Fundamentals of Radar Signal Processing

Fundamentals of Radar Signal Processing

Mark A. Richards
Georgia Institute of Technology

McGraw-Hill

New York Chicago San Francisco Lisbon London Madrid
Mexico City Milan New Delhi San Juan Seoul
Singapore Sydney Toronto

Library of Congress Cataloging-in-Publication Data

Richards, M. A. (Mark A.), date.
  Fundamentals of radar signal processing / Mark A. Richards.
    p.   cm.
  Includes bibliographical references and index.
  ISBN 0-07-144474-2 (alk. paper)
  1. Radar.   2. Signal processing.   I. Title.

TK6575.R47   2005
621.3848—dc22
                                                        2005047891

Copyright © 2005 by The McGraw-Hill Companies, Inc. All rights reserved. Printed in the United States of America. Except as permitted under the United States Copyright Act of 1976, no part of this publication may be reproduced or distributed in any form or by any means, or stored in a data base or retrieval system, without the prior written permission of the publisher.

2 3 4 5 6 7 8 9 0   DOC/DOC   0 1 0 9 8 7 6

ISBN 0-07-144474-2

The sponsoring editor for this book was Stephen S. Chapman and the production supervisors were Sherri Souffrance and Rick Ruzycka. It was set in Century Schoolbook by International Typesetting and Composition. The art director for the cover was Handel Low.

Printed and bound by RR Donnelley.

 This book was printed on recycled, acid-free paper containing a minimum of 50% recycled, de-inked fiber.

McGraw-Hill books are available at special quantity discounts to use as premiums and sales promotions, or for use in corporate training programs. For more information, please write to the Director of Special Sales, McGraw-Hill Professional, Two Penn Plaza, New York, NY 10121-2298. Or contact your local bookstore.

Information contained in this work has been obtained by The McGraw-Hill Companies, Inc. ("McGraw-Hill") from sources believed to be reliable. However, neither McGraw-Hill nor its authors guarantee the accuracy or completeness of any information published herein, and neither McGraw-Hill nor its authors shall be responsible for any errors, omissions, or damages arising out of use of this information. This work is published with the understanding that McGraw-Hill and its authors are supplying information but are not attempting to render engineering or other professional services. If such services are required, the assistance of an appropriate professional should be sought.

To my family: Theresa, Jessica, and Benjamin;
And in memory of my parents

Contents

Preface xiii
Acknowledgments xv
List of Symbols xvii
List of Acronyms xxv

Chapter 1. Introduction to Radar Systems — 1

1.1 History and Applications of Radar — 1
1.2 Basic Radar Functions — 3
1.3 Elements of a Pulsed Radar — 6
    1.3.1 Transmitter and waveform generator — 7
    1.3.2 Antennas — 10
    1.3.3 Receivers — 16
1.4 Review of Selected Signal Processing Concepts and Operations — 21
    1.4.1 Resolution — 21
    1.4.2 Spatial frequency — 23
    1.4.3 Fourier transforms — 24
    1.4.4 The sampling theorem and spectrum replication — 27
    1.4.5 Vector representation of signals — 32
    1.4.6 Data integration — 33
    1.4.7 Correlation — 35
1.5 A Preview of Basic Radar Signal Processing — 37
    1.5.1 Radar time scales — 39
    1.5.2 Phenomenology — 40
    1.5.3 Signal conditioning and interference suppression — 41
    1.5.4 Imaging — 45
    1.5.5 Detection — 47
    1.5.6 Postprocessing — 49
1.6 Radar Literature — 49
    1.6.1 Radar systems and components — 50
    1.6.2 Radar signal processing — 50
    1.6.3 Advanced radar signal processing — 50
    1.6.4 Current radar research — 51
    References — 51

Chapter 2. Signal Models — 53

2.1 Components of a Radar Signal — 53
2.2 Amplitude Models — 54

		2.2.1 Simple point target radar range equation	54
		2.2.2 Distributed target forms of the range equation	57
		2.2.3 Radar cross section	64
		2.2.4 Radar cross section for meteorological targets	66
		2.2.5 Statistical description of radar cross section	67
		2.2.6 Swerling models	79
	2.3	Clutter	82
		2.3.1 Behavior of σ^0	83
		2.3.2 Signal-to-clutter ratio	84
		2.3.3 Temporal and spatial correlation of clutter	85
		2.3.4 Compound models of radar cross section	86
	2.4	Noise Model and Signal-to-Noise Ratio	88
	2.5	Jamming	92
	2.6	Frequency Models: The Doppler Shift	92
		2.6.1 Doppler shift	92
		2.6.2 Simplified approach to Doppler shift	95
		2.6.3 The "stop-and-hop" assumption and spatial Doppler	97
	2.7	Spatial Models	100
		2.7.1 Variation with angle or cross-range	103
		2.7.2 Variation with range	106
		2.7.3 Projections	107
		2.7.4 Multipath	108
	2.8	Spectral Model	109
	2.9	Summary	111
		References	112

Chapter 3. Sampling and Quantization of Pulsed Radar Signals   115

	3.1	Domains and Criteria for Sampling Radar Signals	115
		3.1.1 Time and frequency samples	116
		3.1.2 Spatial samples	118
		3.1.3 Sampling criteria	119
	3.2	Sampling in the Fast Time Dimension	121
	3.3	Sampling in Slow Time: Selecting the Pulse Repetition Interval	123
	3.4	Sampling the Doppler Spectrum	128
		3.4.1 The Nyquist rate in Doppler	129
		3.4.2 Straddle loss	131
	3.5	Sampling in the Spatial and Angle Dimensions	136
		3.5.1 Phased array element spacing	137
		3.5.2 Antenna beam spacing	138
	3.6	Quantization	140
	3.7	I/Q Imbalance and Digital I/Q	145
		3.7.1 I/Q imbalance and offset	145
		3.7.2 Correcting I/Q errors	149
		3.7.3 Digital I/Q	152
		References	157

Chapter 4. Radar Waveforms   159

	4.1	Introduction	159
	4.2	The Waveform Matched Filter	161
		4.2.1 The matched filter	161

	4.2.2	Matched filter for the simple pulse	163
	4.2.3	All-range matched filtering	165
	4.2.4	Range resolution of the matched filter	166
4.3	Matched Filtering of Moving Targets		167
4.4	The Ambiguity Function		169
	4.4.1	Definition and properties of the ambiguity function	169
	4.4.2	Ambiguity function of the simple pulse	173
4.5	The Pulse Burst Waveform		176
	4.5.1	Matched filter for the pulse burst waveform	177
	4.5.2	Pulse-by-pulse processing	178
	4.5.3	Range ambiguity	180
	4.5.4	Doppler response of the pulse burst waveform	181
	4.5.5	Ambiguity function for the pulse burst waveform	183
	4.5.6	Relation of slow-time spectrum to ambiguity function	187
4.6	Frequency-Modulated Pulse Compression Waveforms		188
	4.6.1	Linear frequency modulation	188
	4.6.2	The principle of stationary phase	192
	4.6.3	Ambiguity function of the LFM waveform	194
	4.6.4	Range-Doppler coupling	197
	4.6.5	Stretch processing	198
4.7	Range Side Lobe Control for FM Waveforms		201
	4.7.1	Matched filter frequency response shaping	202
	4.7.2	Waveform spectrum shaping	204
4.8	The Stepped Frequency Waveform		206
4.9	Phase-Modulated Pulse Compression Waveforms		211
	4.9.1	Biphase codes	212
	4.9.2	Polyphase codes	218
4.10	Costas Frequency Codes		222
	References		223

Chapter 5. Doppler Processing   225

5.1	Alternate Forms of the Doppler Spectrum		226
5.2	Moving Target Indication (MTI)		228
	5.2.1	Pulse cancellers	230
	5.2.2	Vector formulation of the matched filter	234
	5.2.3	Matched filters for clutter suppression	235
	5.2.4	Blind speeds and staggered PRFs	239
	5.2.5	MTI figures of merit	244
	5.2.6	Limitations to MTI performance	251
5.3	Pulse Doppler Processing		253
	5.3.1	The discrete time Fourier transform of a moving target	255
	5.3.2	Sampling the DTFT: the discrete Fourier transform	258
	5.3.3	Matched filter and filterbank interpretations of pulse Doppler processing with the DFT	261
	5.3.4	Fine Doppler estimation	264
	5.3.5	Modern spectral estimation in pulse Doppler processing	270
	5.3.6	Dwell-to-dwell stagger	272
5.4	Pulse Pair Processing		273
5.5	Additional Doppler Processing Issues		279
	5.5.1	Combined MTI and pulse Doppler processing	279
	5.5.2	Transient effects	279
	5.5.3	PRF Regimes and Ambiguity Resolution	280

5.6	Clutter Mapping and the Moving Target Detector	284
	5.6.1 Clutter mapping	284
	5.6.2 The moving target detector	286
5.7	MTI for Moving Plateforms: Adaptive Displaced Phase Center Antenna Processing	287
	5.7.1 The DPCA concept	287
	5.7.2 Adaptive DPCA	289
	References	293

Chapter 6. Detection Fundamentals — 295

6.1	Radar Detection as Hypothesis Testing	296
	6.1.1 The Neyman-Pearson detection rule	297
	6.1.2 The likelihood ratio test	298
6.2	Threshold Detection in Coherent Systems	308
	6.2.1 The Gaussian case for coherent receivers	308
	6.2.2 Unknown parameters and threshold detection	312
	6.2.3 Linear and square-law detectors	319
	6.2.4 Other unknown parameters	319
6.3	Threshold Detection of Radar Signals	321
	6.3.1 Coherent, noncoherent, and binary integration	322
	6.3.2 Nonfluctuating targets	324
	6.3.3 Albersheim's equation	329
	6.3.4 Fluctuating targets	331
	6.3.5 Shnidman's equation	336
6.4	Binary integration	338
6.5	Useful Numerical Approximations	342
	6.5.1 Approximations to the error function	342
	6.5.2 Approximations to the magnitude function	344
	References	345

Chapter 7. Constant False Alarm Rate (CFAR) Detection — 347

7.1	The Effect of Unknown Interference Power on False Alarm Probability	347
7.2	Cell-Averaging CFAR	349
	7.2.1 The effect of varying P_{FA}	349
	7.2.2 The cell-averaging CFAR concept	350
	7.2.3 CFAR reference windows	352
7.3	Analysis of Cell-Averaging CFAR	353
	7.3.1 Derivation of CA CFAR threshold	353
	7.3.2 Cell-averaging CFAR performance	354
	7.3.3 CFAR loss	357
7.4	CA CFAR Limitations	358
	7.4.1 Target masking	359
	7.4.2 Clutter edges	363
7.5	Extensions to Cell-Averaging CFAR	364
7.6	Order Statistic CFAR	370
7.7	Additional CFAR Topics	374
	7.7.1 Adaptive CFAR	374
	7.7.2 Two-parameter CFAR	375
	7.7.3 Clutter map CFAR	377
	7.7.4 Distribution-free CFAR	379
	7.7.5 System-level control of false alarms	381
	References	382

Chapter 8. Introduction to Synthetic Aperture Imaging — 385

- 8.1 Introduction to SAR Fundamentals — 390
  - 8.1.1 Cross-range resolution in radar — 390
  - 8.1.2 The synthetic aperture viewpoint — 392
  - 8.1.3 Doppler viewpoint — 399
  - 8.1.4 SAR coverage and sampling — 401
- 8.2 Stripmap SAR Data Characteristics — 404
  - 8.2.1 Stripmap SAR Geometry — 404
  - 8.2.2 Stripmap SAR data set — 407
- 8.3 Stripmap SAR Image Formation Algorithms — 410
  - 8.3.1 Doppler beam sharpening — 411
  - 8.3.2 Quadratic phase error effects — 416
  - 8.3.3 Range-Doppler algorithms — 421
  - 8.3.4 Depth of focus — 426
- 8.4 Spotlight SAR Data Characteristics — 428
- 8.5 The Polar Format Image Formation Algorithm for Spotlight SAR — 433
- 8.6 Interferometric SAR — 436
  - 8.6.1 The effect of height on a SAR image — 436
  - 8.6.2 IFSAR processing steps — 439
- 8.7 Other Considerations — 444
  - 8.7.1 Motion compensation and autofocus — 444
  - 8.7.2 Autofocus — 447
  - 8.7.3 Speckle reduction — 455
- References — 458

Chapter 9. Introduction to Beamforming and Space-Time Adaptive Processing — 461

- 9.1 Spatial Filtering — 461
  - 9.1.1 Conventional beamforming — 461
  - 9.1.2 Adaptive beamforming — 465
  - 9.1.3 Adaptive beamforming with preprocessing — 469
- 9.2 Space-Time Signal Environment — 471
- 9.3 Space-Time Signal Modeling — 475
- 9.4 Processing the Space-Time Signal — 479
  - 9.4.1 Optimum matched filtering — 479
  - 9.4.2 STAP metrics — 480
  - 9.4.3 Relation to displaced phase center antenna processing — 484
  - 9.4.4 Adaptive matched filtering — 488
- 9.5 Computational Issues in STAP — 491
  - 9.5.1 Power domain solution — 492
  - 9.5.2 Computational load of the power domain solution — 493
  - 9.5.3 Voltage domain solution and computational load — 495
  - 9.5.4 Conversion to computational rates — 496
- 9.6 Reduced-Dimension STAP — 497
- 9.7 Advanced STAP Algorithms and Analysis — 499
- 9.8 Limitations to STAP — 501
- References — 502

Index 505

Preface

The goal of this book is to provide in-depth coverage of fundamental topics in radar signal processing from a digital signal processing perspective. The techniques of linear systems, filtering, sampling, and Fourier analysis techniques and interpretations are used throughout to provide a modern and unified tutorial approach. The coverage includes a full range of the basic signal processing techniques on which all radar systems rely, including topics such as target and interference models, matched filtering, waveform design, Doppler processing, and threshold detection and CFAR. In addition, introductions are provided to the advanced topics of synthetic aperture imaging and space-time adaptive array processing.

This book is intended to fill what I perceive to be a void in the technical literature on radar. The literature offers a number of excellent books on radar systems in general. Recent examples include the books by Edde and Skolnik. These books provide an excellent qualitative and descriptive introduction to radar systems as a whole and are recommended as first texts for anyone interested in the topic. The text by Peebles provides greater quantitative depth on basic radar systems and some signal processing topics. However, the goal of this text is to delve more deeply into the signal processing aspects of radar in particular. A number of good quality texts on advanced topics in radar signal processing, principally synthetic aperture imaging and space-time adaptive processing, have appeared in recent years. Synthetic aperture imaging examples include the books by Jakowatz et al., Carrara et al., Soumekh, and Cumming and Wong; space-time adaptive processing examples include the books by Klemm and Guerci. The book by Sullivan spans both areas. However, there is a substantial gap between the qualitative systems books and the advanced signal processing books. Specifically, I believe the radar community lacks a current text providing a concise, unified, and modern treatment of the basic radar signal processing techniques on which these more advanced methods are founded: signal modeling, matched filtering and pulse compression, Doppler processing, and threshold detection. It is my hope that this book will fill that gap.

This book has been developed and used over several years in support of two courses at Georgia Tech. It was primarily developed as a product of ECE 6272,

Fundamentals of Radar Signal Processing, a semester-length first-year graduate course in which these notes first served as supplemental material and more recently as the text for the course. Elements of this book have also been used in abbreviated and simplified form in the one-week professional education course of the same name taught annually through Georgia Tech's Distance Learning and Professional Education division.

A one-semester course in radar signal processing can cover virtually all of the text. Such a course provides a solid foundation for more advanced work in detection theory, adaptive array processing, synthetic aperture imaging, and more advanced radar concepts such as passive and bistatic systems. A quarter-length course could cover Chapters 1 through 7 reasonably thoroughly, perhaps also skipping some of the later sections of Chapters 2 and 3 for additional time savings. In either case, a firm background in basic continuous and discrete signal processing and at least an introductory exposure to random processes is advisable.

A web site is available to provide additional support information related to this book, such as lists of errata and software tools. The URL is:

http://users.ece.gatech.edu/~mrichard/FundRadarSigProc.htm

Acknowledgments

I am indebted to many colleagues and students who have helped me to learn this material, write it, and (try to) debug it once written. Dr. Edward Reedy hired me into my first job after graduation, at the Georgia Tech Research Institute (GTRI), converting me from a speech processor to a radar signal processor. Dr. Jim Echard had the unenviable task of actually overseeing that conversion as my first supervisor and thus my first teacher of the basics of radar. In later years, we jointly developed and taught the course that eventually became ECE 6272 and led to this book. GTRI colleagues too numerous to mention helped me along the way, but a few on whom I have relied particularly merit special mention. Dr. Byron Keel shared his expertise in waveforms and CFAR. Dr. Christopher Barnes (now at Georgia Tech's Savannah campus) and Dr. Gregory Showman both led me to a greater understanding of SAR. Dr. William Melvin introduced me to STAP. I am grateful to each of them for their knowledge and friendship. I am also indebted to numerous students who have soldiered through the development of the textbook chapters as part of ECE 6272. Their contributions to correcting errors and improving the presentation are greatly appreciated. Special mention here goes to Brian Mileshosky and Anders Roos, who provided especially thorough readings and numerous valuable corrections and clarifications. Any errors that remain are strictly my responsibility.

Mark A. Richards

List of Symbols

The following definitions and relations between symbols are used throughout this text except as otherwise specifically noted. Some symbols, such as θ, have more than one usage; their meaning is generally clear from the context.

$*$	Convolution operator
$\otimes$	Kronecker product operator
$\odot$	Hadamard product operator
(x)	Continuous variable x
$[x]$	Discrete variable x
$((\cdot))_x$	Modulo x
$\sim$	"Is distributed as"
$\mathbf{x}$	Vector variable
$\mathbf{X}$	Matrix variable
$\mathbf{x}'$	Matrix or vector transpose
x^*	Conjugate
$\mathbf{x}^H$	Hermitian transpose
α_q	Clutter temporal fluctuation vector
α_{GO}	Threshold multiplier, "greatest-of" CFAR
$\alpha_{\log}$	Threshold multiplier, log CFAR
α_{SO}	Threshold multiplier, "smallest-of" CFAR
β	Bandwidth; Standard deviation of interference process
β_D	Doppler bandwidth
β_L	Linear term of nonlinear FM waveform bandwidth
β_C	Nonlinear term of nonlinear FM waveform bandwidth
β_r	Rayleigh bandwidth
β_n	Noise-equivalent receiver bandwidth
β_{nn}	Null-to-null bandwidth
$\beta_x, \beta_y, \beta_z$	Spatial bandwidth

List of Symbols

χ	Signal-to-noise ratio
χ_1	Single sample signal-to-noise ratio
χ_N	N-sample signal-to-noise ratio
χ_∞	Signal-to-noise ratio with perfect noise level estimate
δ	Grazing angle
$\delta[\cdot]$	Discrete impulse function
$\delta_D(\cdot)$	Dirac impulse ("delta") function
δR	Range error
δt	Differential delay
$\Delta\theta$	Angular resolution
$\Delta\psi$	Change in squint angle
ΔCR	Cross-range resolution
ΔF	Frequency step size
ΔF_D	Doppler frequency resolution
Δh	Height displacement
ΔR	Range resolution
ΔR_b	Range relative to central reference point
ΔR_c	Range curvature
ΔR_s	Range bin spacing
ΔR_w	Range walk
Δt_b	Time relative to central reference point delay
ε	I/Q amplitude mismatch
ϕ	Elevation angle; Phase; Baseband received signal phase
$\bar{\phi}$	Nonbaseband received signal phase
ϕ_3	3-dB elevation beamwidth
ϕ_{fg}	Interferometric phase difference
$\phi_{\max}$	Maximum quadratic phase error
ϕ_n	Subpulse phase in phase-coded waveform
ϕ_{nn}	Null-to-null elevation beamwidth
γ	Q channel DC offset
η	Volume reflectivity
κ	I channel DC offset; Doppler spectrum oversampling factor; Adaptive filter scale factor
λ	Wavelength
Λ	Likelihood ratio
$\bar{\mu}$	Estimated mean
θ	Azimuth angle; Phase; Baseband transmitted signal phase
$\theta(t)$	Phase modulation of waveform
$\bar{\theta}$	Nonbaseband transmitted signal phase

List of Symbols

θ	Vector of random phases
θ_3	3-dB azimuth beamwidth
θ_{az}	Azimuth beamwidth
θ_{el}	Elevation beamwidth
θ_{nn}	Null-to-null azimuth beamwidth
θ_{SAR}	Effective beamwidth of synthetic aperture radar
θ_t	Target angle of arrival
$\rho = \zeta \exp[j\psi]$ $= \rho_I + j\rho_Q$	Complex baseband reflectivity
ρ_I	Baseband reflectivity in-phase (I) component
ρ_Q	Baseband reflectivity quadrature-phase (Q) component
ρ_f, ρ_{fg}	Normalized autocorrelation or cross-correlation function
ρ'	Effective baseband complex reflectivity
$\bar{\rho} = \bar{\zeta} \exp[j\bar{\psi}]$ $= \bar{\rho}_I + j\bar{\rho}_Q$	Complex nonbaseband reflectivity
$\bar{\rho}_I$	Nonbaseband reflectivity in-phase (I) component
$\bar{\rho}_Q$	Nonbaseband reflectivity quadrature-phase (Q) component
$\tilde{\rho}$	Cross-range averaged effective baseband complex reflectivity
$\tilde{P}$	Range spatial spectrum (Fourier transform of $\tilde{\rho}$)
$\sigma = \|\rho\|^2 = \zeta^2$	Radar cross section (RCS)
σ^0	Area reflectivity
σ_x^2	Variance of random variable x
Σ	Diagonal matrix of clutter powers
ζ	Baseband reflectivity amplitude ($\zeta \geq 0$)
$\bar{\zeta}$	Nonbaseband reflectivity amplitude ($\bar{\zeta} \geq 0$)
τ	Pulse length
τ_c	Subpulse length in phase coded waveform
Υ	Sufficient statistic
ω	Normalized frequency (radians/sample, radians)
ω_D	Normalized Doppler frequency shift (radians)
ω_s	Sampling interval in normalized frequency ω (samples/radian)
Ω	Frequency (radians/second)
Ω_θ	Azimuth rotation rate (radians/second)
Ω_D	Doppler frequency shift (radians/second)
Ω_{diff}	Doppler frequency mismatch (radians/second)
Ω_i	Matched Doppler frequency shift (radians/second)
ψ	Baseband reflectivity phase; Squint angle
$\bar{\psi}$	Nonbaseband reflectivity phase
a	Baseband transmitted signal amplitude

List of Symbols

$\mathbf{a}_s(\theta)$	Spatial steering vector
$\mathbf{a}_t(\theta)$	Temporal steering vector
$\bar{a}$	Nonbaseband transmitted signal amplitude
$A, \hat{A}, \tilde{A}$	Signal amplitude
A_e	Effective antenna aperture size
$\mathbf{A}_q$	Covariance matrix of clutter temporal fluctuations
A_n	Complex amplitude of subpulse in phase coded waveform
$AF(\theta), AF(\theta, \phi)$	Phased array antenna array factor
b	Amplitude
$\bar{b}$	Nonbaseband received signal amplitude
B	Number of bits; Interferometric baseline
B_N	Length of Barker phase code
c	Speed of light
$\mathbf{c}_q$	Clutter space-time steering vector for patch q
CA	Clutter attenuation
$C_x(\cdot)$	Characteristic function of random variable x
d	Phased array element spacing
d_{pc}	Phase center spacing
D	Antenna aperture size
D_{az}	Antenna size, azimuth dimension
D_{el}	Antenna size, elevation dimension
D_{SAR}	Synthetic aperture size
DOF	Degrees of freedom
DR	Dynamic range
D_x, D_y, D_z	Antenna aperture size in x, y, or z dimension
E	Pulse energy
$\mathbf{E}(\cdot)$	Expected value
$E(\theta, \phi)$	Electric field amplitude
$E_{el}(\theta), E_{el}(\theta, \phi)$	Phased array antenna element pattern
f	Normalized frequency (cycles/sample, cycles)
$\tilde{f}$	Quantized version of a function f
f_D	Normalized Doppler frequency shift (cycles)
f_{Dt}	Target normalized Doppler frequency shift (cycles)
F	Fourier transform of a function f
F	Frequency (cycles/second, hertz)
F_θ	Spatial frequency (cycles)
F_b	Beat frequency (hertz)
F_c	Corner frequency (hertz)
F_D	Doppler frequency shift (hertz)
F_g	Greatest common divisor of a set of staggered PRFs

Symbol	Description
F_i	Instantaneous frequency (hertz)
F_n	Noise figure
F_s	Sampling frequency (samples/second)
F_t	Transmitted frequency (hertz)
F_{us}	Unstaggered blind Doppler frequency
G	Antenna power gain
G_{nc}	Noncoherent integration gain
G_s	Maximum receiver gain
h	Height
$\mathbf{h}$	Filter weight vector; Beamformer weight vector
$h(t)$	Impulse response (continuous time)
$h[n]$	Impulse response (discrete time)
H_0	Null hypothesis (interference only)
H_1	Nonnull hypothesis (target plus interference)
$H(f), H(F), H(\omega), H(\Omega)$	Frequency response in various units
$H(z)$	Discrete-time system function
$H_N(z)$	System function of N-pulse canceller
$H_{N,P}(F)$	Frequency response of N-pulse canceller with P staggered PRFs
I	Interference power; improvement factor
I_{opt}	Improvement factor for matched filter
$\mathbf{I}_N$	Nth-order identity matrix
I_{sub}	Suboptimum improvement factor
I	In-phase channel
I_k	In-phase component, sample k
$I(\cdot, \cdot)$	Incomplete gamma function
$J_n(t)$	Jammer signal
$\mathbf{J}$	Jammer signal sample vector
k_p	Stagger ratio
K	Spatial frequency (cycles/meter); DFT size; Normalized quantizer step size
K_R	Range spatial frequency (cycles/meter)
K_u	Cross-range spatial frequency (cycles/meter)
K_x, K_y, K_z	Spatial frequency in x, y, or z dimensions
K_θ	Spatial frequency corresponding to AOA θ
$\tilde{K}_\theta$	Normalized spatial frequency corresponding to AOA θ
L	Number of fast time samples per pulse
L_a	Atmospheric loss factor
L_0	Maximum acceptable signal-to-interference ratio loss
L_s	System loss factors; Synthetic aperture radar swath length

Symbol	Description
L_{SIR}	Signal-to-interference ratio loss
LPG	Loss in processing gain
m	Mean of random variable
m_n	Mean of nth element of a random vector
$\bar{\mathbf{m}} = \tilde{\mathbf{m}}\exp(j\theta)$	Mean of random vector
$\tilde{\mathbf{m}}$	Nonrandom portion of mean of random vector random
M	Number of slow time samples per coherent processing interval
M_{DD}	Minimum detectable Doppler shift
M_{DD_+}, M_{DD_-}	Minimum detectable positive, negative Doppler shift
M_{opt}	Optimum value of M in "M of N" detection rule
M_s	DPCA time slip
n_P	Matched filter output noise power
N	Noise power; number of phase centers
N_γ	Number of spotlight SAR radial slices
N_R	Number of spotlight SAR range samples
N_{spot}	Number of spotlight SAR iamges per unit time
$p_x(\cdot)$	Probability density function for a random variable x
P	Power
$P(\theta,\phi)$	Antenna one-way power pattern
$P_\theta(\theta)$	Azimuth one-way antenna power pattern
$P_\phi(\phi)$	Elevation one-way antenna power pattern
P_b	Backscattered power
P_{CD}	Cumulative probability of detection
P_{CFA}	Cumulative probability of false alarm
P_D	Probability of detection
$\overline{P}_D$	Expected value of probability of detection
P_{FA}	Probability of false alarm
$\overline{P}_{FA}$	Expected value of probability of false alarm
P_M	Probability of miss
P_r	Received power; relative power of I/Q mismatch image
P_t	Transmitted power
PL	Processing loss
PRF	Pulse repetition frequency
q	Quantizer step size
Q	Power density
Q	Quadrature channel
$\mathbf{Q}$	$\mathbf{Q}$ matrix in $\mathbf{Q}$-$\mathbf{R}$ decomposition
Q_b	Backscattered power density
Q_k	Quadrature component, sample k

Symbol	Description
Q_M	Marcum Q function
Q_t	Transmitted power density
R, R_0	Range
R	**R** matrix in **Q-R** decomposition
R_a	Apparent range
R_P	Range coordinate of synthetic aperture radar scatterer
R_t	True range
R_{ua}	Unambiguous range
R_w	Range window
RFLOP	Real floating point operations
RFLOPS	Real floating point operations per second
$\hat{s}$	Estimated standard deviation
s_A	Autocorrelation of phase code complex amplitude sequence
s_f	Autocorrelation of a function f
s_{fg}	Cross-correlation of functions f and g
$s_p(t)$	Output of filter matched to single pulse in pulse train waveform
S	Polarization scattering matrix
$S_f(\omega)$	Power spectrum of a function f
$S_{fg}(\omega)$	Cross-power spectrum of functions f and g
$\mathbf{S_x}$	Covariance matrix for a random vector **x**
$\tilde{\mathbf{S}}_x$	Transformed covariance matrix
$\hat{\mathbf{S}}_x$	Estimated covariance matrix
SIR	Signal-to-interference ratio
SQNR	Signal-to-quantization noise ratio
t, t_0	Time
t	Target model vector
$\tilde{\mathbf{t}}$	Transformed target model vector
T	Pulse repetition interval
T	Transformation matrix
T'	Equivalent receiver temperature; Detection threshold
$\hat{T}$	Estimated threshold
$\hat{T}_{\log}$	Estimated threshold, log CFAR
T_θ	Sampling interval in θ
T_a	Aperture time
T_{avg}	Average PRI of a set of staggered PRFs
T_M	Time of matched filter output peak
T_p	pth PRI in set of staggered PRFs
T_s	Fast time sampling interval; sampling interval in $s = \sin\theta$
T_{tot}	Sum of PRIs corresponding to set of staggered set of PRFs

List of Symbols

Symbol	Description
T_w	Time corresponding to swath width
u	Along-track coordinate of synthetic aperture radar platform
UDSF	Usable Doppler space fraction
$\text{var}(x)$	Variance of random variable x
v	Platform velocity
v_{blind}	Blind speed
v_{ua}	Unambiguous velocity
$\mathbf{w}_f$	Temporal weight vector
$\mathbf{w}_\theta$	Spatial weight vector
$\bar{x}$	Mean of random variable x
$x = a\exp[j\theta] = x_I + jx_Q$	Transmitted signal, baseband
$\tilde{x} = \tilde{a}\exp[j\tilde{\theta}] = \tilde{x}_I + j\tilde{x}_Q$	Transmitted signal, nonbaseband
x_I	Baseband transmitted signal in-phase (I) component
$\tilde{x}_I$	Nonbaseband transmitted in-phase (I) component
x_P	Along-track coordinate of synthetic aperture radar scatterer
x_Q	Baseband transmitted signal quadrature-phase (Q) component
$\tilde{x}_Q$	Nonbaseband transmitted quadrature-phase (Q) component
$x_p(t)$	Subpulse of phase coded waveform
$x_p(t)$	Single pulse of pulse train waveform
$y = b\exp[j\phi] = y_I + jy_Q$	Received signal, baseband
$\mathbf{y}$	Baseband received signal sample vector
$\tilde{\mathbf{y}}$	Transformed baseband received signal sample vector
$\tilde{y} = \tilde{b}\exp[j\tilde{\phi}] = \tilde{y}_I + j\tilde{y}_Q$	Received signal, nonbaseband
y_I	Baseband received signal in-phase (I) component
$\tilde{y}_I$	Nonbaseband received signal in-phase (I) component
y_Q	Baseband received signal quadrature-phase (Q) component
$\tilde{y}_Q$	Nonbaseband received signal quadrature-phase (Q) component
$y[l, m, n]$	Datacube for one coherent processing interval
$y[l, m]$	Fast time/slow time data matrix for one CPI
$y_s[m]$	Slow time sequence for one CPI
z	Detected output
$\tilde{z}$	Transformed detected output
Z	Meteorological reflectivity; altitude

List of Acronyms

The following acronyms are used throughout this text.

1D	One-Dimensional
2D	Two-Dimensional
AC	Alternating Current
A/D	Analog-to-Digital
AF	Ambiguity Function
AGC	Automatic Gain Control
AMF	Adaptive Matched Filter
AMTI	Airborne Moving Target Indication
AOA	Angle of Arrival
AR	Autoregressive
ASR	Airport Surveillance Radar
BPF	Bandpass Filter
BSR	Beam Sharpening Ratio
BT	Time-Bandwidth Product
CA	Clutter Attenuation
CA-CFAR	Cell-Averaging Constant False Alarm Rate
CAT	Computerized Axial Tomography
CCD	Coherent Change Detection
CFAR	Constant False Alarm Rate
CMT	Covariance Matrix Taper
CNR	Clutter-to-Noise Ratio
CRP	Central Reference Point
CRT	Chinese Remainder Theorem
CPI	Coherent Processing Interval
CUT	Cell Under Test
CW	Continuous Wave

D/A	Digital-to-Analog
dB	Decibel
dBsm	Decibels relative to 1 square meter
DBS	Doppler Beam Sharpening
DC	Direct Current
DCT	Discrete Cosine Transform
DF CFAR	Distribution-Free Constant False Alarm Rate
DFT	Discrete Fourier Transform
DOF	Degrees of Freedom
DPCA	Displaced Phase Center Antenna
DSP	Digital Signal Processing
DTED	Digital Terrain Elevation Data
DTFT	Discrete Time Fourier Transform
ECM	Electronic Countermeasures
ECCM	Electronic Counter-Countermeasures
EM	Electromagnetic
EMI	Electromagnetic Interference
ERIM	Environmental Research Institute of Michigan
FFT	Fast Fourier Transform
FFTW	Fastest Fourier Transform in the West
FIR	Finite Impulse Response
FLOP	Floating Point Operation
FLOPS	Floating Point Operations per Second
FM	Frequency Modulation
FMCW	Frequency Modulated Continuous Wave
GLRT	Generalized Likelihood Ratio Test
GMTI	Ground Moving Target Indication
GOCA CFAR	Greatest-Of Cell Averaging Constant False Alarm Rate
GPS	Global Positioning System
HF	High Frequency
HPRF	High Pulse Repetition Frequency
I	In-phase
ICM	Internal Clutter Motion; Intrinsic Clutter Motion
IDFT	Inverse Discrete Fourier Transform
IF	Intermediate Frequency
IFFT	Inverse Fast Fourier Transform
IFSAR	Interferometric Synthetic Aperture Radar
IIR	Infinite Impulse Response
IMU	Inertial Measurement Unit

INS	Inertial Navigation System
ISAR	Inverse Synthetic Aperture Radar
ISL	Interference Subspace Leakage
JNR	Jammer-to-Noise Ratio
KA	Knowledge-Aided
LC	Least Common Multiple
LEO	Low Earth Orbit
LFM	Linear Frequency Modulation
LNA	Low Noise Amplifier
LO	Local Oscillator
LPF	Lowpass Filter
LPG	Loss in Processing Gain
LPRF	Low Pulse Repetition Frequency
LRT	Likelihood Ratio Test
LSB	Least Significant Bit
MDD	Minimum Detectable Doppler
MDV	Minimum Detectable Velocity
MMW	Millimeter Wave
MPRF	Medium Pulse Repetition Frequency
MSB	Most Significant Bit
MTD	Moving Target Detector
MTI	Moving Target Indication
NASA	National Aeronautics and Space Agency
NEXRAD	Next Generation Radar
NLFM	Nonlinear Frequency Modulation
NRL	Naval Research Laboratory
OS CFAR	Order Statistic Constant False Alarm Rate
PC	Principal Components
PDF	Probability Density Function
PFA	Polar Format Algorithm
PGA	Phase Gradient Algorithm
PL	Processing Loss
PPP	Pulse Pair Processing
PRF	Pulse Repetition Frequency
PRI	Pulse Repetition Interval
PSL	Peak Sidelobe Level
PSM	Polarization Scattering Matrix
PSP	Principle of Stationary Phase
PSR	Point Spread Response

Q	Quadrature	
RCS	Radar Cross Section	
RCSR	Radar Cross Section Reduction	
RD	Range-Doppler	
RF	Radar Frequency	
RFLOP	Real Floating Point Operations	
RFLOPS	Real Floating Point Operations per Second	
RMB	Reed-Mallet-Brennan	
ROI	Region of Interest	
RSS	Root Sum Square	
RVP	Residual Video Phase	
SAR	Synthetic Aperture Radar	
SQNR	Signal-to-Quantization Noise Ratio	
S-CFAR	Switching Constant False Alarm Rate	
SCR	Signal-to-Clutter Ratio	
SIR	Signal-to-Interference Ratio; Shuttle Imaging Radar	
SMI	Sample Matrix Inverse	
SMTI	Surface Moving Target Indication	
SNR	Signal-to-Noise Ratio	
SOCA CFAR	Smallest-Of Cell Averaging Constant False Alarm Rate	
STALO	Stable Local Oscillator	
STAP	Space-Time Adaptive Processing	
STC	Sensitivity Time Control	
T/R	Transmit/Receive	
UDSF	Usable Doppler Space Fraction	
UHF	Ultra-High Frequency	
UMP	Uniformly Most Powerful	
VHF	Very High Frequency	

Fundamentals of Radar Signal Processing

Chapter 1

Introduction to Radar Systems

1.1 History and Applications of Radar

The word "radar" was originally an acronym, RADAR, for "*r*adio *d*etection *a*nd *r*anging." Today, the technology is so common that the word has become a standard English noun. Many people have direct personal experience with radar in such applications as measuring fastball speeds or, often to their regret, traffic control.

The history of radar extends to the early days of modern electromagnetic theory (Swords, 1986; Skolnik, 2001). In 1886, Hertz demonstrated reflection of radio waves, and in 1900 Tesla described a concept for electromagnetic detection and velocity measurement in an interview. In 1903 and 1904, the German engineer Hülsmeyer experimented with ship detection by radio wave reflection, an idea advocated again by Marconi in 1922. In that same year, Taylor and Young of the U.S. Naval Research Laboratory (NRL) demonstrated ship detection by radar, and in 1930 Hyland, also of NRL, first detected aircraft (albeit accidentally) by radar, setting off a more substantial investigation that led to a U.S. patent for what would now be called a *continuous wave* (CW) radar in 1934.

The development of radar accelerated and spread in the middle and late 1930s, with largely independent developments in the United States, Britain, France, Germany, Russia, Italy, and Japan. In the United States, R. M. Page of NRL began an effort to develop pulsed radar in 1934, with the first successful demonstrations in 1936. The year 1936 also saw the U.S. Army Signal Corps begin active radar work, leading in 1938 to its first operational system, the SCR-268 antiaircraft fire control system and in 1939 to the SCR-270 early warning system, the detections of which were tragically ignored at Pearl Harbor. British development, spurred by the threat of war, began in earnest with work by Watson-Watt in 1935. The British demonstrated pulsed radar that year, and by 1938 established the famous Chain Home surveillance radar network that remained active until the end of World War II. They also built the first airborne interceptor radar in 1939. In 1940, the United States and Britain began to exchange information on radar development. Up to this time, most radar work was conducted at *high frequency* (HF) and *very high frequency* (VHF) wavelengths; but with the British disclosure of the critical cavity magnetron microwave power tube and the United States formation of the Radiation Laboratory at the Massachusetts Institute of Technology, the groundwork was laid for the successful development of radar at the microwave frequencies that have predominated ever since.

Each of the other countries mentioned also carried out CW radar experiments, and each fielded operational radars at some time during the course of World War II. Efforts in France and Russia were interrupted by German occupation. On the other hand, Japanese efforts were aided by the capture of U.S. radars in the Philippines and by the disclosure of German technology. The Germans themselves deployed a variety of ground-based, shipboard, and airborne systems. By the end of the war, the value of radar and the advantages of microwave frequencies and pulsed waveforms were widely recognized.

Early radar development was driven by military necessity, and the military is still the dominant user and developer of radar technology. Military applications include surveillance, navigation, and weapons guidance for ground, sea, and air vehicles. Military radars span the range from huge ballistic missile defense systems to fist-sized tactical missile seekers.

Radar now enjoys an increasing range of applications. One of the most common is the police traffic radar used for enforcing speed limits (and measuring the speed of baseballs and tennis serves). Another is the "color weather radar" familiar to every viewer of local television news. The latter is one type of meteorological radar; more sophisticated systems are used for large-scale weather monitoring and prediction and atmospheric research. Another radar application that affects many people is found in the air traffic control systems used to guide commercial aircraft both en route and in the vicinity of airports. Aviation also uses radar for determining altitude and avoiding severe weather, and may soon use it for imaging runway approaches in poor weather. Radar is commonly used for collision avoidance and buoy detection by ships, and is now beginning to serve the same role for the automobile and trucking industries. Finally, spaceborne (both satellite and space shuttle) and airborne radar is an important tool in mapping earth topology and environmental characteristics such as water and ice conditions, forestry conditions, land usage, and pollution. While this sketch of radar applications is far from exhaustive, it does indicate the breadth of applications of this remarkable technology.

This text tries to present a thorough, straightforward, and consistent description of the signal processing aspects of radar technology, focusing primarily on the more fundamental functions common to most radar systems. Pulsed radars are emphasized over CW radars, though many of the ideas are applicable to both. Similarly, *monostatic* radars, where the transmitter and receiver antennas are collocated (and in fact are usually the same antenna), are emphasized over *bistatic* radars, where they are significantly separated, though again many of the results apply to both. The reason for this focus is that the majority of radar systems are monostatic, pulsed designs. Finally, the subject is approached from a digital signal processing (DSP) viewpoint as much as practicable, both because most new radar designs rely heavily on digital processing and because this approach can unify concepts and results often treated separately.

1.2 Basic Radar Functions

Most uses of radar can be classified as *detection, tracking*, or *imaging*. In this text, the emphasis is on detection and imaging, as well as the techniques of signal acquisition and interference reduction necessary to perform these tasks.

The most fundamental problem in radar is detection of an object or physical phenomenon. This requires determining whether the receiver output at a given time represents the echo from a reflecting object or only noise. Detection decisions are usually made by comparing the amplitude $A(t)$ of the receiver output (where t represents time) to a threshold $T(t)$, which may be set a priori in the radar design or may be computed adaptively from the radar data; in Chap. 6 it will be seen why this detection technique is appropriate. The time required for a pulse to propagate a distance R and return, thus traveling a total distance $2R$, is just $2R/c$; thus, if $y(t) > T(t)$ at some time delay t_0 after a pulse is transmitted, it is assumed that a target is present at range

$$R = \frac{ct_0}{2} \qquad (1.1)$$

where c is the speed of light.[†]

Once an object has been detected, it may be desirable to track its location or velocity. A monostatic radar naturally measures position in a spherical coordinate system with its origin at the radar antenna's phase center, as shown in Fig. 1.1. In this coordinate system, the antenna look direction, sometimes called the *boresight* direction, is along the $+x$ axis. The angle θ is called *azimuth* angle, while ϕ is called *elevation* angle. Range R to the object follows directly from the elapsed time from transmission to detection as just described. Elevation and azimuth angle ϕ and θ are determined from the antenna orientation, since the target must normally be in the antenna main beam to be detected. Velocity is estimated by measuring the Doppler shift of the target echoes. Doppler shift provides only the radial velocity component, but a series of measurements of position and radial velocity can be used to infer target dynamics in all three dimensions.

Because most people are familiar with the idea of following the movement of a "blip" on the radar screen, detection and tracking are the functions most commonly associated with radar. Increasingly, however, radars are being used to

---

[†]$c = 2.99792458 \times 10^8$ m/s in a vacuum. A value of $c = 3 \times 10^8$ m/s is normally used except where very high accuracy is required.

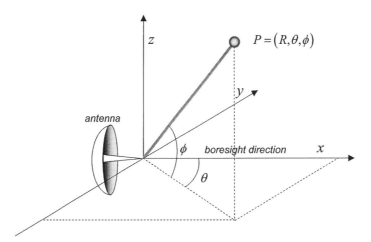

Figure 1.1 Spherical coordinate system for radar measurements.

generate two-dimensional images of an area. Such images can be analyzed for intelligence and surveillance purposes, for elevation/topology mapping, or for analysis of earth resources issues such as mapping, land use, ice cover analysis, deforestation monitoring, and so forth. They can also be used for "terrain following" navigation by correlating measured imagery with stored maps. While radar images have not achieved the resolution of optical images, the very low attenuation of electromagnetic waves at microwave frequencies gives radar the important advantage of "seeing" through clouds, fog, and precipitation very well. Consequently, imaging radars generate useful imagery when optical instruments cannot be used at all.

The quality of a radar system is quantified with a variety of figures of merit, depending on the function being considered. In analyzing detection performance, the fundamental parameters are the *probability of detection* P_D and the *probability of false alarm* P_{FA}. If other system parameters are fixed, increasing P_D always requires accepting a higher P_{FA} as well. The achievable combinations are determined by the signal and interference statistics, especially the *signal-to-interference ratio* (SIR). When multiple targets are present in the radar field of view, additional considerations of resolution and side lobes arise in evaluating detection performance. For example, if two targets cannot be resolved by a radar, they will be registered as a single object. If side lobes are high, the echo from one strongly reflecting target may mask the echo from a nearby but weaker target, so that again only one target is registered when two are present. Resolution and side lobes in range are determined by the radar waveform, while those in angle are determined by the antenna pattern.

In radar tracking, the basic figure of merit is *accuracy* of range, angle, and velocity estimation. While resolution presents a crude limit on accuracy, with appropriate signal processing the achievable accuracy is ultimately limited in each case by the SIR.

In imaging, the principal figures of merit are spatial resolution and dynamic range. Spatial resolution determines what size objects can be identified in the final image, and therefore to what uses the image can be put. For example, a radar map with 1 km by 1 km resolution would be useful for land use studies, but useless for military surveillance of airfields or missile sites. Dynamic range determines image contrast, which also contributes to the amount of information that can be extracted from an image.

The purpose of signal processing in radar is to improve these figures of merit. SIR can be improved by pulse integration. Resolution and SIR can be jointly improved by pulse compression and other waveform design techniques, such as frequency agility. Accuracy benefits from increased SIR and "filter splitting" interpolation methods. Side lobe behavior can be improved with the same windowing techniques used in virtually every application of signal processing. Each of these topics are discussed in the chapters that follow.

Radar signal processing draws on many of the same techniques and concepts used in other signal processing areas, from such closely related fields as communications and sonar to very different applications such as speech and image processing. Linear filtering and statistical detection theory are central to radar's most fundamental task of target detection. Fourier transforms, implemented using *fast Fourier transform* (FFT) techniques, are ubiquitous, being used for everything from fast convolution implementations of matched filters, to Doppler spectrum estimation, to radar imaging. Modern model-based spectral estimation and adaptive filtering techniques are used for beamforming and jammer cancellation. Pattern recognition techniques are used for target/clutter discrimination and target identification.

At the same time, radar signal processing has several unique qualities that differentiate it from most other signal processing fields. Many modern radars are coherent, meaning that the received signal, once demodulated to baseband, is complex-valued rather than real-valued. Radar signals have very high dynamic ranges of several tens of decibels, in some extreme cases approaching 100 dB. Thus, gain control schemes are common, and side lobe control is often critical to avoid having weak signals masked by stronger ones. SIR ratios are often relatively low. For example, the SIR at the point of detection may be only 10 to 20 dB, while the SIR for a single received pulse prior to signal processing is frequently less than 0 dB.

Especially important is the fact that, compared to most other DSP applications, radar signal bandwidths are large. Instantaneous bandwidths for an individual pulse are frequently on the order of a few megahertz, and in some high-resolution radars may reach several hundred megahertz and even as high as 1 GHz. This fact has several implications for digital signal processing. For example, very fast *analog-to-digital* (A/D) converters are required. The difficulty of designing good converters at megahertz sample rates has historically slowed the introduction of digital techniques into radar signal processing. Even now, when digital techniques are common in new designs, radar word lengths in high-bandwidth systems are usually a relatively short 8 to 12 bits, rather

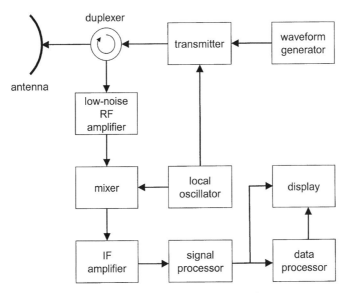

Figure 1.2 Block diagram of a pulsed monostatic radar.

than the 16 bits common in many other areas. The high data rates have also historically meant that it has often been necessary to design custom hardware for the digital processor in order to obtain adequate throughput, that is, to "keep up with" the onslaught of data. This same problem of providing adequate throughput has resulted in radar signal processing algorithms being relatively simple compared to, say, sonar processing techniques. Only in the late 1990s has Moore's Law[†] provided us enough computing power to host radar algorithms for a wide range of systems on commercial hardware. Equally important, this same technological progress has allowed the application of new, more complex algorithms to radar signals, enabling major improvements in detection, tracking, and imaging capability.

1.3 Elements of a Pulsed Radar

Figure 1.2 is one possible block diagram of a simple pulsed monostatic radar. The waveform generator output is the desired pulse waveform. The transmitter modulates this waveform to the desired *radio frequency* (RF) and amplifies it to a useful power level. The transmitter output is routed to the antenna through a *duplexer*, also called a *circulator* or *T/R switch* (for transmit/receive). The returning echoes are routed, again by the duplexer, into the radar receiver. The receiver is usually a superheterodyne design, and often the first stage is

---

[†]Gordon Moore's famous 1965 prediction was that the number of transistors on an integrated circuit would double every 18 to 24 months. This prediction has held remarkably true for nearly 40 years, enabling the computing and networking revolutions that began in the 1980s.

a low-noise RF amplifier. This is followed by one or more stages of modulation of the received signal to successively lower *intermediate frequencies* (IFs) and ultimately to *baseband*, where the signal is not modulated onto any carrier frequency. Each modulation is carried out with a *mixer* and a *local oscillator* (LO). The baseband signal is next sent to the signal processor, which performs some or all of a variety of functions such as pulse compression, matched filtering, Doppler filtering, integration, and motion compensation. The output of the signal processor takes various forms, depending on the radar purpose. A tracking radar would output a stream of detections with measured range and angle coordinates, while an imaging radar would output a two- or three-dimensional image. The processor output is sent to the system display, the data processor, or both as appropriate.

The configuration of Fig. 1.2 is not unique. For example, many systems perform some of the signal processing functions at IF rather than baseband; matched filtering, pulse compression, and some forms of Doppler filtering are very common examples. The list of signal processing functions is redundant as well. For example, pulse compression and Doppler filtering can both be considered part of the matched filtering process. Another characteristic which differs among radars is at what point in the system the analog signal is digitized. Older systems are, of course, all analog, and many currently operational systems do not digitize the signal until it is converted to baseband. Thus, any signal processing performed at IF must be done with analog techniques. Increasingly, new designs digitize the signal at an IF stage, thus moving the A/D converter closer to the radar front end and enabling digital processing at IF. Finally, the distinction between signal processing and data processing is sometimes unclear or artificial.

In the next few subsections, the major characteristics of these principal radar subsystems are briefly discussed.

1.3.1 Transmitter and waveform generator

The transmitter and waveform generator play a major role in determining the sensitivity and range resolution of radar. Radar systems have been operated at frequencies as low as 2 MHz and as high as 220 GHz (Skolnik, 2001); laser radars operate at frequencies on the order of 10^{12} to 10^{15} Hz, corresponding to wavelengths on the order of 0.3 to 30 µm (Jelalian, 1992). However, most radars operate in the microwave frequency region of about 200 MHz to about 95 GHz, with corresponding wavelengths of 0.67 m to 3.16 mm. Table 1.1 summarizes the letter nomenclature used for the common nominal radar bands (IEEE, 1976). The millimeter wave band is sometimes further decomposed into approximate sub-bands of 36 to 46 GHz (Q band), 46 to 56 GHz (V band), and 56 to 100 GHz (W band) (Eaves and Reedy, 1987).

Within the HF to K_a bands, specific frequencies are allocated by international agreement to radar operation. In addition, at frequencies above X band, atmospheric attenuation of electromagnetic waves becomes significant.

TABLE 1.1 Letter Nomenclature for Nominal Radar Frequency Bands

Band	Frequencies	Wavelengths
HF	3–30 MHz	100–10 m
VHF	30–300 MHz	10–1 m
UHF	300 MHz–1 GHz	1–30 cm
L	1–2 GHz	30–15 cm
S	2–4 GHz	15–7.5 cm
C	4–8 GHz	7.5–3.75 cm
X	8–12 GHz	3.75–2.5 cm
K_u	12–18 GHz	2.5–1.67 cm
K	18–27 GHz	1.67–1.11 cm
K_a	27–40 GHz	1.11 cm–7.5 mm
mm	40–300 GHz	7.5–1 mm

Consequently, radar in these bands usually operates at an "atmospheric window" frequency where attenuation is relatively low. Figure 1.3 illustrates the atmospheric attenuation for one-way propagation over the most common radar frequency ranges under one set of atmospheric conditions. Most K_a band radars operate near 35 GHz and most W band systems operate near 95 GHz because of the relatively low atmospheric attenuation at these wavelengths.

Lower radar frequencies tend to be preferred for longer range surveillance applications because of the low atmospheric attenuation and high available powers. Higher frequencies tend to be preferred for higher resolution, shorter range applications due to the smaller achievable antenna beamwidths for a given antenna size, higher attenuation, and lower available powers.

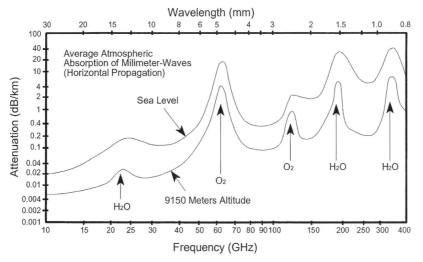

Figure 1.3 One-way atmospheric attenuation of electromagnetic waves. (*Source: EW and Radar Systems Engineering Handbook, Naval Air Warfare Center, Weapons Division, http://ewhdbks.mugu.navy.mil/*)

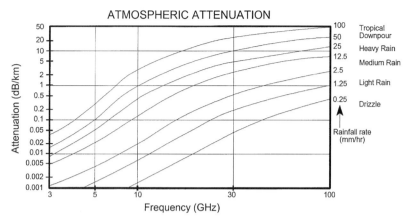

Figure 1.4 Effect of different rates of precipitation on one-way atmospheric attenuation of electromagnetic waves. (Source: *EW and Radar Systems Engineering Handbook*, Naval Air Warfare Center, Weapons Division, http://ewhdbks.mugu.navy.mil/)

Weather conditions can also have a significant effect on radar signal propagation. Figure 1.4 illustrates the additional one-way loss as a function of RF frequency for rain rates ranging from a drizzle to a tropical downpour. X-band frequencies (typically 10 GHz) and below are affected significantly only by very severe rainfall, while millimeter wave frequencies suffer severe losses for even light-to-medium rain rates.

Radar transmitters operate at peak powers ranging from milliwatts to in excess of 10 MW. One of the more powerful existing transmitters is found in the AN/FPS-108 COBRA DANE radar, which has a peak power of 15.4 MW (Brookner, 1988). The interval between pulses is called the *pulse repetition interval* (PRI), and its inverse is the *pulse repetition frequency* (PRF). PRF varies widely but is typically between several hundred pulses per second (pps) and several tens of thousands of pulses per second. The duty cycle of pulsed systems is usually relatively low and often well below 1 percent, so that average powers rarely exceed 10 to 20 kW. COBRA DANE again offers an extreme example with its average power of 0.92 MW. Pulse lengths are most often between about 100 ns and 100 μs, though some systems use pulses as short as a few nanoseconds while others have extremely long pulses, on the order of 1 ms.

It will be seen (Chap. 6) that the detection performance achievable by a radar improves with the amount of energy in the transmitted waveform. To maximize detection range, most radar systems try to maximize the transmitted power. One way to do this is to always operate the transmitter at full power during a pulse. Thus, radars generally do not use amplitude modulation of the transmitted pulse. On the other hand, the nominal range resolution ΔR is determined by the waveform bandwidth β according to

$$\Delta R = \frac{c}{2\beta} \qquad (1.2)$$

(Chap. 4). For an unmodulated pulse, the bandwidth is inversely proportional to its duration. To increase waveform bandwidth for a given pulse length without sacrificing energy, many radars routinely use phase or frequency modulation of the pulse.

Desirable values of range resolution vary from a few kilometers in long-range surveillance systems, which tend to operate at lower RFs, to a meter or less in very high-resolution imaging systems, which tend to operate at high RFs. Corresponding waveform bandwidths are on the order of 100 kHz to 1 GHz, and are typically 1 percent or less of the RF. Few radars achieve 10 percent bandwidth. Thus, most radar waveforms can be considered narrowband, bandpass functions.

1.3.2 Antennas

The antenna plays a major role in determining the sensitivity and angular resolution of the radar. A wide variety of antenna types are used in radar systems. Some of the more common types are parabolic reflector antennas, scanning feed antennas, lens antennas, and phased array antennas.

From a signal processing perspective, the most important properties of an antenna are its gain, beamwidth, and side lobe levels. Each of these follows from consideration of the antenna *power pattern*. The power pattern $P(\theta, \phi)$ describes the radiation intensity during transmission in the direction (θ, ϕ) relative to the antenna boresight. Aside from scale factors, which are unimportant for normalized patterns, it is related to the radiated electric field intensity $E(\theta, \phi)$, known as the antenna *voltage pattern*, according to

$$P(\theta, \phi) = |E(\theta, \phi)|^2 \quad (1.3)$$

For a rectangular aperture with an illumination function that is separable in the two aperture dimensions, $P(\theta, \phi)$ can be factored as the product of separate one-dimensional patterns (Stutzman and Thiele, 1998):

$$P(\theta, \phi) = P_\theta(\theta) P_\phi(\phi) \quad (1.4)$$

For most radar scenarios, only the *far-field* (also called *Fraunhofer*) power pattern is of interest. The far-field is conventionally defined to begin at a range of D^2/λ or $2D^2/\lambda$ for an antenna of aperture size D. Consider the azimuth (θ) pattern of the one-dimensional linear aperture geometry shown in Fig. 1.5. From a signal processing viewpoint, an important property of aperture antennas (such as flat plate arrays and parabolic reflectors) is that the electric field intensity as a function of azimuth $E(\theta)$ in the far field is just the inverse Fourier transform of the distribution $A(y)$ of current across the aperture in the azimuth plane (Bracewell, 1999; Skolnik, 2001):

$$E(\theta) = \int_{-D_y/2}^{D_y/2} A(y) e^{\left[j \frac{2\pi y}{\lambda} \sin\theta\right]} dy \quad (1.5)$$

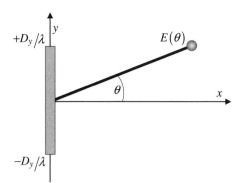

Figure 1.5 Geometry for one-dimensional electric field calculation on a rectangular aperture.

where the "frequency" variable is $(2\pi/\lambda)\sin\theta$ and is in radians per meter. The idea of spatial frequency will be expanded in Sec. 1.4.2.

To be more explicit about this point, define $s = \sin\theta$ and $\zeta = y/\lambda$. Substituting these definitions in Eq. (1.5) gives

$$\int_{-D_y/2\lambda}^{D_y/2\lambda} A(\lambda\zeta)e^{j2\pi\zeta s}d\zeta = \hat{E}(s) \qquad (1.6)$$

which is clearly of the form of an inverse Fourier transform. (The finite integral limits are due to the finite support of the aperture.) Note that $\hat{E}(s) = E(\sin^{-1}\theta)$, that is, $\hat{E}(s)$ and $E(\theta)$ are related by a nonlinear mapping of the θ or s axis. It of course follows that

$$A(\lambda\zeta) = \int_{-\infty}^{+\infty} \hat{E}(s)e^{j2\pi\zeta s}ds \qquad (1.7)$$

The infinite limits in Eq. (1.7) are misleading, since the variable of integration $s = \sin\theta$ can only range from -1 to $+1$. Because of this, $\hat{E}(s)$ is zero outside of this range on s.

Equation (1.5) is a somewhat simplified expression, which neglects a range-dependent overall phase factor and a slight amplitude dependence on range (Balanis, 1982). This Fourier transform property of antenna patterns will, in Chap. 2, allow the use of linear system concepts to understand the effects of the antenna on cross-range resolution and the pulse repetition frequencies needed to avoid spatial aliasing.

An important special case of Eq. (1.5) occurs when the aperture current illumination is a constant, $A(z) = A_0$. The normalized far field voltage pattern is then the familiar sinc function,

$$E(\theta) = \frac{\sin[\pi(D_y/\lambda)\sin\theta]}{\pi(D_y/\lambda)\sin\theta} \qquad (1.8)$$

If the aperture current illumination is separable, then the far field is the product of two Fourier transforms, one in azimuth (θ) and one in elevation (ϕ).

The magnitude of $E(\theta)$ is illustrated in Fig. 1.6, along with the definitions for two important figures of merit of an antenna pattern. The angular resolution of the antenna is determined by the width of its main lobe, and is conventionally expressed in terms of the *3-dB beamwidth*. This can be found by setting $E(\theta) = 1/\sqrt{2}$ and solving for the argument $\alpha = \pi(D_y/\lambda)\sin\theta$. The answer can be found numerically to be $\alpha = 1.4$, which gives the value of θ at the -3-dB point as $\theta_0 = 0.445\lambda/D_y$. The 3-dB beamwidth extends from $-\theta_0$ to $+\theta_0$ and is therefore

$$\text{3-dB beamwidth} \equiv \theta_3 = 2\sin^{-1}\left(\frac{1.4\lambda}{\pi D_y}\right) \approx 0.89\frac{\lambda}{D_y} \text{ radians} \qquad (1.9)$$

Thus, the 3-dB beamwidth is 0.89 divided by the aperture size in wavelengths. Note that a smaller beamwidth requires a larger aperture or a shorter wavelength. Typical beamwidths range from as little as a few tenths of a degree to several degrees for a *pencil beam antenna* where the beam is made as narrow as possible in both azimuth and elevation. Some antennas are deliberately designed to have broad vertical beamwidths of several tens of degrees for convenience in wide area search; these designs are called *fan beam antennas*.

The *peak side lobe* of the pattern affects how echoes from one scatterer affect the detection of neighboring scatterers. For the uniform illumination pattern, the peak side lobe is 13.2 dB below the main lobe peak. This is often considered too high in radar systems. Antenna side lobes can be reduced by use of a nonuniform aperture distribution (Skolnik, 2001), sometimes referred to as *tapering* of

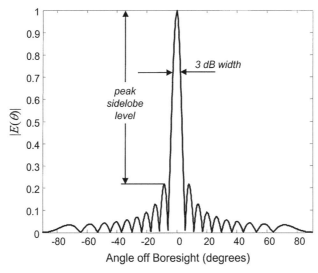

Figure 1.6 One-way radiation pattern of a uniformly illuminated aperture. The 3-dB beamwidth and peak side lobe definitions are illustrated.

shading the antenna. In fact, this is no different from the window or weighting functions used for side lobe control in other areas of signal processing such as digital filter design, and peak side lobes can easily be reduced to around 25 to 40 dB at the expense of an increase in main lobe width. Lower side lobes are possible, but are difficult to achieve due to manufacturing imperfections and inherent design limitations.

The factor of 0.89 in Eq. (1.9) is often dropped, thus roughly estimating the 3-dB beamwidth of the uniformly illuminated aperture as just λ/D_y radians. In fact, this is the 4-dB beamwidth, but since aperture weighting spreads the main lobe it is a good rule of thumb.

The antenna *power gain* G is the ratio of peak radiation intensity from the antenna to the radiation that would be observed from a lossless, isotropic (omnidirectional) antenna if both have the same input power. Power gain is determined by both the antenna pattern and by losses in the antenna. A useful rule of thumb for a typical antenna is (Stutzman, 1998)

$$G \approx \frac{26,000}{\theta_3 \phi_3} \quad (\theta_3, \phi_3 \text{ in degrees})$$
$$= \frac{7.9}{\theta_3 \phi_3} \quad (\theta_3, \phi_3 \text{ in radians}) \qquad (1.10)$$

Though both higher and lower values are possible, typical radar antennas have gains from about 10 dB for a broad fan-beam search antenna to approximately 40 dB for a pencil beam that might be used for both search and track.

Effective aperture A_e is an important characteristic in describing the behavior of an antenna being used for reception. If a wave with power density W W/m² is incident on the antenna, and the power delivered to the antenna load is P, the effective aperture is defined as the ratio (Balanis, 1982)

$$A_e = \frac{P}{W} \quad \text{m}^2 \qquad (1.11)$$

Thus, the effective aperture is the area A_e such that, if all of the power incident on the area was collected and delivered to the load with no loss, it would account for all of the observed power output of the actual antenna. (Note, however, that A_e is *not* the actual physical area of the antenna. It is a fictional area that accounts for the amount of incident power density captured by the receiving antenna.) Effective aperture is directly related to antenna directivity, which in turn is related to antenna gain and efficiency. For most antennas, the efficiency is near unity and the effective aperture and gain are related by (Balanis, 1982)

$$G = \frac{4\pi}{\lambda^2} A_e \qquad (1.12)$$

Two more useful antenna concepts are the antenna *phase front* (or *wave front*) and *phase center* (Balanis, 1982; Sherman, 1984). A phase front of a radiating

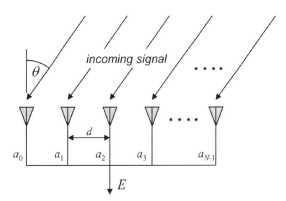

Figure 1.7 Geometry of the uniform linear array antenna.

antenna is any surface on which the phase of the field is a constant. In the far field, the phase fronts are usually approximately spherical, at least over localized regions. The phase center of the antenna is the center of curvature of the phase fronts. Put another way, the phase center is the point at which an isotropic radiator should be located so that the resulting phase fronts best match those of the actual antenna. The phase center concept is useful because it defines an effective location of the antenna, which can in turn be used for analyzing effective path lengths, Doppler shifts, and so forth. For symmetrically illuminated aperture antennas, the phase center will be centered in the aperture plane, but may be displaced forward or backward from the actual aperture. Referring to Fig. 1.5, the phase center would occur at $y = 0$ and $z = 0$, but possibly $z \neq 0$, depending on the detailed antenna shape.

Another important type of antenna is the *array antenna*. An array antenna is one composed of a collection of individual antennas called array *elements*. The elements are typically identical dipoles or other simple antennas with very broad patterns. Usually, the elements are evenly spaced to form a *uniform linear array* as shown in Fig. 1.7. Figure 1.8 illustrates examples of real array and aperture antennas.

The voltage pattern for the linear array is most easily arrived at by considering the antenna in its receive, rather than transmit mode. Suppose the leftmost element is taken as a reference point, there are N elements in the array, and the elements are isotropic (unity gain for all θ). The signal in branch n is weighted with the complex weight a_n. For an incoming electric field $E_0 \exp(j\Omega t)$ at the reference element, the total output voltage E can easily be shown to be (Stutzman and Thiele, 1998; Skolnik, 2001)

$$E(\theta) = E_0 \sum_{n=0}^{N-1} a_n e^{j(2\pi/\lambda)nd\sin\theta} \qquad (1.13)$$

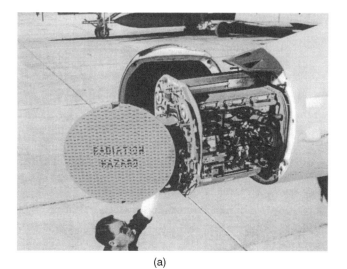

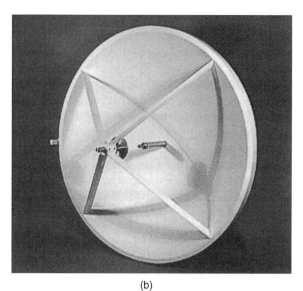

Figure 1.8 Examples of typical array and aperture antennas. (*a*) Slotted phased array in the nose of an F/A-18 aircraft. This antenna is part of the AN/APG-73 radar system. (*b*) A Cassegrain reflector antenna. (Image (*a*) courtesy of Raytheon Corp. Image (*b*) courtesy of Quinstar Corp. Used with permission.)

This is similar in form to the *discrete Fourier transform* (DFT) of the weight sequence $\{a_n\}$. Like the aperture antenna, the antenna pattern of the linear array thus involves a Fourier transform, this time of the weight sequence (which determines the current distribution in the antenna). For the case where all the

$a_n = 1$, the pattern is the familiar "aliased sinc" function, whose magnitude is

$$|E(\theta)| = E_0 \left| \frac{\sin[N(\pi d/\lambda)\sin\theta]}{\sin[(\pi d/\lambda)\sin\theta]} \right| \qquad (1.14)$$

This function is very similar to that of Eq. (1.8) and Fig. 1.6. If the number of elements N is reasonably large (nine or more) and the product Nd is considered to be the total aperture size D, the 3-dB beamwidth is $0.89\lambda/D$ and the first side lobe is 13.2 dB below the main lobe peak; both numbers are the same as those of the uniformly illuminated aperture antenna. Of course, by varying the amplitudes of the weights a_n, it is possible to reduce the side lobes at the expense of a broader main lobe. The phase center is at the center of the array.

Actual array elements are not isotropic radiators. A simple model often used as a first-order approximation to a typical element pattern $E_{el}(\theta)$ is

$$E_{el}(\theta) \approx \cos\theta \qquad (1.15)$$

The right-hand side of Eq. (1.13) is then called the *array factor* $AF(\theta)$, and the composite radiation pattern becomes

$$E(\theta) = AF(\theta)E_{el}(\theta) \qquad (1.16)$$

Because the cosine function is slowly varying in θ, the beamwidth and first side lobe level are not greatly changed by including the element pattern for signals arriving at angles near broadside (near $\theta = 0°$). The element pattern does reduce distant side lobes, thereby reducing sensitivity to waves impinging on the array from off broadside.

The discussion so far has been phrased in terms of the transmit antenna pattern (for aperture antennas) or the receive pattern (for arrays), but not both. The patterns described have been *one-way antenna patterns*. The reciprocity theorem guarantees that the receive antenna pattern is identical to the transmit antenna pattern (Balanis, 1982). Consequently, for a monostatic radar, the *two-way antenna pattern* (power or voltage) is just the square of the corresponding one-way pattern. It also follows that the antenna phase center is the same in both transmit and receive modes.

1.3.3 Receivers

It was shown in Sec. 1.3.1 that radar signals are usually narrowband, bandpass, phase or frequency modulated functions. This means that the echo waveform $r(t)$ received from a single scatterer is of the form

$$r(t) = A(t)\sin[\Omega t + \theta(t)] \qquad (1.17)$$

where the amplitude modulation $A(t)$ represents only the pulse envelope. The major function of the receiver processing is demodulation of the information-bearing part of the radar signal to baseband, with the goal of measuring $\theta(t)$.

Figure 1.9 illustrates the conventional approach to receiver design used in most classical radars.

The received signal is split into two channels. One channel, called the *in-phase* or "I" channel of the receiver (the lower branch in Fig. 1.9) mixes the received signal with an oscillator, called the *local oscillator* (LO), at the radar frequency. This generates both sum and difference frequency components:

$$2\sin(\Omega t)A(t)\sin[\Omega t + \theta(t)] = A(t)\cos[\theta(t)] - A(t)\cos[2\Omega t + \theta(t)] \quad (1.18)$$

The sum term is then removed by the lowpass filter, leaving only the modulation term $A(t)\cos[\theta(t)]$. The other channel, called the *quadrature* phase or "Q" channel, mixes the signal with an oscillator having the same frequency but a 90° phase shift from the I channel oscillator. The Q channel mixer output is

$$2\cos(\Omega t)A(t)\sin[\Omega t + \theta(t)] = A(t)\sin[\theta(t)] + A(t)\sin[2\Omega t + \theta(t)] \quad (1.19)$$

which, after filtering, leaves the modulation term $A(t)\sin[\theta(t)]$. If the input $r(t)$ is written as $A(t)\cos[\Omega t + \theta(t)]$ instead, the upper channel of Fig. 1.9 becomes the I channel and the lower the Q channel, with outputs $A(t)\cos[\theta(t)]$ and $-A(t)\sin[\theta(t)]$, respectively. In general, the I channel is the one where the oscillator function (sine or cosine) is the same as that used in modeling the signal.

The reason that both the I and Q channels are needed is that either one alone does not provide sufficient information to determine the phase modulation $\theta(t)$ unambiguously. Figure 1.10 illustrates the problem. Consider the case shown in Fig. 1.10a. The signal phase $\theta(t)$ is represented as a solid black phasor in the complex plane. If only the I channel is implemented in the receiver, only the cosine of $\theta(t)$ will be measured. In this case, the true phasor will be indistinguishable from the gray phasor $-\theta(t)$. Similarly, if only the Q channel is implemented so that only the sine of $\theta(t)$ is measured, then the true phasor will be indistinguishable from the gray phasor of Fig. 1.10b, which corresponds to $\pi - \theta(t)$. When both the I and Q channels are implemented, the phasor quadrant

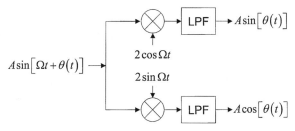

Figure 1.9 Conventional quadrature channel receiver model. In this illustration, the lower channel is the in-phase ("I") channel, and the upper is the quadrature phase ("Q") channel.

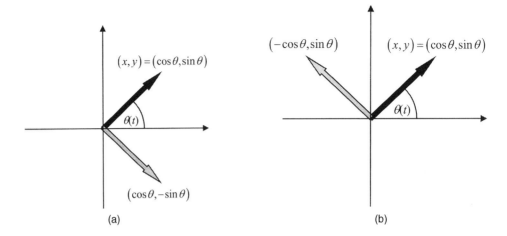

Figure 1.10 (*a*) The I channel of the receiver in Fig. 1.9 measures only the cosine of the phasor $\theta(t)$. (*b*) The Q channel measures only the sine of the phasor.

is determined unambiguously.[†] In fact, the signal processor will normally assign the I signal to be the real part of a complex signal and the Q signal to be the imaginary part, forming a single complex signal

$$x(t) = \mathrm{I}(t) + j\mathrm{Q}(t) = e^{j\theta(t)} \tag{1.20}$$

Equation (1.20) implies a more convenient way of representing the effect of an ideal coherent receiver on a transmitted signal. Instead of representing the transmitted signal by a sine function, an equivalent complex exponential function is used instead. The echo signal of (1.17) is thus replaced by

$$r(t) = A(t)e^{j[\Omega t + \theta(t)]} \tag{1.21}$$

The receiver structure of Fig. 1.9 is then replaced with the simplified model of Fig. 1.11, where the echo is demodulated by multiplication with a complex reference oscillator $\exp(-j\Omega t)$. This technique of assuming a complex transmitted signal and corresponding complex demodulator produces exactly the same result obtained in Eq. (1.20) by explicitly modeling the real-valued signals and the I and Q channels, but is much simpler and more compact. This complex exponential analysis approach is used throughout the remainder of the book. It is important to remember that this is an analysis technique; actual analog hardware must still operate with real-valued signals only. However, once signals are digitized, they may be treated explicitly as complex signals in the digital processor.

[†]This is analogous to the use of the two-argument atan2( ) function instead of the single-argument atan( ) function in many programming languages such as FORTRAN or C.

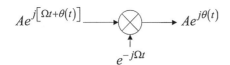

Figure 1.11 Simplified transmission and receiver model using complex exponential signals.

Figure 1.9 implies several requirements on a high-quality receiver design. For example, the local oscillator and the transmitter frequencies must be identical. This is usually ensured by having a single *stable local oscillator* (STALO) in the radar system that provides a frequency reference for both the transmitter and the receiver. Furthermore, many types of radar processing require *coherent* operation. The IEEE *Standard Radar Definitions* defines "coherent signal processing" as "echo integration, filtering, or detection using amplitude *and phase* of the signal referred to a coherent oscillator" (emphasis added) (IEEE, 1982). Coherency is a stronger requirement than frequency stability. In practice, it means that the transmitted carrier signal must have a fixed phase reference for several, perhaps many, consecutive pulses. Consider a pulse transmitted at time t_1 of the form $a(t - t_1) \sin[\Omega(t - t_1) + \phi]$, where $a(t)$ is the pulse shape. In a coherent system, a pulse transmitted at time t_2 will be of the form $a(t - t_2) \sin[\Omega(t - t_1) + \phi]$. Note that both pulses have the same argument $\Omega(t - t_1) + \phi$ for their sine term; only the envelope term changes location on the time axis. Thus, both sinusoids are referenced to the same absolute starting time and phase. This is as opposed to the second pulse being of the form $a(t - t_2) \sin[\Omega(t - t_2) + \phi']$, which is nonzero over the same time interval as the coherent pulse $a(t - t_2) \sin[\Omega(t - t_1) + \phi]$ and has the same frequency, but has a different phase at any instant in time. Figure 1.12 illustrates the difference visually. In the coherent case, the two pulses appear as if they were excised from the same continuous, stable sinusoid; in the noncoherent case, the second pulse is not in phase with the extension of the first pulse. Because of the phase ambiguity discussed earlier, coherency also implies a system having both I and Q channels.

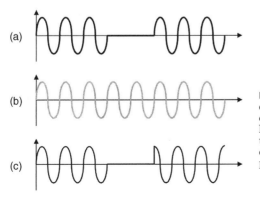

Figure 1.12 Illustration of the concept of a fixed phase reference in coherent signals. (*a*) Coherent pulse pair generated from the reference sinusoid. (*b*) Reference sinusoid. (*c*) Noncoherent pulse pair.

Another requirement is that the I and Q channels have perfectly matched transfer functions over the signal bandwidth. Thus, the gain through each of the two signal paths must be identical, as must be the phase delay (electrical length) of the two channels. Of course, real receivers do not have perfectly matched channels. The effect of gain and phase imbalances will be considered in Chap. 3. Finally, a related requirement is that the oscillators used to demodulate the I and Q channels must be exactly in quadrature, that is, 90° out of phase with one another.

In the receiver structure shown in Fig. 1.9, the information-bearing portion of the signal is demodulated from the carrier frequency to baseband in a single mixing operation. While convenient for analysis, pulsed radar receivers are virtually never implemented this way in practice. One reason is that active electronic devices introduce various types of noise into their output signal, such as *shot noise* and *thermal noise*. One noise component, known as *flicker noise* or $1/F$ *noise*, has a power spectrum that behaves approximately as F^{-1} and is therefore strongest near zero frequency. Since received radar signals are very weak, they can be corrupted by $1/F$ noise if they are translated to baseband before being amplified.

Figure 1.13 shows a more representative *superheterodyne* receiver structure. The received signal, which is very weak, is amplified immediately upon reception using a *low-noise amplifier* (LNA). The LNA, more than any other component, determines the *noise figure* of the overall receiver. It will be seen in Sec. 2.3 that this is an important factor in determining the radar's *signal-to-noise ratio* (SNR), so good design of the LNA is important. The key feature of the superheterodyne receiver is that the demodulation to baseband occurs in two or more stages. First, the signal is modulated to an IF, where it receives additional amplification. Amplification at IF is easier because of the greater percentage bandwidth of the signal and the lower cost of IF components compared to microwave components. In addition, modulation to IF rather than to baseband incurs a lower conversion loss, improving the receiver sensitivity, and the extra IF amplification also reduces the effect of flicker noise (Eaves and Reedy, 1987). Finally, the amplified signal is demodulated to baseband. Some receivers may use more than two demodulation stages (so that there are two or more IF frequencies), but two stages is the most common choice. One final advantage of the superheterodyne configuration is its adaptability. The same IF stages can

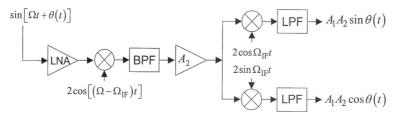

Figure 1.13 Structure of a superheterodyne radar receiver.

be used with variable RFs simply by tuning the LO so as to track changes in the transmitted frequency.

1.4 Review of Selected Signal Processing Concepts and Operations

A few basic concepts and signal processing operations appear over and over again, even if sometimes in disguise, in radar signal processing. In this section, the most frequently encountered concepts and operations are previewed to establish a simple basis for understanding the effect of many radar signal processing steps, and to point out a few aspects not often emphasized in some other application fields.

1.4.1 Resolution

The concept of a *resolution cell* will arise frequently. A resolution cell is the volume in space that contributes to the echo received by the radar at any one instant. Figure 1.14 illustrates resolution in the range dimension for a simple, constant-frequency pulse. If a pulse transmitted at time $t = 0$ has duration τ seconds, then at time t_0 the echo of the leading edge of the pulse will be received from scatterers at range $ct_0/2$. At the same time, echoes of the trailing edge of the pulse from scatterers at range $c(t_0 - \tau)/2$ are also received. Thus, scatterers distributed over $c\tau/2$ in range contribute simultaneously to the received voltage. In order to separate the contributions from two scatterers into different time samples, they must be spaced by more than $c\tau/2$ meters. The quantity $c\tau/2$ is called the *range resolution* ΔR.

This description of range resolution applies only to unmodulated, constant-frequency pulses. As will be seen in Chap. 4, pulse modulation combined with

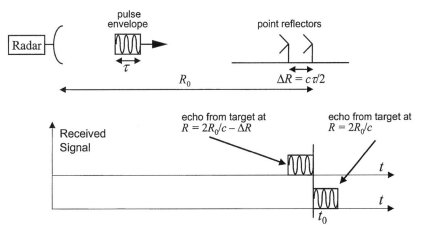

Figure 1.14 Geometry for describing conventional pulse range resolution. (See text for explanation).

matched filtering can be used to obtain range resolution finer than $c\tau/2$. However, this simplest case is adequate for introducing the concept.

Angular resolution in the azimuth and elevation dimensions is determined by the antenna beamwidths in the same planes. Two scatterers at the same range but different azimuth (or elevation) angles will contribute simultaneously to the received signal if they are within the antenna main lobe and thus are both illuminated at the same time. For the purpose of estimating angular resolution, the main lobe width is taken to be the 3-dB beamwidth θ_3 of the antenna. Thus, the two point scatterers in Fig. 1.15 located at the 3-dB edges of the beam define the angular resolution of the radar. The figure illustrates the relation between the angular resolution in radians and the equivalent resolution in units of distance, which will be called the *cross-range resolution* to denote resolution in a dimension orthogonal to range. The arc length at a radius R for an angle subtending θ_3 radians is exactly $R\theta_3$. The cross-range resolution ΔCR is the distance between two scatterers located at the 3-dB edges of the beam, corresponding to the dashed line in Fig. 1.15, and is given by

$$\Delta CR = 2R \sin\left(\frac{\theta_3}{2}\right) \approx R\theta_3 \qquad (1.22)$$

where the approximation holds when the 3-dB beamwidth is small, which is usually the case for pencil beam antennas. This result is applicable in either the azimuth or elevation dimension.

Three details bear mentioning. First, the literature frequently fails to specify whether one- or two-way 3-dB beamwidth is required or given. The two-way beamwidth should be used for monostatic radar. Second, note that cross-range resolution increases linearly with range, whereas range resolution was a constant. Finally, as with range resolution, it will be seen later (Chap. 8) that signal processing techniques can be used to improve resolution far beyond the conventional $R\theta$ limit and to make it independent of range as well.

The radar resolution cell volume ΔV is approximately the product of the total solid angle subtended by the 3-dB antenna main lobe, converted to units of area, and the range resolution. For an antenna having an elliptical beam

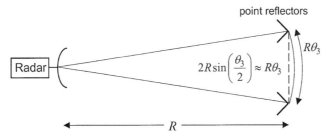

Figure 1.15 The angular resolution is determined by the 3-dB antenna beamwidth θ_3.

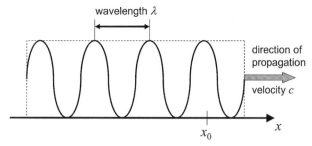

Figure 1.16 Temporal and spatial frequency for a sinusoidal pulse.

with azimuth and elevation beamwidths θ_3 and ϕ_3, this is

$$\Delta V = \pi \left(\frac{R\theta_3}{2}\right)\left(\frac{R\phi_3}{2}\right)\Delta R = \frac{\pi}{4}R^2\theta_3\phi_3\Delta R$$
$$\approx R^2\theta_3\phi_3\Delta R \qquad (1.23)$$

The approximation in the second line of Eq. (1.23) is 27 percent larger than the expression in the first line, but is widely used. Note that resolution cell volume increases with the square of range, because of the two-dimensional spreading of the beam at longer ranges.

1.4.2 Spatial frequency

In the discussion of antennas in Sec. 1.3.2 the idea of "spatial frequency" was introduced. This is an important concept in any study involving propagating waves, and it will be needed to analyze spatial sampling and space-time adaptive processing. A simplified, intuitive introduction to the concept follows in this section. For a more complete discussion, see Johnson and Dudgeon (1993).

Consider the sinusoidal pulse propagating in the $+x$ direction with wavelength λ and velocity c as shown in Fig. 1.16. An observer at a fixed spatial position x_0 will see successive crests of the electric field at a time interval (period) of $T = \lambda/c$ seconds; thus the temporal frequency of the wave is the usual $F = 1/T = c/\lambda$ hertz or $2\pi c/\lambda$ radians per second.

A spatial period can also be defined; it is simply the interval between successive crests in space for a fixed observation time. From Fig. 1.16, the spatial period of the pulse is obviously λ meters. The spatial frequency is therefore $1/\lambda$ cycles per meter or $2\pi/\lambda$ radians per meter. It is common to call the latter quantity the *wavenumber* of the pulse and to denote it with the symbol K.[†]

[†]Most radar and array processing literature uses a lower case k for spatial frequency. Here upper case K is used in keeping with the convention to use uppercase letters for analog quantities and lower case for quantities normalized by a sampling interval, such as F and f for analog and normalized frequency.

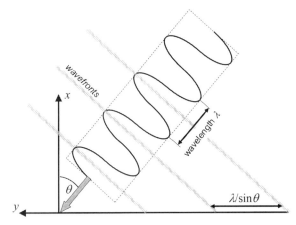

Figure 1.17 Spatial frequency and wavelength in the y direction, expressed in terms of the spatial frequency in the direction of propagation and the incidence angle with respect to the x axis.

Because position in space and velocity are three-dimensional vector quantities in general, so is the wavenumber. For simplicity of illustration, consider the two-dimensional version of Fig. 1.16 shown in Fig. 1.17. The pulse, now propagating at an angle in an xy plane, still has a wavenumber $K = 2\pi/\lambda$ in the direction of propagation. However, measured in the $+x$ direction, the wavenumber is $K_x = (2\pi/\lambda)\cos\theta$, where θ is the angle of incidence of the pulse, measured relative to the $+x$ axis. Similarly, the same signal has a wavenumber in the y direction of $K_y = (2\pi/\lambda)\sin\theta$.[†] Note that as $\theta \to 0$, the wavelength in the y dimension tends to ∞ so that $K_y \to 0$.

The extension to three dimensions of space is straightforward. The total wavenumber is related to the components in the obvious way

$$K = \sqrt{K_x^2 + K_y^2 + K_z^2} \qquad (1.24)$$

and always equals $2\pi/\lambda$. Note that the temporal frequency remains c/λ hertz regardless of the direction of propagation.

1.4.3 Fourier transforms

The Fourier transform is as ubiquitous in radar signal processing as in most other signal processing fields. Frequency domain representations are often used to separate desired signals from interference; the Doppler shift is a frequency

---

[†]Incidence angle will often be measured with respect to the normal to the y axis, i.e., the x axis, because this is convenient and conventional in analyzing antenna patterns. If the antenna aperture lies in the y dimension, then an incidence angle of $\theta = 0$ indicates a wave propagating normal to the aperture, i.e., in the boresight (x) direction.

domain phenomenon of critical importance; and it will be seen that in some radar systems, especially imaging systems, the collected data are related to the desired end product by a Fourier transform.

Both continuous and discrete signals are of interest, and therefore Fourier transforms are required for both. Consider a signal $x(u)$ that is a function of a continuous variable in one dimension called the *signal domain*.[†] Its Fourier transform, denoted as $X(\Omega)$, is given by

$$X(\Omega) = \int_{-\infty}^{\infty} x(u) e^{-j\Omega u} du \qquad \Omega \in (-\infty, \infty) \qquad (1.25)$$

and is said to be a function in the *transform domain*. The inverse transform is

$$x(u) = \frac{1}{2\pi} \int_{-\infty}^{\infty} X(\Omega) e^{+j\Omega u} d\Omega \qquad u \in (-\infty, \infty) \qquad (1.26)$$

In Eqs. (1.25) and (1.26), the frequency variable Ω is in radians per unit of u. For example, if $u = t$, that is, u is in units of seconds, then Ω is the usual radian frequency in units of radians per second; if u is a spatial variable in meters, then Ω is spatial frequency in units of radians per meter.

An equivalent transform pair using a cyclical frequency variable $F = \Omega/2\pi$ is

$$X(F) = \int_{-\infty}^{\infty} x(u) e^{-j 2\pi F u} du \qquad F \in (-\infty, \infty) \qquad (1.27)$$

$$x(u) = \int_{-\infty}^{\infty} X(F) e^{+j 2\pi F u} dF \qquad u \in (-\infty, \infty) \qquad (1.28)$$

If the signal domain is time ($u = t$), then F is in cycles per second, or hertz. There are many excellent textbooks on continuous-variable Fourier transforms and their properties. Two classics are Papoulis (1987) and Bracewell (1999). Papoulis uses primarily the radian frequency notation, while Bracewell uses cyclical frequency.

The continuous-variable Fourier transform is important for some analyses, particularly those relating to establishing sampling requirements; but most actual processing will be performed with discrete-variable signals. There are two classes of Fourier transforms for discrete-variable signals. Directly analogous to the continuous-variable case is the following transform pair for a

---

[†]To unify discussion of sampling in time, frequency, and space, in this discussion the signal to be sampled is referred to as a function in the *signal domain*, and its Fourier transform as a function in the *Fourier domain* so as to maintain generality.

discrete-variable signal $x[n]$:[†]

$$X(\omega) = \sum_{n=-\infty}^{\infty} x[n]e^{-j\omega n} \qquad \omega \in (-\infty, \infty) \qquad (1.29)$$

$$x[n] = \frac{1}{2\pi} \int_{-\pi}^{\pi} X(\omega)e^{+j\omega n}\,d\omega \qquad n \in (-\infty, \infty) \qquad (1.30)$$

In this pair, ω is a normalized *continuous* frequency variable with units of just radians (not radians per second or radians per meter). The normalized frequency ω is related to the conventional frequency Ω according to

$$\omega = \Omega T_s \qquad (1.31)$$

where T_s is the sampling interval. A related Fourier transform pair using a normalized cyclical frequency $f = (\Omega/2\pi)T_s = FT_s$ in cycles is given by

$$X(f) = \sum_{n=-\infty}^{\infty} x[n]e^{-j2\pi f n} \qquad f \in (-\infty, \infty) \qquad (1.32)$$

$$x[n] = \int_{-0.5}^{0.5} X(f)e^{+j2f n}df \qquad n \in (-\infty, \infty) \qquad (1.33)$$

The function $X(\omega)$ is called the *discrete-time Fourier transform* (DTFT) of $x[n]$. It is readily seen from Eq. (1.29) that $X(\omega)$ is continuous in the frequency variable ω with a period of 2π radians; that is, the DTFT repeats itself every 2π radians. Though it is defined for all ω, usually only one period of the DTFT, normally the principal period $-\pi \leq \omega < \pi$, is discussed and illustrated. Similarly, $X(f)$ has a period of 1 cycle, and the principal period is $-0.5 \leq f < 0.5$. The properties of DTFTs are described in most modern digital signal processing textbooks, for example Oppenheim and Schafer (1999).

Of course, $X(\omega)$ or $X(f)$ cannot be computed for all of the uncountably infinite possible values of the continuous frequency variable ω or f; a finite, discrete set of frequency values is needed. The *discrete Fourier transform* (DFT, not to be confused with the DTFT) is a computable Fourier transform defined for *finite-length* discrete-variable signals (Oppenheim and Schafer, 1999). The DFT and its inverse for an N-point signal $x[n]$ are given by

$$X[k] = \sum_{n=0}^{N-1} x[n]e^{-j2\pi nk/N} \qquad k \in [0, N-1] \qquad (1.34)$$

$$x[n] = \frac{1}{N}\sum_{k=0}^{N-1} X[k]e^{+j2\pi nk/N} \qquad n \in [0, N-1] \qquad (1.35)$$

[†]The fairly common convention in DSP texts of enclosing the independent variable in square brackets when it is discrete, and parentheses when it is continuous, will be followed. Thus $x(t)$ is a function of a continuous time variable, while $x[n]$ is a function of a discrete time variable.

Inspection shows that, provided the signal $x[n]$ is of finite duration N samples, $X[k]$ is simply a sampled version of $X(\omega)$ or $X(f)$, with N samples distributed uniformly across the principal period:

$$X[k] = X(\omega)|_{\omega=\frac{2\pi}{N}k} = X(f)|_{f=\frac{1}{N}k} \qquad k \in [0, N-1] \qquad (1.36)$$

In continuous-variable units, the DFT frequency sample locations are equivalent to frequencies of k/NT_s cycles per unit (hertz if T_s is in seconds) or $2\pi k/NT_s$ radians per unit. Of course, $X[k]$ is also periodic, with period N samples.

1.4.4 The sampling theorem and spectrum replication

The subject of this text is *digital* processing of radar signals. A digital signal is one that is discretized in two ways. First, it is a function of a discrete, rather than continuous, variable; one discrete variable for one-dimensional signals, two variables for two-dimensional signals, and so forth. Discrete-variable representations of what are usually continuous-variable physical quantities are therefore of interest. An example is a discrete-time sampled version of the continuous-time output of an antenna and receiver, but sampled functions of frequency and spatial position are of concern as well. The second discretization is quantization of the signal's value. Even when sampled on a discrete grid, each sample can still take on an infinity of possible values. *Quantization* is the process of mapping the continuous amplitude of a signal to one of a finite set of values. The number of permissible values is determined by the number of bits available in the quantized signal representation. Figure 1.18 illustrates the distinction among analog, quantized, and digital (sampled *and* quantized) signals. Not shown in Fig. 1.18 is a "sampled" or discrete-time signal, in which continuous values of amplitude are allowed but the time axis is sampled.

The most fundamental question in developing a sampled-variable representation is choosing the sampling interval. As usual in signal processing, the Nyquist sampling theorem, discussed at length in the book by Oppenheim and Schafer (1999), provides the needed guidance. In summary, it states that if the Fourier transform $X(U)$ of a signal $x(u)$ is bandlimited to an interval in the Fourier domain of total width β_F cyclical frequency units (equivalently, β_Ω radian frequency units), then the signal can be recovered from a set of samples taken at an interval

$$T_s < \frac{1}{\beta_F} = \frac{2\pi}{\beta_\Omega} \qquad (1.37)$$

by an appropriate interpolation filtering operation. Equivalently, the sampling rate $F_s = 1/T$ must satisfy the simple relation

$$F_s > \beta_F \qquad (1.38)$$

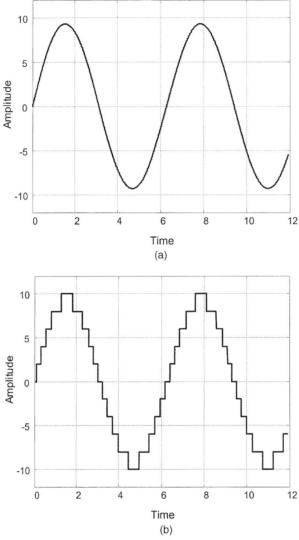

Figure 1.18 (a) Continuous-time ("analog") sinusoid. (b) Quantized sinusoid with 16 amplitude values allowed. (c) Digital sinusoid (quantized in amplitude and sampled in time).

This formula will be used to establish sampling rates for pulse echoes, Doppler processing, and phased array antenna element spacing, among other things.

The Nyquist theorem is easy to derive. Consider a possibly complex-valued signal $x(u)$ strictly bandlimited to $\pm \beta_F/2$ in cyclical frequency units:

$$X(U) \equiv 0 \quad |U| > \frac{\beta_F}{2} \tag{1.39}$$

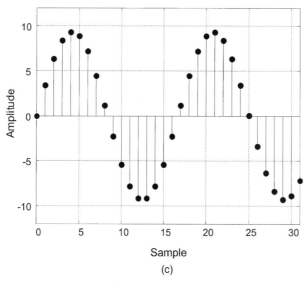

Figure 1.18 (*Continued*).

A bandlimited Fourier transform $X(U)$ is shown in Fig. 1.19a. Because $x(u)$ is complex valued, $X(U)$ does not have Hermitian symmetry in general.

The process of sampling $x(u)$ is modeled as multiplication by an infinite impulse train:

$$x_s(u) = x(u)\left[\sum_{n=-\infty}^{+\infty} \delta_D(u - nT_s)\right] \quad (1.40)$$

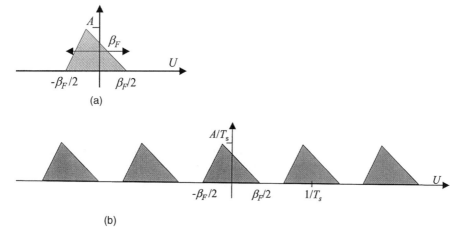

Figure 1.19 (*a*) Fourier spectrum of a bandlimited signal of total width β_F cyclical frequency units. If the original signal variable is time, then β_F is in Hertz. (*b*) Fourier spectrum of signal sampled at an interval T_s showing spectral replication.

The sampled signal $x_s(u)$ then has the Fourier transform, in cyclical frequency units, given by (Oppenheim and Schafer, 1999)

$$X_s(U) = \sum_{k=-\infty}^{+\infty} X\left(U - \frac{k}{T_s}\right) = \sum_{n=-\infty}^{+\infty} X(U - kF_s) \qquad (1.41)$$

where $F_s = 1/T$ is the sampling rate. This spectrum is shown in Fig. 1.19b, which illustrates a very important effect of sampling: infinite *replication* of the original spectrum, at an interval of $1/T_s$ units in cyclical frequency ($2\pi/T_s$ in radian units). This Fourier domain replication occurs any time a signal is subsampled at regular intervals. Because the forward and inverse Fourier transforms are dual operations, this also occurs in reverse. That is, if one samples the Fourier domain signal, then the signal is replicated in the original signal domain.

Form a discrete-time signal from the sampled data by the simple assignment

$$x[n] = x(nT_s) \qquad n \in [-\infty, +\infty] \qquad (1.42)$$

The DTFT of $x[n]$, computed using Eq. (1.29) or (1.32), is the same as $X_s(U)$, but simply expressed on a normalized frequency scale (Oppenheim and Schafer, 1999):

$$X(f) = \frac{1}{T_s} \sum_{k=-\infty}^{+\infty} X((f + k)F_s) \qquad (1.43)$$

in cyclical units, or in radian units

$$X(\omega) = \frac{1}{T_s} \sum_{k=-\infty}^{+\infty} X((\omega + 2\pi k)F_s) \qquad (1.44)$$

Figure 1.20 illustrates this using the cyclical frequency units of Eq. (1.43). Note that $X(f)$ is periodic with a period of 1, and $X(\omega)$ with period 2π, as required for spectra of discrete-variable signals. The normalized frequency variables can be translated back to the original frequency units according to

$$\omega = 2\pi U T_s = \frac{2\pi U}{F_s} \qquad f = U T_s = \frac{U}{F_s} \qquad (1.45)$$

Figure 1.20 makes intuitively clear the sampling rate necessary to enable reconstruction of the original signal $x(u)$ from its samples $x[n]$. Since a signal and its Fourier transform form a unique one-to-one pair, reconstruction of the original spectrum $X(U)$ is equivalent to reconstructing $x(u)$. Inspection of Fig. 1.20 shows that it is possible to reconstruct $X(U)$ by lowpass filtering to remove all of the spectrum replicas, provided the replica centered at $U = 0$ is undistorted and separate from the others. The strict bandlimited condition of Eq. (1.39) guarantees the replicas do not overlap if the sampling frequency

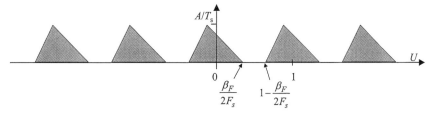

Figure 1.20 Spectrum of discrete-variable signal illustrating replication and normalization of the frequency axis.

is high enough. By inspection, the condition needed is $\beta_F/2F_s < 1 - \beta_F/2F_s$, which gives the simple result $F_s > \beta_F$ already quoted in Eq. (1.38).

Since β_F is twice the highest positive frequency $\beta_F/2$ contained in the spectrum of the original signal $x(u)$, Eq. (1.38) expresses the conventional wisdom that the sampling frequency should be at least twice the highest frequency component in the signal. However, a more direct interpretation is that the sampling frequency should be greater than the total spectral width of the signal. This interpretation is more easily generalized to nonbaseband signals.

Note that no assumption has been made that $x(u)$ is real-valued. The Nyquist theorem applies equally well to real or complex signals. One should realize, however, that in the case of a complex signal of total spectral width β_F hertz, the Nyquist criterion implies collecting at least β_F *complex* samples per second, equivalent to $2\beta_F$ real samples per second.

The fact that the original spectrum was centered at the origin in the Fourier domain was also not used; it can be offset to any location without changing the required sampling rate. This fact is the basis of the so-called "RF sampling schemes;" a related "digital IF" sampling scheme will be considered in Chap. 3. However, if the original spectrum is not centered at zero, then none of the replicas will necessarily be centered at zero in the discrete-variable signal's spectrum unless appropriate relationships between the sampling frequency, spectrum bandwidth, and spectrum offset are maintained. In the end, it is usually desirable to have the information-bearing portion of the spectrum (the gray shaded part of Fig. 1.19a) end up centered at the origin for ease of processing.

It is useful to make one observation about the implications of Eq. (1.38) for a baseband signal, meaning one whose spectrum $X(U)$ has support on $(-\beta_F/2, +\beta_F/2)$. In this case, the magnitude of the frequency of the most rapidly varying signal component present is $\beta_F/2$ hertz. The period of this signal is therefore $2/\beta_F$ seconds. Equation (1.38) equivalently requires the sampling period $T_s \leq 1/\beta_F$ seconds. Thus, the Nyquist criterion requires at least two samples per period of the highest frequency component for baseband signals.

Note also that the Nyquist theorem is not peculiar to sampling of time domain signals. It is a consequence of the replication effect of sampling on the Fourier transform of the sampled signal, and can equally well be applied to the sampling of frequency spectra or spatial signals. For example, when one samples a frequency spectrum, the corresponding time domain signal is replicated.

So long as those time domain replicas are of finite duration and do not overlap, the original spectrum can be recovered, meaning that the frequency spectrum is adequately sampled.

Finally, note that replication also occurs when "sampling" a discrete-variable signal by decimation. Consider $x[n]$, with DTFT $X(f)$. Define the *decimated* signal

$$y[n] = x[nM] \qquad -\infty < n < +\infty \qquad (1.46)$$

The signal $y[n]$ is said to be decimated by the factor M. The DTFT of $y[n]$ is related to that of $x[n]$ according to (Oppenheim and Schafer, 1999)

$$Y(f) = \frac{1}{M} \sum_{k=0}^{M-1} X\left(\frac{f-k}{M}\right) \qquad (1.47)$$

(The summation is finite because the DTFT already repeats periodically.) This is exactly the same type of scaling and replication of the spectrum seen in the sampling of a continuous-variable signal.

1.4.5 Vector representation of signals

It will sometimes be convenient to represent a finite-length signal as a vector, rather than in indexed sequence notation. That is, if the signal $x[n]$ is defined for $0 \leq n \leq N-1$, it can be denoted in vector form as

$$\mathbf{x} = [x[0] \quad x[1] \cdots x[N-2] \quad x[N-1]]' \qquad (1.48)$$

where the apostrophe denotes the matrix transpose.[†] Signal vectors will usually be defined as column vectors, and boldface notation will be used to indicate vectors and matrices.

Many important signal processing operations can be expressed in matrix-vector form. Of particular importance is the calculation of a single output sample of a finite-impulse response (FIR) linear filter. Suppose the filter impulse response is denoted as $h[n]$, $0 \leq n \leq L-1$, and $L \leq N$. Then the filter output is given by the convolution sum:

$$y[n] = \sum_{l=0}^{L-1} h[l]\, x[n-l] \qquad (1.49)$$

Vector notation can be used to represent $h[l]$ by the L-element column vector $\mathbf{h}$; this is just the coefficients of the filter impulse response

$$\mathbf{h} = [h[0] \quad h[1] \cdots h[L-1]]' \qquad (1.50)$$

[†]In the MATLAB™ computer language, care must be taken because the apostrophe (') operator represents a conjugate transpose, not just the transpose. The MATLAB™ operator '.' performs a nonconjugate transpose.

Now define the L-element signal vector $\mathbf{x}_n$ according to

$$\mathbf{x}_n = [x[n] \quad x[n-1] \cdots x[n-L+1]]' \tag{1.51}$$

Equation (1.49) can then be written as the vector inner product

$$y[n] = \mathbf{h}'\mathbf{x}_n \tag{1.52}$$

or simply as $y = \mathbf{h}'\mathbf{x}_n$. This notation will be convenient in discussing matched filters and array processing in later chapters.

1.4.6 Data integration

Another fundamental operation in radar signal processing is *integration* of samples to improve the SIR. Both *coherent integration* and *noncoherent integration* are of interest. The former refers to integration of complex (i.e., magnitude and phase) data, while the latter refers to integration based only on the magnitude (or possibly the squared or log magnitude) of the data.

Suppose a pulse is transmitted, reflects off a target, and at the appropriate time the receiver output signal is measured, consisting of a complex echo amplitude $Ae^{j\phi}$ corrupted by additive noise w. The noise is assumed to be a sample of a random process with power σ^2. The single-pulse SNR is defined as

$$\chi_1 = \frac{\text{signal energy}}{\text{noise power}} = \frac{A^2}{\sigma^2} \tag{1.53}$$

Now suppose the measurement is repeated $N-1$ more times. One expects to measure the same deterministic echo response, but with an independent noise sample each time. Form a single measurement z by integrating (summing) the individual measurements; this sum of complex samples, retaining the phase information, is a coherent integration:

$$\begin{aligned} z &= \sum_{n=0}^{N-1} \{Ae^{j\phi} + w[n]\} \\ &= NAe^{j\phi} + \sum_{n=0}^{N-1} w[n] \end{aligned} \tag{1.54}$$

The energy in the integrated signal component is clearly N^2A^2. Provided the noise samples $w[n]$ are independent of one another and zero mean, the power in the noise component is the sum of the power in the individual noise samples. Further assuming each has the same power σ^2, the total noise power is now $N\sigma^2$. The integrated SNR becomes

$$\chi_N = \frac{N^2A^2}{N\sigma^2} = N\left(\frac{A^2}{\sigma^2}\right) = N\chi_1 \tag{1.55}$$

Coherently integrating N measurements has improved the SNR by a factor of N; this increase is called the *integration gain*. Later chapters show that, as one would expect, increasing the SNR improves detection and parameter estimation performance. The cost is the extra time, energy, and computation required to collect and combine the N pulses of data.

In coherent integration, the signal components added in phase, i.e., coherently. This is often described as adding on a *voltage* basis, since the amplitude of the integrated signal component increased by a factor of N, with the result that signal energy increased by N^2. The noise samples, whose phases varied randomly, added on a *power* basis. It is the alignment of the signal component phases that allowed the signal power to grow faster than the noise power.

Sometimes the data must be preprocessed to ensure that the signal component phases align so that a coherent integration gain can be achieved. If the target had been moving in the previous example, the signal component of the measurements would have exhibited a Doppler shift, and Eq. (1.54) would instead become

$$z = \sum_{n=0}^{N-1} \{Ae^{j(2\pi f_D n + \phi)} + w[n]\} \tag{1.56}$$

for some value of normalized Doppler frequency f_D. The signal power in this case will depend on the particular Doppler shift, but except in very fortunate cases will be less than $A^2 N^2$. However, if the Doppler shift is known in advance, the phase progression of the signal component can be compensated before summing:

$$z' = \sum_{n=0}^{N-1} e^{-j2\pi f_D n} \{Ae^{j(2\pi f_D n + \phi)} + w[n]\}$$

$$= NAe^{j\phi} + \sum_{n=0}^{N-1} e^{-j2\pi f_D n} w[n] \tag{1.57}$$

The phase correction aligns the signal component phases so that they add coherently. The noise phases are still random with respect to one another. Thus, the integrated signal power is again $N^2 A^2$, while the integrated noise power is again $N\sigma^2$, and therefore an integration gain of N is again achieved. Compensation for the phase progression so that the compensated samples add in phase can be considered a form of *motion compensation*: an estimate of the Doppler shift is obtained, its effect on the data predicted, and the data modified to reverse that effect.

In noncoherent integration, the phases are discarded and the magnitudes of the measured data samples are added:

$$z = \sum_{n=0}^{N-1} \left| Ae^{j(2\pi f_D n + \phi)} + w[n] \right| \tag{1.58}$$

Alternatively, any function of the magnitude can be used; the squared-magnitude and log-magnitude are common. The important fact is that phase information in the received signal samples is discarded. Because of the non-linear operation, noncoherent integration is much more difficult to analyze, typically requiring derivation of the probability density functions of the noise-only and signal-plus-noise cases in order to determine the effect on detection and parameter estimation. Chap. 6 will show that in many useful cases, the effect of noncoherent integration is approximately equivalent to an integration gain that increases as N^α, where α ranges from about 0.7 or 0.8 for small N to about 0.5 ($\sqrt{N}$) for large N, rather than in direct proportion to N. Thus, noncoherent integration is less efficient than coherent integration. This should not be surprising, since not all of the signal information is used.

1.4.7 Correlation

Correlation is an operation that compares one signal against a reference signal to determine their degree of similarity. *Cross-correlation* is the correlation of two different signals; *autocorrelation* is the correlation of a signal with itself. Correlation is frequently defined in both a statistical sense, as a descriptive property of random signals, and as a deterministic operation performed on actual digital signals. If a random process is *ergodic*, the two interpretations are closely linked; see Oppenheim and Schafer (1999) or any text on random signals and processes for an introduction to these concepts. The deterministic processing operation is of concern here.

Consider two signals $x[n]$ and $y[n]$ with DTFTs $X(f)$ and $Y(f)$. Their deterministic cross-correlation is defined as

$$s_{xy}[k] = \sum_{n=-\infty}^{+\infty} x[n]y^*[n+k] \qquad -\infty < k < +\infty \qquad (1.59)$$

If $x[n] = y[n]$ this is the autocorrelation of $x[n]$, denoted $s_x[k]$. The value $s_x[k]$ is called the kth autocorrelation *lag*. It is straightforward to show that the cross-correlation of $x[n]$ and $y[n]$ is identical to the convolution of $x[n]$ and $y^*[-n]$. The Fourier transform of the cross-correlation function is called the *cross-power spectrum*, and can be expressed in terms of the individual spectra:

$$S_{xy}(f) = \mathrm{F}\{s_{xy}[k]\} = X(f)Y^*(f) \qquad (1.60)$$

The Fourier transform of the autocorrelation function is usually called simply the *power spectrum*. Notice that the power spectrum S_x is the squared-magnitude of the Fourier transform of the underlying signal x:

$$S_x(f) = \mathrm{F}\{s_x[k]\} = X(f)X^*(f) = |X(f)|^2 \qquad (1.61)$$

Thus, the power spectrum is not dependent on the phase of the signal spectrum.

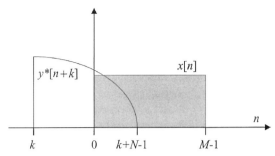

Figure 1.21 Illustration of cross-correlation $s_{xy}[k]$.

The two-dimensional versions of Eqs. (1.59) and (1.60) will also prove useful. The extensions to two dimensions are obvious:

$$s_{xy}[l,k] = \sum_{m=-\infty}^{+\infty} \sum_{n=-\infty}^{+\infty} x[m,n]y^*[m+l,n+k] \qquad -\infty < l,k < +\infty \qquad (1.62)$$

$$S_{xy}(f,g) = \mathrm{F}\{s_{xy}[l,k]\} = X(f,g)\,Y^*(f,g) \qquad (1.63)$$

Graphically, correlation corresponds to overlaying the two constituent signals; multiplying them sample by sample; and adding the results to get a single value. One of the two signals is then shifted and the process repeated, creating a series of output values which form the correlation sequence $s_{xy}[k]$. This is illustrated notionally in Fig. 1.21, which shows the cross-correlation of two functions: $x[n]$, which is nonzero for $0 \leq n \leq M - 1$, and $y[n]$, nonzero for $0 \leq n \leq N - 1$. Note that $s_{xy}[k]$ will be nonzero only for $1 - N \leq m \leq M - N$.

It is sometimes convenient to define the *normalized correlation* function

$$\rho_x[k] \equiv \frac{s_x[k]}{s_x[0]} \qquad (1.64)$$

Normalized versions of the cross-correlation function, and of the two-dimensional auto- and cross-correlation functions, are defined similarly.

The properties of correlation functions are described in many standard texts. Here only two properties of particular importance are presented, without proof

$$s_x[k] \leq s_x[0] \qquad (1.65)$$

$$s_{xy}[k] = s_{xy}^*[-k] \qquad (1.66)$$

The first property states that the zero lag ($k = 0$) of an autocorrelation function is always the peak value. In terms of Fig. 1.21, this corresponds to the case where the signal is being correlated with itself, and the two replicas are completely overlapping. In this case Eq. (1.59) specializes to

$$s_x[0] = \sum_{n=-\infty}^{+\infty} x[n]x^*[n] = \sum_{n=-\infty}^{+\infty} |x[n]|^2 \qquad (1.67)$$

Note that $s_x[0]$ is necessarily real, even if $x[n]$ is complex. Furthermore, $s_x[0]$ is the total energy in $x[n]$.

The second property establishes the symmetry of any auto- or cross-correlation function. This property is sometimes called *Hermitian* symmetry. From the properties of the Fourier transform, it follows that the power spectrum will have even symmetry if the correlation function is also real-valued.

The computation of the zero autocorrelation lag can be used as a kind of weighted integration of the signal samples. In Sec. 1.4.5 it was noted that in coherent integration, it is necessary to align the phases of the signal samples to maximize the coherent sum. This is exactly what happens in Eq. (1.67). If $x[n]$ is represented in the magnitude-phase form $Ae^{j\phi}$, multiplication by $x^*[n] = Ae^{-j\phi}$ cancels the phase component so that the products are all real, positive values and therefore add in phase. Thus when performing an autocorrelation, when the signal perfectly overlays itself, the resulting zero lag term is the weighted coherent integration of the signal samples. This observation will be useful in understanding matched filtering and synthetic aperture imaging.

Figure 1.22 illustrates one of the important uses of correlation in radar signal processing: detecting and locating desired signals in the presence of noise. The sequence in part *a* of the figure is the sum of a zero mean, unit variance Gaussian random noise signal with the finite duration pulse $y[n] = 2, 9 \leq n \leq 38$ and zero otherwise. The SNR is 6 dB. While there is some evidence of the presence of $y[n]$ in the interval $9 \leq n \leq 38$ in Fig. 1.22*a*, it does not stand out well above the noise. However, the cross-correlation of the signal with $y[n]$, shown in Fig. 1.22*b*, displays a clear peak at lag $k = 9$, indicating the presence of the signal $y[n]$ in the noisy signal of Fig. 1.22*a* and furthermore indicating that it begins nine samples from the origin. Thus, correlation can be used to aid in identification and location of known signals in noise, i.e., as a way to "look for" a signal of interest in the presence of interference.

1.5 A Preview of Basic Radar Signal Processing

There are a number of instances where the design of a component early in the radar signal processing chain is driven by properties of some later component. For example, in Chap. 4 it will be seen that the matched filter maximizes SNR; but it is not until the performance curves for the detectors that follow the matched filter are derived that it will be seen that maximizing SNR also optimizes detection performance. Until the detector is considered, it is hard to understand why it is important to maximize SNR. Having seen the major components of a typical pulsed coherent radar system, the most common signal processing operations in the radar signal processing chain are now described heuristically. By sketching out this preview of the "big picture" from beginning to end, it will be easier to understand the motivation for, and interrelation of, many of the processing operations to be described in later chapters.

Figure 1.23 illustrates one possible sequence of operations in a generic radar signal processor. The sequence shown is not unique, nor is the set of operations exhaustive. In addition, the point in the chain at which the signal is digitized

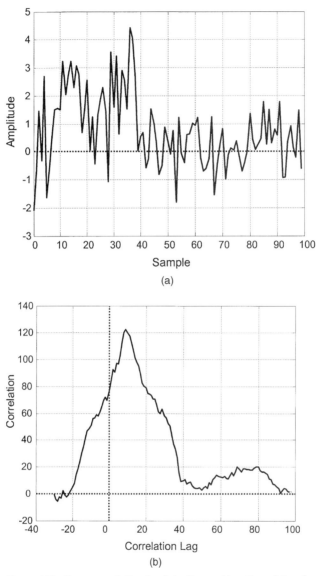

Figure 1.22 Cross-correlation for detection and location of signals corrupted by noise. (a) Signal $w[n]$ composed of unit-variance white Gaussian noise containing a 30-sample pulse $x[n]$ with amplitude 2 in the region $9 \leq n \leq 38$. (b) $s_{xw}[k]$ showing peak correlation at lag $k = 9$.

varies in different systems; it might occur as late as the output of the clutter filtering step. The operations can be generally grouped into *signal conditioning and interference suppression; imaging; detection;* and *postprocessing*. Radar signal *phenomenology* must also be considered. In the next few subsections the basic purpose and operation of each block in this signal processing chain is described.

Introduction to Radar Systems 39

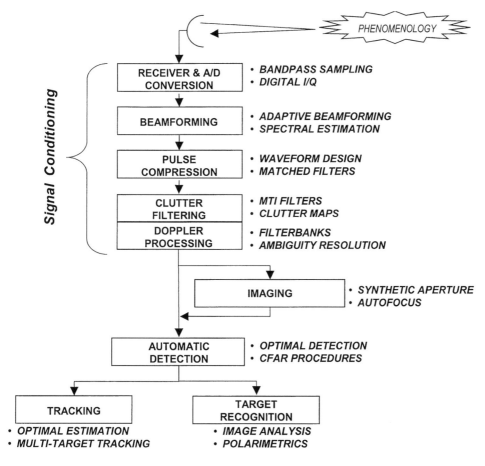

Figure 1.23 One example of a generic radar signal processor flow of operations.

1.5.1 Radar time scales

Radar signal processing operations take place on time scales ranging from less than a nanosecond to tens of seconds or longer, a range of 10 to 12 orders of magnitude. Different classes or levels of operations tend to operate on significantly different time scales. Figure 1.24 illustrates one possible association of operations and time scale.

Operations that are applied to data from a single pulse occur on the shortest time scale, often referred to as *fast time* because the sample rate, determined by the instantaneous pulse bandwidth (see Chap. 2), is on the order of hundreds of kilohertz (kHz) to as much as a few gigahertz in some cases. Corresponding sampling intervals range from a few microseconds down to a fraction of a nanosecond, and signal processing operations on these samples therefore tend to act over similar time intervals. Typical fast time operations are digital I/Q signal formation, beamforming, pulse compression or matched filtering, and sensitivity time control.

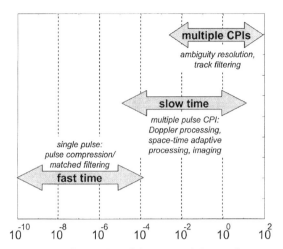

Figure 1.24 Illustration of the range of time scales over which radar signal processing

The next level up in signal processing operations operates on data from multiple pulses. The sampling interval between pulses, that is, the PRI, is typically on the order of tens of microseconds to hundreds of milliseconds, so again operations that involve multiple pulses occupy similar time scales. Due to the much slower sampling rate compared to single-pulse operations, such operations are said to act in *slow time*. Typical operations include coherent and noncoherent integration, Doppler processing of all types, synthetic aperture imaging, and space-time adaptive processing. The idea of slow and fast time will be revisited in the discussion of the data organizational concept of the datacube in Chap. 3.

A group of pulses that are to be somehow combined coherently, for example via Doppler processing or *synthetic aperture radar* (SAR) imaging, are said to form a *coherent processing interval* (CPI). A still higher level of radar processing acts on data from multiple CPIs and therefore acts on even longer time scales, typically milliseconds to ones or tens of seconds. Operations on this scale include multiple-CPI ambiguity resolution techniques, multilook SAR imaging, and track filtering. Most operations on this scale are beyond the scope of this book.

1.5.2 Phenomenology

To design a successful signal processor, the characteristics of the signals to be processed must be understood. *Phenomenology* refers to the characteristics of the signals received by the radar. Relevant characteristics include signal power, frequency, polarization, or angle of arrival; variation in time; and randomness. The received signal phenomenology is determined by both intrinsic features of the physical object(s) giving rise to the radar echo, such as their physical size, or orientation and velocity relative to the radar; and the characteristics of the radar itself, such as its transmitted waveform or polarization or antenna gain. For example, if more power is transmitted, a more powerful received echo is expected, all other things being equal.

In Chap. 2, models of the behavior of typical measured signals that are relevant to the design of signal processors is developed. The radar range equation will give a means of predicting nominal signal power. The Doppler phenomenon will predict received frequency. It will be seen that the complexity of the real world gives rise to very complex variations in radar signals; this will lead to the use of random processes to model the signals, and to particular probability density functions that match measured behavior well. It will also be shown that measured signals can be represented as the convolution of the "true" signal representing the ideal measurement with the radar waveform (in the range dimension) or its antenna pattern (in the azimuth or elevation dimension, both also called cross-range dimension). Thus, a combination of random process and linear systems theory will be used to describe radar signals and to design and analyze radar signal processors.

1.5.3 Signal conditioning and interference suppression

The first several blocks after the antenna in Fig. 1.23 can be considered as signal conditioning operations, whose purpose is to improve the SIR of the data prior to detection, parameter measurement, or imaging operations. That is, the intent of these blocks is to "clean up" the radar data as much as possible. This is done in general with a combination of fixed and adaptive *beamforming, pulse compression, clutter filtering*, and *Doppler processing*.

Beamforming is applicable when the radar antenna is an array, i.e., when there are multiple phase center signals, or *channels*, available to the signal processor. Fixed beamforming is the process of combining the outputs of the various available phase centers to form a directive gain pattern, similar to that shown in Fig. 1.6. The high-gain main lobe and low side lobes selectively enhance the echo strength from scatterers in the antenna look direction while suppressing the echoes from scatterers in other directions, typically clutter. The side lobes also provide a measure of suppression of jamming signals so long as the jammer is not in the main lobe of the antenna. By proper choice of the weights used to combine the channels, the main lobe of the beam can be steered to various look directions, and the tradeoff between the side lobe level and the main lobe width (angular resolution) can be varied.

Adaptive beamforming takes this idea a step further. By examining the correlation properties of the received data across channels, it is possible to recognize the presence of jamming and clutter entering the antenna pattern side lobes, and to design a set of weights for combining the channels such that the antenna not only has a high-gain main lobe and generally low side lobes, but also has a null in the antenna pattern at the angle of arrival of the jammer. Much greater jammer suppression can be obtained in this way. Similarly, it is also possible to increase clutter suppression by this technique. Figure 1.25 illustrates the suppression of ground clutter and jammer signals that can be obtained by this technique, allowing the previously invisible target signal to be seen and perhaps detected. The two vertical bands in Fig. 1.25a represent jammer energy, which comes from a fixed angle of arrival but is usually in the form of relatively wideband noise; thus it is present at all Doppler frequencies observed by the radar. The diagonal band in Fig. 1.25a is due to ground clutter, for which the Doppler shift depends on the angle from the radar to the ground patch contributing energy. Figure 1.25b shows that the adaptive filtering has created nulls along the loci of the jammer and clutter energy, making the target at 0° angle of arrival and 400 Hz Doppler shift apparent. Adaptive interference suppression will be introduced in Chap. 9.

Pulse compression is a special case of *matched filtering*. Many radar system designs strive for both high sensitivity in detecting targets, and high range resolution (the ability to distinguish closely spaced targets). Upcoming chapters show that target detectability improves as the transmitted energy increases, and that range resolution improves as the transmitted waveform's instantaneous bandwidth increases. If the radar employs a simple, constant-frequency rectangular envelope pulse as its transmitted waveform, then the pulse must be lengthened to increase the transmitted energy for a given power level. However, lengthening the pulse also decreases its instantaneous bandwidth, degrading the range resolution. Thus sensitivity and range resolution appear to be in conflict with one another.

Pulse compression provides a way out of this dilemma by decoupling the waveform bandwidth from its duration, thus allowing both to be independently

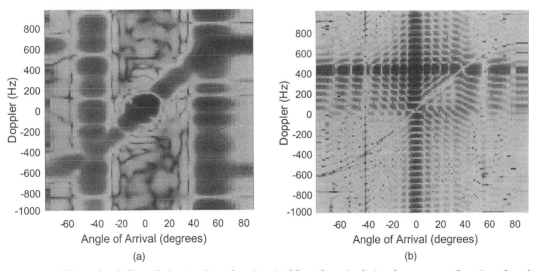

Figure 1.25 Example of effect of adaptive beamforming. (a) Map of received signal power as a function of angle of arrival and signal Doppler shift. (b) Angle-Doppler map after adaptive processing. (Images courtesy of Dr. W. L. Melvin. Used with permission.)

specified. This is done by abandoning the constant-frequency pulse and instead designing a modulated waveform. A very common choice is the linear frequency modulated (linear FM, LFM, or "chirp") waveform, shown in Fig. 1.26a. The instantaneous frequency of an LFM pulse is swept over the desired bandwidth during the pulse duration; the frequency may be swept either up or down, but the rate of frequency change is constant.

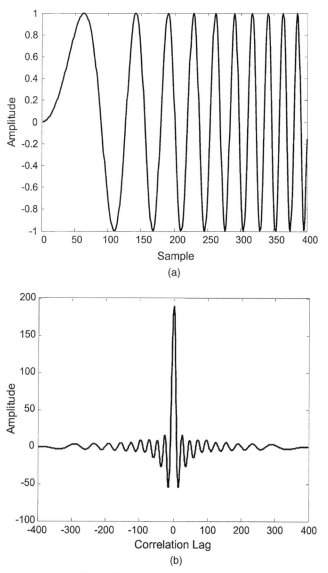

Figure 1.26 (a) Linear FM waveform modulation function, showing an increasing instantaneous frequency. (b) Output of the matched filter for the LFM waveform of (a).

The matched filter is by definition a filter in the radar receiver designed to maximize the SNR at its output. The impulse response of the filter having this property turns out to be a replica of the transmitted waveform's modulation function that has been reversed in time and conjugated; thus the impulse response is "matched" to the particular transmitted waveform modulation. Pulse compression is the process of designing a waveform and its corresponding matched filter so that the matched filter output in response to the echo from a single point scatterer concentrates most of its energy in a very short duration, thus providing good range resolution while still allowing the high transmitted energy of a long pulse. Figure 1.26b shows the output of the matched filter corresponding to the LFM pulse of Fig. 1.26a; note that the main lobe of the response is much narrower than the duration of the original pulse. The concepts of matched filtering, pulse compression, and waveform design, as well as the properties of linear FM and other common waveforms, are described in Chap. 4. There it is seen that the 3-dB width of the main lobe in time is approximately $1/\beta$ seconds, where β is the instantaneous bandwidth of the waveform used. This width determines the ability of the waveform to resolve targets in range. Converted to equivalent range units, the range resolution is given by

$$\Delta R = \frac{c}{2\beta} \qquad (1.68)$$

[This is the same as Eq. (1.2), presented earlier.]

Clutter filtering and Doppler processing are closely related. Both are techniques for improving the detectability of moving targets by suppressing interference from clutter echoes, usually from the terrain in the antenna field of view, based on differences in the Doppler shift of the echoes from the clutter and from the targets of interest. The techniques differ primarily in whether they are implemented in the time or frequency domain and in historical usage of the terminology.

Clutter filtering usually takes the form of *moving target indication*, or MTI, which is simply pulse-to-pulse highpass filtering of the radar echoes at a given range to suppress constant components, which are assumed to be due to nonmoving clutter. Extremely simple, very low-order (most commonly first- or second-order) digital filters are applied in the time domain to samples taken at a fixed range but on successive transmitted pulses.

The term "Doppler processing" generally implies the use of the fast Fourier transform algorithm, or occasionally some other spectral estimation technique, to explicitly compute the spectrum of the echo data for a fixed range across multiple pulses. Due to their different Doppler shifts, energy from moving targets is concentrated in different parts of the spectrum from the clutter energy, allowing detection and separation of the targets. Doppler processing obtains more information from the radar signals, such as number and approximate velocity of moving targets, than does MTI filtering. The cost is more required radar pulses, thus consuming energy and timeline, and greater processing complexity.

Many systems use both techniques in series. Clutter filtering and Doppler processing are the subjects of Chap. 5.

1.5.4 Imaging

Most people are familiar with the idea of a radar producing "blips" on a screen to represent targets, and in fact systems designed to detect and track moving targets may do exactly that. However, radars can also be designed to compute high-resolution images of a scene. Figure 1.27 compares the quality routinely obtainable in SAR imagery in the mid-1990s to that of an aerial photograph of the same scene; close examination reveals many similarities and many significant differences in the appearance of the scene at radar and visible wavelengths. Not surprisingly, the photograph is easier for a human to interpret and analyze, since the imaging wavelengths (visible light) and phenomenology are the same as the human visual system. In contrast, the radar image, while remarkable, is monochromatic, offers less detail, and exhibits a "speckled" texture as well as some seemingly unnatural contrast reversals. Given these drawbacks, why is radar imaging of interest?

While radars do not obtain the resolution or image quality of photographic systems, they have two powerful advantages. First, they can image a scene through clouds and inclement weather due to the superior propagation of RF wavelengths. Second, they can image equally well 24 hours a day, since they do not rely on the sun for illumination; they provide their own "light" via the transmitted pulse. Figure 1.28 is a notional example of the radar and photographic images of the Albuquerque airport that might be obtained on a rainy night.

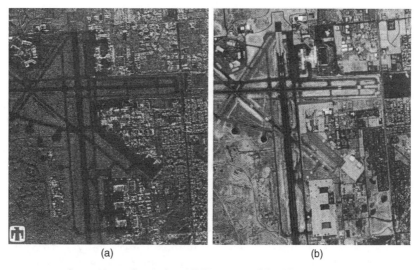

Figure 1.27 Comparison of optical and SAR images of the Albuquerque airport. (*a*) K_u band (15 GHz) SAR image, 3-m resolution. (*b*) Aerial photograph. (Images courtesy of Sandia National Laboratories. Used with permission.)

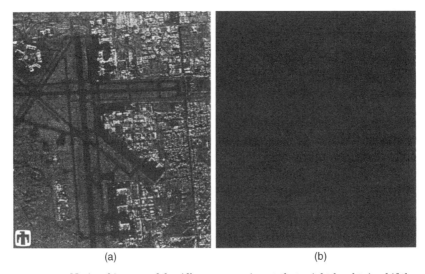

Figure 1.28 Notional images of the Albuquerque airport that might be obtained if the experiment of Fig. 1.27 were repeated on a rainy night. (a) Radar image. (b) Simulated aerial photograph. (Radar image courtesy of Sandia National Laboratories. Used with permission.)

The example is not authentic: the "visible image" is simply a black rectangle, not an actual photograph taken at night in inclement weather. Nonetheless, the point is valid: photography and other optical sensors would fail in those conditions, while radar imaging at typical ranges and wavelengths would not be affected in any noticeable way

To obtain high-resolution imagery, radars use a combination of high-bandwidth waveforms to obtain good resolution in the range dimension, and the synthetic aperture radar technique to obtain good resolution in the cross-range dimension. The desired range resolution is obtained while maintaining adequate signal energy by using pulse compression waveforms, usually linear FM. A long pulse that is nonetheless swept over a large enough bandwidth β and processed using a matched filter can provide very good range resolution according to Eq. (1.68). For example, range resolution of 1 m can be obtained with a waveform swept over 150 MHz. Depending on their applications, modern imaging radars usually have range resolution of 30 m or better; many systems have 10 m or better resolution, and some advanced systems have resolution under 1 m.

For a conventional nonimaging radar, referred to as a *real aperture* radar, the resolution in cross-range is determined by the width of the antenna beam at the range of interest and is given by $R\theta_3$ as shown in Eq. (1.22). Realistic antenna beamwidths for narrow-beam antennas are typically $1°$ to $3°$, or about 17 to 52 mrad. Even at a relatively short imaging range of 10 km, the cross-range resolution that results would be 170 to 520 m, much worse than typical range resolutions and too coarse to produce useful imagery. This poor cross-range resolution is overcome by using SAR techniques.

The synthetic aperture technique refers to the concept of synthesizing the effect of a very large antenna by having the actual physical radar antenna move in relation to the area being imaged. Thus, SAR is most commonly associated with moving airborne or space-based radars, rather than with fixed ground-based radars. Figure 1.29 illustrates the concept for the airborne case. By transmitting pulses at each indicated location, collecting all of the resulting data, and properly processing it together, a SAR system creates the effect of a large phased array antenna extending over the distance flown while collecting data. As suggested by Eq. (1.9) (though some details differ in the SAR case), a very large aperture size produces a very narrowly focused effective antenna beam, thus making possible very fine cross-range resolution. The SAR concept is explained more fully, and a modern analysis of the technique provided, in Chap. 8.

1.5.5 Detection

The most basic function of a radar signal processor is detection of the presence of one or more targets of interest. Information about the presence of targets is contained in the echoes of the radar pulses. These echoes compete with receiver noise, undesired echoes from clutter signals, and possibly intentional or unintentional jamming. The signal processor must somehow analyze the total received signal and determine whether it contains a desirable target echo and, if so, at what range, angle, and velocity.

Because the complexity of radar signals leads us to employ statistical models, detection of target echoes in the presence of competing interference signals is a

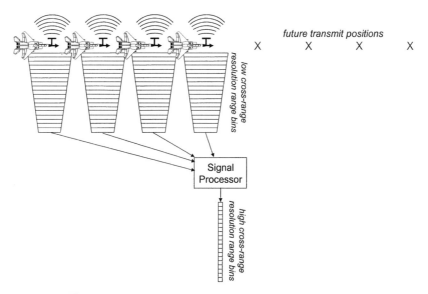

Figure 1.29 The concept of synthetic aperture radar.

problem in statistical decision theory. The theory as applied to radar detection will be developed in Chap. 6. There it will be seen that in most cases optimal performance can be obtained using the technique of *threshold detection*. In this method, the magnitude of each complex sample of the radar echo signal, possibly after signal conditioning and interference suppression, is compared to a precomputed threshold. If the signal amplitude is below the threshold, it is assumed to be due to interference signals only. If it is above the threshold, it is assumed that the stronger signal is due to the presence of a target echo in addition to the interference, and a detection or "hit" is declared. In essence, the detector makes a decision as to whether the energy in each received signal sample is too large to likely have resulted from interference alone; if so, it is assumed a target echo contributed to that sample. Figure 1.30 illustrates the concept. The "clutter + target" signal might represent the variation in received signal strength versus range (fast time) for a single transmitted pulse. It crosses the threshold at three different times, suggesting the presence of three targets at different ranges.

Because they are the result of a statistical process, threshold detection decisions have a finite probability of being wrong. For example, a noise spike could cross the threshold, leading to a false target declaration, commonly called a *false alarm*. These errors are minimized if the target spikes stand out strongly from the background interference, i.e., if the SIR is as large as possible; in this case the threshold can be set relatively high, resulting in few false alarms while still detecting most targets. This conclusion is deduced mathematically in Chap. 6. This fact also accounts for the importance of matched filtering in radar systems. The matched filter maximizes the SIR, thus providing the best threshold detection performance. Furthermore, the achievable SIR is monotonically increasing with the transmitted pulse energy E, thus encouraging the use of longer pulses to get more energy on the target. Since longer simple pulses reduce range resolution (see Sec. 1.4.1), the technique of pulse compression is also important so that high resolution can be obtained while maintaining good detection performance.

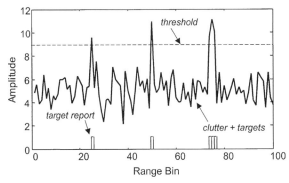

Figure 1.30 Illustration of threshold detection.

The concept of threshold detection can be applied to many different radar signal processing systems. Figure 1.30 illustrated its application to a fast-time (range) signal trace, but it can be equally well applied to a signal composed of measurements at different Doppler frequencies for a fixed range, or in a two-dimensional form to SAR imagery.

There are numerous significant details in implementing threshold detection. Various detector designs work on the magnitude, squared-magnitude, or even log-magnitude of the complex signal samples. The threshold is computed from knowledge of the interference statistics so as to limit false alarms to an acceptable rate. However, in real systems the interference statistics are rarely known accurately enough to allow for precomputing a fixed threshold. Instead, the required threshold is estimated using interference statistics estimated from the data itself, a process called *constant-false-alarm rate* (CFAR) detection. Detection processing is described in detail in Chap. 6 and 7.

1.5.6 Postprocessing

Radar systems employ a wide variety of postprocessing operations after the point of detection. Often these are referred to as *data processing* rather than signal processing operations. While these techniques are beyond the scope of this book, two of the most common postprocessing methods can be mentioned here.

Tracking is an essential component of many systems. The radar signal processor detects the presence of targets using threshold detection methods. Once detected, the signal processor may also estimate the range to the target, which is determined by the time delay after pulse transmission at which the threshold crossing occurred, the angle of the target relative to the antenna look direction, and its radial velocity using Doppler measurements. The angle measurements are obtained using angle tracking techniques, especially *monopulse tracking* (Sherman, 1984). These signal processor measurements provide a snapshot of the target location at one instant in time. The term *track filtering* describes a higher-level process of integrating a series of such measurements to compute a complete trajectory of the target motion over time. The individual position measurements will have some error due to interference, and there may also be multiple targets with crossing or closely spaced trajectories. Consequently, track filtering must also deal with the problems of determining which measurements to associate with which targets being tracked, and with correctly resolving nearby and crossing trajectories. A variety of optimal estimation techniques have been developed to perform track filtering. An excellent reference in this area is Bar-Shalom (1988).

1.6 Radar Literature

This text covers a middle ground in radar technology. It focuses on basic radar signal processing from a digital signal processing point of view. It does not address radar systems, components, or phenomenology in any great depth except where needed to explain the signal processing aspects; nor does it provide in-depth coverage of advanced radar signal processing specialties. Fortunately, there are many excellent radar reference books that address both needs. Good books appear every year; those listed in the paragraphs that follow are current as of the year 2005.

1.6.1 Radar systems and components

Probably the most classic introductory text to radar systems, now in its third edition, is by Skolnik (2001). Another classic that grew from one of the most successful continuing education courses in radar technology is by Eaves and Reedy (1987). The 1990s saw the introduction of several general radar system textbooks. The text by Edde (1995) also has an associated self-study course. Peebles (1998) provides a recent, comprehensive introduction, while Mahafza (2000) provides a number of useful MATLAB™ files to aid in simulation and experimentation. Morris and Harkness (1996) provides a good introduction to airborne pulsed Doppler systems specifically.

1.6.2 Radar signal processing

It is this author's opinion that there are a number of excellent books about radar systems in general, including coverage of components and system designs; and several on advanced radar signal processing topics, especially in the area of synthetic aperture imaging. There have been few books that address the middle ground of basic radar signal processing, such as pulse compression, Doppler filtering, and CFAR detection. Such books are needed to provide greater quantitative depth than is available in the radar systems books without restricting themselves to in-depth coverage of a single advanced application area, and this text aims to fill that gap. Nonetheless, there are a few texts that fit somewhat into this middle area. Nathanson (1991) wrote a classic book, now in its second edition, that covers radar systems in general but in fact concentrates on signal processing issues, especially RCS and clutter modeling, waveforms, MTI, and detection. Probably the closest text in intent to this one is by Levanon (1988), which provides excellent analyses of many basic signal processing functions. The new text by Levanon and Mozeson (2004) addresses the widening variety of radar waveforms in detail. A recent text by Sullivan (2000) is interesting especially for its introductory coverage of both SAR and space-time adaptive processing (STAP), thus providing a bridge between basic signal processing and more advanced texts specializing in SAR and STAP.

1.6.3 Advanced radar signal processing

Two very active areas of advanced radar signal processing research are SAR imaging, and STAP. SAR research extends back to 1951, but only in the 1990s did open literature textbooks begin to appear in the market. There are now at least 10 good textbooks on SAR. The first comprehensive text was by Curlander and McDonough (1991). Based on experience gained at the NASA Jet Propulsion Laboratory, it emphasizes space-based SAR and includes a strong component of scattering theory as well. Cumming and Wong (2005), the newest SAR text at this writing, also emphasizes spaced-based SAR. The spotlight SAR mode received considerable development in the 1990s, and two major groups published competing texts in the mid-1990s. Carrara, Goodman, and Majewski (1995) represented the work of the group at the Environmental Research Institute of Michigan (ERIM, now a part of general Dynamics, Inc.); Jakowatz, Jr., et al. (1996) represented the work of a group at Sandia National Laboratories, a unit of the U.S. Department of Energy. Franceschetti and Lanari (1999) provide a compact, unified treatment of both major modes of SAR imaging, namely stripmap and spotlight. The book by Soumekh (1999) is the most complete academic reference on synthetic aperture imaging and includes a number of MATLAB™ simulation resources.

STAP, one of the most active radar signal processing research areas, began in earnest in 1973 and is correspondingly less mature than SAR processing. Klemm (1998) wrote the first significant open literature text on the subject. Just as with the Curlander and McDonough book in the SAR community, this book was the first sign that a series of STAP texts can be expected as that research topic matures and reaches mainstream use. The book by Guerci (2003) is the newest primer on this subject at this writing, while Van Trees (2002) prepared a detailed text that continues his classic series on detection and estimation. Additionally, there are other texts on more limited forms of adaptive interference rejection. A good example is the one by Nitzberg (1999), which discusses several forms of side lobe cancellers.

1.6.4 Current radar research

Current radar research appears in a number of scientific and technical journals. The most important in the United States are the Institute of Electrical and Electronics Engineers (IEEE) *Transactions on Aerospace and Electronic Systems, Transactions on Geoscience and Remote Sensing*, and *Transactions on Image Processing*. Radar-related material in the latter is generally limited to papers related to SAR processing, especially interferometric three-dimensional SAR. In the United Kingdom, radar technology papers are often published in the Institution of Electrical Engineers (IEE) *Proceedings: Radar, Sonar, and Navigation*.

References

Balanis, C. A., *Antenna Theory*. Harper & Row, New York, 1982.
Bar-Shalom, Y., and T. E. Fortmann, *Tracking and Data Association*. Academic Press, Boston, MA, 1988.
Bracewell, R. N., *The Fourier Transform and Its Applications*, 3d ed., McGraw-Hill, New York, 1999.
Brookner, E. (ed.), *Aspects of Modern Radar*. Artech House, Boston, MA, 1988.

Carrara, W. G., R. S. Goodman, and R. M. Majewski, *Spotlight Synthetic Aperture Radar*. Artech House, Norwood, MA, 1995.

Curlander, J. C., and R. N. McDonough, *Synthetic Aperture Radar*. J. Wiley, New York, 1991.

Cumming, I. G., and F. N. Wong, *Digital Processing of Synthetic Aperture Radar Data*. Artech House, Norwood, MA, 2005.

Eaves, J. L., and E. K. Reedy, *Principles of Modern Radar*. Van Nostrand Reinhold, New York, 1987.

Edde, B., *Radar: Principles, Technology, Applications*. Prentice Hall PTR, Upper Saddle River, NJ, 1995.

EW and Radar Systems Engineering Handbook, Naval Air Warfare Center, Weapons Division., Available at http://ewhdbks.mugu.navy.mil/.

Franceschetti, G., and R. Lanari, *Synthetic Aperture Radar Processing*. CRC Press, New York, 1999.

Guerci, J. R., *Space-Time Adaptive Processing for Radar*. Artech House, Norwood, MA, 2003.

Institute of Electrical and Electronics Engineers, "IEEE Standard Letter Designations for Radar-Frequency Bands," Standard 521-1976, Nov. 30, 1976.

Institute of Electrical and Electronics Engineers, "IEEE Standard Radar Definitions," Standard 686-1982, Nov. 30, 1982.

Jakowatz, C. V., Jr., et al., *Spotlight-Mode Synthetic Aperture Radar: A Signal Processing Approach*. Kluwer, Boston, MA, 1996.

Jelalian, A. V., *Laser Radar Systems*. Artech House, Boston, MA, 1992.

Johnson, D. H., and D. E. Dudgeon, *Array Signal Processing: Concepts and Techniques*. Prentice Hall, Englewood Cliffs, NJ, 1993.

Klemm, R., *Space-Time Adaptive Processing: Principles and Applications*. INSPEC/IEEE, London, 1998.

Levanon, N., *Radar Principles*. J. Wiley, New York, 1988.

Levanon, N., and E. Mozeson, *Radar Signals*. Wiley, New York, 2004.

Mahafza, B. R., *Radar Systems Analysis and Design Using MATLAB*. Chapman & Hall/CRC, New York, 2000.

Morris, G. V., and L. Harkness (ed.), *Airborne Pulsed Doppler Radar*, 2d ed. Artech House, Boston, MA, 1996.

Nathanson, F. E., (with J. P. Reilly and M. N. Cohen), *Radar Design Principles*, 2d edition. McGraw-Hill, New York, 1991.

Nitzberg, R., *Radar Signal Processing and Adaptive Systems*, 2d ed. Artech House, Boston, MA, 1999.

Oppenheim, A. V., and R. W. Schafer, *Discrete-Time Signal Processing*, 2d ed. Prentice Hall, Englewood Cliffs, NJ, 1999.

Papoulis, A., *The Fourier Integral and its Applications*, 2d ed. McGraw-Hill, New York, 1987.

Peebles, Jr., P. Z., *Radar Principles*. Wiley, New York, 1998.

Sherman, S. M., *Monopulse Principles and Techniques*. Artech House, Boston, MA, 1984.

Skolnik, M. I., *Introduction to Radar Systems*, 3d ed. McGraw-Hill, New York, 2001.

Soumekh, M., *Synthetic Aperture Radar Signal Processing with MATLAB Algorithms*. Wiley, New York, 1999.

Stutzman, W. L., "Estimating Gain and Directivity of Antennas," IEEE *Transactions on Antennas and Propagation*, vol. 40(4), pp. 7–11, Aug. 1998.

Stutzman, W. L., and G. A. Thiele, *Antenna Theory and Design*. Wiley, New York, 1998.

Sullivan, R. J., *Microwave Radar: Imaging and Advanced Concepts*. Artech House, Boston, MA, 2000.

Swords, S. S., *Technical History of the Beginnings of RADAR*. Peter Peregrinus Ltd., London, 1986.

Van Trees, H. L., *Optimum Array Processing: Part IV of Detection, Estimation, and Modulation Theory*. Wiley, New York, 2002.

Chapter 2

Signal Models

2.1 Components of a Radar Signal

While a radar transmits a controlled, well-defined signal, the signal measured at the receiver output in response is the superposition of several major components, none of them entirely under the control of the designer. The major components are the *target, clutter, noise*, and, in some cases, *jamming*. These signals are sometimes subdivided further. For instance, clutter can be separated into ground clutter and weather clutter (such as rain), while jamming can be separated into active jamming (noise transmitters) and passive jamming (such as chaff clouds). Signal processing is applied to this composite signal; the goal is to extract useful information regarding the presence of targets and their characteristics, or to form a radar image. Noise and jamming are interference signals; they degrade the ability to measure targets. Clutter may be interference in some cases, such as detecting aircraft, or may be the desired signal itself, as with a ground imaging radar. The effectiveness of the signal processing is measured by the improvement it provides in the various figures of merit discussed in Chap. 1, such as detection probability, *signal-to-interference ratio* (SIR), or angle accuracy.

It was shown in Chap. 1 that conventional pulsed radars transmit narrowband, bandpass signals. Transmitted energy is maximized by restricting amplitude modulation to on-off pulsing; phase modulation is used to expand the instantaneous bandwidth when needed to improve resolution. Thus, an individual transmitted radar pulse can be written as

$$\tilde{x}(t) = a(t)\sin[2\pi F_t t + \theta(t)] \qquad (2.1)$$

where $a(t)$ is the constant amplitude pulse envelope, F_t is the radar carrier frequency, and $\theta(t)$ may be a constant or may represent phase modulation of the pulse. It will usually be assumed that $a(t)$ is an ideal, square pulse envelope of amplitude A and duration τ seconds. The instantaneous power of this signal is just $P_s = A^2/2$. The signal at the receiver output will be a combination of echoes of $\tilde{x}(t)$ from targets and clutter, noise, and possibly jamming.

Because the target and clutter components are delayed echoes of the transmitted pulse, they are also narrowband signals, although their amplitude and phase modulation will in general be altered, e.g., by propagation loss and Doppler shift. Receiver noise appears as an additive random signal. Thus, the received signal resulting from a single pulse echoing from a scatterer at range $R_0 = ct_0/2$ can be modeled as

$$\tilde{y}(t) = b(t - t_0) e^{j[2\pi F_t (t - t_0) + \phi(t)]} + n(t) \quad (2.2)$$

where $n(t)$ = receiver noise
$b(t)$ = echo amplitude
$\phi(t)$ = echo phase modulation.

The important parameters of $\tilde{y}(t)$ are the delay time t_0, echo component amplitude $b(t)$, its power relative to the noise component, and the echo phase modulation function $\phi(t)$. These characteristics are used to estimate target range, scattering strength, and radial velocity.

The amplitude and phase modulation functions also determine the range resolution ΔR of a measurement. For example, $\Delta R = c\tau/2$ if $\theta(t)$ is a constant. Resolution in angle and cross-range is determined by the 3-dB width of the antenna pattern.

In order to design good signal processing algorithms, good models of the signals to be processed are needed. In this chapter, an understanding of common radar signal characteristics pertinent to signal processing is developed by presenting models of the effect of the scattering process on the amplitude, phase, and resolution properties of radar measurements. While deterministic models suffice for simple scatterers, it will be seen that complicated real targets require statistical descriptions of the scattering process.

2.2 Amplitude Models

2.2.1 Simple point target radar range equation

The *radar range equation* (Eaves and Reedy, 1987; Skolnik, 2001) is a deterministic model that relates received echo power to transmitted power in terms of a variety of system design parameters. It is a fundamental relation used for basic system design and analysis. Since the received signals are narrowband pulses of the form of Eq. (2.2), the received power P_r estimated by the range equation can be directly related to the received pulse amplitude.

To derive the range equation, assume that an isotropic radiating element transmits a waveform of power P_t watts into a lossless medium. Because the transmission is isotropic and no power is lost in the medium, the power density at a range R is the total power P_t divided by the surface area of a sphere of radius R, which is

$$\text{Isotropic transmitted power density} = \frac{P_t}{4\pi R^2} \quad \text{W/m}^2 \quad (2.3)$$

Of course, real radars use directive antennas to focus the outgoing energy, rather than isotropic radiators. As described in Chap. 1, the antenna gain G is the ratio of maximum power density to isotropic density. Thus, in the direction of maximum radiation intensity, the power density at range R becomes

$$\text{Peak transmitted power density} = Q_t = \frac{P_t G}{4\pi R^2} \quad \text{W/m}^2 \quad (2.4)$$

This is the power density incident upon the target if it is aligned with the antenna's axis of maximum gain.

When the electromagnetic wave with power density given by Eq. (2.4) is incident upon a single discrete scattering object, or *point target*, at range R the incident energy is scattered in various directions; some of it may also be absorbed by the scatterer itself. In particular, some of the incident power is reradiated, or *backscattered*, toward the radar. Imagine that the target collects all of the energy incident upon a collector of area σ square meters and reradiates it isotropically. The reradiated power is then

$$\text{Backscattered power} = P_b = \frac{P_t G \sigma}{4\pi R^2} \quad \text{W} \quad (2.5)$$

The quantity σ is called the *radar cross section* (RCS) of the target. One important fact about RCS is that σ is *not* equal to the physical cross-sectional area of the target; it is an equivalent area that can be used to relate incident power density at the target to the reflected power density that results at the receiver. RCS will be discussed further in Sec. 2.2.3.

Because RCS is defined under the assumption that the backscattered power is reradiated isotropically, the density of the backscattered power at a range R is found by dividing the power of Eq. (2.5) by the surface area of a sphere of radius R as was done in Eq. (2.3), giving the backscattered power density at the radar receiver as

$$\text{Backscattered power density} = Q_b = \frac{P_t G \sigma}{(4\pi)^2 R^4} \quad \text{W/m}^2 \quad (2.6)$$

If the effective aperture size of the radar antenna is A_e square meters, the total backscattered power collected by the receiving antenna will be

$$\text{Received power} = P_r = \frac{P_t G A_e \sigma}{(4\pi)^2 R^4} \quad \text{W} \quad (2.7)$$

It was shown in Chap. 1 that the effective aperture of an antenna is related to its gain and operating wavelength according to $A_e = \lambda^2 G / 4\pi$. Thus

$$P_r = \frac{P_t G^2 \lambda^2 \sigma}{(4\pi)^3 R^4} \quad \text{W} \quad (2.8)$$

Equation (2.8) describes the power that would be received if an ideal radar operated in free space and used no signal processing techniques to improve sensitivity. Various additional "loss" and "gain" factors are customarily added to the formula to account for a variety of additional considerations. For example, losses incurred in various components such as the duplexers, power dividers, waveguide, and radome (a protective covering over the antenna), and propagation effects not found in free space propagation, can be lumped into a *system loss factor* L_s that reduces the received power. System losses are typically in the range of 3 to 10 dB but can vary widely. One of the most important loss factors, particularly at X band and higher frequencies, is atmospheric attenuation $L_a(R)$. Unlike system losses, atmospheric losses are a function of range. If the one-way loss in decibels per kilometer of Fig. 1.3 is denoted by α, the loss in decibels for a target at range R meters (not kilometers) is

$$L_a(R)(\text{dB}) = 2\alpha(R/1000) = \alpha R/500 \qquad (2.9)$$

In linear units, the loss is therefore

$$L_a(R) = 10^{\alpha R/5000} \qquad (2.10)$$

Atmospheric loss can be inconsequential at 10 GHz and moderate ranges, or tens of decibels at 60 GHz and a range of a few kilometers (the reason why 60 GHz is not a popular radar frequency). This example also shows that, like system losses, atmospheric loss is a strong function of radar frequency.

Incorporating atmospheric and system losses in Eq. (2.8) finally gives

$$P_r = \frac{P_t G^2 \lambda^2 \sigma}{(4\pi)^3 R^4 L_s L_a(R)} \quad \text{W} \qquad (2.11)$$

Equation (2.11) is one simple form of the *radar range equation*. It relates received echo power to fundamental radar system and target parameters such as transmitted power, operating frequency, and antenna gain; radar cross section; and range. Because the power of the radar signal is proportional to the square of the electric field amplitude, the range equation also serves as a model of the amplitude of the target and clutter components of the signal. Note that all variables in Eq. (2.11) are in linear units, not decibels, even though several of the parameters are often specified in decibels; frequent examples include the atmospheric losses, antenna gain, and RCS. Also note that P_r is instantaneous, not average, received power.

As an example, consider an X-band (10 GHz) radar with a peak transmitted power of 1 kW and a pencil beam antenna with a 1° beamwidth, and suppose an echo is received from a jumbo jet aircraft with an RCS of 100 m² at a range of 10 km. The received power can be determined using Eq. (2.11). The antenna gain can be estimated from Eq. (1.10) to be $G = 26{,}000/(1)(1) = 26{,}000 = 44$ dB. The wavelength is $\lambda = c/F = 3 \times 10^8/10 \times 10^9 = 3 \times 10^{-2}$ m = 3 cm.

Assuming atmospheric and system losses are negligible, the received power is

$$P_r = \frac{(1000)(26,000)^2(0.03)^2(100)}{(4\pi)^3(10,000)^4} = 3.066 \times 10^{-9} \quad \text{W} \quad (2.12)$$

The received power is only 1.8 nW, 12 orders of magnitude less than the transmitted power. Nonetheless, it will be seen that this signal level is adequate for reliable detection in many cases. This example illustrates the huge dynamic ranges observed in radar between transmitted and received signal powers.

An important consequence of Eq. (2.11) is that, for a point target, the received power decreases as the fourth power of range from the radar to the target. Thus, the ability to detect a target of a given radar cross section decreases rapidly with range. Range can be increased by increasing transmitted power, but because of the R^4 dependence, the power must be raised by a factor of 16 (12 dB) just to double the effective range. Alternatively, the antenna gain can be increased by a factor of 4 (6 dB), implying an increase in antenna area by a factor of 4. On the other hand, designers of "stealth" aircraft and other vehicles must reduce the RCS σ by a factor of 16 in order to halve the range at which they can be detected by a given radar system.

The range equation is a fundamental radar system design and analysis tool. More elaborate or specialized versions of the equation can be formulated to show the effect of other variables, such as pulse length, *intermediate frequency* (IF) bandwidth, or signal processing gains. The range equation also provides the basis for calibrating a radar system. If the system power, gain, and losses are carefully characterized, then the expected received power of echoes from test targets of known RCS can be computed. Calibration tables equating receiver voltage observed due to those same echoes to incident power density can then be constructed.

Signal processing techniques can increase the effective received power, and therefore increase the obtainable range. The effect of each technique on received power is discussed as they are introduced in later chapters.

2.2.2 Distributed target forms of the range equation

Not all scattering phenomena can be modeled as a reflection from a single point scatterer. Ground clutter, for example, is best modeled as distributed scattering from a surface, while meteorological phenomena such as rain or hail are modeled as distributed scattering from a three-dimensional volume. The radar range equation can be rederived in a generalized way that accommodates all three cases.

Equation (2.3) is still applicable as a starting point. To consider distributed scatterers, and because the gain of the antenna varies with azimuth and elevation angle, Eq. (2.4) must be replaced with an equation that accounts for the effect of the antenna power pattern $P(\theta, \phi)$ on the power density radiated in a particular direction (θ, ϕ)

$$Q_t(\theta, \phi) = \frac{P_t P(\theta, \phi)}{4\pi R^2} \tag{2.13}$$

Assume that the antenna boresight corresponds to $\theta = \phi = 0$. The antenna boresight is normally the axis of maximum gain so that $P(0, 0) = G$.

Now consider the scattering from an incremental volume dV located at range and angle coordinates (R, θ, ϕ). Suppose the incremental RCS of the volume element is $d\sigma$ square meters, and that $d\sigma$ in general varies with position in space. The incremental backscattered power from dV is

$$dP_b(\theta, \phi) = \frac{P_t P(\theta, \phi) \, d\sigma(R, \theta, \phi)}{4\pi R^2} \tag{2.14}$$

As before, $d\sigma$ is defined such that it is assumed this power is reradiated isotropically, and then collected by the antenna effective aperture, adjusted for the angle of arrival. After substituting for effective aperture and accounting for losses, this results in an incremental received power of

$$dP_r = \frac{P_t P^2(\theta, \phi) \lambda^2 \, d\sigma(R, \theta, \phi)}{(4\pi)^3 R^4 L_s L_a(R)} \tag{2.15}$$

The total received power is obtained by integrating over all space to obtain the *generalized radar range equation*

$$P_r = \frac{P_t \lambda^2}{(4\pi)^3 L_s} \int_V \frac{P^2(\theta, \phi)}{R^4 L_a(R)} d\sigma(R, \theta, \phi) \tag{2.16}$$

In Eq. (2.16), the volume of integration V is all of three-dimensional space. However, as discussed in Sec. 1.4.1, only scatterers within a single resolution cell volume ΔV contribute significantly to the radar receiver output at any given instant. Thus, a more appropriate form of the generalized radar range equation is

$$P_r = \frac{P_t \lambda^2}{(4\pi)^3 L_s} \int_{\Delta V(R, \theta, \phi)} \frac{P^2(\theta, \phi)}{R^4 L_a(R)} d\sigma(R, \theta, \phi) \tag{2.17}$$

where $\Delta V(R_0, \theta_0, \phi_0)$ is the volume of the resolution cell at nominal coordinates (R_0, θ_0, ϕ_0).

By integrating power, it is being assumed that the backscatter from each volume element adds *noncoherently* rather than *coherently*. This means that the power of the composite electromagnetic wave formed from the backscatter of two or more scattering centers is the sum of the individual powers, as opposed to the voltage (electric field amplitude) being the sum of the individual amplitudes, in which case the power would be the square of the voltage sum. Noncoherent addition occurs when the phases of the individual contributors are random and uncorrelated with one another, as opposed to the coherent case when they are in phase.

The general result of Eq. (2.17) is more useful if evaluated for the special cases of point, volume, and area scatterers. Beginning with the point scatterer, the differential RCS in the resolution cell volume is represented by a Dirac impulse function of weight σ:

$$d\sigma(R, \theta, \phi) = \sigma \delta_D(R - R_0, \theta - \theta_0, \phi - \phi_0) \, dV \text{ (point scatterer)} \quad (2.18)$$

Using Eq. (2.18) in Eq. (2.17) gives the range equation for a point target

$$P_r = \frac{P_t P^2(\theta_0, \phi_0) \lambda^2 \sigma}{(4\pi)^3 R_0^4 L_s L_a(R_0)} \quad (2.19)$$

If the point scatterer is located on the antenna boresight $\theta_0 = \phi_0 = 0$, $P(\theta_0, \phi_0) = G$ and Eq. (2.19) is identical to Eq. (2.11).

Next consider the *volume scattering* case where the RCS seen by the radar is presumed to be due to a distribution of scatterers evenly distributed throughout the volume, rather than associated with a single point. In this case, σ is expressed in terms of RCS per cubic meter, or *volume reflectivity*, denoted as η. The units of reflectivity are $m^2/m^3 = m^{-1}$. The RCS of a differential volume element dV is just

$$d\sigma = \eta \, dV = \eta R^2 \, dR \, d\Omega \text{ (volume scatterer)} \quad (2.20)$$

where $d\Omega$ is a differential solid angle element. The range equation becomes

$$P_r = \frac{P_t \lambda^2 \eta}{(4\pi)^3 L_s} \int_{\Delta V(R,\theta,\phi)} \frac{P^2(\theta, \phi)}{R^2 L_a(R)} dR \, d\Omega \quad (2.21)$$

Consider integration over the range coordinate first. If it is assumed that atmospheric loss is very slowly varying over the extent of a range resolution cell, then $L_a(R)$ can be replaced by $L_a(R_0)$ and removed from the integral. The integral over range that remains is

$$\int_{R_0 - \frac{\Delta R}{2}}^{R_0 + \frac{\Delta R}{2}} \left(\frac{dR}{R^2} \right) = \frac{\Delta R}{R_0^2 - (\Delta R/2)^2} \approx \frac{\Delta R}{R_0^2} \quad (2.22)$$

provided the range resolution is small compared to the absolute range, which is usually the case. Using Eq. (2.22) in Eq. (2.21) gives

$$P_r = \frac{P_t \lambda^2 \eta \, \Delta R}{(4\pi)^3 R_0^2 L_s L_a(R_0)} \int_{\Delta \Omega} P^2(\theta, \phi) \, d\Omega \quad (2.23)$$

Integration over the angular coordinates requires knowledge of the antenna pattern. The main lobe of many antennas can be reasonably well approximated by a Gaussian function (Sauvageot, 1992). It can be shown that a good approximation to the integral in Eq. (2.23) over the cross-range variables for the Gaussian case is (Probert-Jones, 1962)

$$\iint P^2(\theta,\phi)\,d\theta\,d\phi \approx \frac{\pi\theta_3\phi_3}{8\ln 2}G^2 = 0.57\theta_3\phi_3 G^2 \tag{2.24}$$

where θ_3 and ϕ_3 are the 3-dB beamwidths in azimuth and elevation. For first-order calculations, the even simpler assumption is frequently made that the antenna power pattern $P(\theta,\phi)$ is a constant equal to the gain G over the 3-dB beamwidths and zero elsewhere, so that the integral reduces to $G^2\theta_3\phi_3$, a value 2.5 dB higher than that of Eq. (2.24). Using this simpler approximation, Eq. (2.23) reduces to the range equation for volume scatterers:

$$P_r = \frac{P_t G^2 \lambda^2 \eta \Delta R \theta_3 \phi_3}{(4\pi)^3 R_0^2 L_s L_a(R_0)} \quad \text{(volume scatterers)} \tag{2.25}$$

Unlike the point scatterer case described by Eq. (2.11) or (2.19), the received power in the volume scattering case of Eq. (2.25) decreases only as R^2 instead of R^4. The reason is that the size of the radar resolution cell, which determines the extent of the scatterers contributing to the received power at any one instant, increases as R^2 due to the spreading of the antenna beam in angle at longer ranges.

Finally, the *area scattering* case will be considered. This model is used for the RCS of electromagnetic scatter from the ground, forest, ocean, and other surfaces that have a large area relative to the antenna beam main lobe. The area scattering case must further be divided into two subcases depending on whether the range extent of the scatterers contributing to the echo is limited by the antenna elevation beamwidth or by the range resolution.

First assume that the scattering surface is represented by a flat plane[†] and consider the extent of the main lobe on the surface. The cross-range extent is simply $R_0\theta_3$, where R_0 is the nominal range to the center of the illuminated area. To estimate the down-range extent, consider Fig. 2.1 that shows the boresight vector intersecting the scattering plane at a *grazing angle* of δ radians. The extent of the beam "footprint" in the down-range dimension is therefore $R_0\phi_3/\sin\delta$ meters.

Now suppose a pulse of range resolution ΔR is transmitted as shown in Fig. 2.2. Regardless of the antenna footprint, the range extent of scatterers within the resolution cell, and therefore emitting backscattered energy at any instant, is $\Delta R/\cos\delta$ meters.

Scatterers will not contribute to the received signal unless they are both illuminated (so that there is some backscatter) and within the main lobe of the antenna (so that their backscatter is not excessively attenuated). Consequently,

---

[†]This ignores earth curvature effects that are significant in very long range or spaceborne radars. See the books by Nathanson (1991) or Skolnik (2001) for additional details.

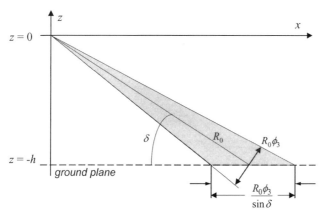

Figure 2.1 Projection of elevation beamwidth onto a horizontal plane at a slant range R_0 and grazing angle δ.

the effective down-range extent of the resolution cell is the lesser of the range resolution and the elevation beamwidth as each is projected onto the scattering surface. Depending on the relative values of range, range resolution, and grazing angle, either could be the limiting factor. If the range resolution limits the effective extent, the resolution cell is said to be *pulse limited*; if the main lobe extent is the limiting factor, it is said to be *beam limited*. These two cases are shown in Fig. 2.3.

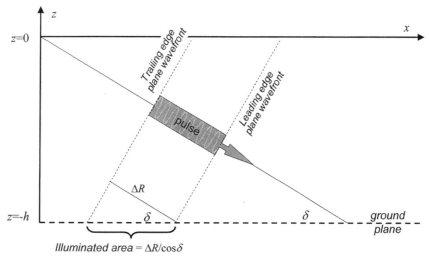

Figure 2.2 Projection of range resolution onto a horizontal plane at a slant range R_0 and grazing angle δ.

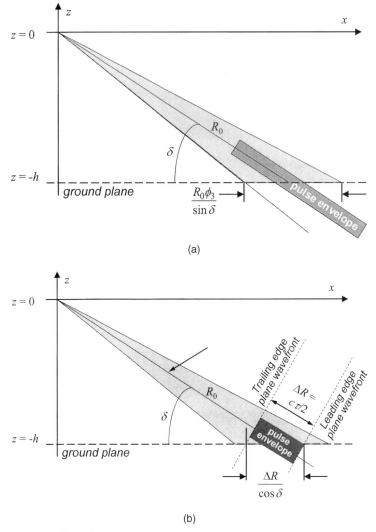

Figure 2.3 Relative geometry of antenna footprint and pulse envelope. (a) Beam-limited case. (b) Pulse-limited case.

The boundary between the two cases is

$$\text{Beam-limited:} \quad \frac{\Delta R}{R_0} \tan \delta > \phi_3$$

$$\text{Pulse-limited:} \quad \frac{\Delta R}{R_0} \tan \delta < \phi_3 \qquad (2.26)$$

In area scattering, all of the differential RCS is due to backscatter from a surface at a particular range within the resolution cell, say R_0; the remainder of the cell contributes no scattering. In this case, the differential volume RCS of Eq. (2.20) is proportional to the differential area of the scattering surface and can be represented as

$$d\sigma = \delta_D(R - R_0)\sigma^0 dA \qquad (2.27)$$

where σ^0 (called "sigma-nought") is the RCS in m²/m² and is therefore dimensionless. The generalized range equation becomes

$$P_r = \frac{P_t \lambda^2 \sigma^0}{(4\pi)^3 L_s} \int_{\Delta A(R_0,\theta_0,\phi_0)} \frac{P^2(\theta,\phi)}{R^4 L_a(R)} \delta(R-R_0) dA$$
$$= \frac{P_t \lambda^2 \sigma^0}{(4\pi)^3 R_0^4 L_s L_a(R_0)} \int_{\Delta A(R_0,\theta,\phi)} P^2(\theta,\phi) dA \qquad (2.28)$$

where $\Delta A(R_0, \theta, \phi)$ is the illuminated area at range R_0.

If the illuminated area is beam-limited, applying the geometry of Fig. 2.3a to the differential scattering element at range R_0 shows that the area contributing to the backscatter at one instant is $R^2 \phi_3 \theta_3 / \sin \delta$. Thus, a differential area contributing to the received power is of the form

$$dA = R_0 d\theta \frac{R_0}{\sin \delta} d\phi = \frac{R_0^2}{\sin \delta} d\theta \, d\phi \quad \text{(beam-limited case)} \qquad (2.29)$$

Applying this to Eq. (2.28) and again using the constant-gain approximation to the antenna 3-dB beamwidth gives the beam-limited range equation for area scatterers:

$$P_r = \frac{P_t G^2 \lambda^2 \phi_3 \theta_3 \sigma^0}{(4\pi)^3 R_0^2 L_s L_a(R_0) \sin \delta} \quad \text{(area scatterers, beam-limited case)} \qquad (2.30)$$

If the illuminated area is pulse limited, the geometry of Fig. 2.3b shows that the area contributing to the backscatter at one instant is $R\theta_3 \Delta R / \cos \delta$. The differential contribution is thus

$$dA = R_0 \, d\theta \frac{\Delta R}{\cos \delta} d\phi = \frac{R_0 \Delta R}{\cos \delta} d\theta \, d\phi \quad \text{(pulse-limited case)} \qquad (2.31)$$

The first-order approximation of constant gain over the main lobe can be used again, though the integral over ϕ is now limited to the range that covers the extent of the pulse on the ground. Equation (2.28) becomes

$$P_r = \frac{P_t G^2 \lambda^2 \sigma^0 \Delta R \theta_3}{(4\pi)^3 R_0^3 L_s L_a(R_0) \cos \delta} \quad \text{(area scatterers, pulse-limited case)} \qquad (2.32)$$

Note that power varies as R^{-2} in the beam-limited case because, as with the volume scattering, the resolution cell size grows in both cross-range and down-range extent with increasing range. In the pulse-limited case, power varies as R^{-3} because the resolution cell extent increases in only the cross-range dimension with increasing range.

2.2.3 Radar cross section

Section 2.2.1 introduced the radar cross section to heuristically account for the amount of power reradiated by the target back toward the radar transmitter. To restate the concept, assume that the incident power density at the target is Q_t and the backscattered power density at the transmitter is Q_b. If that backscattered power density resulted from isotropic radiation from the target, it would have to satisfy

$$Q_b = \frac{P_b}{4\pi R^2} \tag{2.33}$$

for some total backscattered power P_b. RCS is the *fictional* area over which the transmitted power density Q_t must be intercepted to collect a total power P_b that would account for the received power density, that is, σ must satisfy

$$P_b = \sigma Q_t \tag{2.34}$$

Combining Eqs. (2.33) and (2.34) gives

$$\sigma = 4\pi R^2 \frac{Q_b}{Q_t} \tag{2.35}$$

This definition is usually written in terms of electric field amplitude. Also, in order to make the definition dependent only on the target characteristics, range is eliminated by taking the limit as R tends to infinity. Thus, the formal definition of radar cross section becomes (Knott, Shaeffer, and Tuley, 1985)

$$\sigma = 4\pi \lim_{R \to \infty} \left[R^2 \frac{|\mathbf{E}^b|^2}{|\mathbf{E}^t|^2} \right] \tag{2.36}$$

where $|\mathbf{E}^b|^2$ and $|\mathbf{E}^t|^2$ are the backscattered and transmitted electric field squared magnitudes, respectively.

The RCS just defined is a single real scalar number. Implicit in the definition is the use of a single polarization of the transmitted wave and a single receiver polarization, usually the same as the transmitted polarization. However, the polarization state of a transverse electromagnetic plane wave is a two-dimensional vector, and therefore two orthogonal polarization basis vectors are required to fully describe the wave. The most common basis choices are linear (horizontal and vertical polarizations) and circular (left and right rotating polarizations). Furthermore, a general target will modify the polarization of an incident wave, so that the energy backscattered from, say, the vertical component of the incident wave may have both vertical and horizontal components. To account fully for polarization effects, RCS must be generalized to the *polarization scattering matrix* (PSM) **S**, which relates the complex amplitudes of the incident and backscattered fields. For a radar using a linear polarization basis, e.g., this relation is (Knott, Shaeffer, and Tuley, 1985; Mott, 1986; Holm, 1987)

$$\begin{bmatrix} \mathbf{E}_H^b \\ \mathbf{E}_V^b \end{bmatrix} = \begin{bmatrix} S_{HH} & S_{HV} \\ S_{VH} & S_{VV} \end{bmatrix} \begin{bmatrix} \mathbf{E}_H^t \\ \mathbf{E}_V^t \end{bmatrix}$$

$$= \mathbf{S} \begin{bmatrix} \mathbf{E}_H^b \\ \mathbf{E}_V^b \end{bmatrix} \quad (2.37)$$

Instead of a single real number, the target backscattering characteristics are now described by four complex numbers. If the radar transmitted and received, say, only the vertical component, then the RCS σ would be related to $\mathbf{S}$ by

$$\sigma = |S_{VV}|^2 \quad (2.38)$$

Radars can be designed to measure the full PSM. Other designs measure the magnitudes of the elements of the PSM, but not the phases, or the magnitudes of two of the PSM elements. These *polarimetric* measurements can be used for a variety of target analysis purposes. However, a discussion of polarimetric techniques is beyond the scope of this book. Henceforth, it will be assumed that only a single fixed polarization is transmitted and a single fixed polarization received, and consequently that RCS is described by a scalar, rather than matrix, function. The reader is referred to the works by Holm (1987) and Mott (1986) for discussions of polarimetric radars and polarimetric signal processing.

Typical values of RCS for targets of interest range from 0.01 m² (−20 dB with respect to 1 m², or −20 dBsm) to hundreds of square meters (≥ +20 dBsm). Both larger and smaller values are also observed. Table 2.1 lists representative RCS values for various types of targets.

TABLE 2.1 Typical RCS Values at Microwave Frequencies

Target	RCS, m²	RCS, dBsm
Conventional unmanned winged missile	0.5	−3
Small single-engine aircraft	1	0
Small fighter aircraft or 4-passenger jet	2	3
Large fighter aircraft	6	8
Medium bomber or jet airliner	20	13
Large bomber or jet airliner	40	16
Jumbo jet	100	20
Small open boat	0.02	−17
Small pleasure boat	2	3
Cabin cruiser	10	10
Large ship at zero grazing angle	10,000+	40+
Pickup truck	200	23
Automobile	100	20
Bicycle	2	3
Man	1	0
Bird	0.01	−20
Insect	0.00001	−50

SOURCE: After Skolnik (2001).

2.2.4 Radar cross section for meteorological targets

The field of radar meteorology expresses the reflectivity of weather targets such as rain or snow in terms of a normalized factor called the *reflectivity* (here called the *volume reflectivity*) and usually represented with the symbol Z (Sauvageot, 1992; Doviak and Zrnic, 1993). Weather targets are an example of volume clutter. The actual observed echo is the composite backscatter of many raindrops, suspended water particles, or snowflakes in the radar's resolution cell.

Suppose the RCS of the ith individual scatterer is σ_i. Then the total RCS of a volume V containing N such scatterers is $\sum \sigma_i$ and the volume reflectivity is

$$\eta = \frac{1}{\Delta V} \sum_{i=1}^{N} \sigma_i \tag{2.39}$$

Water droplets are often modeled as small conducting spheres. When the ratio of the sphere radius a to the radar wavelength λ is small, specifically $2\pi a/\lambda \ll 1$, the radar cross section associated with the ith scatterer can be expressed as

$$\sigma_i = \frac{\pi^5 |K|^2 D_i^6}{\lambda^4} \tag{2.40}$$

where D_i is the drop diameter, usually given in millimeters and

$$K = \frac{m^2 - 1}{m^2 + 1} \tag{2.41}$$

and m is the complex index of refraction. The index of refraction is a function of both the temperature and wavelength. However, for wavelengths between 3 and 10 cm (radar frequencies between X band (10 GHz) and C band (3 GHz)) and temperatures between 0°C and 20°C, the value of $|K|^2$ is approximately a relatively constant 0.93 for scatterers composed of water and 0.197 for ice. Substituting Eq. (2.40) in Eq. (2.39) gives

$$\eta = \frac{1}{\Delta V} \sum_{i=1}^{N} \frac{\pi^5 |K|^2 D_i^6}{\lambda^4} = \frac{\pi^5 |K|^2}{\lambda^4} \frac{1}{\Delta V} \sum_{i=1}^{N} D_i^6 \tag{2.42}$$

Now define the quantity

$$Z \equiv \frac{1}{\Delta V} \sum_{i=1}^{N} D_i^6 \tag{2.43}$$

Z is called the *volume reflectivity* and is usually expressed in units of mm^6/m^3. Due to the large range of values observed for Z, it is commonly expressed on a decibel scale and denoted as dBz. Using this definition in Eq. (2.42) gives the following expression for the observed RCS

$$\eta = \frac{\pi^5 |K|^2}{\lambda^4} Z \tag{2.44}$$

TABLE 2.2 Correspondence between dBz Reflectivity and Rain Rate

Level	Rain fall rate (mm/hr)	Reflectivity dBz	Category
1	0.49 to 2.7	18 to <30	Light mist
2	2.7 to 13.3	30 to <41	Moderate
3	13.3 to 27.3	41 to <46	Heavy
4	27.3 to 48.6	46 to <50	Very Heavy
5	48.6 to 133.2	50 to <57	Intense
6	133.2 and greater	57 and above	Extreme

Thus, given a measured echo power, the radar range equation can be used to estimate η, and then Eq. (2.44) can be used to convert η to Z.

Because it is related only to the volume density and size of scatterers, meteorologists prefer to express radar echo strength in terms of the reflectivity Z rather than the RCS η. The value of Z can then be related to the amount of water in the air or the precipitation rate. A number of models are used to relate the observed values of Z to rain rates. These models depend on the type of precipitation, e.g., snow versus thunderstorm rain versus "orographic"[†] rain. The most common model is that of Table 2.2, which shows the six-level equivalence between observed Z values (in dBz) and rainfall rates used in the U.S. NEXRAD national weather radar system. Very similar scales are used in the commercial "Doppler weather radar" systems familiar to every watcher of television weather forecasts.

It is important to note that the dBz values in Table 2.2 are 10 times the base 10 logarithm of Z in mm^6/m^3. When Z is given in $m^6/m^3 = m^3$, it must be multiplied by 10^{18} to convert it to units of mm^6/m^3 before converting to a decibel scale and using Table 2.2.

2.2.5 Statistical description of radar cross section

The radar cross section of real targets cannot be effectively modeled as a simple constant. In general, RCS is a complex function of aspect angle, frequency, and polarization, even for relatively simple scatterers. For example, the conducting trihedral corner reflector of Fig. 2.4 is often used as a calibration target in field measurements. Its RCS when viewed along its axis of symmetry (looking "into the corner") can be determined theoretically; it is (Knott, Shaeffer, and Tuley, 1985)

$$\sigma = \frac{12\pi a^4}{\lambda^2} \quad (2.45)$$

Thus the RCS increases with increasing frequency. On the other hand, at least one frequency- and aspect-independent scatterer exists. The RCS of a conducting sphere of radius a is a constant πa^2, provided $a \gg \lambda$. It is independent of aspect angle because of the spherical symmetry.

[†]A form of rain that occurs when moist air is lifted over an obstacle such as a mountain range, cooling as it rises and condensing into rainfall.

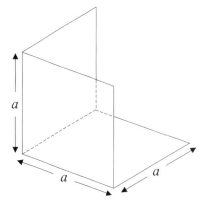

Figure 2.4 Square trihedral corner reflector.

A simple example of frequency and aspect dependence is the two-scatterer "dumbbell" target of Fig. 2.5. If the nominal range R is much greater than the separation D, the range to the two scatterers is approximately

$$R_{1,2}(\theta) \approx R \pm \frac{D}{2} \sin \theta \tag{2.46}$$

If the signal $a e^{j 2\pi F t}$ is transmitted, the echo from each scatterer will be proportional to $a e^{j 2\pi F(t - 2R_{1,2}(\theta)/c)}$. The voltage $\bar{y}(t)$ of the composite echo is therefore proportional to

$$\begin{aligned}
\bar{y}(t) &= a e^{j 2\pi F(t - 2R_1(\theta)/c)} + a e^{j 2\pi F(t - 2R_2(\theta)/c)} \\
&= a e^{j 2\pi F(t - 2R/c)} [e^{-j \pi F D \sin \theta / c} + e^{+j \pi F D \sin \theta / c}] \\
&= 2a e^{j 2\pi F(t - 2R/c)} \cos(\pi F D \sin \theta / c)
\end{aligned} \tag{2.47}$$

RCS is proportional to the power of the composite echo. Taking the squared magnitude of Eq. (2.47) and simplifying leads to the result

$$\sigma = 4a^2 |\cos(\pi F D \sin \theta / c)|^2 = 4a^2 |\cos(\pi D \sin \theta / \lambda)|^2 \tag{2.48}$$

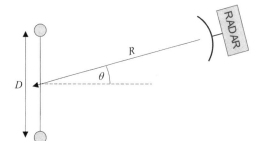

Figure 2.5 Geometry for determining relative RCS of a "dumbbell" target.

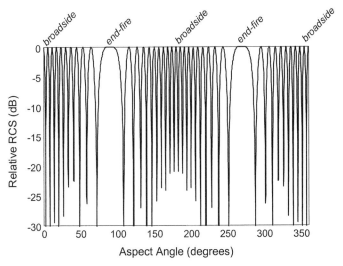

Figure 2.6 Relative radar cross section of the "dumbbell" target of Fig. 2.5 when $D = 10\lambda$ and $R = 10,000D$.

Equation (2.48) shows that the RCS is a periodic function of both radar frequency and aspect angle. The larger the scatterer separation in terms of wavelengths, the more rapidly the RCS varies with angle or frequency. An exact calculation of the variation in RCS of the dumbbell target is plotted in Fig. 2.6 for the case $D = 10\lambda$ and $R = 10,000D$. The plot has been normalized so that the maximum value corresponds to 0 dB. Notice the multilobed structure as the varying path lengths traversed by the echoes from the two scatterers cause their echoes to shift between constructive and destructive interference. Also note that the maxima at aspect angles of 90° and 270° (the two "end fire" cases) are the broadest, while the maxima at the two "broadside" cases of 0° and 180° are the narrowest. Figure 2.7 plots the same data in a more traditional polar format.

The relative RCS of a target with multiple scatterers can be computed using a generalization of Eq. (2.47). Suppose there are N scatterers, each with its own RCS σ_i, located at ranges $R_i(\theta)$ from the radar. Note that the ranges R_i vary with aspect angle θ. The complex voltage of the echo will be, to within a proportionality constant

$$\bar{y}(t) = \sum_{i=1}^{N} \sqrt{\sigma_i}\, e^{j 2\pi F(t - 2R_i(\theta)/c)}$$

$$= e^{j 2\pi F t} \sum_{i=1}^{N} \sqrt{\sigma_i}\, e^{-j 4\pi F R_i(\theta)/c}$$

$$= e^{j 2\pi F t} \sum_{i=1}^{N} \sqrt{\sigma_i}\, e^{-j 4\pi R_i(\theta)/\lambda} \qquad (2.49)$$

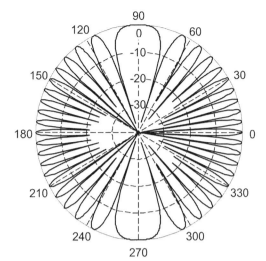

Figure 2.7 Polar plot of the data of Fig. 2.6.

The RCS σ is proportional to $|\bar{y}|^2$. Define

$$\zeta \equiv |\bar{y}| = \left| \sum_{i=1}^{N} \sqrt{\sigma_i}\, e^{-j4\pi R_i(\theta)/\lambda} \right| \quad (2.50)$$

and

$$\sigma = \zeta^2 = \left| \sum_{i=1}^{N} \sqrt{\sigma_i}\, e^{-j4\pi R_i(\theta)/\lambda} \right|^2 \quad (2.51)$$

RCS variations like those of Fig. 2.6 become very complicated for complex targets having many scatterers of varying individual RCS. Figure 2.8 shows a "target" consisting of 50 point scatterers randomly distributed within a rectangle 5 m wide and 10 m long. The RCS of each individual point scatterer is a

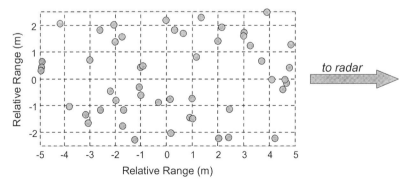

Figure 2.8 Random distribution of 50 scatterers used to obtain Fig. 2.9. (See text for additional details.)

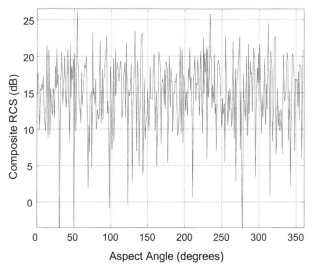

Figure 2.9 Relative RCS of the complex target of Fig. 2.8 at a range of 10 km and radar frequency of 10 GHz.

constant, $\sigma_i = 1.0$. Figure 2.9 shows the relative RCS, computed at $0.2°$ increments using Eq. (2.51), which results when this target is viewed 10 km from its center at a frequency of 10 GHz. The dynamic range is similar to that of the simple dumbbell target, but the lobing structure is much more complicated.

The complicated behavior observed for even moderately complex targets leads us to use a statistical description for radar cross section (Levanon, 1988; Nathanson, 1991; Skolnik, 2001). This means that the RCS σ of the scatterers within a single resolution cell is considered to be a random variable with a specified *probability density function* (pdf). The radar range equation is used to estimate the mean RCS, and one of a variety of pdfs are used to describe the statistical behavior of the RCS.

Consider first a target consisting of a large number of individual scatterers (similar to that of Fig. 2.8), each with its own individual but fixed RCS and randomly distributed in space. It can be assumed that the phase of the echoes from the various scatterers is a random variable distributed uniformly on $(0, 2\pi]$. Under these circumstances, the central limit theorem guarantees that the real and imaginary parts of the composite echo can each be assumed to be independent, zero mean Gaussian random variables with the same variance, say α^2 (Papoulis, 1984; Beckmann and Spizzichino, 1963). In this case, the squared-magnitude σ has an exponential pdf (Papoulis, 1984):

$$p_\sigma(\sigma) = \begin{cases} \dfrac{1}{\bar{\sigma}} \exp\left[\dfrac{-\sigma}{\bar{\sigma}}\right] & \sigma \geq 0 \\ 0 & \sigma < 0 \end{cases} \qquad (2.52)$$

where $\bar{\sigma} = 2\alpha^2$ is the mean value of the RCS σ. The voltage or magnitude ζ has a Rayleigh pdf

$$p_\zeta(\zeta) = \begin{cases} \dfrac{2\zeta}{\bar{\sigma}} \exp\left[\dfrac{-\zeta^2}{\bar{\sigma}}\right] & \zeta \geq 0 \\ 0 & \zeta < 0 \end{cases} \qquad (2.53)$$

While the Rayleigh/exponential model is only strictly accurate in the limit of a very large number of scatterers, in practice it can be a good model for a target having as few as 10 or 20 significant scatterers. Figure 2.10 compares a histogram of the RCS values from Fig. 2.9 to an exponential curve of the form of Eq. (2.52) having the same mean $\bar{\sigma}$. Even though only 50 scatterers are used, the fit of the total RCS histogram to the Rayleigh/exponential distribution is quite good. This same effect is observed when the randomly distributed scatterers also have random individual cross sections drawn from the same Gaussian distribution, a somewhat more general and plausible situation than the fixed-RCS case.

Many radar targets are not well modeled as an ensemble of equal-strength scatterers, so many other pdfs have been advocated and used. Table 2.3 summarizes several of the more common RCS models. The mean value $\bar{\sigma}$ of RCS is given for each case in which the pdf is not written explicitly in terms of $\bar{\sigma}$. The variance var(σ) is given for each case. The naming terminology can be confusing, because in some cases the name traditionally applied to the distribution of RCS σ is actually that of the density function of the corresponding amplitude ζ. For example, the exponential RCS distribution of Eq. (2.52) is frequently referred to as the Rayleigh model.

The shape of the pdf of RCS directly affects detection performance, as is seen in Chap. 6. Figure 2.11a compares the Rayleigh/exponential, fourth-degree chi-square, Rice, Weibull, and log-normal density functions when all have an

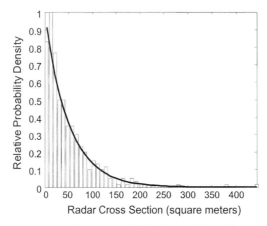

Figure 2.10 Histogram of RCS data of Fig. 2.9 and exponential pdf with the same mean.

TABLE 2.3 Common Statistical Models for Radar Cross Section

Model name	pdf for RCS σ	Comment
Nonfluctuating, Marcum, Swerling 0, or Swerling 5	$p_\sigma(\sigma) = \delta_D(\sigma - \bar{\sigma})$ $\text{var}(\sigma) = 0$	Constant echo power, e.g., calibration sphere or perfectly stationary reflector with no radar or target motion.
Rayleigh/exponential, chi-square of degree 2	$p_\sigma(\sigma) = \dfrac{1}{\bar{\sigma}} \exp\left[\dfrac{-\sigma}{\bar{\sigma}}\right]$ $\text{var}(\sigma) = \bar{\sigma}^2$	Many scatterers, randomly distributed, none dominant. Used in Swerling case 1 and 2 models.
Chi-square of degree 4	$p_\sigma(\sigma) = \dfrac{4\sigma}{\bar{\sigma}^2} \exp\left[\dfrac{-2\sigma}{\bar{\sigma}}\right]$ $\text{var}(\sigma) = \bar{\sigma}^2/2$	Approximation to case of many small scatterers + one dominant, with RCS of dominant equal to $1 + \sqrt{2}$ times the sum of RCS of others. Used in Swerling case 3 and 4 models.
Chi-square of degree $2m$, Weinstock	$p_\sigma(\sigma) = \dfrac{m}{\Gamma(m)\bar{\sigma}} \left[\dfrac{m\sigma}{\bar{\sigma}}\right]^{m-1} \exp\left[\dfrac{-m\sigma}{\bar{\sigma}}\right]$ $\text{var}(\sigma) = \bar{\sigma}^2/m$	Generalization of the two preceding cases. Weinstock cases correspond to $0.6 \leq 2m \leq 4$. Higher degrees correspond to presence of a more dominant single scatterer.
Rice or Ricean, noncentral chi-square of degree 2	$p_\sigma(\sigma) = \dfrac{1}{\bar{\sigma}}(1+a^2)\exp\left[-a^2 - \dfrac{\sigma}{\bar{\sigma}}(1+a^2)\right]$ $\times I_0\left[2a\sqrt{(1+a^2)(\sigma/\bar{\sigma})}\right]$ $\text{var}(\sigma) = \dfrac{(1+2a^2)}{(1+a^2)^2}\bar{\sigma}^2$	Exact solution for one dominant scatterer plus many small ones. Ratio of dominant RCS to sum of small RCS is a^2.
Weibull	$p_\sigma(\sigma) = CB\sigma^{C-1}\exp[-B\sigma^C]$ $\bar{\sigma} = \Gamma(1+1/C)B^{-1/C}$ $\text{var}(\sigma) = B^{-2/C}[\Gamma(1+2/C) - \Gamma^2(1+1/C)]$	Empirical fit to many measured target and clutter distributions. Can have longer "tail" than previous cases.
Log-normal	$p_\sigma(\sigma) = \dfrac{1}{\sqrt{2\pi}\,s\sigma}\exp\left[-\ln^2(\sigma/\sigma_m)/2s^2\right]$ $\bar{\sigma} = \sigma_m \exp(s^2/2)$ $\text{var}(\sigma) = \sigma_m^2 \exp(s^2)[\exp(s^2) - 1]$	Empirical fit to many measured target and clutter distributions. "Tail" is longest of previous cases. σ_m is the median value of σ.

RCS variance of 0.5. The exponential distribution then necessarily has a mean of $\sqrt{0.5}$. The fourth-degree chi-square necessarily has a mean of 1.0, and the parameters of the remaining density functions have been chosen to give them a mean of 1.0 as well. Figure 2.11b repeats the same data on a semilogarithmic scale so that the behavior of the pdf "tails" is more evident. Note that the Weibull and Rice distributions are very similar. The chi-square is also similar, but has a somewhat less extensive tail to the distribution. The log-normal has both the narrowest peak and the longest tail of any of the distributions shown. Unlike all of the others, the exponential does not peak near the mean RCS. Each of the others does have a distinct peak, making them suitable for distributions with one or a few dominant scatterers.

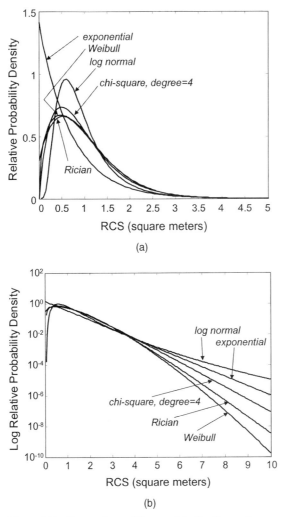

Figure 2.11 Comparison of five models for the probability density function of radar cross section. (See text for additional details.) (a) Linear scale. (b) Log scale.

One fundamental difference among the various RCS models of Table 2.3 is whether the probability density function has one or two free parameters. The nonfluctuating, Rayleigh/exponential, and all chi-square (once the order is stated) are all *one-parameter distributions*. The one parameter in the form given earlier is the mean RCS, $\bar{\sigma}$. The Rice, Weibull, and log-normal are *two-parameter distributions*. The parameters are $\bar{\sigma}$ and a^2 for the Rice, B and C for the Weibull, and σ_m and s for the log-normal in the forms given. For a one-parameter distribution, estimating the mean also provides an estimate of the variance. For the two-parameter case, separate estimates of mean and variance must be computed. This distinction is important in the design of automatic detection algorithms in Chap. 7.

Most radar analysis and measurement programs emphasize RCS measurements, which are proportional to received power. Sometimes ζ, the corresponding voltage, is of interest, particularly for use in simulations where Eq. (2.49) is used explicitly to model the composite echo from a multiple scatterer target. The probability density function for the voltage is required in order to properly model the probabilistic variations of the complex sum. The pdf of $|\zeta|$ is easily derived from the pdf of σ using basic results of random variables (Papoulis, 1984). Because RCS is nonnegative, the transformation[†]

$$\zeta = \sqrt{\sigma} \qquad (2.54)$$

from RCS to voltage has only one real solution for σ, namely $\sigma = \zeta^2$. It then follows that the pdf of ζ is given by

$$p_\zeta(\zeta) = \frac{p_\sigma(\zeta^2)}{d\zeta/d\sigma}$$
$$= 2\zeta p_\sigma(\zeta^2) \qquad (2.55)$$

Equation (2.55) can be used to write the voltage pdfs by inspection from Table 2.3. The results, given in Table 2.4, are expressed in terms of the parameters of the corresponding RCS distribution from Table 2.3. Note that the nonfluctuating, Weibull, and log-normal RCS distributions all result in distributions of the same type (but with one or more parameters changed) for the voltage. Also note that the voltage in the Rayleigh/exponential case is Rayleigh distributed, explaining the name.

As has been seen, the RCS of a complex target varies with both transmitted frequency and aspect angle. Another important characteristic of a target's signature is the *correlation "length"* in time, frequency, and angle. This is the change in time, frequency, or angle required to cause the echo amplitude to decorrelate to a specified degree. If a rigid target such as a building is illuminated with a series of identical radar pulses and there is no motion between the radar and target, one expects the same received complex voltage y from each pulse (ignoring receiver noise). If motion between the two is allowed, however, the relative path length between the radar and the various scatterers comprising the target will change, causing the composite echo amplitude to fluctuate. Thus, for rigid targets, decorrelation of the RCS is induced by changes in range and aspect angle. On the other hand, if natural clutter such as the ocean surface or a stand of trees is illuminated, the signature will decorrelate even if the radar and target do not move relative to one another. In this case the decorrelation

---

[†]Because the power P of a real sinusoid is related to its amplitude A according to $P = A^2/2$ instead of just $P = A^2$, some authors (e.g., Levanon (1988)) present a slightly different form for the voltage distributions.

TABLE 2.4 Voltage Distributions Corresponding to Common Statistical Models of Radar Cross Section

RCS model name	pdf for voltage ζ	Comment
Nonfluctuating, Marcum, Swerling 0, or Swerling 5	$p_\zeta(\zeta) = \delta_D(\zeta - \sqrt{\bar{\sigma}})$ $\bar{\zeta} = \sqrt{\bar{\sigma}}, \quad \text{var}(\zeta) = 0$	Also nonfluctuating model.
Rayleigh/exponential, chi-square of degree 2	$p_\zeta(\zeta) = \dfrac{2\zeta}{\bar{\sigma}} \exp\left[\dfrac{-\zeta^2}{\bar{\sigma}}\right]$ $\bar{\zeta} = \dfrac{1}{2}\sqrt{\pi\bar{\sigma}}, \quad \text{var}(\zeta) = \bar{\sigma}(1 - \pi/4)$	Rayleigh distribution. RCS has exponential distribution.
Chi-square of degree 4	$p_\zeta(\zeta) = \dfrac{8\zeta^3}{\bar{\sigma}^2} \exp\left[\dfrac{-2\zeta^2}{\bar{\sigma}}\right]$ $\bar{\zeta} = \dfrac{3}{4}\sqrt{\pi\bar{\sigma}}, \quad \text{var}(\zeta) = \left(1 - \dfrac{9}{32}\pi\right)\bar{\sigma}$	Central Rayleigh distribution of degree 4
Chi-square of degree $2m$, Weinstock	$p_\zeta(\zeta) = \dfrac{2\zeta m}{\Gamma(m)\bar{\sigma}} \left(\dfrac{m\zeta^2}{\bar{\sigma}}\right)^{m-1} \exp\left[-\dfrac{m\zeta^2}{\bar{\sigma}}\right]$ $\bar{\zeta} = \sqrt{\dfrac{\bar{\sigma}}{m}}, \quad \text{var}(\zeta) = \bar{\sigma}\left\{1 - \dfrac{1}{m}[\Gamma(m+0.5)/\Gamma(m)]^2\right\}$	Central Rayleigh distribution of degree $2m$
Rice or Rician, noncentral chi-square of degree 2	$p_\zeta(\zeta) = \dfrac{2\zeta(1+a^2)}{\bar{\sigma}} \exp\left[-a^2 - \dfrac{\zeta^2}{\bar{\sigma}}(1+a^2)\right] I_0\left(2a\zeta\sqrt{(1+a^2)/\bar{\sigma}}\right)$ $\bar{\zeta} = \dfrac{1}{2}\sqrt{\dfrac{\pi\bar{\sigma}}{1+a^2}} \, e^{-a^2} {}_1F_1\left[1.5, 1; a^2\right]$ $\text{var}(\zeta) = \left(\dfrac{\bar{\sigma}}{1+a^2}\right) e^{-a^2}\left[{}_1F_1(2,1;a^2) - \dfrac{\pi}{4}e^{-a^2}{}_1F_1^2(1.5,1;a^2)\right]$	Noncentral Rayleigh distribution of degree 2. ${}_1F_1(\cdot)$ is the confluent hypergeometric function.
Weibull	$p_\zeta(\zeta) = 2CB\zeta^{2C-1}\exp[-B\zeta^{2C}]$ $\bar{\zeta} = \Gamma(1 + 1/2C)B^{-1/2C}$ $\text{var}(\zeta) = [\Gamma(1+1/C) - \Gamma^2(1+1/2C)]B^{-1/2C}$	Also Weibull, one parameter (C) changed.
Log-normal	$p_\zeta(\zeta) = \dfrac{2}{\sqrt{2\pi}\,s\zeta}\exp\left[-2\ln^2\left(\zeta/\sqrt{\sigma_m}\right)^2/s^2\right]$ $\bar{\zeta} = \sqrt{\sigma_m}\exp(s^2/8)$ $\text{var}(\zeta) = \sigma_m\exp(s^2/4)[\exp(s^2/4) - 1]$	Also log-normal, both parameters (s, σ_m) changed.

is caused by the "internal motion" of the clutter, such as the wave motion on the sea surface or the blowing leaves and limbs of the trees. Decorrelation is therefore unavoidable, and the rate of decorrelation is influenced by factors external to the radar such as wind speed. Range or aspect changes also induce decorrelation of clutter signatures.

Although the behavior of real targets can be quite complex, a sense of the change in frequency and angle required to decorrelate a target or clutter patch can be obtained by the following simple argument.

Consider a target consisting of a uniform line array of point scatterers tilted at an angle θ with respect to the antenna boresight and separated by Δx from one another, as shown in Fig. 2.12. For simplicity, assume an odd number $2M+1$ of scatterers indexed from $-M$ to $+M$ as shown. If the nominal distance to the radar R_0 is much larger than the target extent (that is, $R_0 \gg (2M+1)\Delta x$), then the range to the nth scatterer is approximately

$$R_n \approx R_0 + n\,\Delta x\,\sin\theta \qquad (2.56)$$

If the target is illuminated with the waveform $Ae^{j\Omega t}$, the received signal is

$$\bar{y}(t) = \sum_{n=-M}^{M} Ae^{j\Omega(t-2R_n/c)}$$

$$= Ae^{j\Omega(t-2R_0/c)} \sum_{n=-M}^{M} e^{-j4\pi n \Delta x \sin\theta\, F/c} \qquad (2.57)$$

To simplify the notation, define

$$z = F\sin\theta \qquad \alpha = 4\pi\,\Delta x/c \qquad (2.58)$$

Then $\bar{y}(t)$ can be considered as a function $\bar{y}(t;z)$ of both t and z. The correlation in the variable z is of interest, which includes both aspect angle and

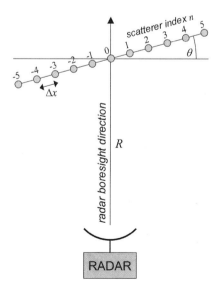

Figure 2.12 Geometry for calculation of RCS correlation length in frequency and aspect angle.

frequency. Note that $\bar{y}(t;z)$ is periodic in z with period $2\pi/\alpha$. The deterministic autocorrelation function is therefore

$$s_{\bar{y}}(\Delta z) = \int_{-\pi/\alpha}^{\pi/\alpha} \bar{y}(t;z)\bar{y}^*(t;z+\Delta z)\,dz$$

$$= \int_{-\pi/\alpha}^{\pi/\alpha} \left[A^* e^{j\Omega(t-2R_0/c)} \sum_{n=-M}^{M} e^{-j\alpha z n} \right] \left[e^{-j\Omega(t-2R_0/c)} \sum_{l=-M}^{M} e^{+j\alpha(z+\Delta z)l} \right] dz \quad (2.59)$$

The complex exponential terms outside the summations cancel. Interchanging integration and summation and collecting terms then gives

$$s_{\bar{y}}(\Delta z) = |A|^2 \sum_{l=-M}^{M} e^{+j\alpha\Delta z l} \sum_{n=-M}^{M} \left[\int_{z=-\pi/\alpha}^{\pi/\alpha} e^{-j\alpha(n-l)z}\,dz \right] \quad (2.60)$$

A change of variables $z' = \alpha z$ makes it clear that the integral has the form of the inverse discrete-time Fourier transform of a constant spectrum $S(\Omega) = 2\pi/\alpha$. Therefore, the integral is just the discrete impulse function $(2\pi/\alpha)\delta_D[n-l]$. Using this fact reduces Eq. (2.60) to a single summation over l that can be evaluated to give

$$s_{\bar{y}}(\Delta z) = \frac{2\pi|A|^2}{\alpha} \frac{\sin[\alpha(2M+1)\Delta z/2]}{\sin[\alpha\Delta z/2]} \quad (2.61)$$

The correlation length can now be determined by evaluating Eq. (2.61) to find the value of Δz which reduces s to a given level. This value of Δz can then be converted into equivalent changes in frequency or aspect angle.

One criterion is to choose the value of Δz corresponding to the first zero of the correlation function. This occurs when the argument of the numerator equals π. Using Eq. (2.58) and defining the target length $L = (2M+1)\Delta x$,

$$\pi = \frac{4\pi \,\Delta x(2M+1)\Delta z}{2c} \Rightarrow \Delta z = \frac{c}{2L} \quad (2.62)$$

Recall that $z = F\sin\theta$. To determine the decorrelation length in angle, fix the transmitted frequency F so that $\Delta z = F(\Delta\sin\theta)$. Assuming θ is small (i.e., the radar is near broadside), $\Delta\sin\theta \approx \Delta\theta$. Equation (2.62) then becomes the desired result for the change in angle required to decorrelate the echo amplitude:

$$\Delta\theta = \frac{c}{2LF} \quad (2.63)$$

The frequency step required to decorrelate the target is obtained by fixing the aspect angle θ so that $\Delta z = \Delta F\sin\theta$. The result is

$$\Delta F = \frac{c}{2L\sin\theta} \quad (2.64)$$

This is minimum when $\theta = 90°$. Note that $L\sin\theta$ is the length of the target projected along the radar boresight.

As an example, consider a target the size of an automobile, about 5 m long. At L band (1 GHz), the target signature can be expected to decorrelate in $(3 \times 10^8)/(2 \times 5 \times 10^9) = 30$ mrad of aspect angle rotation, about $1.7°$, while at W band (95 GHz), this is reduced to only $0.018°$. The frequency step required for decorrelation with an aspect angle of $45°$ is 42.4 MHz. This result does not depend on the transmitted frequency.

The results of Eqs. (2.63) and (2.64) are based on a highly simplified target model and an assumption about what constitutes "decorrelation." For example, defining "decorrelation" to be the point at which the correlation function s first drops to $1/2$ or $1/e$ of its peak results in a smaller estimate of the required change in angle or frequency to decorrelate the target. Also, as will be seen in Chap. 6, many radars operate on the magnitude-squared of the echo amplitude, rather than the magnitude as has been assumed in this derivation. A square law detector produces a correlation function proportional to the square of Eq. (2.61) (Birkmeier and Wallace, 1963). The first zero therefore occurs at the same value of Δz, and the previous conclusions still apply. However, if a different definition of decorrelation is used (such as the 50 percent decorrelation point), the required change in Δz is less for the square law than for the linear detector.

For rigid targets, the amount of aspect angle rotation required to decorrelate the target echoes can be estimated from Eq. (2.63). The corresponding amount of time required to decorrelate successive measurements depends on the geometry of the radar/target encounter and their relative velocity. It is seen in the next section that target echo amplitudes are represented by statistical models that involve a choice of both a pdf for the target echoes and an assumption about the degree of correlation between successive measurements. It is shown in Chap. 6 that the correlation time assumption significantly affects estimates of detection performance.

It is also seen in Chap. 6 that in most cases, detection performance is improved when successive target echoes are uncorrelated. For this reason, some radars use a technique called *frequency agility* to force decorrelation of successive measurements (Ray, 1966). In this process, the radar frequency is increased by ΔF hertz or more between successive pulses, where ΔF is given by Eq. (2.64), ensuring that the target echo decorrelates from one pulse to the next. Once the desired number of samples is obtained, the cycle of increasing frequencies is repeated for the next set of measurements.

2.2.6 Swerling models

An extensive body of radar detection theory results have been built up using the four *Swerling models* of target RCS fluctuation (Swerling, 1960; Meyer and Mayer, 1973; Nathanson, 1991; Skolnik, 2001). Swerling models are intended to address the common problem of making a detection decision based not on one, but on a block of M echo samples from a given resolution cell. One motivation for considering detection based on a block of M samples may have originally been based on a simplified model of the operation of a surveillance radar, such

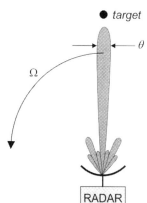

Figure 2.13 Rotating antenna rationale for Swerling model decorrelation assumptions.

as might be used to detect aircraft. Consider a radar with an antenna that rotates at a constant angular velocity Ω radians per second with an azimuth beamwidth of θ radians and a pulse repetition frequency of PRF hertz. Suppose that a target is present at a particular location. The geometry is shown in Fig. 2.13. Although some echo energy from the target is received on every pulse through the antenna side lobes, significant returns are received only when the target is in the antenna main lobe. Every complete 360° sweep of the antenna results in a new set of $M = (\theta/\Omega)PRF$ mainbeam pulses.

This is not the only way a block of related pulse echoes can arise. Many modern systems are designed to transmit bursts of pulses at a constant PRF, often with the antenna staring in a fixed direction. The time interval required for this measurement, which is simply M/PRF, is called a *coherent processing interval* (CPI).[†] The system may then repeat the entire measurement in the same or a different look direction, change the PRF to make a related measurement, or make any of a number of other changes in collecting the next CPI of data. As is seen in Chap. 4, the M-pulse burst is a common waveform well-suited to Doppler measurements, adaptive interference suppression, and imaging applications.

Each Swerling model is a combination of a probability density function and a decorrelation time for the RCS σ. They are formed from the four combinations of two choices for the pdf and two for the decorrelation time. Swerling considered two extreme cases for the correlation properties of this block of M measurements of σ. The first assumes they are all perfectly correlated, so that all M pulses collected on one sweep have the same value. The M new pulses collected on the next antenna sweep all have the same value as one another also, but their value is independent of the value measured on the first sweep.

[†]The term "CPI" is often used to refer to the block of data samples collected within the time interval, as well as to the time interval itself.

This case is referred to as *scan-to-scan decorrelation*. The second case assumes that each individual pulse on each sweep results in an independent value for σ. This case is referred to as *pulse-to-pulse decorrelation*.

The two density functions used by Swerling to describe RCS are the Rayleigh/exponential and the chi-square of degree four (see Table 2.3). The Rayleigh model describes the behavior of a complex target consisting of many scatterers, none of which is dominant. The fourth-degree chi-square models targets having many scatterers of similar strength with one dominant scatterer. Although the Rice distribution with $a^2 = 1$ is the exact pdf for this case, the chi-square is an approximation that is more analytically tractable. The approximation of the Rice distribution by the fourth-degree chi-square is based on matching the first two moments of the two pdfs (Meyer and Mayer, 1973). These moments match when the RCS of the dominant scatterer is $1+\sqrt{2} = 2.414$ times that of the sum of the RCS of the small scatterers, so the fourth-degree chi-square model fits best for this case. More generally, a chi-square of degree $2m = 1 + [a^2/(1+2a)]$ is a good approximation to a Rice distribution with a ratio of a^2 of the dominant scatterer to the sum of the small scatterers; this is readily seen using the mean and variance formulas for the two distributions given in Table 2.3. However, only the specific case of the fourth-degree chi-square is considered a Swerling model.

The Swerling models are the four combinations of the two choices for the pdf of σ and the two choices for the decorrelation characteristics. The models are denoted as "Swerling 1," "Swerling 2," and so forth. Table 2.5 defines the four cases. In some sources, the nonfluctuating target is identified as the "Swerling 0" or "Swerling 5" model.

Figures 2.14 and 2.15 illustrate the difference in the behavior of two of the Swerling models. In both cases, the received power from a single point scatterer for 500 samples of a unit mean Swerling RCS is plotted, and in both it is assumed that 10 samples are obtained on each scan of the radar, while no echo is received when the point target is outside the beam. Figure 2.14 is a sample Swerling 1 (exponential pdf, scan-to-scan decorrelation) series. The scan-to-scan decorrelation implies that all 10 samples within a single scan are identical, but independent of the 10 received on the next scan. In contrast, Fig. 2.15 illustrates a Swerling 4 case (chi-square pdf, pulse-to-pulse decorrelation) in which each individual sample is independent of the others.

TABLE 2.5 Swerling Models

Probability density function of RCS	Decorrelation	
	Scan-to-scan	Pulse-to-pulse
Rayleigh/exponential	Case 1	Case 2
Chi-square, degree 4	Case 3	Case 4

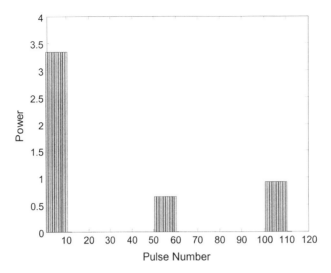

Figure 2.14 500 samples of a unit mean Swerling 1 power sequence with a 10-sample scan time.

2.3 Clutter

The term *clutter* implies an interference signal, but in radar it refers to a component of the received signal due to echoes from volume or surface scatterers. Such scatterers include the earth's surface, including both terrain and sea; weather echoes (for example, rain clouds); and man-made distributed clutter, such as

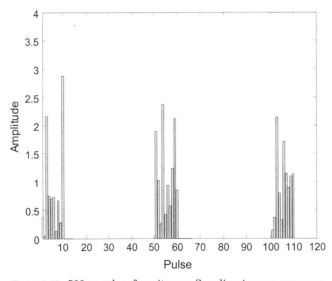

Figure 2.15 500 samples of a unit mean Swerling 4 power sequence.

so-called *chaff* clouds of airborne scatterers, typically made out of lightweight strips of reflecting material. Clutter echoes differ from targets and noise in that they are sometimes interference and sometimes the desired signal. For instance, synthetic aperture imaging radars are designed to image the earth surface, thus the terrain is the target in a SAR radar. For an airborne or spaceborne surveillance radar trying to detect moving vehicles on the ground, the surrounding terrain echo is an interference signal.

From a signal processing point of view, the major concern is how to model clutter echoes. As with man-made targets, terrain is a complex target and the echoes are highly sensitive to radar parameters and aspect angle. Thus, clutter is also modeled as a random process. Clutter differs from noise in two major ways: its power spectrum is not white (i.e., it is correlated interference), and, since it is the result of echo, the power is affected by such radar and scenario parameters as the antenna gain, transmitted power, signal processing gain, and the range from the radar to the terrain. In addition to temporal correlation, clutter can also exhibit spatial correlation: the reflectivity samples from adjacent resolution cells may be correlated. Two excellent general references on land and sea clutter phenomenology are books by Ulaby and Dotson (1989) and Long (2001).

2.3.1 Behavior of σ^0

Area clutter (land and sea surface) is of the most interest. Area clutter reflectivity is characterized by its mean or median value of radar cross section, σ^0 (dimensionless) and the probability density function of the reflectivity variations. For a given value of mean or median σ^0, many of the same pdfs described in Sec. 2.2.4 are applied to modeling clutter as well. Popular examples include Rayleigh/exponential, log-normal, and Weibull distributions.

The nature of terrain observed by the radar varies with spatial location, weather, engagement geometry, and other factors. Consequently, selection of a pdf is not sufficient to model clutter. It is also necessary to model the dependence of σ^0 on these parameters. First consider land clutter. σ^0 is a strong function of terrain type, wavelength, polarization, grazing angle, surface roughness, and moisture, to name a few parameters. Values of σ^0 commonly range from -60 to -10 dB. Extensive measurement programs over the years have collected statistics of land clutter under various conditions and resulted in many tabulations of σ^0 for various terrain types and conditions, as well as models for the variation of σ^0. One particular concern has been characterizing the variation of σ^0 with grazing angle δ. Generally, σ^0 decreases rapidly at very low grazing angles, and increases rapidly at very high grazing angles (radar look direction normal to the clutter surface), with a milder variation of σ^0 with grazing angle in a middle "plateau region." Figure 2.16 is a notional diagram of this behavior.

One of the most popular models for the behavior of σ^0 is the "constant gamma" model (Long, 2001):

$$\sigma^0 = \gamma \sin \delta \qquad (2.65)$$

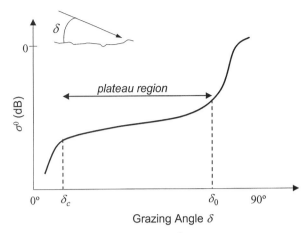

Figure 2.16 General behavior of σ^0 with grazing angle for land clutter. (After Long, 2001.)

where γ is a characteristic of the particular clutter type at the radar frequency and polarization of interest. This model predicts a maximum value of σ^0 at normal incidence and σ^0 vanishing as the grazing angle tends to zero. However, it does not accurately reflect the sharp increase in σ^0 at near-normal incidence angles, and additional models are often used at the two extremes of near-grazing and near-normal incidence.

2.3.2 Signal-to-clutter ratio

In many scenarios, the dominant interference is not noise, but clutter. Consequently, the *signal-to-clutter* ratio (SCR) is often of more importance than the *signal-to-noise* (SNR) ratio. The SCR is easily obtained as the ratio of the received signal power, given by Eq. (2.11) to the received clutter power, given by Eq. (2.25), (2.30), or (2.32) for the volume clutter, beam-limited area clutter, and pulse-limited area clutter cases, respectively. The resulting equations are

$$\begin{aligned}
\text{SCR} &= \frac{\sigma}{R^2 \eta \Delta R \theta_3 \phi_3} \quad \text{(volume clutter case)} \\
&= \frac{\sigma \sin \delta}{R^2 \phi_3 \theta_3 \sigma^0} \quad \text{(beam-limited area clutter case)} \\
&= \frac{\sigma \cos \delta}{R^3 \sigma^0 \Delta R \theta_3} \quad \text{(pulse-limited area clutter case)} \quad (2.66)
\end{aligned}$$

In each case, such system parameters as the transmitted power and the antenna gain cancel out. This occurs because both the clutter and target signals are echoes of the radar pulse; increasing power or antenna gain increases the strength of both types of echoes equally. Thus, the SCR just becomes the ratio of the target RCS to the total RCS of the contributing clutter.

2.3.3 Temporal and spatial correlation of clutter

Clutter decorrelation in time is induced by internal motion for clutter, such as leaves moving in the wind on trees or waves on the sea surface, and by changes in radar-target geometry for both clutter and targets. Various investigators have experimentally characterized the decorrelation characteristics of clutter echoes due to internal motion, or equivalently, their power spectrum. For example, one model suggested to estimate the power spectrum of the RCS of foliated trees or rain uses a cubic spectrum (Currie, 1987):

$$S_\sigma(F) = \frac{A}{1 + (F/F_c)^3} \quad (2.67)$$

The corner frequency F_c is a function of the wavelength and either wind speed (for trees) or rain rate (for rain). Some sample measured values are given in Table 2.6. Another model frequently used to model generic power spectra is the Gaussian given by

$$S_\sigma(F) = A \exp\left[-\alpha \left(\frac{F}{F_0}\right)^2\right] \quad (2.68)$$

The Gaussian model is very commonly used in weather radar, and is the basis of the pulse pair Doppler velocity estimation technique discussed in Chap. 5.

Haykin et al. show that both the cubic and Gaussian power spectral models can be well matched by a low-order autoregressive (AR, or all-pole) spectrum model of the form (Haykin, Currie, and Kessler, 1982)

$$S_\sigma(F) = \frac{A}{1 + \sum_{k=1}^{N} \alpha_k F^{2k}} \quad (2.69)$$

Real clutter measured from ground-based radars appears to be well matched using an order N of only two to four (Haykin, Currie, and Kessler, 1982). Other studies of clutter measured by airborne radars in a landing scenario indicate that orders up to 10 may be required (Baxa, 1991). The AR clutter spectrum model has the advantage that its parameters can be computed directly from

TABLE 2.6 Cubic Power Spectrum Corner Frequencies (Hz) for Rain and Tree Clutter

Target	Radar frequency, GHz		
	10	35	95
Rain, 5 mm/hr	35	80	140
Rain, 100 mm/hr	70	120	500
Trees, 6-15 mph wind	9	21	35

SOURCE: Currie, N. C. "Clutter Characteristics and Effects," chapter 10 in J. L. Eaves and K. E. Reedy (eds.), *Principles of Modern Radar*. Van Nostrand Reinhold, New York, 1987.

measured data and adapted in real time using the Levinson-Durbin or similar algorithms (Kay, 1988). Furthermore, the AR parameters can be used to construct optimal adaptive clutter suppression filters, as is seen in Chap. 5. The disadvantage is that the calculations rapidly become computationally intensive as the model increases.

2.3.4 Compound models of radar cross section

As is seen in Chap. 6, radar detection performance predictions depend strongly on the details of target and clutter RCS models. Furthermore, it is well known that RCS statistics vary significantly with a host of factors such as geometry, resolution, wavelength, and polarization. Consequently, the development of good statistical RCS models is a very active area of empirical and analytical research. Following are three brief examples of an extension to the basic modeling approach described earlier, all motivated by the complexities of modeling clutter.

Some RCS probability density functions are physically motivated, especially the Rayleigh/exponential model (which follows from a central limit theorem argument) and the Rice model (which corresponds to a Rayleigh model with an additional dominant scattering source). Others, such as the lognormal or Weibull, have been developed empirically by fitting distributions to measured data. One attempt to provide a physical justification for a non-Rayleigh model abandons the single-pdf approach, instead assuming that the random variable representing echo amplitude (voltage) can be written as the product of two independent random variables:

$$\zeta = xy \qquad (2.70)$$

The pdf of ζ can then be represented in a Bayesian formulation as

$$p_\zeta(\zeta) = p_x(x) p_{\zeta|x}(\zeta|x) \qquad (2.71)$$

Ward uses this model to describe sea clutter (Ward, 1981). He identifies the random variable x with a slowly decorrelating component having a voltage distribution following a chi-square of degree $2m$ with $m \geq 2.5$. This component is introduced to account for "bunching" of scatterers due to ocean swell structure and radar geometry, and represents variation in the mean of the voltage over time. The distribution $p_{\zeta|x}(\zeta|x)$ is assumed to represent the composite of a large number of independent scatterers. Its voltage distribution is therefore Rayleigh. The overall pdf $p_\zeta(\zeta)$ can be shown to be the *K-distribution*, which had previously been proposed as a model of sea clutter (Jakeman and Pusey, 1976). Thus, the product formulation suggests that modulation of a standard Rayleigh variable by a chi-square distributed geometric term can account for observed sea clutter distributions.

More recent research has begun to bridge the gap between the physics of scattering and the apparent success of compound clutter models of the type promoted by Ward and Jakeman and Pusey. Sangston summarizes the work that considers an extension of the "many scatterer" physical model that leads to the Rayleigh distribution (Sangston, 1997). Specifically, consider the model of Eq. (2.50), but with the fixed number of scatterers N replaced by a random variable N

$$\zeta = \left| \sum_{i=1}^{N} \sqrt{\sigma_i}\, e^{-j 4\pi R_i / \lambda} \right| \qquad (2.72)$$

This representation is referred to as a "number fluctuations" model. Depending on the choice of the statistics of the number N of scatterers contributing to the return at any given time, the model of Eq. (2.72) can result in K, Weibull, gamma, Nakagami-m, or any of a number of other distributions in the class of so-called Rayleigh mixtures (Sangston, 1997).

Much of the work in compound RCS models has been performed in the context of sea clutter analysis, and empirical sea clutter data have often been observed to exhibit non-Rayleigh statistics such as Weibull, K, and log-normal distributions. The number fluctuation model is intuitively appealing in this case, because it can be related to the physical behavior of waves. Specifically, scattering theory suggests that the principal scatterers on the ocean surface are the small capillary waves, as opposed to the large swells, and that these small scattering centers tend to cluster near the crest of the swells, with fewer of them in between. In other words, they are nonuniformly distributed over the sea surface (Sangston, 1997). Consequently, a radar illuminating the sea will receive echoes from a variable number N of scatterers as the crests of the swells move into and out of a given resolution cell. Thus the success of the number fluctuation model, which sums echoes from a variable number of scatterers, in predicting Weibull and K distributions provides a link between a phenomenological model of sea scatter and empirically observed statistics.

All of the statistical models described in Sec. 2.2.4 apply to the scattering observed from a single resolution cell, i.e., they represent the variations in RCS observed by measuring the same region of physical space multiple times, e.g., by transmitting multiple pulses in the same direction and measuring the received power at the same delay after each transmission. Another use of the product model of Eq. (2.70) and (2.71) is to describe the spatial variation of clutter reflectivity. If the scene being viewed by the radar is nonhomogenous, then the characteristics of the RCS observed in one resolution cell might vary significantly from those of another. For example, the dominant clutter observed by a scanning radar at a coastal site might be an urban area in one look direction and the sea in another. Another example occurs when scattered rain cells occupy only part of the scanned region, so that some resolution cells contain rain while others are clear.

This situation can be modeled by letting the slowly decorrelating term x in the product model represent spatial variations in the local mean of the received voltage. If the pdf of x is log-normal with a large variance and the pdf of ζ conditioned on x is gamma distributed (which includes Rayleigh as a special case), then the overall pdf of the product ς has a log-normal distribution (Lewinski, 1983). Consequently, the product model implies that log-normal variations of the local mean from one resolution cell to another could account for the log-normal variation often used to model ground clutter returns. A similar argument can be used to justify the log-normal model for target RCS by modeling the variation of RCS with aspect angle as a log-normal process.

2.4 Noise Model and Signal-to-Noise Ratio

The echo signal received from a target or clutter inevitably competes with noise. There are two sources of noise: that received through the antenna from external sources, and that generated in the radar receiver itself.

External noise is a strong function of the direction in which the radar antenna is pointed. The primary contributor is the sun. If the antenna is directed toward the night sky and there are no interfering microwave sources, the primary source is *galactic* (also called *cosmic*) *noise*. Internal noise sources include *thermal noise* (also called *Johnson noise*) due to ohmic losses, *shot noise* and *partition noise* due to the quantum nature of electric current, and *flicker noise* due to surface leakage effects in conducting and semiconducting devices (Carlson, 1976).

Of these various sources, thermal noise is normally dominant. The theories of statistical and quantum mechanics dictate that thermal noise voltage in an electronic circuit is a zero-mean Gaussian random process (Curlander and McDonough, 1991). The mean energy of the random process is $kT/2$ joules, where T is the temperature of the noise source in degrees Kelvin (absolute temperature) and $k = 1.38 \times 10^{-23}$ J/K is *Boltzmann's constant*. The power spectrum $S_n(F)$ of the thermal noise delivered to a matched load is

$$S_n(F) = \frac{hF}{\exp(hF/kT) - 1} \quad \text{W/Hz} \tag{2.73}$$

(Ziemer and Tranter, 1976) where $h = 6.6254 \times 10^{-34}$ J/s is *Planck's constant*. If $hF/kT \ll 1$, a series approximation gives $\exp(hF/kT) \approx 1 + hF/kT$ so that Eq. (2.73) reduces to the white noise spectrum

$$S_n(F) = kT \quad \text{W/Hz} \tag{2.74}$$

Note that Eq. (2.74), when integrated over frequency, implies infinite power in the white noise process. In reality, however, the noise is not white (Eq. (2.73)) and, in any event, it is observed in any real system only over a finite bandwidth. For frequencies below 100 GHz, the approximation of Eq. (2.74) requires the effective noise temperature T (to be defined below) to be larger than about 50K, which is almost always the case. Consequently, thermal noise has a white power spectrum. For many practical systems, it is reasonable to choose the temperature of the system to be the "standard" temperature $T_0 = 290$K. In this case, $kT_0 \approx 4 \times 10^{-21}$ W/Hz.

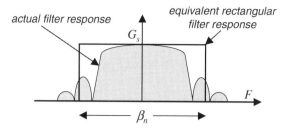

Figure 2.17 Illustration of the concept of noise equivalent bandwidth of a filter.

In a coherent radar receiver, the noise present at the front end of the system contributes noise to both the I and Q channels after the quadrature demodulation. It is easy to show (Ziemer and Tranter, 1976) that the I and Q channel noises are both zero-mean Gaussian random processes with equal power. Since the total noise spectral density is kT watts per hertz, the noise density in each channel individually is $kT/2$ watts per hertz. Furthermore, if the power spectrum of the input noise is white, then the I and Q noise processes are uncorrelated and their power spectra are also white. Since the I and Q noise processes are Gaussian and uncorrelated, it follows that they are also independent (Papoulis, 1984). Finally, since the I and Q signals are independent zero-mean Gaussian processes, it also follows that the magnitude of the complex signal $I + jQ$ is Rayleigh distributed, the magnitude-squared is exponentially distributed, and the phase angle $\tan^{-1}(Q/I)$ is uniformly distributed over $(0, 2\pi]$ as described in Sec. 2.2.5.

Radar receivers do not have infinite bandwidth. The bandwidth of the various components of a receiver varies, but the narrowest bandwidth is generally approximately equal to the bandwidth of the transmitted pulse. If the receiver contains any component of narrower bandwidth, signal energy will be lost, reducing sensitivity. If the most narrowband component has a bandwidth appreciably wider than the pulse bandwidth, the signal will have to compete against more noise power than necessary, again reducing sensitivity. Thus the frequency response of the receiver can be approximated as a bandpass filter centered at the transmit frequency, with a bandwidth equal to the waveform bandwidth.

Real filters do not have perfectly rectangular passbands. For analyzing noise power, the *noise-equivalent bandwidth* β_n of a filter described by the transfer function $H(F)$ is used. Figure 2.17 illustrates the concept. The noise equivalent bandwidth is the width an ideal rectangular filter, with gain equal to the peak gain of the actual filter, must have so that the area under the two squared frequency responses are equal. This condition guarantees that given a white noise input, both filters exhibit the same output noise power. Thus

$$\beta_n = \frac{\int_{-\infty}^{\infty} |H(F)|^2 dF}{\max[|H(F)|^2]} = \frac{1}{G_s} \int_{-\infty}^{\infty} |H(F)|^2 dF \qquad (2.75)$$

where G_s is defined as the maximum gain of $H(F)$. The total noise power N present at the output of the filter $H(F)$ is then given by

$$N = \int_{-\infty}^{\infty} |H(F)|^2 S_n(F) \, dF$$
$$= kT \int_{-\infty}^{\infty} |H(F)|^2 dF$$
$$= kT \beta_n G_s \tag{2.76}$$

White noise passed through a filter $H(F)$ is no longer white, but instead has the power spectrum $|H(F)|^2$. If $|H(F)|^2$ is modeled as a rectangular filter of two-sided bandwidth β_n hertz, the autocorrelation function of the noise at the filter output is a sinc function with its first zero at delay $1/\beta_n$ seconds. Thus, the receiver output noise is not white, and its decorrelation time is on the order of $1/\beta_n$ seconds. However, it will be seen in Chap. 3 that the receiver output is normally sampled at intervals of approximately $1/\beta_n$ seconds. Consequently, the noise component of the successive receiver output samples are still uncorrelated with one another.

The power spectral density at the output of any noise source or circuit can be described as the product of Boltzmann's constant and some equivalent temperature T', mimicking the simple formulation of Eq. (2.74). Source noise power is usually referenced to the input of a system, so that the power gain G_s (or loss, in which case $G_s < 1$) of the system must also be taken into account. That is, if the observed output power spectral density (still assumed white) is some value S_n, then an equivalent temperature T' of the noise source is defined to be

$$T' \equiv \frac{S_n}{kG_s} \tag{2.77}$$

so that $S_n = kT'G_s$ and the total noise power, using Eq. (2.76) is

$$N = kT'\beta_n G_s \tag{2.78}$$

The total output noise power at the receiver output is the primary quantity of interest. In a radar system, the contributors to this noise include the external noise, the intrinsic $kT_0\beta_n$ thermal noise, and additional thermal noise due to losses in the antenna structure and nonideal receivers. Detailed noise analyses assign individual equivalent noise temperatures to each stage in the system; good introductory description was given by Curlander and McDonough (1991). When considering the system as a whole, it is common to express the total output noise power as the sum of the power that would be observed due to the minimum noise density kT_0 at the input and a second term that accounts for the additional noise due to the nonideal system

$$N = kT_0\beta_n G_s + kT_e\beta_n G_s \tag{2.79}$$

In this equation, G_s is now the power gain of the complete receiver system, including antenna loss effects. The equivalent temperature T_e used to account for noise above the theoretical minimum is called the *effective temperature* of the system.

The noise temperature description of noise power is most useful for low-noise receivers. An alternative description common in radar is based on the idea of *noise figure* F_n, which is the ratio of the actual noise power at the output of a system to the minimum power $kT_0\beta_n G_s$ (Skolnik, 1980). As with noise temperatures, various noise figures can be defined to include the effects of just the receiver, or of the entire antenna and receiver system, and so forth. Here, the term *noise figure* used without qualification will mean the noise figure of the complete receiver system, so that

$$F_n = \frac{N}{kT_0\beta_n G_s} \qquad (2.80)$$

Equation (2.80) implies that knowledge of the noise equivalent bandwidth and gain of the receiver system are sufficient to calculate the output noise power using $N = kT_0\beta_n F_n G_s$. It also follows from using Eq. (2.79) in Eq. (2.80) that $T_e = (F_n - 1)T_0$. Typical noise figures for radars can be as low as 2 or 3 dB, and as high as 10 dB or more. Corresponding effective temperatures range from about 170K to over 2600K.

In Sec. 2.2, the term "radar range equation" was applied to Eqs. (2.11), (2.25), (2.30), and (2.32). These expressions described the power received by the radar given various system and propagation conditions. As will be seen in Chap. 6, the detection performance of a radar depends not on the received power per se but on the SNR at the point of detection. The earlier results can be used to convert the power range equations to SNR range equations.

To illustrate, consider the point target range Eq. (2.11), which expresses the power P_r of the signal available at the input to the receiver. The signal power at the output will be $P_o = G_s P_r$ provided the signal bandwidth is entirely contained within the receiver bandwidth B_n. From Eq. (2.80), the output noise power is $N_o = kT_0\beta_n F_n G_s$. The signal to noise ratio is therefore

$$\begin{aligned}\chi &= \frac{P_o}{N_o} \\ &= \frac{G_s P_t G^2 \lambda^2 \sigma}{(4\pi)^3 R^4 L_s L_a(R)} \cdot \frac{1}{kT_0\beta_n F_n G_s} \\ &= \frac{P_t G^2 \lambda^2 \sigma}{(4\pi)^3 R^4 kT_0\beta_n F_n L_s L_a(R)}\end{aligned} \qquad (2.81)$$

The last expression in Eq. (2.81) gives the SNR in terms of transmitter and receiver characteristics, target RCS, range, and loss factors. Modifications of Eqs. (2.25), (2.30), and (2.32) for volume and area scatterers to express them in terms of signal to noise ratio are obtained in the same manner by simply adding the term $kT_0\beta_n F_n$ to their denominators.

Like Eq. (2.11), Eq. (2.81) is also often called the radar range equation. In the remainder of this text, the term "range equation" or "radar range equation" refers to the SNR form of Eq. (2.81) and its analogues for volume and area scatterers.

2.5 Jamming

Jamming refers to intentional interference directed at the radar system from a hostile emitter. Jamming is an example of *electronic countermeasures* (ECM) or *electronic attack* (EA). As noted earlier, the purpose of most radar signal processing is to improve the SIR of the data before a detection test is carried out. The purpose of jamming is just the opposite: to reduce the SIR so that the detection performance of the radar is degraded.

The most basic form of jamming is simple noise jamming. A hostile emitter directs an amplified noise waveform at the victim radar, essentially increasing the noise level out of the receiver. If the noise power spectrum fills the entire radar receiver bandwidth, then the noise out of the receiver will appear like any other white noise process and is modeled in the same way. More advanced forms of noise jamming use various amplitude and frequency modulations. Instead of noise, other jamming techniques use waveforms designed to mimic target echoes and fool the radar into detecting and tracking non-existent targets.

Even a limited discussion of ECM is out of the scope of this text, due both to the breadth of the topic and the limited amount of material publishable in the open literature. The reader is referred to the book by Lothes et al., (1990) for a good general reference on jamming signals in radar.

2.6 Frequency Models: The Doppler Shift

2.6.1 Doppler shift

If the radar and scatterer are not at rest with respect to one another, the frequency F_r of the received echo will differ from the transmitted frequency F_t due to the Doppler effect. A correct description of the Doppler shift for electromagnetic waves requires the theory of special relativity. Consider a monostatic radar, where the transmitter and receiver are at the same location and do not move with respect to one another. Suppose a scatterer in the radar field of view is moving with a velocity component v toward the radar. If the transmitted radar frequency is F_t, then the theory of special relativity predicts that the received frequency F_r will be (Temes, 1959; Gill, 1965)

$$F_r = \left(\frac{1+v/c}{1-v/c}\right)F_t \qquad (2.82)$$

Thus, an approaching target causes an increase in the received frequency. Substituting $v = -v$ shows that a receding target decreases the received frequency. This is in keeping with our experience with the whistles of passing trains.

Equation (2.82) can be simplified without significant loss of precision because the velocity of actual radar targets is a small fraction of c. For example, the value of v/c for a supersonic aircraft traveling at Mach 2 (about 660 m/s) is only 2.2×10^{-6}. Expand the denominator of Eq. (2.82) in a binomial series:

$$\begin{aligned} F_r &= (1+v/c)(1-v/c)^{-1}F_t \\ &= (1+v/c)[1+(v/c)+(v/c)^2+\cdots]F_t \\ &= [1+2(v/c)+2(v/c)^2+\cdots]F_t \end{aligned} \quad (2.83)$$

Discarding all second-order and higher terms in (v/c) leaves

$$F_r = [1 + 2(v/c)]F_t \quad (2.84)$$

The difference F_D between the transmitted and received frequencies is called the *Doppler frequency* or *Doppler shift*. For this case of an approaching target it is

$$F_D = +\frac{2v}{c}F_t = +\frac{2v}{\lambda_t} \quad (2.85)$$

where λ_t is the transmitted wavelength; for a receding target the Doppler shift would be equal in magnitude to Eq. (2.85), but negative. The magnitude of Eq. (2.85) is the expression conventionally used in radar to evaluate Doppler shift, with the sign determined by whether the radar and target are closing or opening in relative range.

The numerical values of Doppler shift are small compared to the RF frequencies. Table 2.7 gives the magnitude of the Doppler shift corresponding to a velocity of 1 m/s at various typical radar frequencies. The Mach 2 aircraft, observed with the L band radar, would cause a Doppler shift of only 4.4 kHz in a 1 GHz carrier frequency.

For a monostatic radar, the observed Doppler shift is proportional to the component of velocity in the direction of the radar, called the *radial velocity*. If the angle between the velocity vector of target traveling at v meters per second and the radar boresight is ψ, the radial velocity is $v \cos \psi$ meters per second. The geometry is illustrated in two dimensions in Fig. 2.18. The magnitude of the Doppler shift is maximum when the target is traveling directly toward or away from the radar. The Doppler shift is zero, regardless of the target velocity, when the target is crossing orthogonally to the radar boresight.

TABLE 2.7 Doppler Shift Resulting from a Velocity of 1 m/s

Band	Frequency (GHz)	Doppler shift (Hz) for $v = 1$ m/s
L	1	6.67
C	6	40.0
X	10	66.7
K_a	35	233
W	95	633

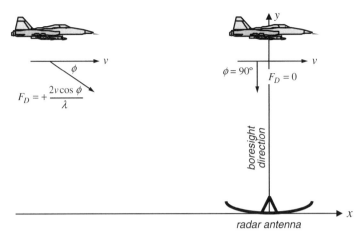

Figure 2.18 Doppler shift is determined by the radial component of relative velocity between the target and radar.

Radar waveforms are not pure monochromatic sinusoids, but instead have a finite bandwidth β_t. Applying Eq. (2.85) to the upper and lower frequency components shows that the Doppler shifted received bandwidth β_r is

$$\beta_r = [1 + 2(v/c)]\beta_t \tag{2.86}$$

Because the bandwidth is almost always 10 percent or less of the carrier frequency (and usually 1 percent or less), the change in bandwidth of $2(v/c)\beta_t$ is usually completely insignificant.

As mentioned earlier, the values of Doppler shift are quite small. As an example, consider a target moving at a velocity of 100 m/s (224 mi/h) with respect to an X band (10 GHz) radar. Using Eq. (2.85) or Table 2.7, the resulting Doppler shift will be 6.67 kHz. This Doppler shift is too small to be measured in a single pulse echo. Suppose the pulse length is 10 μs. The frequency resolution available with a 10 μs observation time is 1/(10 μs) = 100 kHz, much larger than that required to estimate the 6.67 kHz shift. Put another way, in 10 μs a 6.67 kHz sinusoid progresses through only 6.67 percent of a cycle. For this reason, most radars cannot measure Doppler shifts of interest from a single pulse.

Despite the small values of Doppler shift, it will be seen in Sec. 2.6.3 that Doppler shifts are large enough to be detected. This will be accomplished by measuring phase shifts over multiple pulse repetition intervals (PRIs), giving a much longer observation timeline and thus much better Doppler resolution. These Doppler shifts can be used to advantage to detect echoes from moving targets in the presence of much stronger echoes from clutter or to drastically improve cross-range resolution. Uncompensated Doppler shifts can also have harmful effects, particularly a loss of sensitivity for some types of waveforms.

2.6.2 Simplified approach to Doppler shift

As argued in deriving Eq. (2.84), the relativistic effect on Doppler shift is insignificant for any realistic target velocity because v/c is so small. Classical derivations can therefore be used with no significant error (Gill, 1965). In the analysis so far, Doppler shift has been considered only for a sinusoid of infinite duration and a constant radial velocity v. A simple generalization of the Doppler shift formula can be developed that is easily applied to arbitrary motion between the radar and target and any waveform using a classical physics approach.

Suppose the one-way range from the radar to the target as a function of time is $R(t)$. If the transmitter radiates a signal $\tilde{x}(t)$, the echo received will simply be (ignoring amplitude scaling factors) a delayed replica of the transmitted signal:

$$\tilde{y}(t) = \tilde{x}\left(t - \frac{2R(t)}{c}\right) \tag{2.87}$$

Though simple, Eq. (2.87) leads to correct descriptions of time-varying Doppler shifts. For example, suppose the signal $\tilde{x}(t)$ is a simple pulse of the form

$$\tilde{x}(t) = a(t)\exp(j2\pi F_t t) \tag{2.88}$$

where $a(t)$ is the pulse envelope. Now suppose the target is moving toward the radar at a constant radial velocity v. This means that

$$R(t) = R_0 - vt \tag{2.89}$$

where R_0 is the range at some time $t = 0$. Combining Eq. (2.87) through (2.89) gives the received signal as

$$\begin{aligned}\tilde{y}(t) &= a\left(t - \frac{2(R_0 - vt)}{c}\right)\exp\left[j2\pi F_t\left(t - \frac{2(R_0 - vt)}{c}\right)\right]\\ &\approx a\left(t - \frac{2R_0}{c}\right)\exp\left(-j\frac{4\pi c}{\lambda_t}R_0\right)\exp\left[+j2\pi\left(\frac{2v}{\lambda_t}\right)t\right]\exp(j2\pi F_t t)\end{aligned} \tag{2.90}$$

In Eq. (2.90), the first exponential term is a constant phase factor of no consequence, while the last is the carrier frequency. The middle exponential is a sinusoid of frequency $+2v/\lambda_t$, which is exactly the Doppler shift of Eq. (2.85). Thus the analysis approach of Eq. (2.87) is consistent with the earlier results.

In deriving Eq. (2.90), the time delay for the one-way propagation from the radar to the target has been effectively ignored; this assumption will be examined more closely in the next subsection. The approximation in going from the first line of Eq. (2.90) to the second is due to ignoring the term $2vt/c$ in the argument of the pulse envelope. This represents time compression of the envelope due to the moving target. In evaluating this term, the largest value of t that need be considered is the pulse length τ. The ratio of this time shift to the pulse length is therefore $(2v\tau/c)/\tau = 2v/c$. Even for the fastest targets, this compression in the envelope duration is insignificant.

For a more interesting example of the use of Eq. (2.87), consider Fig. 2.18 again. Let a sidelooking radar be located at (x, y) coordinates $(x_r = 0, y_r = 0)$, and let the coordinates of the target aircraft be $(x_t = vt, y_t = R_0)$. This means that the target aircraft is initially directly abreast of the radar platform at a range R_0 at time $t = 0$, and is crossing orthogonal to the radar line of sight at a velocity v meters per second. The range between radar and aircraft is the Euclidean distance

$$R(t) = \sqrt{R_0^2 + (vt)^2}$$
$$= R_0\sqrt{1 + \left(\frac{vt}{R_0}\right)^2} \qquad (2.91)$$

While it is possible to work with Eq. (2.91) directly, it is customary to expand the square root in a power series:

$$R(t) = R_0\left[1 + \frac{1}{2}\left(\frac{vt}{R_0}\right)^2 - \frac{3}{8}\left(\frac{vt}{R_0}\right)^4 - \cdots\right] \qquad (2.92)$$

In evaluating this expression, the range of t that must be considered may be limited by any of several factors, such as the time the aircraft is within the radar main beam, or the coherent processing interval duration over which pulses will be collected for subsequent combining in the signal processor. In any event, it is almost always true that the distance traveled by the target within the time of interest vt is much less than the nominal range R_0 so that higher order terms can be neglected. Thus

$$R(t) \approx R_0 + \left(\frac{v^2}{2R_0}\right)t^2 \qquad (2.93)$$

Equation (2.93) shows that the range is approximately a quadratic function of time for the crossing target scenario of Fig. 2.18. Using this truncated series in Eqs. (2.87) and (2.88) will give a result similar to Eq. (2.90)

$$\tilde{y}(t) \approx a\left(t - \frac{2R_0}{c}\right)\exp\left(-j\frac{4\pi}{\lambda_t}R_0\right)\exp\left[-j2\pi\left(\frac{v^2}{R_0\lambda_t}\right)t^2\right]\exp(j2\pi F_t t) \qquad (2.94)$$

All of the terms are the same as in Eq. (2.90) except for the middle exponential. Because the phase function varies with the square of time, this is not a Doppler shift in the conventional sense of the term. Since instantaneous frequency is proportional to the time derivative of phase

$$F(t) = \frac{1}{2\pi}\frac{d\varphi(t)}{dt} \qquad (2.95)$$

this quadratic phase function represents a frequency component that varies linearly with time:

$$F_D(t) = -\frac{2v^2}{R_0 \lambda_t} t \qquad (2.96)$$

Thus, the "Doppler shift" of the received signal varies linearly due to the changing radar-target geometry. As the target aircraft approaches from the left in Fig. 2.18 ($t < 0$), the instantaneous Doppler shift is positive. When the aircraft is abreast of the radar ($t = 0$), the Doppler shift is zero because the radial component of velocity is zero. Finally, as the aircraft passes by the radar ($t > 0$), the Doppler shift becomes negative, as would be expected for a receding target.

This example shows how relative motion between radar and target can create time-varying Doppler shifts, and how the simple approach of Eq. (2.87) can be used to compute these shifts. The quadratic range case is important in synthetic aperture radar, and it will be revisited in Chap. 8 in more detail.

2.6.3 The "stop-and-hop" assumption and spatial Doppler

In this section, the model for an echo from a moving target is considered more closely. The one-way range from the radar to the target is again $R(t)$ along the line-of-sight axis, denoted here as the z axis. Suppose the transmitter is located at $z = 0$ and radiates the signal $\bar{x}(t)$ at time $t = 0$; the leading edge of the pulse is then at $z = ct$ at time t. The outgoing pulse will strike the target at some time t_1 such that $ct_1 = R(t_1)$. The echo will then be received another t_1 seconds later, so that the total two-way trip time is $t_d = 2t_1$. The echo received will simply be (ignoring amplitude scaling factors) a delayed replica of the transmitted signal

$$\bar{y}(t) = \bar{x}(t - t_d) \qquad (2.97)$$

Consider again the constant velocity case where $R(t) = R_0 - vt$. The one-way travel time is given by

$$ct_1 = R_0 - vt_1 \Rightarrow t_1 = \frac{R_0}{c+v} \Rightarrow t_d = \frac{2R_0}{c+v} \qquad (2.98)$$

Now consider the mth pulse in an M-pulse train. The result of Eq. (2.98) can be modified by assuming that the mth pulse is transmitted at time $t = mT$, where T is the radar PRI. The pulse and target are then located at z coordinates $c(t - mT)$ and $R_0 - vt = R_0 - vmT - v(t - mT)$ at time t. The one-way travel time after the pulse is transmitted is

$$t_1 = t - mT = \frac{R_0 - vmT}{c+v} \qquad (2.99)$$

and the total round-trip delay t_d is twice this. It is convenient to form the following approximation to t_d

$$\begin{aligned} t_d &= \frac{2(R_0 - vmT)}{c + v} = \frac{2(R_0 - vmT)}{c\left(1 + \frac{v}{c}\right)} \\ &= \frac{2(R_0 - vmT)}{c}\left[1 - \frac{v}{c} + \left(\frac{v}{c}\right)^2 - \cdots\right] \\ &\approx \frac{2(R_0 - vmT)}{c}\left(1 - \frac{v}{c}\right) \end{aligned} \quad (2.100)$$

The transmitted signal for pulse m is

$$\tilde{x}_m(t) = a(t - mT)\exp[j2\pi F_t(t - mT)] \quad (2.101)$$

and, using Eq. (2.100), the received signal becomes

$$\begin{aligned} \tilde{y}_m(t) &= a\left(t - \frac{2(R_0 - vmT)}{c + v}\right)\exp\left[j2\pi F_t\left(t - \frac{2(R_0 - vmT)}{c + v}\right)\right] \\ &\approx a\left[t - mT - \frac{2(R_0 - vmT)}{c}\left(1 - \frac{v}{c}\right)\right] \\ &\quad \times \exp\left\{j2\pi F_t\left[t - mT - \frac{2(R_0 - vmT)}{c}\left(1 - \frac{v}{c}\right)\right]\right\} \\ &\approx a\left(t - mT - \frac{2(R_0 - vmT)}{c}\right) \\ &\quad \times \exp\left\{j2\pi F_t\left[t - mT - \frac{2(R_0 - vmT)}{c}\left(1 - \frac{v}{c}\right)\right]\right\} \end{aligned} \quad (2.102)$$

where the last line takes advantage of the fact that the additional time shift of the signal envelope due to the v/c term is insignificant.

Now sample this echo at a time delay after transmission that corresponds to the nominal range R_0. The sampled echo is

$$\begin{aligned} \tilde{y}_m\left(mT + \frac{2R_0}{c}\right) &= a\left(\frac{v}{c}mT\right)\exp\left[j2\pi F_t\left(\frac{2v}{c}mT\right)\left(1 - \frac{v}{c}\right)\right] \\ &= a\left(\frac{v}{c}mT\right)\exp\left[j\frac{4\pi vmT}{\lambda_t}\left(1 - \frac{v}{c}\right)\right] \\ &= a\left(\frac{v}{c}mT\right)\exp\left[j2\pi\left(\frac{2v}{\lambda_t}\right)mT\right]\exp\left[-j\frac{4\pi v^2 mT}{\lambda_t c}\right] \end{aligned} \quad (2.103)$$

Denote the maximum value of mT as the *aperture time*[†] T_a; thus $T_a = (M-1)T \approx MT$. The envelope term $a(vmT/c)$ is a constant equal to the

---
[†]This terminology is adopted from synthetic aperture imaging and will be explained in Chap. 8.

envelope amplitude so long as $vT_a < c\tau$, which states that the target motion during the aperture time T_a is less than the pulse span $c\tau$, that is, the target does not move more than a pulse length. This is normally the case for Doppler processing, which uses short aperture times. When synthetic aperture imaging is considered in Chap. 8, this condition may be violated, introducing the complication of *range migration*.

Now consider the two phase terms in Eq. (2.103). The first term is a discrete time complex sinusoid $\exp(j\omega m)$ with $\omega = 2\pi(2v/\lambda_t)T$, corresponding again to an analog Doppler frequency of $2v/\lambda_t$ hertz. The second term is another discrete sinusoid with a frequency of $-2v^2T/\lambda_t c$ hertz. It is more useful to consider the total phase excursion of this term. It is seen in Chap. 8 that this term has no significant impact on the spectrum of the measured data sequence provided that the maximum phase change over the aperture time is less than a fraction of π radians. Choosing a standard of $\pi/4$ and recalling that the maximum value of mT is no more than T_a, this requires that

$$\frac{4\pi v^2 T_a}{\lambda_t c} < \frac{\pi}{4} \Rightarrow T_a < \frac{\lambda_t c}{16v^2} \qquad (2.104)$$

This inequality must be evaluated for any particular situation, but is easily satisfied for a vast majority of cases.[†] In this event, the received and sampled signal becomes simply

$$\bar{y}_m\left(mT + \frac{2R_0}{c}\right) = a\left(\frac{v}{c}mT\right)\exp\left[j2\pi\left(\frac{2v}{\lambda_t}\right)mT\right] \qquad (2.105)$$

Now repeat the development of Eqs. (2.99) through (2.103) under the assumption that there is no range change between the target and radar during the one-way travel time of a pulse. This will result in the $(c+v)$ term in Eq. (2.99) being replaced by simply c. This in turn will replace the $[1+(v/c)]$ terms with simply 1, and the model of the received signal will then be exactly that of Eq. (2.105). This assumption is often referred to as the *stop-and-hop assumption*. In synthetic aperture imaging, where the "target" is stationary ground scatterers, it effectively assumes that the radar, which is on a moving platform, stops and hovers as it transmits and receives each pulse, and then "hops" forward by vT meters before stopping to transmit the next pulse. In Doppler processing, where the target is moving and the radar platform may or may not be in motion, it is tantamount to assuming that both the target and radar stop during pulse transmission and reception, and then move the appropriate amounts during the pulse repetition interval before stopping for the next pulse. The stop-and-hop assumption allows the second phase term in Eq. (2.103) to be neglected, permitting use of the simple $2v/\lambda$ model for Doppler shift.

It was pointed out previously that typical values of Doppler shift are too small to be detected within a typical pulse length. How then is Doppler shift

---

[†]While almost always valid in radar, the stop-and-hop assumption is often not valid in sonar signal processing, requiring extra phase corrections in sonar Doppler processing and imaging not used in radar algorithms.

measured? Equation (2.105) showed that the sequence of coherent samples taken from the *same* range bin (same time delay $2R/c$ after pulse transmission) over multiple pulses forms a discrete-time sinusoid. The normalized radian frequency is $4\pi vT/\lambda_t$ radians, corresponding to $2v/\lambda_t$ hertz. Note that this series of samples is simply the slow-time sequence $y[l_0, m]$, where l_0 is the range bin corresponding to range R_0. Thus, the phase progression of a slow-time data sequence for a range bin containing a moving target provides a measure of the Doppler shift of the target. The longer observation time T_a corresponding to having M pulses provides the resolution necessary to measure small Doppler shifts. Recall that it is assumed that the relationship between the target velocity, total observation time, and range bin spacing is such that the target remains within a single range bin over M pulses, i.e., for MT seconds.

The manifestation of the target Doppler in the slow-time phase progression is sometimes referred to as *spatial Doppler*. This terminology emphasizes the fact that the Doppler shift is measured not from intrapulse frequency changes, but rather from the change of absolute phase of the echoes at a given range bin over a series of pulses. Because of the inability to measure intrapulse Doppler frequency shifts in most systems, the term Doppler processing in radar refers to sensing and processing this spatial Doppler information. This concept extends to synthetic aperture imaging where, as shown previously, the Doppler shift is not constant.

2.7 Spatial Models

Previous sections have dealt with models of Doppler shift and the received power (both mean value and statistical fluctuations) of radar echoes from a single resolution cell. In this section, the variation in received power or complex voltage as a function of the spatial dimensions of range and angle will be considered. It will be seen that the observed complex voltage can be viewed as the output of a linear filter with the "true" variation in reflectivity over range or angle as its input. These relationships will lay the groundwork for an analysis of data sampling requirements and range and angle resolution in subsequent chapters.

Consider a stationary pulsed radar. On the mth pulse it transmits the equivalent complex signal.

$$\tilde{x}(t; m) = \sqrt{P_t}\, x(t - t_m) \exp(j 2\pi F_t t) \qquad (2.106)$$

In Eq. (2.106), $x(t)$ represents the modulation of the sinusoidal carrier, including both amplitude (typically on-off pulsing) and any phase modulation, and t_m is the time of transmission of the mth pulse. Assume that $x(t)$ has unit amplitude, so that the transmitted signal amplitude is represented by the term $\sqrt{P_t}$.

This signal echoes off a differential scatterer of cross section $d\sigma(R, \theta, \phi)$ at coordinates (R, θ, ϕ). The corresponding differential contribution to the complex voltage is, from Eq. (2.50), $d\zeta(R,\theta,\phi)\exp[j\psi(R,\theta,\phi)]$, so that $d\sigma = |d\zeta \exp(j\psi)|^2$.[†] The antenna may be scanning in either or both angle coordinates, so that at time t_m it is steered in the direction (θ_m, ϕ_m). Then from Eq. (2.16) and (2.2) the differential received signal is

$$d\bar{y}(\theta_m,\phi_m,t-t_m;R,\theta,\phi) = \sqrt{\frac{P_t P^2(\theta-\theta_m,\phi-\phi_m)\lambda^2}{(4\pi)^3 R^4 L_s L_a(R)}} d\zeta(R,\theta,\phi) \cdots$$

$$\times \exp[j\psi(R,\theta,\phi)]x\left(t-t_m-\frac{2R}{c}\right)\exp\left[j2\pi F_t\left(t-\frac{2R}{c}\right)\right] \quad (2.107)$$

Equation (2.107) can be simplified by separating the reflectivity terms and the terms which depend on spatial location, collapsing all of the other system-dependent amplitude terms into a single constant A_r. The term $d\zeta \exp(j\psi)$ is termed the *baseband complex reflectivity* or just *reflectivity* of the differential scatterer, and will be denoted as $d\rho$. Making these substitutions gives

$$d\bar{y}(\theta_m,\phi_m,t-t_m;R,\theta,\phi)$$
$$= A_r\, d\rho(R,\theta,\phi)\left[\frac{P(\theta-\theta_m,\phi-\phi_m)}{\sqrt{L_a(R)}\,R^2}x\left(t-t_m-\frac{2R}{c}\right)\exp\left(j\frac{4\pi}{\lambda_t}R\right)\right]$$
$$\times \exp(j2\pi F_t t) \qquad (2.108)$$

Coherent demodulation removes the carrier term, leaving only the baseband complex received voltage dy for the single differential scatterer

$$dy(\theta_m,\phi_m,t-t_m;R,\theta,\phi)$$
$$= A_r\, d\rho(R,\theta,\phi)\left[\frac{P(\theta-\theta_m,\phi-\phi_m)}{\sqrt{L_a(R)}\,R^2}x\left(t-t_m-\frac{2R}{c}\right)\exp\left(j\frac{4\pi}{\lambda_t}R\right)\right]$$
$$(2.109)$$

Equation (2.109) gives the contribution to the received voltage of the echo of the mth pulse from a differential scatterer element at coordinates (R,θ,ϕ). The total received voltage is obtained by integrating these differential contributions over all space:

$$y(\theta_m,\phi_m,t-t_m;R,\theta,\phi) = \int_{\phi=-\frac{\pi}{2}}^{\frac{\pi}{2}}\int_{\theta=-\pi}^{\pi}\int_{R=0}^{\infty} dy(\theta_m,\phi_m,t-t_m;R,\theta,\phi)$$
$$(2.110)$$

[†]The term involving ψ accounts for a possible constant phase shift on reflection at the scatterer surface.

Now write $d\rho(R,\theta,\phi) = \rho(R,\theta,\phi)\,dR\,d\theta\,d\phi$, $t'_m = t - t_m$ and use those results in Eq. (2.109) and (2.110) to obtain

$$y(\theta_m,\phi_m,t'_m;R,\theta,\phi) = A_r \int_{\phi=\frac{\pi}{2}}^{\frac{\pi}{2}} \int_{\theta=-\pi}^{\pi} \int_{R=0}^{\infty} \left\{ \frac{\exp[j(4\pi/\lambda_t)R]}{\sqrt{L_a(R)}R^2} \rho(R,\theta,\phi) \right\} \cdots$$
$$\times \left[P(\theta-\theta_m,\phi-\phi_m) x\left(t'_m - \frac{2R}{c}\right) dR\,d\theta\,d\phi \right] \quad (2.111)$$

Define the *effective reflectivity* ρ' to include the amplitude modulation due to wave spreading and atmospheric loss, and also the phase rotation due to two-way propagation range

$$\rho'(R,\theta,\phi) = \frac{\exp[j(4\pi/\lambda_t)R]}{R^2\sqrt{L_a(R)}}\rho(R,\theta,\phi) \quad (2.112)$$

Applying Eq. (2.112) to Eq. (2.111), the received signal is recognized as a *three-dimensional convolution* of the effective reflectivity with a convolution kernel comprising the antenna power pattern in the angle coordinates and the pulse modulation function in the range coordinate. Specifically

$$y(\theta_m,\phi_m,t'_m) = A_r \rho'(ct'_m/2,\theta_m,\phi_m) *_{t'_m} *_{\theta_m} *_{\phi_m} [P(-\theta_m,-\phi_m) x(t'_m)] \quad (2.113)$$

where the symbols $*_{t'_m}$, $*_{\theta_m}$, and $*_{\phi_m}$ denote convolution over the indicated coordinate. Now assume the antenna pattern is symmetric in the two angular coordinates, as is often the case; rescale the time variable to units of range; and replace θ_m and ϕ_m with general angular variables θ and ϕ. These substitutions finally give

$$y(\theta,\phi,R) = A_r \rho'(R,\theta,\phi) *_R *_\theta *_\phi \left[P(\theta,\phi) x\left(\frac{2R}{c}\right) \right] \quad (2.114)$$

Strictly speaking, Eq. (2.111) can be interpreted as the three-dimensional convolution of Eq. (2.114) only with the aid of some additional assumptions. These are discussed in Sec. 2.7.1 and 2.7.2, but do not detract from our main argument.

Equation (2.114) is a fundamental result. It shows that the transmitted waveform and antenna power pattern act as linear filters in the range and angle coordinates, respectively. The observed reflectivity distribution, as represented by the receiver output voltage, is a filtered and weighted version of the actual reflectivity distribution. The range resolution of the reflectivity function is therefore limited by the pulse modulation function $x(t)$ while the angular resolution is limited by the antenna beamwidth. Since most pulsed radars use pulse lengths between perhaps 50 ns and 100 μs, the corresponding range resolution of the measured reflectivity profile varies from 7.5 m to 15 km and is determined by the pulse length. (In Chap. 4 it will be seen that the introduction of matched filtering will significantly change this statement.) Antennas have been designed for a wide range of beamwidths, with the higher resolution "pencil beam" antennas typically between 1/3° and 10° beamwidths. The corresponding angular resolution in meters is usually coarser than the range resolution, and furthermore increases with range. (In Chap. 7 it will be seen that the introduction of synthetic aperture techniques also significantly changes this statement.)

It also follows from the filtering action of $x(t)$ and $P(\theta, \phi)$ that the bandwidth of the measured reflectivity function is limited by the bandwidth of the waveform modulation function and antenna power pattern. This observation will be used in Chap. 3 to determine the range and angle sampling requirements.

2.7.1 Variation with angle or cross-range

Now consider the variation in reflectivity with angle for a fixed range, say R_0. Define the range-averaged effective reflectivity

$$\hat{\rho}(\theta, \phi; R_0) = \int_R x\left[\frac{2}{c}(R_0 - R)\right] \rho'(R, \theta, \phi)$$
$$= \left[\rho'(R, \theta, \phi) *_R x\left(\frac{2R}{c}\right)\right]_{R=R_0} \quad (2.115)$$

This is the reflectivity variation in angle, taking into account the range averaging at each angle due to the finite pulse length. Note that in the limit of very fine range resolution, i.e., if the pulse modulation $x(2R/c) \to \delta_D(R - R_0)$, then $\hat{\rho}(\theta, \phi; R_0) \to \rho'(R_0, \theta, \phi)$, that is, the "range-averaged" reflectivity would exactly equal the effective reflectivity evaluated at the range of interest R_0.

Applying Eq. (2.115) to Eq. (2.114) gives

$$y(\theta, \phi; R_0) = A_r \int_{\xi=-\frac{\pi}{2}}^{\frac{\pi}{2}} \int_{\zeta=-\pi}^{\pi} P(\zeta - \theta, \xi - \phi)\hat{\rho}(\theta, \phi; R_0)\, d\zeta\, d\xi$$
$$= \hat{\rho}(\theta, \phi; R_0) *_\theta *_\phi P(\theta, \phi) \quad (2.116)$$

where again symmetry of the antenna pattern has been assumed in the second line. Equation (2.116) is a special case of Eq. (2.114) showing that the complex voltage at the output of a coherent receiver for a fixed range and a scanning antenna is the convolution in the angle dimensions of the range-averaged effective reflectivity function evaluated at the range R_0, $\hat{\rho}(\theta, \phi; R_0)$, with the antenna power pattern $P(\theta, \phi)$.

As mentioned earlier, some care must be taken in interpreting the relation of Eq. (2.116) as a convolution. Suppose that the elevation angle ϕ is fixed, and consider only the variation in azimuth angle θ. Because the integration is over a full 2π radians and the integrand is periodic in θ with period 2π, the integration over azimuth is a true convolution of periodic functions.

This would not appear to be the case if instead θ is fixed and ϕ varies because the integrand is over a range of only π radians. However, one could equally well write Eq. (2.116) as

$$y(\theta, \phi; R_0) = A_r \int_{\xi=-\pi}^{\pi} \int_{\zeta=-\frac{\pi}{2}}^{\frac{\pi}{2}} P(\zeta - \theta, \xi - \phi)\hat{\rho}(\theta, \phi; R_0) \, d\zeta \, d\xi$$
$$= \hat{\rho}(\theta, \phi; R_0) *_\theta *_\phi P(\theta, \phi) \tag{2.117}$$

For fixed azimuth, this is now a convolution of periodic functions in elevation. Taken together, there is a two-dimensional averaging over the (θ, ϕ) space. Finally, note that Eq. (2.116) can be recast from angle units into units of cross range X and Y using the transformations $X = R_0\theta$ and $Y = R_0\phi$:

$$y(X, Y; R_0) = A_r \int_{X=-\frac{\pi R}{2}}^{\frac{\pi R}{2}} \int_{Y=-\pi R}^{\pi R} P\left[\frac{1}{R_0}(X - \alpha, Y - \beta)\right] \hat{\rho}\left(\frac{X}{R_0}, \frac{Y}{R_0}; R_0\right) d\alpha \, d\beta \tag{2.118}$$

Figure 2.19 illustrates intuitively in one angle dimension the process described by Eq. (2.116). Assume that the elevation angle is fixed at $\phi = 0°$ and consider only the azimuth variation. An array of ideal point scatterers is illuminated by a radar that scans in azimuth across the target field. The response to any one scatterer is maximum when the radar boresight is aimed at that scatterer; as the radar boresight moves away, the strength of the echo declines because less energy is directed to the scatterer on transmission, and the antenna is also less sensitive to echoes from directions other than the boresight on reception. For an isolated scatterer, the amplitude of the coherent baseband received signal $y(\theta, 0; R_0)$ at the receiver output will be proportional

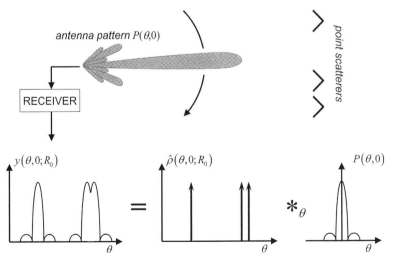

Figure 2.19 When scanning past an array of point scatterers, the receiver output is a superposition of replicas of the antenna pattern.

to $P(\theta, 0)$. Thus, a graph of the received signal mimics the antenna azimuth pattern.

Assuming a linear receiver so that superposition applies, the response to two closely spaced point scatterers is proportional to two replicas of the antenna pattern, overlapped and added to get a composite response. If the two scatterers are close enough together, the individual responses are not resolved, but instead blur together into a single peak as illustrated in Fig. 2.19. Clearly, the separation at which scatterers are resolved depends on the antenna pattern $P(\theta, 0)$.

The spatial Fourier transform of the observed signal is the input spatial Fourier transform multiplied by the Fourier transform of the antenna pattern. Practical antenna patterns have lowpass spectra. Equation (2.119) gives the ideal two-way azimuth voltage patterns for circular and rectangular apertures of width D (Balanis, 1982):

$$P(\theta, 0) = \left[\frac{J_1(\pi D \sin\theta/\lambda)}{\pi D \sin\theta/\lambda}\right]^2 \quad \text{(circular aperture)}$$

$$P(\theta, 0) = \left[\frac{\sin(\pi D \sin\theta/\lambda)}{\pi D \sin\theta/\lambda}\right]^2 \quad \text{(rectangular aperture)}$$
(2.119)

Figure 2.20 plots these patterns on a decibel scale for the case $D = 40\lambda$.

The corresponding spatial spectra are shown in Fig. 2.21; for the rectangular aperture, it is a triangle function with a support of twice the aperture width. The reason is easy to see: the one-way voltage pattern is just the inverse Fourier transform of the aperture function, which for uniform illumination is a rectangular pulse of the width of the aperture. When that pattern is squared to get the two-way pattern, the Fourier transform of the squared pattern is the self-convolution of the Fourier transform of the unsquared pattern. Thus, the rectangular aperture function is convolved with itself to give a triangle of twice the aperture width. The spectrum for the circular aperture has the same width but is somewhat smoother.

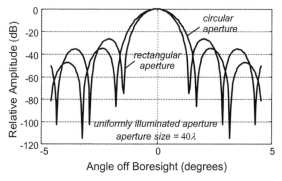

Figure 2.20 Two-way antenna gain patterns for ideal, uniformly illuminated circular and rectangular apertures.

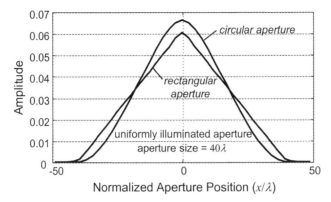

Figure 2.21 Spatial spectra corresponding to the antenna patterns of Fig. 2.20.

The spatial spectra of these idealized, but typical, antenna patterns are lowpass functions. Thus, the upper frequencies in the spatial spectrum of the observed data will be strongly attenuated and in fact effectively removed. Since resolution is proportional to bandwidth, Eq. (2.118) and Fig. 2.21 show that the antenna pattern reduces resolution because it has a strongly low-pass spatial spectrum.

2.7.2 Variation with range

A development similar to that in Sec. 2.7.1 can be carried out to specialize Eq. (2.115) for the variation of received voltage in the range dimension along the boresight look direction (θ_m, ϕ_m).[†] First interchange the order of integration in Eq. (2.111) so that the outer integral is over range. Next, define the new quantity

$$\tilde{\rho}(R; m) = \iint_{\theta,\phi} P(\theta - \theta_m, \phi - \phi_m)\rho'(R, \theta, \phi)\,d\theta\,d\phi$$

$$= \rho'(R, \theta, \phi) *_\theta *_\phi P(\theta, \phi)|_{\theta=\theta_m, \phi=\phi_m} \quad (2.120)$$

This is the reflectivity variation in range, taking into account the azimuth and elevation averaging at each range due to the nonideal antenna power pattern. Note that in the limit as the antenna power pattern tends to the ideal $P(\theta, \phi) \to G\delta_D(\theta, \phi)$, then $\tilde{\rho}(R; m) \to \rho'(R, \theta_m, \phi_m)$, that is, the "angle-averaged" reflectivity exactly equals the effective reflectivity along the antenna look direction, as expected.

[†]The analysis can be carried out equally easily for an off-boresight look direction. The only difference is to substitute an antenna gain value other than the peak gain G.

Applying Eq. (2.120) to Eq. (2.113) leaves (Munson and Visentin, 1989)

$$y\left(\frac{2R}{c}; m\right) = A_r \tilde{\rho}(R; m) *_R \left[x\left(\frac{2R}{c}\right)\right]$$

$$= A_r \int_{R'=0}^{\infty} x\left[\frac{2}{c}(R-R')\right] \tilde{\rho}(R'; m) \, dR' \quad (2.121)$$

or an equivalent equation, using time units instead of range units

$$y(t; m) = A_r \tilde{\rho}\left(\frac{ct}{2}; m\right) *_t [x(t)]$$

$$= A_r \int_{t'=0}^{\infty} x(t'-t) \tilde{\rho}\left(\frac{ct'}{2}; m\right) dt' \quad (2.122)$$

Equation (2.121) or (2.122), which are simply special cases of Eq. (2.113), make clear that the complex voltage at the output of a coherent receiver is the convolution in the range dimension of the angle-averaged effective reflectivity function in the radar look direction, $\tilde{\rho}(R; m)$ with the waveform modulation function $x(t)$.

Equation (2.112) expressed the relationship between the amplitude of the effective reflectivity ρ' and the actual reflectivity ρ resulting from weighting due to atmospheric loss and R^2 spreading. Whether this term is significant depends on the ratio of the maximum and minimum ranges of interest. For example, if a surveillance radar is mapping a 20-km swath between 80 and 100 km range, the variation in the R^2 term over the swath is only 1.94 dB. While significant, this is probably much less than the variations in ρ itself and may often be ignored. On the other hand, a millimeter wave seeker searching a footprint from 1 to 3 km downrange sees a variation of 9.54 dB in this term, which can probably not be ignored. In either case, the variation in amplitude due to R^2 is predictable and can be compensated in the signal processor by applying a range-dependent gain to the data. The compensation can be applied as the data are received by using a rapidly time-varying gain proportional to t^2. This procedure is called *sensitivity time control* (STC). Alternatively, the correction can be applied in the signal or data processor.

The R^2 voltage gain (equivalent to R^4 power gain) correction assumes a range dependence of the received voltage corresponding to the point target form of the radar range equation (Eq. (2.19)). If the radar is illuminating a volume or area scatterer, a different correction factor is required because of the different range dependence of the received power. Thus, for volume scatterers or beam-limited area scatterers (Eq. (2.25) or (2.30)) the *voltage* gain should be varied as R, while for pulse-limited area scatterers Eq. (2.31) the voltage gain should be varied as $R^{3/2}$.

2.7.3 Projections

The range-averaged reflectivity $\hat{\rho}(\theta, \phi; R_0)$ of Eq. (2.115) and the angle-averaged reflectivity $\tilde{\rho}(R; m)$ of Eq. (2.120) are examples of *projections*. In each case, the three-dimensional reflectivity is reduced in dimension by integrating over one or more dimensions. The angle-averaged reflectivity is reduced to a two-dimensional function by integrating over range, while the range-averaged reflectivity is reduced to a one-dimensional function by integrating over both angle coordinates.

The idea of projections, particularly the angle-averaged projection $\tilde{\rho}(R; m)$, will be important in deriving the polar format spotlight SAR algorithm in Chap. 8. The projections that will be needed are integrals over straight lines or planar surfaces. The averaging in Eq. (2.120) is over the surface of a sphere. However, for small beamwidths only a region of θ_3 radians in azimuth and ϕ_3 radians in elevation contributes significantly to the integral, and at long ranges this limited region is nearly planar.

2.7.4 Multipath

The convolutional model of the measured range profile is based on the assumption of superposition of backscattered fields and a one-to-one mapping of echo arrival time to range (Eq. (2.107)). The superposition of electric fields is a valid assumption while the mapping of time to range may not be. To illustrate, consider Fig. 2.22, which diagrams two phenomena that violate this assumption. Figure 2.22a illustrates the problem of *multipath*, in which echoes from the same target arrive at the radar receiver via two different paths. The first is the direct path of total length $2R_0$. The second is the "multipath" or "ground bounce" path with length $R_0 + R_1 + R_2 > 2R_0$. Though not shown,

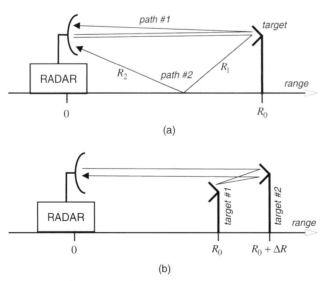

Figure 2.22 Illustration of two multiple-bounce scattering phenomena which violate the one-to-one mapping of time to range.

it is also possible for a portion of the transmitted wave to arrive at the target via the ground bounce and be scattered back along both of the paths in Fig. 2.22a, meaning that there may also be an echo with a time delay corresponding to a two-way path length of $2(R_0 + R_1 + R_2)$. Consequently, one scatterer may produce echoes at three different apparent ranges if multipath is present. Whether these appear as distinct echoes depends on the relationship between the path length difference and the pulse length. The ground bounce echoes are often, but not always, significantly attenuated with respect to the direct path echo. The degree of attenuation depends on the bistatic scattering characteristics of the surface, the antenna pattern characteristics (because the multipath bounce is not on the peak of the main lobe) and the problem geometry. As the range between target and radar varies, the path length difference also varies, so that the direct and multipath bounces may alternately add in and out of phase, provided the path length difference is such that the two received echoes overlap. Multipath is generally most significant for targets located at low altitude over a good reflecting surface such as a relatively smooth terrain or calm ocean and at long range, so that the grazing angles involved are small.

Figure 2.22b illustrates the effect of *multiple bounce* echoes in a situation involving two scatterers. A portion of the energy reflected from the more distant scatterer bounces off the nearer scatterer, then reflects a second time off the distant scatterer and returns to the radar. Obviously additional multiple bounces are also possible. For the situation sketched, three apparent echoes will again result, with the third due to a phantom scatterer $2\Delta R$ behind the second actual scatterer. As with multipath, the amplitude of multiple bounce echoes often falls off rapidly, and the same considerations of in- and out-of-phase superposition apply.

These possible differences in the measured and actual reflectivity distributions do not mean that range profile measurements are not useful. They do mean that in situations where significant multipath or multiple bounce phenomena are possible, the range profiles must be interpreted with care.

2.8 Spectral Model

There is one more interpretation of the received radar signal that proves useful in subsequent chapters. The preceding two sections have emphasized linear filtering models of the spatial reflectivity distribution as observed through the received complex baseband signals. However, it was pointed out previously that radar cross section is a function of, among many other things, the radar frequency. Thus, it is useful to investigate the significance of the radar transmitted frequency F_t on the reflectivity measurements.

To understand the role of transmitted frequency, it is necessary to deal with the radar signal while it is still at the radar frequency F_t. If the development from Eq. (2.106) to Eq. (2.122) is repeated without demodulating the signals to baseband and the range variation is considered, the RF version of Eq. (2.122) can be obtained

$$\bar{y}(t; m) = A_r \int_{t=0}^{\infty} x(t' - t) \tilde{\rho}\left(\frac{ct'}{2}; m\right) \exp(j 2\pi F_t t') \, dt'$$

$$= A_r \left[\tilde{\rho}\left(\frac{ct}{2}; m\right) \exp(j 2\pi F_t t) \right] *_t [x(t)] \quad (2.123)$$

Now consider the Fourier transform of $\bar{y}(t; m)$ with respect to the time (range) variable t. Using simple properties of Fourier transforms gives

$$\bar{Y}(F; k) = \frac{2A_r}{c} X(F) \tilde{P}\left[\frac{2(F - F_t)}{c}\right] \quad (2.124)$$

Figure 2.23 provides a pictorial interpretation of this equation under the assumption that the transmitted waveform $x(t)$ is a narrowband waveform. In this case,

$$\bar{Y}(F; k) \approx \frac{2A_r}{c} \tilde{P}\left(\frac{-2F_t}{c}\right) X(F) \quad (2.125)$$

so that the amplitude of the spectrum of the received pulse, and therefore of the pulse itself, is proportional to the amplitude of the spectrum of the angle averaged range profile, evaluated at the transmitted frequency. Since it is the complex spectrum that appears in Eq. (2.125), both the amplitude and phase of the returned pulse are affected by the amplitude and phase of the reflectivity spectral sample.

Equation (2.125) shows that a narrowband radar pulse can be interpreted as measuring a frequency sample of the spectrum of the angle-averaged reflectivity range variation. This concept is important in understanding the behavior of frequency stepped radars in Chap. 4.

Another case of interest occurs when $x(t)$ is a wideband pulse. In this case, the spectrum $X(F)$ is approximately a rectangle as shown in Fig. 2.24. The spectrum of the receiver waveform $\bar{y}(t; m)$ is then approximately that of the

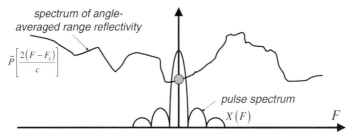

Figure 2.23 Pictorial interpretation of Eq. (2.124), illustrating the spectral windowing effect of a narrowband radar pulse.

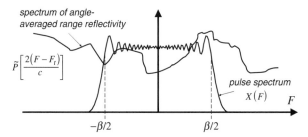

Figure 2.24 Pictorial interpretation of Eq. (2.124) illustrating the spectral sampling effect of a wideband radar pulse.

angle-averaged range profile over the bandwidth of the pulse

$$\bar{Y}(F;m) \approx \frac{2A_r}{c} \tilde{P}\left[\frac{2(F-F_t)}{c}\right] \qquad -\frac{\beta}{2} \leq F \leq +\frac{\beta}{2} \qquad (2.126)$$

In other words, the pulse spectrum acts as a window selecting a portion of the spectrum of the angle-averaged reflectivity. This result will be useful in understanding the use of linear FM and other modulated waveforms to achieve high range resolution in Chap. 4.

2.9 Summary

An understanding of the nature of the signals of interest is prerequisite to the design of successful signal processing systems. This chapter reviews the most common signal models used in designing and analyzing radar signal processors. It has been seen that multiple views of the radar echo are used: its variation in amplitude, space, time, and frequency, and deterministic and statistical interpretations of these variations.

Radar signal modeling traditionally focuses most strongly on amplitude models, that is, on radar cross section. RCS is viewed as a deterministic quantity, predictable in principle through the use of Maxwell's equations if the scattering is modeled accurately enough. The radar range equation in its many forms (only a very small subset of which has been introduced here) is the radar engineer's most fundamental tool for estimating received signal amplitude or, conversely, determining required system characteristics such as transmitted power or antenna gain.

The radar system is a measuring instrument, used to observe the variation of RCS in space. Its pulse function (modulation and carrier term) and antenna power pattern determine its measurement characteristics, which in turn determine the achievable resolution and required sampling rates. The effect of the radar measurement system on the spatial variation of observed RCS is well-modeled by the convolution of the combined pulse-and-antenna pattern measurement kernel with the three-dimensional reflectivity function.

This important observation means that the tools of linear systems analysis can be brought to bear to help analyze and understand the performance of radar systems. The carrier frequency, in combination with any Doppler shifts, determines what portion of the reflectivity frequency spectrum is sampled by the pulse. This observation reinforces the need for frequency domain analyses of radar measurements. Linear systems and frequency domain viewpoints are relied on heavily throughout the remainder of the book.

Even though RCS is a deterministic quantity, its sensitivity to radar frequency, aspect angle, and range coupled with the complexity of typical targets results in very complex behavior of observed amplitude measurements. Statistical models are used to describe this complexity. A variety of statistical models, comprising both probability density functions and correlation properties, have gained acceptance for various scenarios and form the basis for much analysis, particularly in calculations of probabilities of detection and false alarm, two of the most important radar performance measures.

References

Balanis, C. A., *Antenna Theory*. Harper & Row, New York, 1982.

Baxa, E. G., Jr., "Airborne Pulsed Doppler Radar Detection of Low-Altitude Windshear-A Signal Processing Problem," *Digital Signal Processing*, vol. 1, no. 4, pp. 186–197, Oct. 1991.

Beckmann, P., and A. Spizzichino, *The Scattering of Electromagnetic Waves from Rough Surfaces*. MacMillan, New York, 1963.

Birkmeier, W. P., and N. D. Wallace, *AIEE Transactions on Communication Electronics*, vol. 81, pp. 571–575, Jan. 1963.

Carlson, A. B., *Communication Systems*. McGraw-Hill, New York, 1976.

Curlander, J. C., and R. N. McDonough, *Synthetic Aperture Radar*. Wiley, New York, 1991.

Currie, N. C., "Clutter Characteristics and Effects," Chap. 10 in J. L. Eaves and E. K. Reedy (eds.), *Principles of Modern Radar*. Van Nostrand Reinhold, New York, 1987.

Doviak, D. S., and R. J. Zrnic, *Doppler Radar and Weather Observations*, 2d ed. Academic Press, San Diego, CA, 1993.

Eaves, J. L., and E. K. Reedy, *Principles of Modern Radar*. Van Nostrand Reinhold, New York, 1987.

Gill, T. P., *The Doppler Effect*. Logos Press, London, 1965.

Haykin, S., B. W. Currie, and S. B. Kesler, "Maximum Entropy Spectral Analysis of Radar Clutter," *Proceedings of the IEEE*, vol. 70, no. 9, pp. 953–962, Sept. 1982.

Holm, W. A., "MMW Radar Signal Processing Techniques," Chap. 6 in N. C. Currie and C. E. Brown (eds.), *Principles and Applications of Millimeter-Wave Radar*. Artech House, Boston, MA, 1987.

Jakeman, E., and P. N. Pusey, "A Model for Non-Rayleigh Sea Echo," *IEEE Transactions on Antennas and Propagation*, vol. 24, no. 6, pp. 806–814, Nov. 1976.

Kay, S. M., *Modern Spectral Estimation*. Prentice Hall, Englewood Cliffs, NJ, 1988.

Knott, E. F., J. F. Shaeffer, and M. T. Tuley, *Radar Cross Section*. Artech House, Boston, MA, 1985.

Levanon, N., *Radar Principles*. Wiley, New York, 1988.

Lewinski, D. J., "Nonstationary Probabilistic Target and Clutter Scattering Models," *IEEE Transactions on Antennas and Propagation*, vol. AP-31, no. 3, pp. 490–498, May 1983.

Long, M. W., *Radar Reflectivity of Land and Sea*, 3d ed. Artech House, Boston, MA, 2001.

Lothes, R. N., M. B. Szymanski, and R. G. Wiley, *Radar Vulnerability to Jamming*. Artech House, Boston, MA, 1990.

Meyer, D. P., and H. A. Mayer, *Radar Target Detection*. Academic Press, New York, 1973.

Mott, H., *Polarization in Antennas and Radar*. Wiley, New York, 1986.

Munson, D. C., and R. L. Visentin, "A Signal Processing View of Strip-Mapping Synthetic Aperture Radar," *IEEE Transactions on Acoustics, Speech, and Signal Processing*, vol. 27, no. 12, pp. 2131–2147, 1989.

Nathanson, F. E., *Radar Design Principles*, 2d ed. McGraw-Hill, New York, 1991.

Omura, J., and T. Kailath, "Some Useful Probability Distributions," Technical Report No. 7050-6, Stanford Electronics Laboratories, Stanford University, Sep. 1965. Also identified as report no. SU-SEL-65-079.

Papoulis, A., *Probability, Random Variables, and Stochastic Processes*, 2d ed. McGraw-Hill, New York, 1984.

Probert-Jones, J. R., "The Radar Equation in Meteorology," *Quarterly Journal of the Royal Meteorological Society*, vol. 88, pp. 485–495, 1962.

Ray, H., "Improving Radar Range and Angle Detection with Frequency Agility," *Microwave Journal*, p. 64ff, May 1966.

Sangston, K. J., "Toward a Theory of Ultrawideband Sea Scatter," *Proceedings of IEEE National Radar Conference*, pp. 160–165, 1997.

Sauvageot, H., *Radar Meteorology*. Artech House, Boston, MA, 1992.

Skolnik, M. I., *Introduction to Radar Systems*, 3d. ed. McGraw-Hill, New York, 2001.

Swerling, P., "Probability of Detection for Fluctuating Targets," *IRE Transactions on Information Theory*, vol. IT-6, pp. 269–308, April 1960.

Temes, C. L., "Relativistic Consideration of Doppler Shift," *IRE Transactions on Aeronautical and Navigational Electronics*, p. 37, 1959.

Ulaby, F. T., and M. C. Dobson, *Handbook of Radar Scattering Statistics for Terrain*. Artech House, Norwood, MA, 1989.

Ward, K. D., "Compound Representation of High Resolution Sea Clutter," *Electronics Letters*, vol. 17, no. 16, pp. 561–563, Aug. 6, 1981.

Ziemer, R. E., and W. H. Tranter, *Principles of Communications*. Houghton Mifflin, Boston, MA, 1976.

Chapter 3

Sampling and Quantization of Pulsed Radar Signals

3.1 Domains and Criteria for Sampling Radar Signals

As has been seen, a radar measures the spatial distribution of reflectivity in the three-dimensional spherical coordinate system of range, azimuth angle, and elevation angle. Many radars also process the signals in domains corresponding to the Fourier transform of either slow-time (pulse number) or, in the case of array antennas, receiver channel. These spectral domains correspond to Doppler shift and angle of arrival, respectively, and represent the fourth and fifth potential dimensions of a radar signal. If the signals are to be processed digitally, one of the first questions to arise is how the sampling interval should be chosen in each dimension.

A pulsed radar has several distinct and independent sampling intervals; six are identified here. The first two are both time sampling intervals. Consider Fig. 3.1, which illustrates the general data collection strategy of a pulsed radar, possibly with a multi-phase center antenna such as a phased array. The radar emits a periodic series of pulses; the period is denoted as the *pulse repetition interval* (PRI), and its inverse is the *pulse repetition frequency* (PRF).[†] The PRF may range from a few hundred hertz to tens and sometimes a few hundreds of kilohertz. In a portion of the time period between pulses, the received signal from each antenna channel is sampled at a high rate, typically in the range of hundreds of kilohertz to a few tens of megahertz, and sometimes higher. After conversion to baseband, the cluster of high rate samples from one channel and one pulse may be viewed as being stored in a single row and layer of the

---

[†] In this text, the abbreviation "PRF" is used both as an acronym and as a mathematical variable. When used as an acronym, it is not italicized (PRF); when used as a mathematical variable, it is italicized (*PRF*).

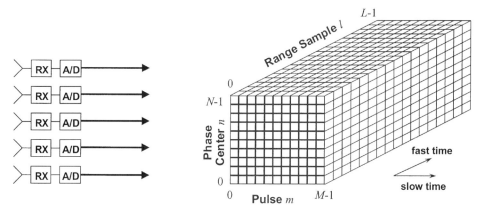

Figure 3.1 General model of data collection and storage strategy for a multichannel pulsed radar.

structure $y[l, m, n]$, called a *datacube*. The cluster of samples taken from the same channel and the next pulse is stored in the second row at the same layer, and so forth. This data collection and storage strategy provides a good conceptual model for understanding most digital radar signal processing operations. It is actually implemented in many processors. The l dimension of $y[l, m, n]$ (out of the page in Fig. 3.1) is sometimes called *fast time* while the horizontal (m) dimension is called *slow-time* due to the usually large difference in sampling intervals between data samples in successive rows of a given column and those in successive columns of a given row. The vertical (n) dimension represents spatial sampling, rather than time sampling.

Many of the basic radar signal processing operations considered in the remainder of this text correspond to processing one-dimensional subvectors or two-dimensional submatrices of the datacube in various dimensions. Figure 3.2 illustrates these relationships. (The particular operations depicted are discussed in upcoming chapters.) For example, pulse compression (see Chap. 4) corresponds to one-dimensional convolution on a single vector in the fast-time (range) dimension. Pulse compression can be performed independently on each such range vector for every pulse and receiver channel.

In many cases only a two-dimensional submatrix of the datacube, typically consisting of the fast- and slow-time (range and pulse number) dimensions for a given receiver channel, is of interest. In systems having only a single receiver channel, the datacube also degenerates into a data matrix. Figure 3.3 illustrates the data matrix $y[l, m]$ obtained by extracting the data for one receiver channel from a datacube.

3.1.1 Time and frequency samples

The first pulsed radar sampling interval of interest, T_s, is that between successive samples of the echoes of a single pulse. This is the range or fast time sampling interval between successive rows of a given column of $y[l, m, n]$, that is,

Sampling and Quantization of Pulsed Radar Signals 117

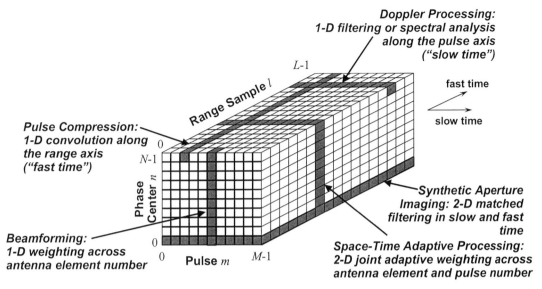

Figure 3.2 Correspondence between key radar signal processing functions and operations on the radar datacube.

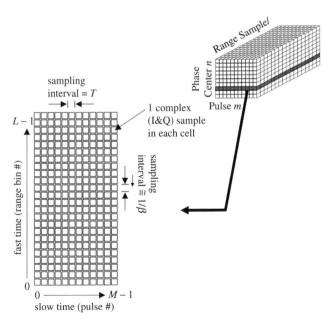

Figure 3.3 Two-dimensional data matrix corresponding to one channel of a datacube.

between $y[l,m,n]$ and $y[l+1,m,n]$. The corresponding sampling rate is $F_s = 1/T_s$. The sampling interval T_s determines the spacing of the radar range samples, or *range bins*,[†] according to

$$\text{Range bin spacing} = \Delta R_s = \frac{cT_s}{2} \qquad (3.1)$$

The second sampling interval is the slow-time sampling rate. As is obvious from Fig. 3.1, this is just the pulse repetition interval, T. The PRI (equivalently, the PRF) determines the unambiguous Doppler spectrum width which can be represented; it is

$$\text{Unambiguous Doppler spectrum width} = \beta_D = PRF \quad \text{Hz} \qquad (3.2)$$

A digitally computed Doppler spectrum must itself be sampled in the Doppler (frequency) variable; this third sampling interval is denoted as ΔF_D. That is, the spectrum is evaluated at frequencies $F = k\Delta F_D, k = 0, \ldots, K-1$.

3.1.2 Spatial samples

Spatial sampling arises in two ways. Consider a system that uses a steerable antenna, whether mechanically steered (typically a parabolic dish or slotted flat-plate array, and others) or electronically steered (phased array). How often must a pulse be transmitted to sample the environment as the antenna is steered in different directions? This question leads to the consideration of sampling intervals in the spatial dimensions of azimuth and elevation angle, $\Delta\theta$ and $\Delta\phi$. These are determined by the interplay of the PRI and the antenna steering behavior. For example, a rotating antenna with a constant angular rate in azimuth of Ω_θ, a fixed elevation angle ϕ, and a pulse repetition interval of T exhibits a sampling interval in azimuth angle of

$$\Delta\theta = \Omega_\theta T \qquad (3.3)$$

Only a single elevation sample is being collected, therefore an elevation sampling interval $\Delta\phi$ cannot be defined in this case.

The second way in which spatial sampling arises occurs when the antenna system is a phased array. In this event, the array elements sample the wavefront that impinges on the array face. Thus, the spacing of the array elements is yet another spatial sampling interval.

[†]Other common names for the range bins include *range cells* and *range gates*. The terms *resolution bins* and *resolution cells* are also sometimes used synonymously with range bins, and in many cases are synonymous; but the sampling interval in range does not always equal the range resolution, so caution should be used in interpreting the latter two terms.

3.1.3 Sampling criteria

The usual approach to choosing a sampling rate is based on the application of the Nyquist criterion, which was reviewed briefly in Chap. 1. The Nyquist criterion chooses the sampling rate such that the original continuous-time (or space) signal can be exactly reconstructed from its samples, provided the signal is bandlimited (Oppenheim and Schafer, 1999). This criterion can, of course, be applied to radar signals. However, in radar the original signal is virtually never reconstructed from its samples; the samples are processed to estimate information about the environment and then discarded. Furthermore, real-world radar pulses are not always particularly well bandlimited. Therefore, the rationale motivating the Nyquist criterion is not directly relevant to the sampling of radar signals. What other criteria might be used?

The criterion for sampling a signal should be appropriate for the use to which the samples will be put. As discussed in Chap. 1, the basic functions of radar are detection, tracking, and imaging. Of these, detection is most fundamental. Detection can be performed in the time domain on the pulse echoes; in the frequency domain on a Doppler spectral estimate; or in the spatial domain on a radar image. Consider detection based on analysis of the time domain echoes from a single pulse. The basic approach to detection is to monitor the amplitude of the received echo signal and compare it to a threshold T. The threshold is chosen to be high enough so that it is very unlikely that noise or other interference alone will produce a return large enough to cross the threshold. Thus, it is assumed that if the signal being monitored does cross the threshold at some point in time t_0, then that high-amplitude echo is caused by a target located at the corresponding range $ct_0/2$.

A discrete time implementation of this procedure will test the echo signal $y(t)$ only at discrete time instants nT_s, $n \in [-\infty, +\infty]$. If none of these samples falls on or near a peak in $y(t)$, a target may escape detection. Figure 3.4 illustrates the concept of threshold detection and the potential to miss a detection in a discrete time implementation. The continuous signal (solid line) crosses the threshold in the vicinity of abscissa values of 155 and 355. The solid circles indicate samples taken every 10 units along the abscissa. The sample sequence crosses the threshold only for the peak at 355; the samples in the vicinity of 155 straddle the small peak and fail to cross the threshold. Thus, one range sampling criterion that is directly relevant to radar detection performance is to sample densely enough to ensure at least one sample "close to" the peak. This reasoning is applicable in the Doppler frequency and spatial domains as well.

A second range sampling criterion can be developed from the viewpoint of resolution. Chapter 8 develops the idea of radar as an imaging technology. An image should be sampled finely enough to preserve all the detail available in the image before sampling. The spatial resolution of the imaging instrument limits the detail in the image. Consequently, a sampling criterion in range and cross-range (or angle) can be developed based on the resolution achieved by the radar. Again, similar resolution-based criteria can be developed in the Doppler frequency domain.

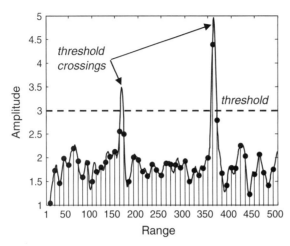

Figure 3.4 The concept of threshold detection and the effect of range sampling rate.

The pulse repetition interval determines the unambiguous interval in range and Doppler. Figure 3.5 illustrates the idea of a range ambiguity in a pulsed radar. An echo of pulse number m from a target at range R_0 is received at the same time as an echo of pulse number $m-1$ from a target at range $R_0 + cT/2$. Thus, the range of a target causing a peak in received echo strength at some time t_0 is *ambiguous* in the sense that the target range is known to be of the form $R_0 + kcT/2$, but k is unknown. There are signal processing methods for resolving range ambiguities (see Chap. 5), but a more direct approach is to choose the PRI such that most or all ambiguities (i.e., values of k) can be ruled out a priori. This approach to selecting the PRI is discussed in Sec. 3.3.

The concept of ambiguities also applies to processing in the Doppler frequency dimension. The range of Doppler shifts present in the received echoes is determined by a variety of factors, several of them external to the radar. For moving targets such as aircraft, the radial component of the relative velocity of the radar and target and the operating wavelength determine the target Doppler shift.

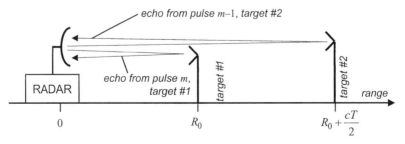

Figure 3.5 The concept of range ambiguities.

In an imaging radar, the imaged terrain exhibits a spread of Doppler frequencies determined by the radar platform motion, the antenna beamwidth and look direction, and the wavelength. While these factors determine the range of Doppler shifts present in the echoes, the PRI (or PRF) determines the unambiguous Doppler spectrum width (Eq. 3.2). Specifically, if the PRF is less than the Doppler spectrum width, the spectrum will be aliased, meaning that targets having Doppler shifts F_D greater in magnitude than $PRF/2$ will appear in the spectrum at an apparent Doppler shift of $F'_D = F_D + kPRF$, where k is the integer that results in $-PRF/2 \leq F'_D \leq +PRF/2$. This incorrect apparent Doppler shift is referred to as *Doppler ambiguity*. As with range, therefore, it is desirable to choose the PRF to minimize or eliminate Doppler ambiguities when possible.

The Doppler spectrum itself is computed only at discrete frequencies separated by some Doppler sampling interval ΔF_D. As with the range dimension, ΔF_D can be chosen based on arguments of peak detection and resolution. Finally, these same ideas can again be used to choose the sampling intervals in the angular spatial coordinates, $\Delta \theta$ and $\Delta \phi$. In the case of the angular dimensions, however, the antenna pattern will be the determinant of the sampling intervals. Sec. 3.5 develops these relationships.

In the end, sampling rules developed from directly relevant concerns such as detection performance or image fidelity are similar to the Nyquist criterion in each dimension. This should not be surprising. While in radar the analog signal is never reconstructed once it has been sampled, it is a goal that all of the useful information in the original signal be retained. The Nyquist criterion guarantees that the original signal can be recovered from its samples, thereby guaranteeing that no meaningful information has been lost in the sampling process.

3.2 Sampling in the Fast Time Dimension

How rapidly should one sample the echo from a single received pulse, i.e., what should the spacing of the range bins be? The Nyquist theorem states that the sampling rate should equal or exceed the bandwidth of the received signal. In Chap. 2 it was shown that the received signal in the range dimension can be modeled as the convolution of the range reflectivity function and the modulation function $x(t)$ of the transmitted waveform. The spectrum of the received signal is thus the product of the spectra of the range reflectivity function and the modulation function. This means that the bandwidth of the received fast time signal will be limited by the bandwidth of the transmitted pulse. Therefore, the Nyquist rate in fast time is simply the bandwidth of the transmitted pulse.

The spectrum of the simple pulse is a sinc function in the frequency domain centered at Ω_0. This spectrum is not strictly bandlimited; however, the 3-dB bandwidth β_3 is $0.89/\tau$ hertz, both the Rayleigh bandwidth β_r and 4 dB bandwidth β_4 are $1/\tau$ hertz, and the null-to-null bandwidth β_{nn} is $2/\tau$ hertz. These approximate bandwidth measures are shown in Fig. 3.6.

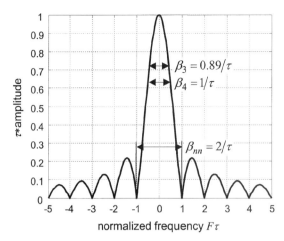

Figure 3.6 Three definitions of bandwidth for the spectrum of a simple rectangular pulse of duration τ.

Since the sinc spectrum is decidedly not bandlimited, a Nyquist bandwidth cannot be unambiguously defined for the simple pulse. An approximate bandwidth can be defined as the two-sided frequency interval beyond which the spectrum amplitude is "insignificant," but this approach is not very useful in this case because of the slow decay of the sinc function. For example, a criterion of 40 dB reduction in the spectrum from peak gives an approximate Nyquist bandwidth of about 66 times the 3-dB bandwidth. In radar, 3 dB bandwidths are commonly used; thus "the bandwidth" of the simple pulse is defined to be $\beta_3 = 0.89/\tau$ hertz. More conservative definitions use the Rayleigh bandwidth of $1/\tau$ hertz and the null-to-null bandwidth $\beta_{nn} = 2/\tau$ Hz. The Rayleigh bandwidth β_r is used here, despite its limitations, to estimate appropriate range sampling rates. In practice, the fast time signal is often sampled at some margin above the Nyquist rate. This compensates both for the transition band of receiver antialiasing filters and for some of the nonbandlimited nature of common pulse waveforms. Sampling rate margins of 20 to 50 percent are common.

It will be seen in Chap. 4 that pulses are often phase modulated in order to increase their bandwidth. The pulse spectrum is then no longer a sinc function. In fact, many phase modulated pulses are designed to have a spectrum that is approximately constant magnitude (but with complicated phase characteristics) over some desired bandwidth β, where β is much larger than the simple pulse bandwidth of approximately $1/\tau$. Thus, an idealized model of the spectrum of the ideal received phase-modulated radar pulse after translation to baseband is

$$Y(\Omega) = \begin{cases} Ae^{j\Phi(\Omega)} & |\Omega| < \dfrac{\beta}{2} \\ 0 & |\Omega| > \dfrac{\beta}{2} \end{cases} \qquad (3.4)$$

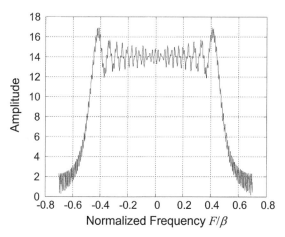

Figure 3.7 Magnitude of the Fourier transform of a linear FM "chirp" waveform with a time-bandwidth product of 100.

where $\Phi(\Omega)$ is some arbitrary phase function. Figure 3.7 shows an example, the spectrum of a linear frequency-modulated, or "chirp," waveform with a time-bandwidth product of 100; this waveform is studied in Chap. 4. On the normalized frequency scale shown, the spectrum is approximately rectangular with support $f \in (-0.5, +0.5)$, corresponding to $\pm \beta/2$ hertz. This case offers a relatively unambiguous definition of the bandwidth of the pulse (namely, β hertz), making application of the Nyquist criterion to range sampling straightforward.

3.3 Sampling in Slow Time: Selecting the Pulse Repetition Interval

The pulse repetition interval is the interval between radar measurements occupying successive columns within a given row of the data matrix $y[l, m]$. These samples represent the echo received after the same delay from the time of transmission for successive pulses. Assuming the antenna boresight is not moving significantly from pulse to pulse, these samples represent the reflectivity from the same range and angle, i.e., the same region in three-dimensional space, measured with a sampling interval equal to the pulse repetition interval T. The slow-time sampling frequency is the pulse repetition frequency *PRF*. As was seen in the discussion of spatial Doppler in Chap. 2, if there is any relative motion from one sample to the next between the radar and the target region being sampled, the phase of successive returns will vary from sample to sample (in addition to the variation due just to noise), and that phase history is equivalent to a Doppler shift. In other words, the signal corresponding to one row of the data matrix, called the *slow-time signal*, will have a nonzero *Doppler bandwidth*. The key requirement in choosing the PRF is to preserve the information in the Doppler spectrum for subsequent processing such as pulse Doppler target detection or synthetic aperture imaging by avoiding aliasing of

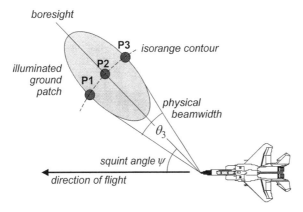

Figure 3.8 Geometry for estimating the Doppler bandwidth component due to radar platform motion.

the spectrum replicas. To determine the required PRF, it is therefore necessary to determine the bandwidth of the slow-time signal so that the Nyquist criterion can be applied.

A nonzero Doppler bandwidth results from two sources: intrinsic motion of the scatterers in the area being measured, and motion of the radar platform. If the area being measured is a target in the conventional sense of a man-made vehicle, its intrinsic motion is simply the motion of the vehicle.[†] If it is clutter, then intrinsic motion can be due to wind blowing the leaves of trees or blades of grass, waves on the ocean, falling and swirling rain, air conditioning fans on tops of buildings, and so forth. For instance, the Doppler power spectrum corner frequencies in Table 2.6 imply an intrinsic Doppler spread on the order of 0.5 to 1.0 m/s for rain. The intrinsic Doppler spread of moving man-made objects can be much larger. Consider an urban clutter scene where a stationary radar observes automobile traffic with a maximum speed of (optimistically) 55 mi/h both toward and away from the radar. The radar therefore sees targets with a velocity spread of 110 mi/h, or about 50 m/s. For a more extreme example, consider a moving radar installed on one of two subsonic (200 m/s) jet aircraft flying in opposite directions. As they approach, the closing rate is 400 m/s; once they pass, they separate at 400 m/s. The change in velocities observed by the radar on one of the aircraft over time is 800 m/s.

A moving radar can also induce a spread in the Doppler bandwidth of stationary objects in the beam. This is most relevant in air-to-ground radars. Figure 3.8 illustrates in two dimensions an approach to estimating the Doppler

---

[†]Note that the velocity of portions of the vehicle may differ from the nominal velocity of the vehicle as a whole. The Doppler frequency corresponding to the nominal velocity is frequently termed the *skin return*. However, the lug nuts on the wheels on a wheeled vehicle have velocities ranging from zero to twice the nominal velocity of the vehicle. Some laser radars even attempt to measure the Doppler shifts due to vehicle vibration.

bandwidth of a patch of terrain induced by radar platform motion. The 3-dB radar beamwidth is θ_3 radians. Recall from Chap. 2 that the Doppler shift for a radar moving at velocity v with its boresight squinted ψ radians off the velocity vector is

$$F_D = \frac{2v}{\lambda} \cos \psi \quad \text{Hz} \tag{3.5}$$

Now consider three point scatterers **P1**, **P2**, and **P3**, each at the same range from the radar. **P1** and **P3** are at the 3-dB edges of the antenna beam, while **P2** is on boresight. Because all three are at the same range, the received echo at a delay corresponding to that range is the superposition of the echoes from all three scatterers. However, each is at a slightly different angle with respect to the aircraft velocity vector. **P2** is on the boresight at the squint angle of ψ, but **P1** and **P3** are at $\psi \pm \theta_3/2$ radians. The *difference* in the Doppler shift of the echoes from **P1** and **P3** is then

$$\begin{aligned}\beta_D &= \frac{2v}{\lambda}[\cos(\psi - \theta_3/2) - \cos(\psi + \theta_3/2)] \\ &= \frac{4v}{\lambda} \sin\left(\frac{\theta_3}{2}\right) \sin \psi\end{aligned} \tag{3.6}$$

Radar antenna beamwidths are small, typically less than five degrees. Applying a small angle approximation to the $\sin(\theta_3/2)$ term in Eq. (3.6) gives a simple expression for Doppler bandwidth due to platform motion

$$\beta_D \approx \frac{2v\theta_3}{\lambda} \sin \psi \quad \text{Hz} \tag{3.7}$$

As can be seen from Fig. 3.8, Eq. (3.6) assumes the radar is squinted sufficiently that the main beam does not include the velocity vector, that is, $|\psi| > \theta_3/2$. If the radar is forward looking or nearly so, then the $\cos(\psi - \theta_3/2)$ term in Eq. (3.6), which represents the largest Doppler shift in the mainbeam, is replaced by 1. A more complete expression for the platform motion-induced Doppler bandwidth is therefore

$$\beta_D \approx \begin{cases} \dfrac{2v\theta_3}{\lambda} \sin \psi & \text{Hz} \quad |\psi| > \dfrac{\theta_3}{2} \quad \text{(squinted)} \\ \dfrac{2v}{\lambda}[1 - \cos(\psi + \theta_3/2)] & \text{Hz} \quad |\psi| < \dfrac{\theta_3}{2} \quad \text{(forward looking)} \end{cases} \tag{3.8}$$

For example, an L band (1 GHz) side-looking ($\psi = 90°$) radar with a beamwidth of 3° traveling at 100 m/s will induce $\beta_D \approx 35$ Hz, while an X band (10 GHz) side-looking radar with a 1° beam flying at 200 m/s will induce $\beta_D \approx 233$ Hz.

In the previous example, the radar was viewing a patch of ground, and the Doppler bandwidth observed by a stationary radar would be 0 Hz. The nonzero Doppler bandwidth β_D is entirely due to the motion of the observing radar, not to the characteristics of the target scene itself.

Although the Doppler spectrum of the illuminated terrain is both shifted according to Eq. (3.5) and broadened according to Eq. (3.8) by relative motion between the terrain and platform, the shift in center frequency is not relevant to selection of the PRF; only the bandwidth determines the Nyquist rate. Also, note that antenna patterns are not strictly limited in spatial extent, and therefore the motion-induced Doppler spectrum is not strictly bandlimited to the value based on the 3-dB bandwidth given in Eq. (3.8). Nonetheless, Eq. (3.8) provides a good basis for estimating motion-induced bandwidth.

The total Doppler bandwidth is approximately the sum of the bandwidth induced by platform motion (Eq. (3.8)) and the intrinsic bandwidth of the scene being measured. The PRF should be chosen equal to or greater than this value to meet the Nyquist sampling criterion for the slow-time signal.

The discussion so far has considered the effect of the PRF selection only on bandwidth of the slow-time signal. However, the PRF also determines range ambiguities, as was shown in Fig. 3.5. In particular, the range extent that can be represented unambiguously given T or PRF is

$$R_{ua} = \frac{cT}{2} = \frac{c}{2PRF} \qquad (3.9)$$

In general, it is desirable to be able to choose the PRF to provide the desired unambiguous range and Doppler bandwidth simultaneously. However, unambiguous range increases with decreasing PRF, while unambiguous Doppler increases with increasing PRF. In fact, since the unambiguous Doppler bandwidth is simply the PRF, Eq. (3.9) can be arranged to make this more explicit

$$R_{ua} PRF = \frac{c}{2} \qquad (3.10)$$

Clearly, increasing R_{ua} requires decreasing PRF and therefore the Doppler bandwidth.

In many situations it is not possible to simultaneously obtain the desired Doppler bandwidth and unambiguous range. To illustrate the result of range ambiguities more fully, consider Fig. 3.9. Part (a) of the figure illustrates a down-looking airborne radar, observing a target (a second aircraft) at a range beyond R_{ua}. A notional sketch of the return power from the ground clutter and the target, as a function of both time delay and range, is also shown. Because of the rising elevation mountain, the clutter power increases with range until the crest of the mountain is reached; clutter past this point will be in the shadow region and there will be no echo. The target return is superimposed at the delay corresponding to its range.

Now suppose a second pulse is transmitted at a PRI $T < 2R_0/c$. The total received power will be the superposition of the returns from each pulse. This is illustrated in Fig. 3.9b. The same clutter and target profile is repeated, simply delayed by T seconds. Because the echoes from the first pulse have not died out

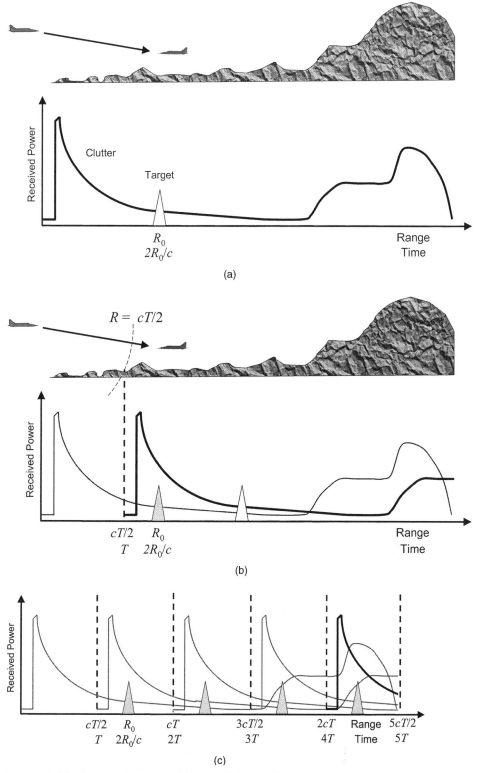

Figure 3.9 Illustration of range-ambiguous clutter and target returns.

128 Chapter Three

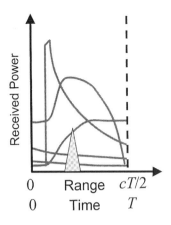

Figure 3.10 Steady-state range-ambiguous return for the scenario of Fig. 3.9.

before those from the second arrive, the actual received power after time T has significant contributions from both pulses. Part (c) of the figure continues this idea for a total of five pulses. In the interval after $t = 5T$, there is overlap of returns from all five pulses, so that the received power simultaneously contains echo from five different ranges. Because the shadowing ensures that the clutter echoes die out after $t = 5T$, steady state has been reached and the composite return after each additional pulse will be similar to that after pulse No. #5. The total return after each pulse from No. #5 onward will be as shown in Fig. 3.10. While the received power actually contains clutter returns from five different range intervals and the return from a target at a range greater that R_{ua}, it appears as if the clutter is all in the range interval $[0, R_{ua}]$ and that the target is at the range $R_0 - R_{ua}$. Thus, once detected, the target's range is ambiguous. Techniques to deal with this problem are discussed in Chap. 5.

This example also illustrates the existence of a start-up transient in processing when the returns are range ambiguous. If the scenario and PRF are such that there will be N range ambiguities in steady state, the received power will not have achieved steady state in the intervals after the first $N-1$ pulses. These first $N-1$ pulses are called *clutter fill* pulses. Processing of this nonstationary data generally gives poor results; it is better to discard the data from the clutter fill pulses and use only the data received after steady state is achieved.

3.4 Sampling the Doppler Spectrum

Selecting a value for the pulse repetition frequency has determined the sampling rate for the slow-time signal. The frequency spectrum of the slow-time signal is traditionally called the *Doppler spectrum*, because the nonzero frequency components are due to the spatial Doppler effect arising from the relative motion between the radar and target scene. Analysis or modification of the information about the target scene contained in the Doppler spectrum will be the processing goal and the subject of Chap. 5. This so-called Doppler processing will sometimes be performed directly in the slow-time domain, that is, directly on the time signal represented by a row of $y[l, m]$; but frequently the spectrum of each row will be explicitly calculated. In a digital processor, this must be done with a DFT or other discrete spectral analysis technique. In this section, it is assumed that the spectrum is computed using conventional discrete Fourier transform techniques; no nonlinear spectral estimation methods or other alternatives are considered. The question then arises as to how closely successive samples of the computed Doppler spectrum should be spaced, i.e., what should be the Doppler sampling interval?

3.4.1 The Nyquist rate in Doppler

The Nyquist sampling rate in the frequency domain can be determined by reviewing the relation between the sampled Doppler spectrum and the slow-time signal. Let a single finite duration slow-time signal (one row of $y[l, m]$) be denoted as $y_s[m]$, $0 \leq m \leq M - 1$, and let the *discrete-time Fourier transform* (DTFT) of $y_s[m]$ be given by (Oppenheim and Schafer, 1999)

$$Y_s(\omega) \equiv \sum_{m=-\infty}^{\infty} y_s[m] e^{-j\omega m} \qquad \omega \in (-\pi, \pi] \qquad (3.11)$$

$Y_s(\omega)$ is a function of a *continuous* frequency variable, despite the fact that the signal $y_s[m]$ is discrete. Furthermore, it is periodic in ω with period 2π.

Consider the K-point discrete spectrum $Y_s[k]$ formed by sampling $Y_s(\omega)$ at K evenly spaced points along the interval $(0, 2\pi]$

$$Y_s[k] = Y_s\left(\frac{2\pi k}{K}\right) \qquad k \in [0, K-1] \qquad (3.12)$$

Interpret $Y_s[k]$ as a discrete Fourier transform. To find the relation between it and the original signal $y_s[m]$, compute its inverse DFT

$$\hat{y}_s[m] = \frac{1}{K} \sum_{k=0}^{K-1} Y_s\left(\frac{2\pi k}{K}\right) e^{j 2\pi mk/K} \qquad m \in [0, K-1]$$

$$= \frac{1}{K} \sum_{k=0}^{K-1} \sum_{p=-\infty}^{\infty} y_s[p] e^{-j 2\pi pk/K} e^{j 2\pi mk/K}$$

$$= \sum_{p=-\infty}^{\infty} y_s[p] \left[\frac{1}{K} \sum_{k=0}^{K-1} e^{j 2\pi (m-p)k/K}\right] \qquad (3.13)$$

The inner sum can be evaluated as

$$\frac{1}{K} \sum_{k=0}^{K-1} e^{j 2\pi (m-p)k/K} = \sum_{q=-\infty}^{\infty} \delta[m - p - qK] \qquad (3.14)$$

where $\delta[\ ]$ is the discrete-time unit impulse function.† Substituting Eq. (3.14) back in Eq. (3.13) gives

$$\hat{y}_s[m] = \sum_{q=-\infty}^{\infty} y_s[m - qK] \qquad m \in [0, K-1] \quad (3.15)$$

Although a finite length signal $y_s[m]$ was assumed in the previous analysis, the result also holds for infinite-length signals.

Equation (3.15) shows that, in the dual of time domain sampling behavior, sampling the frequency spectrum replicates the signal in the time domain with a period proportional to the frequency sampling rate. Specifically, if the slow-time signal spectrum is computed at K frequency points, the time domain signal obtained by an inverse DFT of those frequency samples is the original slow-time signal replicated at intervals of K samples. Recall that $y_s[m]$ is confined to the interval $[0, M-1]$. If $K \geq M$, $y_s[m - qK] = 0$ in the interval of interest $[0, K-1]$ for all $q \neq 0$, so that $\hat{y}_s[m] = y_s[m]$. This is the usual case where the DFT size is at least as long as the data sequence size.

The Nyquist rate for sampling in the frequency dimension is now apparent. If $K \geq M$ the original slow-time signal $y_s[m]$ is not aliased by the frequency domain sampling operation. Consequently, it can be recovered from the replicated signal $\hat{y}_s[m]$ implied by the sampled spectrum by simply excising the principal period $m \in [0, M-1]$. This is the equivalent of the lowpass filter required to reconstruct a sampled time-domain signal; the time and frequency domains have simply been reversed in this discussion of sampling in the frequency domain. The frequency domain sampling interval ω_s must therefore satisfy

$$\omega_s \leq \frac{2\pi}{M} \quad \text{rad} \quad (3.16)$$

The corresponding sampling rate in the frequency domain is

$$K \geq M \quad \text{samples per Doppler spectrum period} \quad (3.17)$$

The input to a K-point DFT must be a K-point discrete sequence, but the available data are the M-point sequence $y_s[m]$. If $K > M$, the data are extended to a K-point sequence by simply appending $K - M$ zeroes to the end of the sequence, an operation called *zero padding*.

In some systems the number of Doppler samples needed is less than the number of data samples available, i.e., $K < M$. One way to compute a K-point DFT from an M-point sequence when $K < M$ is to simply retain only K data samples and compute their K-point DFT. This is not desirable when there are $M > K$ samples available for two reasons. First, the DTFT of the K samples used is not the same as that of the full M-point sequence, so the DFT will give us samples of a reduced-resolution DTFT. Second, by not using all M available samples, the

---

†Not to be confused with the "Dirac delta" impulse function $\delta_D(\cdot)$ used in continuous-time analysis.

signal-to-noise ratio (SNR) of the calculated spectrum is reduced because only K samples instead of all M available samples are integrated. It is rarely a good idea to discard measured data if the highest possible measurement quality is desired.

If the Doppler spectrum samples are still to be equal to samples of the discrete-time Fourier transform of $y_s[m]$ in this case, Eqs. (3.12) to (3.15) imply that it is necessary to form a new, reduced-length K-point sequence $\hat{y}_s[m]$ from the slow-time data sequence $y_s[m]$ by aliasing it according to Eq. (3.15). This operation, depicted pictorially in Fig. 3.11, is sometimes called *data turning*. It maximizes the SNR of the Doppler spectrum samples by using all of the available samples, and is in fact used in some operational radars.

3.4.2 Straddle loss

The previous section established the Nyquist sampling rate in Doppler frequency. When actually computing the sampled spectrum, whether by the DFT or other means, one would like to be confident that the sampled spectrum captures all of the important features of the underlying DTFT. For example, if the DTFT exhibits significant peaks, it is hoped that one of the spectral samples will fall on or very near that peak so that the sampled spectrum captures this feature.

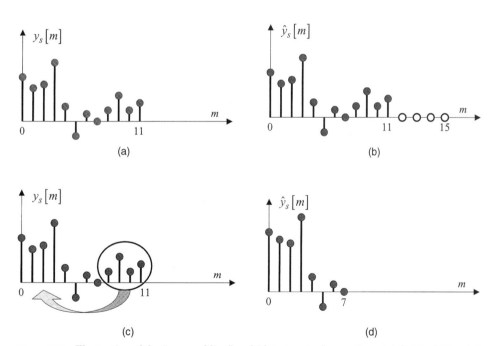

Figure 3.11 Illustration of the "zero padding" and "data turning" operations. (*a*) Original 12-point data sequence. (*b*) Zero-padded to 16 points for use in a 16-point DFT. (*c*) Data turning to create an aliased 8-point sequence shown in (*d*) for use in an 8-point DFT.

An appropriate signal model to consider this issue is a pure complex sinusoid, corresponding for example to a target moving at constant velocity relative to the radar over the observation interval and therefore evincing a constant Doppler shift. Thus, the slow-time signal $y_s[m]$ is modeled as

$$y_s[m] = e^{j\omega_D m} \qquad m \in [0, M-1] \tag{3.18}$$

where ω_D is the Doppler frequency shift in normalized radian frequency units. The DTFT of $y_s[m]$ is

$$Y_s(\omega) = \frac{\sin[(\omega - \omega_D)M/2]}{\sin[(\omega - \omega_D)/2]} e^{-j\left(\frac{M-1}{2}\right)(\omega - \omega_D)} \qquad \omega \in (-\pi, \pi) \tag{3.19}$$

That is, $Y_s(\omega)$ is a so-called *digital sinc, aliased sinc (asinc)*, or *Dirichlet* function, circularly shifted in the frequency domain so that its peak occurs at $\omega = \omega_D$. An example is shown in Fig. 3.12 for the case $\omega_D = \pi/2$ (corresponding to $f_D = \omega_D/2\pi = 0.25$) and $M = 20$. Significant features of this DTFT include the peak amplitude and frequency, the main lobe bandwidth, and the side lobe structure. In particular, the M-point DTFT of a pure complex sinusoid has a peak value of M, with the peak side lobe about 13.2 dB below the peak. The 3-dB width of the main lobe in normalized frequency units is $\beta_3 = 0.89/M$ cycles, while the Rayleigh width is $\beta_r = 1/M$ cycles; thus the null-to-null main lobe width is $\beta_{nn} = 2/M$ cycles. Several of these metrics are illustrated in Fig. 3.12.

The DFT computes samples of this spectrum at normalized frequencies $2\pi k/K$ radians. Figure 3.13 shows the result when $K = M$ and the sinusoid frequency exactly equals one of the DFT frequencies, that is, $\omega_D = 2\pi k_0/K$ for some k_0 ($k_0 = 5$ and $K = 20$ in this example, corresponding to $\omega_D = \pi/2$). One DFT sample falls on the peak of the asinc function, while all of the others

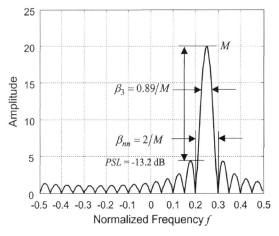

Figure 3.12 The magnitude of the DTFT of a sampled pure complex sinusoid of 20 samples length and normalized frequency 0.25.

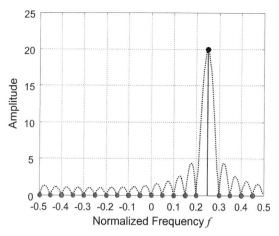

Figure 3.13 The 20-point DFT of a sampled pure complex sinusoid of 20 samples length and normalized frequency 0.25. The dotted line shows the underlying DTFT of the same data from Fig. 3.12.

fall on its zeroes, so that the DFT becomes an impulse function. This could be viewed as an ideal measurement, since the discrete spectrum indicates a single sinusoid at the correct frequency and nothing else; but it does not reveal the main lobe width or side lobe structure of the underlying DTFT.

More importantly, the good result of Fig. 3.13 depends critically on the actual sinusoid frequency exactly matching one of the DFT sample frequencies. If this is not the case, the DFT samples will fall somewhere on the asinc function other than the peak and zeros. Figure 3.14 shows the result when the example

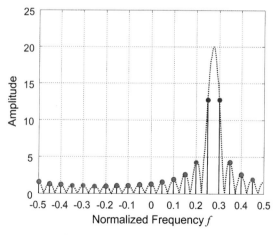

Figure 3.14 Same as Fig. 3.13 except for a frequency shift of the sinusoid by one-half DFT bin to a normalized frequency of 0.275.

of Figs. 3.12 and 3.13 is modified by changing the normalized frequency from 0.25 to 0.275 (equivalently, changing ω_D to 0.55π), exactly halfway between two DFT sample frequencies. Now a pair of DFT samples straddle the actual underlying peak of the asinc function, while the other samples fall near the side lobe peaks. Even though the underlying asinc function is identical in shape in both cases, differing only by a half-bin rotation, the effect on the apparent spectrum measured by the DFT is dramatic: a broadened and attenuated main lobe, and the appearance of significant side lobes where there apparently were none.

Because the DFT sample frequencies straddle the true peak of the underlying DTFT, the apparent peak amplitude of the spectrum in Fig. 3.14 is about 13, whereas the peak amplitude of the underlying DTFT (and thus of the DFT in Fig. 3.13) is 20. This reduction in measured amplitude from the true value is called a *straddle loss* (because the samples straddle the true peak location).[†]

One obvious way to reduce straddle loss is to sample the Doppler frequency axis more densely, i.e., to choose the number of spectrum samples $K > M$. The resulting samples are more closely spaced and thus the maximum amount by which a sample frequency can miss the peak frequency of the DTFT is reduced, and so is the straddle loss. Figure 3.15 shows the result when the sinusoid frequency is again equal to a DFT sampling frequency as in Fig. 3.13, but the sampling density is doubled to $2M$ samples per Doppler spectrum period (40 samples in this case), and then to $25M$ samples per spectrum period (500 samples). Increasing the sample density causes the apparent spectrum measured by the DFT to begin to resemble the underlying asinc of the DTFT even at as little as $2M$ samples per period. At $25M$ samples per spectrum period, the DFT gives an excellent representation of the details of the underlying DTFT.

The off-peak sampling loss (straddle loss) can be limited to a specified value, at least for this idealized signal, by appropriate choice of the sampling rate K (i.e., number of samples in the spectrum). For example, the loss can be limited to 3 dB or less by choosing K such that the interval $2\pi/K$ between samples does not exceed the 3-dB width of the asinc function. The 3-dB width can be found by considering just the magnitude of Eq. (3.19) with $\omega_D = 0$ for convenience. The peak value of the asinc function is M, thus it is necessary to find the value ω_3 of ω such that the asinc function has the value $M/\sqrt{2}$. This is best done numerically. The answer is a strong function of M for small M. The time-bandwidth product $\omega_3 M$ equals π for $M = 2$ but rapidly approaches an asymptotic value of 2.79 for $M \geq 10$. Figure 3.16 shows the variation of $\omega_3 M$ with M.

If the 3-dB point of the asinc function occurs when $\omega_3 M = 2.79$, it follows that the two-sided 3-dB width of the asinc function is $\Delta\omega = 5.58/M$ radians. The sampling interval for a rate of K samples per period is $2\pi/K$ radians. Equating these and solving gives the sampling rate required to limit off-peak sampling

---

[†]Straddle loss is also called *scallop loss* by some authors (Harris, 1978).

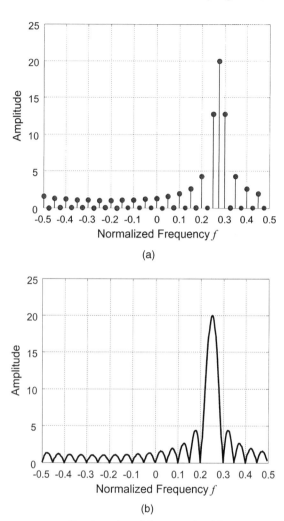

Figure 3.15 Continuation of the example of Fig. 3.13. (*a*) 40-point DFT of the 20-point sinusoid of normalized frequency 0.25. (*b*) 500-point DFT of the same sequence.

attenuation to 3 dB in the Doppler spectrum in terms of the Nyquist rate of M samples per Doppler spectrum period

$$K = \frac{2\pi}{5.58} M = 1.13 M \qquad \text{samples per Doppler spectrum period} \qquad (3.20)$$

which is 13 percent higher than the Nyquist sampling rate in Doppler. If the off-peak sampling loss is to be kept significantly less than 3 dB, the Doppler spectrum must be oversampled still more. Figure 3.17 shows the loss as a function

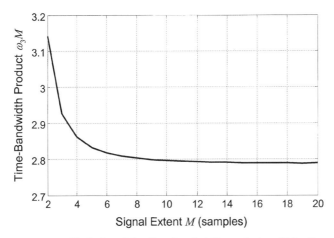

Figure 3.16 Variation of the time-bandwidth product $\omega_3 M$ for the asinc function.

of the oversampling factor κ (i.e., $K = \kappa M$) for the case $M = 100$. Both undersampled ($\kappa < 1$) and oversampled ($\kappa > 1$) cases are shown. The loss is somewhat less for very short duration sequences ($M < 5$).

3.5 Sampling in the Spatial and Angle Dimensions

As discussed earlier, two distinct types of spatial sampling are of concern in a radar system. One type concerns the design of phased array antennas. A phased array samples the incoming wavefront at the individual array element

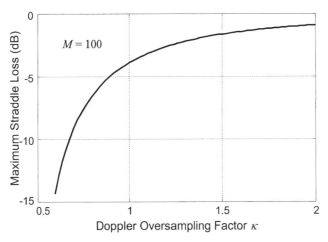

Figure 3.17 Maximum off-peak Doppler spectrum sampling loss for a sinusoidal slow-time signal sampled at κM samples per Doppler spectrum period.

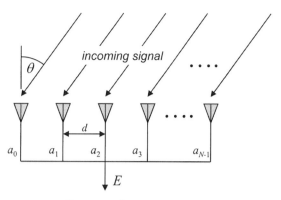

Figure 3.18 Geometry of a uniform linear array antenna.

locations. Thus, the spacing of these elements must be chosen to adequately sample the wavefront for any incident angle. The second concerns beam steering. Mechanically- or electronically-steered antennas can change the pointing direction of their antenna beam. As the beam is scanned to search or map a region in space, a decision must be made as to how far it is permissible to scan before another pulse (or burst of pulses) must be emitted by the radar so that the external environment is adequately sampled. The next two subsections address these two questions.

3.5.1 Phased array element spacing

Chapter 1 introduced the concept of spatial frequency and wavenumber. Consider a uniform linear array with element spacing d, as shown in Fig. 3.18. The wavenumber (spatial radian frequency) of a radio frequency (RF) signal with wavelength λ impinging on the array antenna from a direction of arrival θ radians off the normal to the array, as shown in the figure, is

$$K_x = \frac{2\pi}{\lambda} \sin \theta \quad \text{rad/m} \tag{3.21}$$

The equivalent spatial frequency in cyclical units is just

$$F_x = \frac{1}{\lambda} \sin \theta \quad \text{cycle/m} \tag{3.22}$$

The angle of arrival θ can vary between $-90°$ and $+90°$, so the spatial frequency bandwidth becomes

$$\beta_x = \frac{1}{\lambda} \sin\left(\frac{\pi}{2}\right) - \frac{1}{\lambda} \sin\left(-\frac{\pi}{2}\right) = \frac{2}{\lambda} \quad \text{cycle/m} \tag{3.23}$$

It follows immediately by the Nyquist criterion that the required spatial sampling interval is

$$d \leq \frac{1}{\beta_x} = \frac{\lambda}{2} \quad \text{m} \tag{3.24}$$

Thus, the elements of the array should be spaced no more than $\lambda/2$ meters apart to avoid aliasing of spatial frequencies.[†]

3.5.2 Antenna beam spacing

Consider a steerable or scanning antenna with a 3-dB beamwidth θ_3 radians, or a first null-to-first null beamwidth θ_{nn}. Each pulse transmitted samples the reflectivity of the environment in the direction in which the antenna is pointed. If a region in angular (elevation and azimuth) space is to be searched, the question arises: how densely in angle must the space be sampled? That is, how much can the antenna be steered before another pulse should be transmitted? Smaller angular sampling intervals obviously provide a better representation of the search volume, but they also require more pulses and therefore more time to search a given volume. It was shown in Chap. 1 that the antenna voltage pattern suppresses returns more than about $\pm\theta_3/2$ (or $\pm\theta_{nn}/2$ if preferred) radians from the antenna boresight. Thus, one intuitively expects that to adequately sample the reflectivity of the scene scanned by the antenna, it will be necessary to make a new measurement every time it scans by some angle on the order of θ_3 (or θ_{nn}). The Nyquist criterion can be applied to this spatial sampling problem to quantify our expectation.

It was seen in Chap. 2 that the observed reflectivity in angle, for a constant range, is the convolution of the range-averaged reflectivity with the one-way antenna power pattern. The equivalent expression, in just one angle dimension for simplicity and assuming a symmetric antenna pattern, is

$$\begin{aligned} y(\theta; R_0) &= \hat{\rho}(\theta; R_0) *_\theta P(\theta) \\ &= A_r \int_{-\pi}^{\pi} P(\varsigma - \theta)\hat{\rho}(\varsigma; R_0)\,d\varsigma \end{aligned} \tag{3.25}$$

where $y(\theta; R_0)$ is the complex coherent receiver output as a function of (azimuth or elevation) angle θ at range R_0, $\hat{\rho}(\theta; R_0)$ is the range-averaged reflectivity evaluated at range R_0 and $P(\theta)$ is the one-way power pattern in the angular dimension θ. It follows that the Fourier transform in the angle dimension of y is the product of the Fourier transforms of the antenna pattern and the range-averaged reflectivity.

[†]This same result is often derived in antenna literature by requiring that the antenna pattern not contain *grating lobes*, which are replicas of the antenna pattern caused by sampling of the aperture of a phased array antenna by the elements.

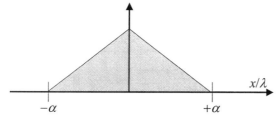

Figure 3.19 Fourier transform of the one-way antenna power pattern for an ideal rectangular antenna aperture with uniform illumination.

Taking the pattern of the ideal rectangular aperture as representative, it was seen in Chap. 1 that the one-way antenna power pattern is

$$P(\theta) = \left[\frac{\sin(\pi D \sin\theta/\lambda)}{\pi D \sin\theta/\lambda} \right]^2 \quad (3.26)$$

Defining $s = \sin\theta$ and $\alpha = D/\lambda$, Eq. (3.26) can be rewritten as

$$P(s) = \left[\frac{\sin(\pi\alpha s)}{\pi\alpha s} \right]^2 \quad (3.27)$$

which is recognizable as a sinc-squared function. It follows immediately from known Fourier transform pairs that its Fourier transform is a triangle function in the normalized variable (x/λ), where x is the spatial dimension of the antenna aperture (Bracewell, 1999). This function is illustrated in Fig. 3.19.

Because the Fourier transform of the antenna pattern has a width of 2α, the Nyquist sampling interval in s must be

$$T_s \le \frac{1}{2\alpha} \quad (3.28)$$

Recall that $s = \sin\theta$. To convert T_s into a sampling interval in θ, consider the differential $ds = \cos\theta \, d\theta$, so that $d\theta = ds/\cos\theta$. Thus, a small interval T_s in s corresponds approximately to an interval $T_\theta = T_s/\cos\theta$ in θ. The minimum value for T_θ occurs when $\theta = 0$ so that $T_\theta = T_s$. Thus, the sampling interval in angle becomes

$$T_\theta \le \frac{1}{2\alpha} \quad (3.29)$$

This is the Nyquist sampling interval in angle for a rectangular aperture of size $D = \alpha\lambda$ with uniform illumination.

As a final step, this result can be expressed in terms of 3-dB beamwidths. The 3-dB beamwidth of an aperture antenna is of the form (Balanis, 1982)

$$\theta_3 = k\frac{\lambda}{D} = \frac{k}{\alpha} \quad \text{rad} \quad (3.30)$$

For the uniformly illuminated case, $k = 0.89$. Combining Eq. (3.29) and Eq. (3.30) gives

$$T_\theta \leq \frac{\theta_3}{2k} \qquad (3.31)$$

For $k = 0.89$, this gives a Nyquist sampling rate of 0.56 times the 3-dB beamwidth, or 1.8 samples per 3 dB beamwidth. In practice, many systems sample in angle at approximately one sample per 3 dB beamwidth. The search space is then undersampled in angle, at least according to the Nyquist criterion.

The previous results, while derived for the uniformly illuminated aperture, in fact apply to all aperture antennas. For a finite aperture of size D, different power patterns (for instance, with lower side lobes at the expense of a wider main lobe) are obtained by changing the aperture illumination function, typically by tapering it in a manner similar to windowing operations in signal processing. The Fourier transform of these antenna power patterns will still be the autocorrelation of the corresponding illumination function. Since the illumination function still has finite support, its autocorrelation will still be limited to a width of 2α in s, as shown in Fig. 3.19; only the detailed shape of the function will change. Thus, Eq. (3.31) applies for any finite aperture antenna. The difference is that the factor k will be different for different illumination functions. Lower side lobe antennas will have values of k in the range of approximately 1.4 to 2.0, giving corresponding Nyquist sampling rates on the order of 2.8 to four samples per 3-dB beamwidth for low side lobe antennas.

3.6 Quantization

Two operations must be performed on an analog signal to convert it to a digital signal. The first is sampling, which has been the subject of the preceding several sections. Each sample $y[n]$ of the resulting discrete-domain (e.g., discrete-time) signal remains continuous in amplitude, however, and so must also be *quantized*, i.e., converted to a quantized value $\tilde{y}[n]$ that takes on one of a finite number of amplitudes. In this section, models for the effects of quantization on the digital signal are discussed.

Suppose that each signal sample is to be represented by a digital word having B bits; then there are 2^B possible digital values. However, the number of distinct amplitudes that can actually be encoded depends on the number representation scheme used. The two most common are two's complement and sign-magnitude encoding (Ercegovac and Lang, 2003). Two's complement allows 2^B distinct amplitudes in the range $(-2^{B-1}, +2^{B-1} - 1)$. Note that this range is nonsymmetric. Sign-magnitude encoding allows $2^B - 1$ distinct amplitudes in the symmetric range $(-2^{B-1} + 1, +2^{B-1} - 1)$. There is one less value available in sign-magnitude because it has codes for both $+0$ and -0; in two's complement, the code for zero is unique. The maximum *dynamic range* supportable by a B-bit quantizer is the ratio of the largest representable magnitude to the

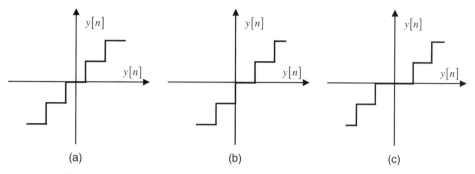

Figure 3.20 Quantizer transfer functions for the combinations of rounding and truncation quantization rules with two's complement and sign-magnitude encoding rules. (*a*) Rounding, two's complement or sign-magnitude. (*b*) Truncation, two's complement. (*c*) Truncation, sign-magnitude.

smallest nonzero representable magnitude. For both two's complement[†] and sign magnitude encoding this is, in dB

$$DR = 20\log_{10}\left(\frac{2^{B-1}-1}{1}\right) \approx 20\log_{10}(2^{B-1})$$
$$= 20(B-1)\log_{10}(2)$$
$$= 6.021(B-1) \text{ dB} \quad (3.32)$$

Thus, available dynamic range increases at about 6 dB per bit, with 8-bit quantization offering about 42 dB while 12-bit quantization gives 66 dB. The error in DR due to the approximation in Eq. (3.32) is less than 1 dB for B equal to 5 bits or more, which is almost always the case.

In addition to the wordlength and encoding, the rule for assigning values of $y[n]$ to permissible values of $\tilde{y}[n]$ must be determined. Two common rules are *rounding* and *truncation*. The effect of these operations depends on whether sign-magnitude or two's complement encoding is used. The details are discussed in the book by Oppenheim and Schafer (1999), but the result can be summarized by the effective quantizer transfer functions relating $\tilde{y}[n]$ to $y[n]$ for the various cases. These are sketched in Fig. 3.20. Note that the transfer functions for rounding of either number representation, and for truncation with sign-magnitude encoding, are symmetric about a line bisecting the second and fourth quadrants of the plot. This means that for an input signal that is symmetrically distributed about $y[n] = 0$, the quantization error will also be symmetric about zero and will therefore have zero mean. The transfer function for truncation with two's complement encoding does not have this symmetry, so that the quantization error will have a positive nonzero mean. In other words, the

---

[†]Although the maximum negative magnitude in two's complement coding is 2^{B-1}, the maximum positive magnitude is only $2^{B-1} - 1$. The tighter constraint is used to compute dynamic range.

quantized signal will include a direct current dc bias not present in the unquantized signal. In some cases, this dc component can manifest itself, particularly in Doppler processing, as an apparent false target or an increase in clutter power. Thus, quantizers that produce zero mean quantization error are preferable.

Once the shape of a quantizer characteristic has been determined, the quantization step size q must be chosen. Heuristically, the choice of q is bounded by two considerations. The maximum value of $y[n]$ should be within the dynamic range of the quantizer. Thus, it is preferable to choose q such that $|y[n]| \leq 2^{B-1}q$. For a given number of bits, this requirement establishes a minimum value of q. In the absence of target or clutter echoes, $y[n]$ will consist entirely of receiver noise. Even when a target signal is present, it will often be below the noise level, particularly if analog-to-digital (A/D) conversion precedes pulse compression (see Chap. 4). If q is so large that the quantizer's least significant bit is not dithered by the noise, the output will be zero, and the signal information is lost entirely. The requirement that noise dither the LSB establishes a maximum value for q. It may not be possible to simultaneously satisfy completely both criteria, so that in many practical instances it is necessary to accept some possibility of quantizer saturation in order to avoid suppressing signals at the noise level.

To put these considerations on a more quantitative basis, consider the case where $y[n]$ is white Gaussian noise with zero mean and a standard deviation of σ. The output of the quantizer can be expressed as the sum of the input and a quantization error $e[n]$

$$\tilde{y}[n] = y[n] + e[n] \tag{3.33}$$

Traditionally, $e[n]$ is modeled as a random variable that is uniformly distributed on $(-q/2, +q/2)$ (and therefore has variance $q^2/12$) and is uncorrelated from sample to sample with both itself and $y[n]$. The *signal-to-quantization noise ratio* (SQNR) is just

$$SQNR = \frac{12\sigma^2}{q^2} \tag{3.34}$$

If the maximum value to be represented by a B-bit quantizer is denoted as $y_{\max}$ and it is assumed for simplicity that B is large enough that $2^{B-1} - 1 \approx 2^{B-1}$, then $q = y_{\max}/2^{B-1}$. The SQNR in decibels then becomes

$$SQNR(\text{dB}) = 10\log_{10}\left(\frac{2^{2B-2}12\sigma^2}{y_{\max}^2}\right) = 10\log_{10}\left(\frac{2^{2B}3\sigma^2}{y_{\max}^2}\right)$$

$$= 6.02B - 10\log_{10}\left(\frac{y_{\max}^2}{3\sigma^2}\right) \tag{3.35}$$

Equation (3.35) expresses the common result that the SQNR improves about 6 dB for every bit added to the quantizer wordlength. This result is called the *linear model* for quantization noise because the SQNR in decibels is a linear function of the number of bits.

The variance of the quantizer output when the input is noise of variance σ^2 is $\tilde{y}[n]$ is $\sigma^2 + q^2/12$. Defining a parameter K as (q/σ), that is, the quantizer step size in multiples of the noise standard deviation, it follows that that noise power at the quantizer output, normalized to that at the input, is

$$\frac{\tilde{\sigma}^2}{\sigma^2} = 1 + \frac{K^2}{12} \qquad (3.36)$$

This equation is an alternate form of the linear model of quantization noise. A desirable outcome is to have $\tilde{\sigma}^2/\sigma^2 \approx 1$, that is, the noise power is not significantly increased by the quantization process.

The linear model is misleading because it predicts continually rising quantization noise in the output signal as K is increased, that is, as the quantizer step size is made larger. As discussed earlier, in reality the output signal will tend toward becoming identically zero when q becomes sufficiently large that the input signal does not toggle the LSB, a condition called *underflow*. A zero output signal has zero variance, an obvious violation of Eq. (3.36).

A more realistic analysis can be developed by examining the exact discrete probability density function of the quantizer output. A result that takes quantizer underflow into account is derived in the chapter by McClellan and Purdy (1978). For a given value of normalized step size K, choose an integer number of quantization levels $N = 2^{B-1} - 1$ such that $(N + 0.5)K \geq 5$; this assures that the probability of the input noise exceeding $(N + 0.5)q$ is negligible, so that potential saturation can be ignored. Then a good approximation to the normalized output power is

$$\frac{\tilde{\sigma}^2}{\sigma^2} \approx 2K^2 \left\{ N^2 \text{erf}\left[\left(N + \frac{1}{2}\right)K\right] - \sum_{i=0}^{N-1}(2i+1)\text{erf}\left[\left(i + \frac{1}{2}\right)K\right] \right\} \qquad (3.37)$$

where
$$\text{erf}(z) \equiv \frac{1}{\sqrt{2\pi}} \int_0^z e^{-x^2/2} \, dx \qquad (3.38)$$

is a standard definition of the error function. Figure 3.21 plots both the McClellan-Purdy model of Eq. (3.37) and the linear model of Eq. (3.36). The McClellan-Purdy model clearly shows the rapid drop in output noise power as the quantizer step size becomes large. A reasonable region of operation from the standpoint of quantization noise is to select K between about -6 dB and $+6$ dB ($1/2 < K < 2$), corresponding to additional quantization noise power of about 0.1 to 1.2 dB. A common choice for the operating point is $K = 1$, that is, $q = \sigma$, in which case the quantizer adds 0.35 dB of noise power.

The analysis so far has effectively assumed an infinite number of quantizer bits. For any finite number B of quantizer bits there is a maximum representable output magnitude $Nq = (2^{B-1} - 1)q = NK\sigma$. Larger values of the input are truncated to Nq, an effect called *saturation*. As K becomes small,

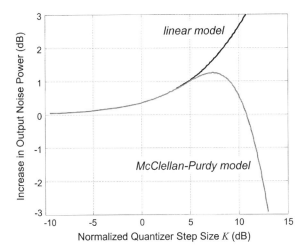

Figure 3.21 Comparison of McClellan-Purdy and linear models of the quantizer output noise power as a function of normalized step size $K = q/\sigma$. (After McClellan and Purdy, 1978.)

the quantization step size becomes small compared to the input noise power, and saturation becomes more likely. Once saturation becomes significant, decreasing K no longer causes $\tilde{\sigma}^2/\sigma^2$ to approach 1. Instead, saturation causes the output noise power to be less than the input noise power, causing a signal suppression effect similar to that observed for large K. The combined effect of

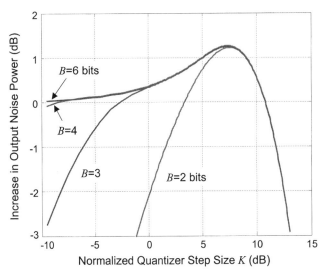

Figure 3.22 Quantizer effect on output noise power when saturation effects are included as a function of normalized step size $K = q/\sigma$ (After Purdy et al., 1974.)

the signal suppression effects at both extremes of quantizer step size on the output noise power can be estimated as (Purdy et al., 1974)

$$\frac{\tilde{\sigma}^2}{\sigma^2} \approx K^2 \left\{ N^2 - \sum_{i=0}^{N-1}(2i+1)\mathrm{erf}\left[\left(i+\frac{1}{2}\right)K\right] \right\} \quad (3.39)$$

This equation is plotted in Fig. 3.22 for B in the range of 2 to 6 bits. Reduced output noise power due to large signal suppression for 2 or 3 bits is obvious for K less than 1 (0 dB). On the other hand, for K equal to as few as four bits, the curve is virtually indistinguishable from the curve in Fig. 3.21, which assumed a large number of bits, over the range of K shown.

Figure 3.22 shows that for small wordlengths there is more than one value of K that produces an output noise power equal to the input noise power. For very small wordlengths, the boundary of the saturation effect region approaches $K = 1$, so that it is better to choose K closer to about 2 (6 dB) and accept the fraction of a decibel increase in noise power that results.

3.7 I/Q Imbalance and Digital I/Q

In Chap. 1, it was shown that the output of a quadrature receiver given a real-valued bandpass signal as input is the same as would be obtained by using the equivalent analytic (one-sided spectrum) complex signal with complex demodulation by the signal $\exp(-j\Omega_0 t)$. In other words, the quadrature receiver acts to select the upper band of the bandpass signal and shift it to baseband. Any system that accomplishes this same result can be used to derive the in-phase and quadrature signals needed for further signal processing.

The quadrature receiver could, in principle, be implemented entirely digitally. The input signal would be converted to a digital signal after the low-noise amplifier. The mixing operations would be replaced by multiplications, and the analog lowpass filters by digital filters. This is not done in practice because a straightforward implementation would require the A/D converter to operate at about twice the carrier frequency rather than twice the information bandwidth of the signal (specifically, $2F_0+\beta$ rather than just β samples/second), a technologically unreasonable requirement. On the other hand, the conventional analog quadrature receiver also has technological limitations, as mentioned briefly in Chap. 1. Correct operation assumes that the two channels are perfectly matched in delay and gain across the frequency band of interest, that there are no dc biases in either channels, and the two reference oscillators are exactly 90° out of phase.

In this section, the effect of I/Q imbalances is investigated, and then two digital I/Q receiver structures that combat imbalance errors are described.

3.7.1 I/Q imbalance and offset

Figure 1.9 describing the conventional quadrature receiver is repeated below as Fig. 3.23, but with the addition of an amplitude mismatch factor $(1 + \varepsilon)$, a

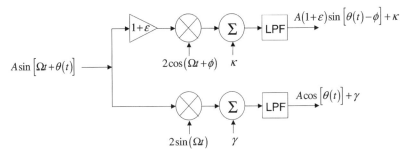

Figure 3.23 Conventional coherent receiver of Fig. 1.9 with amplitude and phase mismatch errors and dc offsets.

phase mismatch ϕ, and dc offsets γ and κ in the in-phase (I) and quadrature (Q) channels, respectively. Take the I channel as the gain and phase reference without loss of generality, so the gain and phase errors are placed entirely in the Q (upper) channel. As shown in the figure, the introduction of these errors is reflected as an undesired gain and phase shift in the Q channel output, along with the dc offset in each channel. For processing, the I and Q channel outputs are combined as usual into a single complex signal, $x(t) = I(t) + jQ(t)$. In the absence of mismatch errors, $x(t) = A\exp[j\theta(t)]$. How are the mismatch errors manifested in $x(t)$?

Inspection of Fig. 3.23 gives

$$x(t) = A\cos\theta + \gamma + j[A(1+\varepsilon)\sin(\theta-\phi) + \kappa]$$
$$= A\{[1 - j(1+\varepsilon)\sin\phi]\cos\theta + j(1+\varepsilon)\cos\phi\sin\theta\} + (\gamma + j\kappa)$$
$$\equiv A(\alpha\cos\theta + j\beta\sin\theta) + (\gamma + j\kappa) \qquad (3.40)$$

where the time dependence of $\theta(t)$ has been dropped to simplify the notation slightly. Note that the constant α is complex but β is not, and that $\alpha = \beta = 1$ in the absence of gain and phase errors, i.e., if $\varepsilon = \phi = 0$. Using the identities

$$\alpha = \frac{\alpha+\beta}{2} + \frac{\alpha-\beta}{2}$$
$$\beta = \frac{\alpha+\beta}{2} - \frac{\alpha-\beta}{2} \qquad (3.41)$$

in Eq. (3.40) and collecting terms of equal amplitude gives

$$x(t) = A\left[\frac{\alpha+\beta}{2}(\cos\theta + j\sin\theta) + \frac{\alpha-\beta}{2}(\cos\theta - j\sin\theta)\right] + (\gamma + j\kappa)$$
$$= A\left[\frac{\alpha+\beta}{2}e^{j\theta} + \frac{\alpha-\beta}{2}e^{-j\theta}\right] + (\gamma + j\kappa) \qquad (3.42)$$

Equation (3.42) shows that in the presence of amplitude or phase errors, the complex signal $x(t)$ will not only contain the desired signal component (with a slightly modified amplitude) $A\exp[j\theta(t)]$, but also an *image* component with a different amplitude and a conjugated phase function, as well as a complex dc term. The image component is an error resulting from the amplitude and phase mismatches; the dc component is the direct result of the individual channel dc offsets.

Recall that the phase function $\exp[j\theta(t)]$ can represent phase modulation of the radar waveform or the effect of the environment on the waveform (such as a phase shift due to spatial Doppler), or both. In the case of a spatial Doppler phase shift, $\theta(t)$ on the mth pulse will be of the form $\Omega_D m$ for some normalized Doppler radian frequency Ω_D. The image component will then have a phase shift of the form $-\Omega_D m$. Thus, over a series of M pulses, the mismatches will give rise to a false signal at the negative of each actual Doppler frequency in addition to the desired signal. Furthermore, the dc component is equivalent to a false signal at a Doppler shift of zero, i.e., clutter, or a stationary target.

As another example, suppose $\theta(t)$ represents the intentional quadratic phase modulation used to construct a linear FM chirp signal, $\theta(t) = \alpha t^2$ (see Chap. 4 for details). Then the image component will have a phase modulation of $-\alpha t^2$, which represents a linear frequency modulation (FM) signal with a slope opposite to the transmitted pulse. This signal will not be properly compressed by the matched filter, instead causing an apparent increase in the noise floor (Sinsky and Wang, 1974).

To judge the significance of the gain and phase mismatch errors, consider the ratio P_r of the power in the image component relative to that in the desired component. From Eq. (3.42), this is

$$P_r = \frac{|(\alpha - \beta)/2|^2}{|(\alpha + \beta)/2|^2}$$

$$= \frac{[1 - (1+\varepsilon)\cos\phi]^2 + [(1+\varepsilon)\sin\phi]^2}{[1 + (1+\varepsilon)\cos\phi]^2 + [(1+\varepsilon)\sin\phi]^2} \quad (3.43)$$

Figure 3.24 illustrates the value of P_r as a function of the phase and amplitude imbalance.

It is also useful to consider simplifications of Eq. (3.43) for the cases of small amplitude mismatch only and small phase mismatch only. First consider the case of small amplitude mismatch only, so that $\phi = 0$. Then

$$P_r = \frac{[1 - (1+\varepsilon)]^2}{[1 + (1+\varepsilon)]^2} = \frac{\varepsilon^2}{(2+\varepsilon)^2}$$

$$\approx \frac{\varepsilon^2}{4} \quad (3.44)$$

where the approximation is valid for small ε.

Amplitude mismatch is often specified in decibels, as is the relative power of the image signal component. A mismatch of k dB implies that $20\log_{10}(1+\varepsilon) = k$.

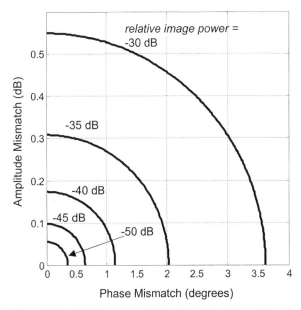

Figure 3.24 Relative power of I/Q mismatch-induced signal image as a function of amount of phase and amplitude mismatch.

Substituting this relation in Eq. (3.44) and expressing the result in decibels gives

$$P_r(\text{dB}) = 20\log_{10}(10^{k/20} - 1) - 6.02 \quad (k \text{ in dB}) \tag{3.45}$$

For example, an amplitude mismatch of 0.1 dB gives rise to an image component 44.7 dB below the desired component of $x(t)$.

A result similar to Eq. (3.44) holds for the case of small phase mismatch only, that is, $\varepsilon = 0$. In this case Eq. (3.43) reduces to

$$P_r = \frac{[1-\cos\phi]^2 + \sin^2\phi}{[1+\cos\phi]^2 + \sin^2\phi} = \frac{1-\cos\phi}{1+\cos\phi}$$

$$\approx \frac{\frac{1}{2}\phi^2}{2+\frac{1}{2}\phi^2} \approx \frac{\phi^2}{4} \tag{3.46}$$

where the second line is obtained using the assumption that ϕ is small and the small angle approximation $\cos\phi \approx 1 - \phi^2/2$. Note that ϕ is in radians. The relative power of the image component in decibels is then

$$P_r(\text{dB}) = 20\log_{10}(\phi) - 6.02 \tag{3.47}$$

As an example, a phase mismatch of $1°$ gives an image component approximately 41.2 dB below the desired response.

3.7.2 Correcting I/Q errors

As shown in Fig. 3.23, the I and Q signals in the presence of mismatch can be modeled as

$$I = A\cos\theta + \gamma \qquad Q = A(1+\varepsilon)\sin(\theta - \phi) + \kappa \qquad (3.48)$$

where the dependence on time t continues to be suppressed to simplify the notation. The desired in-phase signal I is $A\cos\theta$, and in the quadrature channel is $A\sin\theta$. Is it possible to recover the desired outputs from the available measurements of Eq. (3.48)?

Consider forming a new I' and Q' as a linear combination of the measured I and Q. Specifically, require that $I' = A\cos\theta$ and $Q' = A\sin\theta$. Although it is straightforward to solve the general problem, it is obvious that the dc offsets should simply be subtracted, and then a linear combination of the zero-offset data formed

$$\begin{bmatrix} I' \\ Q' \end{bmatrix} = \begin{bmatrix} A\cos\theta \\ A\sin\theta \end{bmatrix} = \begin{bmatrix} a_{11} & a_{12} \\ a_{21} & a_{22} \end{bmatrix} \left[\begin{pmatrix} I \\ Q \end{pmatrix} - \begin{pmatrix} \gamma \\ \kappa \end{pmatrix} \right]$$

$$= \begin{bmatrix} a_{11} & a_{12} \\ a_{21} & a_{22} \end{bmatrix} \left[\begin{pmatrix} A\cos\theta \\ A(1+\varepsilon)\sin(\theta - \phi) \end{pmatrix} \right] \qquad (3.49)$$

By inspection, $a_{11} = 1$ and $a_{12} = 0$. The remaining equation is

$$Q' = A\sin\theta = a_{21}I + a_{22}Q$$
$$= a_{21}A\cos\theta + a_{22}A(1+\varepsilon)\sin(\theta - \phi) \qquad (3.50)$$

Applying a trigonometric identity for $\sin(\theta - \phi)$ and equating terms in $\sin\theta$ and $\cos\theta$ on both sides of Eq. (3.50) leads to the following solution for a_{21} and a_{22}

$$a_{21} = \tan\phi \qquad a_{22} = \frac{1}{(1+\varepsilon)\cos\phi} \qquad (3.51)$$

Using Eq. (3.51) in Eq. (3.49) gives the final transformation required

$$\begin{bmatrix} I' \\ Q' \end{bmatrix} = \begin{bmatrix} 1 & 0 \\ \tan\phi & \frac{1}{(1+\varepsilon)\cos\phi} \end{bmatrix} \left[\begin{pmatrix} I \\ Q \end{pmatrix} - \begin{pmatrix} \gamma \\ \kappa \end{pmatrix} \right] \qquad (3.52)$$

Once the I/Q errors ε, ϕ, γ, and κ are determined, Eq. (3.50) can be used to compute a new value Q' for the quadrature channel sample for each measured I-Q sample pair. The difficulty, of course, is in actually determining the errors; the correction is then easy. The errors are generally estimated by injecting a known *pilot signal*, usually a pure sinusoid, into the receiver and observing the outputs. Details for one specific technique to estimate gain and phase errors are given in the paper by Churchill et al. (1981); that paper also derives limits to mismatch correction (and thus to image suppression) caused by noise, which introduces errors into the estimates of ε and ϕ.

A second method for eliminating I/Q error is based on the idea of transmitting multiple pulses, stepping the starting phase of each pulse through a series of evenly spaced values, and then integrating the measured returns. To see how this technique works, suppose the input signal in Fig. 3.23 is changed to $A\sin[\Omega t+\theta(t)+k(2\pi/N)]$ for some fixed integer N and variable integer k; i.e., the pulse is one of a series of N pulses where the initial phase is increased by $2\pi/N$ radians on each successive pulse. The extra phase shift propagates to the output signals

$$I_k = A\cos\left[\theta(t)+k\frac{2\pi}{N}\right]+\gamma$$
$$Q_k = A(1+\varepsilon)\sin\left[\theta(t)+k\frac{2\pi}{N}-\phi\right]+\kappa \tag{3.53}$$

for $k = 0, 1, \ldots, N-1$. The development leading to Eq. (3.42) can be repeated to obtain the complex signal for this case, which is (again suppressing the t dependence of θ for simplicity)

$$x_k(t) = A\left[\frac{\alpha+\beta}{2}e^{j(\theta+2\pi k/N)}+\frac{\alpha-\beta}{2}e^{-j(\theta+2\pi k/N)}\right]+(\gamma+j\kappa) \tag{3.54}$$

Now integrate the N pulses x_k to form a single composite measurement, applying a counter phase rotation to each to realign their phases

$$x(t) = \frac{1}{N}\sum_{k=0}^{N-1}x_k(t)e^{-jk\frac{2\pi}{N}}$$

$$= \frac{A}{N}\frac{\alpha+\beta}{2}\sum_{k=0}^{N-1}e^{j\theta}+\frac{A}{N}\frac{\alpha-\beta}{2}\sum_{k=0}^{N-1}e^{-j\theta}e^{-j4\pi k/N}+\frac{(\gamma+j\kappa)}{N}\sum_{k=0}^{N-1}e^{-j2\pi k/N}$$

$$= A\frac{\alpha+\beta}{2}e^{j\theta}+A\frac{\alpha-\beta}{2}e^{-j\theta}\sum_{k=0}^{N-1}e^{-j4\pi k/N}+\frac{(\gamma+j\kappa)}{N}\sum_{k=0}^{N-1}e^{-j2\pi k/N} \tag{3.55}$$

The summations in the middle and last terms of Eq. (3.55) can be evaluated in closed form to give

$$\sum_{k=0}^{N-1}e^{-j4\pi k/N} = \begin{cases}N & N=1,2 \\ 0 & N\geq 3\end{cases} \tag{3.56}$$

and

$$\sum_{k=0}^{N-1}e^{-j2\pi k/N} = \begin{cases}1 & N=1 \\ 0 & N\geq 2\end{cases} \tag{3.57}$$

so that

$$x(t) = A\frac{\alpha+\beta}{2}e^{j\theta} \quad (N\geq 3) \tag{3.58}$$

Thus, as long as at least three pulses are used, the process of rotating the transmitted phase, compensating the received measurements, and integrating will suppress both the undesired image component and the dc component!

The algebraic correction technique of Eqs. (3.50) and Eq. (3.51) is applied to individual I/Q sample pairs as a signal processor operation, requiring two real multiplies and one real addition per time sample (assuming a and b have been precomputed). The major advantage of this technique is that it can be applied individually to each pulse of data. Its major disadvantage is that it requires the transmitter/receiver control and analog hardware be augmented to allow pilot signal insertion for determining the correction coefficients. This operation is performed relatively infrequently on the assumption that ε and ϕ vary only slowly.

The phase rotation and integration technique of Eqs. (3.53) to (3.55), in contrast, requires integration of at least three pulses, with the transmitted phase adjusted for each pulse. Thus, the technique requires both high-speed transmitter phase control, and more time to complete a measurement since multiple pulses must be collected. The increase in required time implies also an assumption that the scene being measured does not vary during the time required for the multiple pulses; decorrelation of the scene degrades the effectiveness of the technique. This method also places a heavier load on the signal processor, since the integration requires N complex multiplies and $N-1$ complex additions per time sample, or a total of $4N$ real multiplies and $4N-2$ real additions, with $N \geq 3$. However, the integration method has one very important advantage: it does not require knowledge of any of the errors ε, ϕ, γ, and κ. It also has the side benefit that the integration of multiple pulses increases the signal-to-noise ratio of the final result $x(t)$. Given these considerations, it is often used in instrumentation systems at fixed site installations, such as turntable radar cross section (RCS) measurement facilities. In these systems, N is often on the order of 16 to 64, and may even be as high as 65,536 (64K) in some cases.

Note also that Eqs. (3.53) to (3.55) implicitly assume that the phase modulation $\theta(t)$ is the same for each pulse $x_k(t)$. If $\theta(t)$ represents waveform modulation (e.g., a linear FM chirp), this will be true; but if $\theta(t)$ contains a term representing environmental phase modulation, for example due to Doppler shift, then the technique assumes that the appropriate component of $\theta(t)$ is the same on each of the pulses integrated. This is the case for stationary targets (assuming the radar is also stationary). For constant Doppler targets, the frequency implied by $\theta(t)$ will be the same from pulse to pulse, but the absolute phase will change in general, so that the target response does not integrate properly. For accelerating targets, the assumption will fail entirely. The phase rotation and integration technique is therefore most appropriate for stationary or nearly-stationary (over N PRIs) targets. Note that the algebraic technique does not have this limitation, since it operates on individual pulses only.

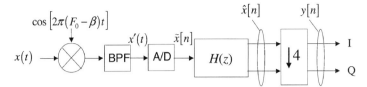

Figure 3.25 Architecture of Rader's system for digital generation of in-phase and quadrature signals. (After Rader, 1984).

3.7.3 Digital I/Q

Digital I/Q or *digital IF* is the name given to a collection of techniques that form the I and Q signals digitally in order to overcome the channel matching limitations of analog receivers. A number of variations have been described in the literature. In general, they all share two characteristics. First, they use analog mixing and filtering to shift the single real-valued input signal to a low intermediate frequency prior to A/D conversion, greatly relaxing the A/D speed requirements compared to RF sampling. Furthermore, the IF is chosen so that required complex multiplications by functions of the form $\exp(j\omega_0 n)$ reduce to particularly simple forms. Second, they use a combination of digital filtering and downsampling to obtain a final output consisting only of the desired sideband of the original spectrum, sampled at the appropriate Nyquist rate of β complex samples per second. Two approaches are briefly described here.

The first method, which is particularly elegant, is described in the paper by Rader (1984). The RF signal is assumed to have a bandpass spectrum with an information bandwidth of β hertz. Figure 3.25 is a block diagram of the system, and Fig. 3.26 sketches the signal spectrum at various points in the system. The first step is an analog frequency shifting operation that translates this spectrum to a low IF of β hertz. The bandpass filter rejects the double frequency terms created by the mixer. The spectrum is therefore bandlimited to $\pm 3\beta/2$ hertz, so the Nyquist rate is 3β samples per second. However, for reasons that will become clear shortly a higher sampling rate of 4β samples/second is used, giving a discrete-time signal with the spectrum shown in Fig. 3.26c.

Remember that the goal of quadrature demodulation is to select one sideband of the bandpass signal and translate it to baseband. Assume that the upper sideband is to be retained. The next step is therefore to filter the real signal $\tilde{x}[n]$ to eliminate the lower sideband. Since the resulting spectrum will not be Hermitian, the output signal must be complex; this is shown in Fig. 3.25 as a one-input, two-output filter. The required frequency response is clear from the spectrum diagrams in Fig. 3.26; it is

$$H(\omega) = \begin{cases} 1 & \dfrac{\pi}{4} < \omega < \dfrac{3\pi}{4} \\ 0 & -\dfrac{3\pi}{4} < \omega < -\dfrac{\pi}{4} \\ \text{don't care} & \text{otherwise} \end{cases} \quad (3.59)$$

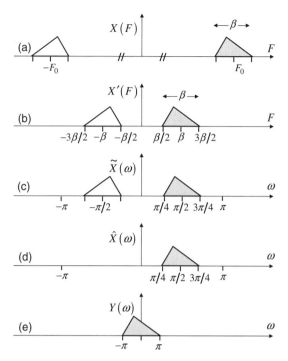

Figure 3.26 Spectra corresponding to successive signals in the digital I/Q system of Fig. 3.25. (a) Spectrum of bandpass input signal with information bandwidth β hertz. (b) Result of translation to an IF frequency also equal to β. (c) One period of spectrum on normalized frequency scale after A/D conversion. (d) Only the upper sideband remains after filtering. (e) A replica of the upper sideband is centered at dc after decimation. (After Rader, 1984).

Obviously, this asymmetric filter frequency response corresponds to a complex-valued impulse response, giving rise to the complex output from the single real input.

While Eq. (3.59) states that the value of $H(\omega)$ around dc is unconstrained, in fact it should be close to zero. The filter will then also suppress any dc component in the signal (not sketched in Fig. 3.26) which may have been introduced by nonideal mixing in the first analog frequency translation. Thus, this digital I/Q architecture also makes it easier to suppress mixer bias terms. Note that this would not be possible if the spectrum had been translated to the lowest possible IF frequency, namely β hertz, since there would then be no region of the spectrum around dc that did not contain signal components of interest.

A particularly efficient design for realizing the filter $H(z)$ as a pair of low-order recursive filters is based on the mathematics of phase-splitting networks; details are given in the paper by Rader (1984). However, the particular design of the filters is not central to the architecture of the approach.

The final step is to translate the remaining spectral sideband, centered at $\omega_0 = \pi/2$, to baseband and to reduce the sampling rate from 4β to the final Nyquist rate β. This can be accomplished by multiplying the complex filter output $\hat{x}[n]$ by the sequence $\exp(-j\pi n/2) = (-j)^n$ and then simply discarding three of every four samples. Because of the special form of the multipliers, the complex multiplications could be implemented simply with sign changes and interchanges of real and imaginary parts, rather than with actual complex multiplications. Note that this is a consequence of having selected the original sampling rate to be 4β instead of 3β.

However, this multiplication is not shown in Fig. 3.25 because, in fact, it is not necessary at all. The spectrum of the decimator output $y[n]$ is related to the spectrum of $\hat{x}[n]$ according to (Oppenheim and Schafer, 1999)

$$Y(\omega) = \frac{1}{4}\sum_{k=0}^{3} \hat{X}\left(\frac{\omega - 2\pi k}{4}\right) = \frac{1}{4}\sum_{k=0}^{3} \hat{X}\left(\frac{\omega}{4} - k\frac{\pi}{2}\right) \quad (3.60)$$

Equation (3.60) states that the decimation process causes the spectrum to replicate at intervals of $\pi/2$ radians. Since the nonzero portion of the spectrum is bandlimited to $\pi/2$ radians, these replications abut but do not alias; furthermore, since the spectrum prior to decimation is centered at $\omega = \pi/2$, one of the replications ($k = 3$, specifically) is centered at $\omega = 2\pi$ radians. The periodicity of the spectrum of a discrete-time signal therefore guarantees that there is a replica centered at $\omega = 0$ as well; this replica is the final desired spectrum. Thus, the real and imaginary outputs of the decimator are the desired I and Q signals. Another digital I/Q system that uses the spectrum replicating properties of decimation to advantage is described in the paper by Rice and Wu (1982).

Again, note that the success of the decimation operation in eliminating the need for a final complex frequency translation depended on the proper relationship between the bandwidth and center frequency of the signal, and the decimation factor. This is the major reason for choosing the IF to be β instead of $\beta/2$ (or some other permissible value), and the sampling frequency as 4β instead of 3β (or some other value).

Rader's digital I/Q architecture has reduced the number of analog signal channels from two to one, making the issues of oscillator quadrature and gain and phase matching completely moot, and providing a natural opportunity to filter out dc biases introduced by the remaining analog mixer. Furthermore, the two A/D converters required at the output of the conventional quadrature receiver to enable subsequent digital processing have been reduced to one. There are two major costs to these improvements. The first is an increase by a factor of four in the A/D converter speed requirement, from β samples per second for conventional baseband sampling to $4B$ samples per second for Rader's system; this may be very difficult at radar signal bandwidths. The second is the introduction of the need for high-rate digital filtering, which is computationally expensive (although Rader's efficient filter design lessens this cost).

Sampling and Quantization of Pulsed Radar Signals   155

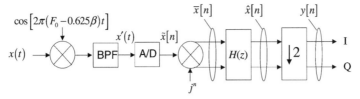

Figure 3.27 Conceptual architecture of Lincoln Laboratory system for digital generation of in-phase and quadrature signals. (After Shaw and Pohlig, 1995).

The second digital I/Q architecture, described in the report by Shaw and Pohlig (1995), is used in a K_a band radar operated by the Massachusetts Institute of Technology's Lincoln Laboratory. Figures 3.27 and 3.28 sketch the processor conceptual architecture and the relevant signals. In this case, analog frequency translation is used to shift the signal spectrum to a lower IF than used by Rader, namely 0.625β. The signal is then A/D converted at a rate of 2.5β samples per second, resulting in the signal $\tilde{x}[n]$ having a spectrum centered at $\omega = \pi/2$ as shown in Fig. 3.28c. An explicit complex modulation by $\exp(+j\pi n/2) = j^n$ then shifts one of the sidebands, in this case the lower one,

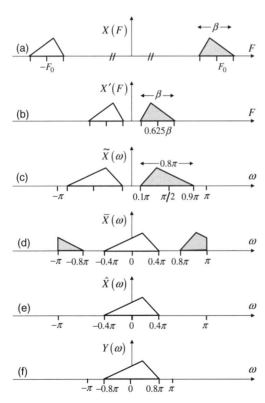

Figure 3.28 Spectra corresponding to successive signals in the digital I/Q system of Fig. 3.27. (a) Spectrum of bandpass input signal with information bandwidth of β hertz. (b) Result of translation to an IF frequency of 0.625β. (c) One period of spectrum on normalized frequency scale after A/D conversion. (d) Digital complex modulation centers the lower sideband at dc. (e) Only the lower sideband remains after lowpass filtering. (f) Decimation by two reduces the sampling rate to 1.25β. (After Shaw and Pohlig, 1995).

to baseband, resulting in the spectrum shown in Fig. 3.28d. Clearly $\bar{x}[n]$ is complex as a result of this complex modulation.

The next step is to lowpass filter $\bar{x}[n]$ to remove the upper sideband, leaving only the baseband portion of the spectrum. The Lincoln Laboratory system uses a 16-point *finite impulse response* (FIR) digital filter for this task. Once the lowpass filtering is completed, the spectrum is nonzero only for $\omega \in (-0.4\pi, +0.4\pi)$. Therefore, the sampling rate is reduced by a factor of 2 by discarding every other output sample. The final result is the desired digital I and Q signals, sampled at a rate of 1.25β samples per second.

As with Rader's system, the computational complexity is actually reduced by taking advantage of the properties of decimation and FIR filters. The decimation is performed immediately after the A/D conversion by splitting the data into even- and odd-numbered sample streams. The complex modulation by j^n, which implies both sign changes and real/imaginary interchanges, then reduces only to sign changes on every other sample in each channel, and the 16-point FIR filters are replaced with 8-point FIR filters in each channel without any reduction in filtering quality. Details are in the report by Shaw and Pohlig (1995).

The chief, and significant, advantage of this system over Rader's is that the A/D converter must operate at only 2.5 times the signal information bandwidth, rather than four times the bandwidth. This is an important savings at high radar bandwidths. There are three disadvantages. The first is that the lower IF and sampling rate require sharper transitions in the digital filter, therefore increasing the filter order necessary to achieve a given stopband suppression and thus the computational complexity of the filter. The second is the requirement for an explicit multiplication by j^n. Although this reduces to sign changes and real/imaginary exchanges, it nonetheless represents extra processing. Third, the final sampling rate exceeds the signal Nyquist rate by 25 percent, whereas in Rader's system it equaled the Nyquist rate. This increases the computational load by 25 percent over the minimum necessary throughout the remainder of the digital processing. This may not be a problem in practice. Sampling rates are usually set somewhat above Nyquist rates anyway to provide a margin of safety, since real signals are never perfectly bandlimited.

Two other details merit mention here. It may appear that modulating the sideband to baseband before filtering eliminates the possibility of using the digital filter to suppress dc bias errors from the analog mixer. However, that same modulation will move any dc term contributed by the mixer to $\omega = \pi/2$, where it can still be removed by the lowpass filter. Finally, in the Rader system the I and Q signals were derived from the upper sideband of the original bandpass signal, while in the Shaw and Pohlig system the lower sideband was used. Because the original signal was real valued, its spectrum was Hermitian, and consequently the spectra of the complex outputs of the two systems, say $Y_1(\omega)$ and $Y_2(\omega)$, are related according to $Y_2(\omega) = Y_1*(-\omega)$ so that $y_2[n] = y_1^*[n]$. Thus the I outputs of the two systems are (ideally) identical, while the Q outputs differ in sign. Clearly, either system could be modified to use the opposite sideband.

References

Balanis, C. A., *Antenna Theory*. Harper & Row, New York, 1982.

Bracewell, R. N., *The Fourier Transform and Its Applications*, 3d ed. McGraw-Hill, New York, 1999.

Churchill, F. E., G. W. Ogar, and B. J. Thompson, "The Correction of I and Q Errors in a Coherent Processor," *IEEE Transactions on Aerospace and Electronic Systems*, vol. 17, no. 1, pp. 131–137, Jan. 1981. See also "Corrections to 'The Correction of I and Q Errors in a Coherent Processor," *IEEE Transactions on Aerospace and Electronic Systems*, vol. 17, no. 2, p. 312, March 1981.

Ercegovac, M., and T. Lang, *Digital Arithmetic*. Morgan-Kauffman, San Francisco, CA, 2003.

Harris, F. J., "On the Use of Windows for Harmonic Analysis with the Discrete Fourier Transform," *Proceedings of the IEEE,* vol. 66, no. 1, pp. 51–83, Jan. 1978.

McClellan, J. H., and R. J. Purdy, "Applications of Digital Signal Processing to Radar," Chap. 5 in *Applications of Digital Signal Processing*, A. V. Oppenheim, ed. Prentice Hall, Englewood Cliffs, NJ, 1978.

Oppenheim, A. V., and R. W. Schafer, *Discrete-Time Signal Processing*, 2d ed. Prentice Hall, Englewood Cliffs, NJ, 1999.

Purdy, R. J., et al., "Digital Signal Processor Designs for Radar Applications," Technical Note 1974-58, vol. 1, Massachusetts Institute of Technology Lincoln Laboratory, Dec. 31, 1974.

Rader, C. M., "A Simple Method for Sampling In-Phase and Quadrature Components," *IEEE Transactions on Aerospace and Electronic Systems*, vol. 20, no. 6, pp. 821–824, Nov. 1984.

Rice, D. W., and K. H. Wu, "Quadrature Sampling with High Dynamic Range," *IEEE Transactions on Aerospace and Electronic Systems*, vol. 18, no. 4, pp. 736–739, Nov. 1982.

Shaw, G. A., and S. C. Pohlig, "I/Q Baseband Demodulation in the RASSP SAR Benchmark," Project Report RASSP-4, Massachusetts Institute of Technology Lincoln Laboratory, Aug. 24, 1995.

Sinsky, A. I., and P. C. P. Wang, "Error Analysis of a Quadrature Coherent Detector Processor," *IEEE Transactions on Aerospace and Electronic Systems*, vol. 10, no. 6, pp. 880–883, Nov. 1974.

Chapter 4

Radar Waveforms

4.1 Introduction

A radar transmits a waveform typically modeled as

$$\bar{x}(t) = a(t)\sin[\Omega t + \theta(t)] \qquad (4.1)$$

The term Ω in the argument of the sine function is the *radio frequency* (RF) carrier frequency in radians per second. The term $a(t)$ represents amplitude modulation of the RF carrier; in a pulsed radar, this is typically just a rectangular function that pulses the waveform on and off. The term $\theta(t)$ models any phase or frequency modulation of the carrier; it can be zero, a nonzero constant, or a nontrivial function. The overbar on $\bar{x}(t)$ denotes that the signal is on a carrier, i.e., it has not yet been demodulated. Figure 4.1 illustrates three example waveform types common in pulsed radar. The simple pulse is simply a constant-amplitude burst at the RF frequency. The frequency of the *linear frequency modulated* (LFM) pulse increases during the time the pulse is on. LFM pulses can also have decreasing frequency during the pulse. The third example is a binary phase-coded pulse. In this waveform, the frequency is constant, but the absolute phase of the waveform changes from zero to π radians several times within the pulse. That is, the value of $\theta(t)$ changes between the constants zero and π at specific times within the pulse.

As discussed in Chap. 1, the real-valued waveform of Eq. (4.1) is more conveniently modeled by its complex equivalent

$$\bar{x}(t) = a(t)e^{j[\Omega t + \theta(t)]} \qquad (4.2)$$

The portion of $\bar{x}(t)$ other than the carrier term, or equivalently the complex baseband signal after demodulation, is called the *complex envelope* of the waveform

$$x(t) = a(t)e^{j\theta(t)} \qquad (4.3)$$

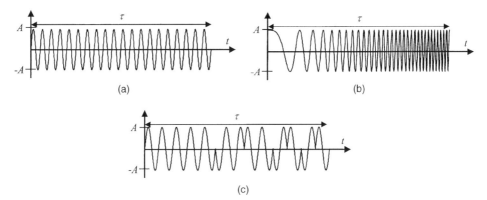

Figure 4.1 Examples of common pulsed radar waveforms. (*a*) Simple pulse. (*b*) Linear frequency modulated (LFM) pulse. (*c*) Binary phase-coded pulse.

It is this function that describes the amplitude and phase or frequency modulation applied to the RF carrier, which is considered to be "the waveform" in this chapter.

Radar waveforms can be characterized in several ways. Perhaps first is whether the waveform is *continuous wave* (CW) or pulsed; sometimes variations such as "interrupted CW" are defined as well. Pulsed waveforms can be defined based on a single pulse, or "the waveform" can be considered to be a multipulse burst. Both pulsed and CW waveforms can be further categorized based on the presence or absence of frequency or phase modulation. If present, the modulation may be intrapulse (applied to individual pulses), interpulse (applied across the pulses of a multipulse waveform), or both. Phase modulation can be biphase (two possible states), or polyphase (more than two phase states); frequency modulation can be linear or nonlinear. Amplitude modulation may be used (but usually is not).

The choice of waveform directly determines or is a major contributor to several fundamental radar system performance metrics. These include the *signal-to-noise ratio* (SNR) χ, the range resolution, the Doppler (velocity) resolution, ambiguities in range and Doppler, range and Doppler side lobes, and range-Doppler coupling. These metrics are determined by such waveform attributes as the pulse duration, bandwidth, amplitude, and phase or frequency modulation. While all of these metrics are discussed, the primary emphasis is on SNR, range resolution, and Doppler resolution, because these are the most fundamental drivers in choosing the waveform. As an example, the simple pulse of Fig. 4.1*a* has a duration of τ seconds and an amplitude of A volts. The SNR will prove to be proportional to the waveform energy, which is the product $A^2\tau$ of its power and duration. As has been seen, the range resolution of $c\tau/2$ is proportional to the pulse duration. It will be shown shortly that both the waveform bandwidth and the Doppler resolution of the simple pulse are inversely proportional to the pulse length.

There are two classic references on radar waveforms, the books by Cook and Bernfeld (1993) and Rihaczek (1996). Most radar systems books cover the fundamentals of radar waveforms (e.g., Nathanson, 1991; Peebles, 1998). An excellent modern reference on radar waveforms is the text by Levanon and Mozeson (2004). Apart from covering the mainstream waveforms such as pulse bursts and LFM, this text covers the many developments in phase codes that have occurred since the Cook and Bernfeld and Rihaczek books.

4.2 The Waveform Matched Filter

4.2.1 The matched filter

So far, it has been implicitly assumed that the overall frequency response of the radar receiver is a bandpass characteristic with a bandwidth equal to or greater than that of the transmitted signal. Equivalently, once the carrier is demodulated out, the effective frequency response is a lowpass filter with a bandwidth equal to that of the complex envelope. It will be shown in Chap. 6 that detection performance improves with increasing signal-to-noise ratio. Thus, it is reasonable to ask what overall receiver frequency response $H(\Omega)$ will maximize the SNR.

To answer this question, note that the spectrum of the receiver output, $y(t)$, will be $Y(\Omega) = H(\Omega)X(\Omega)$, where $X(\Omega)$ is the spectrum of the waveform (and thus, except for an overall delay, of a received target echo). Consider maximizing the SNR at a specific time T_M. The power of the signal component of the output at that instant is

$$|y(T_M)|^2 = \left| \frac{1}{2\pi} \int_{-\infty}^{\infty} X(\Omega) H(\Omega) e^{j\Omega T_M} d\Omega \right|^2 \tag{4.4}$$

To determine the output noise power, consider the case where the interference is white noise with power spectral density $N_0/2$ watts per hertz. The noise power spectral density at the output of the receiver will be $(N_0/2)|H(\Omega)|^2$ watts per hertz. The total output noise power is then

$$n_p = \frac{1}{2\pi} \frac{N_0}{2} \int_{-\infty}^{\infty} |H(\Omega)|^2 d\Omega \tag{4.5}$$

and the SNR measured at time T_M is

$$\chi = \frac{|y(T_M)|^2}{n_p} = \frac{\left| \frac{1}{2\pi} \int_{-\infty}^{\infty} X(\Omega) H(\Omega) e^{j\Omega T_M} d\Omega \right|^2}{\frac{N_0}{4\pi} \int_{-\infty}^{\infty} |H(\Omega)|^2 d\Omega} \tag{4.6}$$

Clearly, χ depends on the receiver frequency response. The choice of $H(\Omega)$ that will maximize χ can be determined via the Schwarz inequality. One of many forms of the Schwarz inequality is

$$\left| \int A(\Omega) B(\Omega) d\Omega \right|^2 \leq \left\{ \int |A(\Omega)|^2 d\Omega \right\} \left\{ \int |B(\Omega)|^2 d\Omega \right\} \tag{4.7}$$

with equality if and only if $B(\Omega) = \alpha A^*(\Omega)$, with α any arbitrary constant. Applying Eq. (4.7) to the numerator of Eq. (4.6) gives

$$\chi \leq \frac{\left[\frac{1}{2\pi}\right]^2 \int_{-\infty}^{\infty} |X(\Omega)e^{j\Omega T_M}|^2 \, d\Omega \int_{-\infty}^{\infty} |H(\Omega)|^2 \, d\Omega}{\frac{N_0}{4\pi} \int_{-\infty}^{\infty} |H(\Omega)|^2 \, d\Omega} \qquad (4.8)$$

The SNR is maximized when

$$H(\Omega) = \alpha X^*(\Omega) e^{-j\Omega T_M} \qquad \text{or}$$

$$h(t) = \alpha x^*(T_M - t) \qquad (4.9)$$

This particular choice of the receiver filter frequency or impulse response is called the *matched filter*, because the response is "matched" to the signal waveform. Thus, the waveform and the receiver filter needed to maximize the output SNR are a matched pair. If the radar changes waveforms, it must also change the receiver filter response in order to stay in a matched condition. The impulse response of the matched filter is obtained by time-reversing and conjugating the complex waveform. The gain constant α is often set equal to unity; it has no impact on the achievable SNR, as seen later in this chapter. The time T_M at which the SNR is maximized is arbitrary. However, $T_M \geq \tau$ is required for $h(t)$ to be causal.

Given an input signal $x'(t)$ consisting of both target and noise components, the output of the filter is given by the convolution

$$y(t) = \int_{-\infty}^{\infty} x'(s) h(t-s) \, ds \qquad h(t) = \alpha x^*(T_M - t)$$

$$= \alpha \int_{-\infty}^{\infty} x'(s) x^*(s + T_M - t) \, ds \qquad (4.10)$$

The second line of Eq. (4.10) is recognized as the cross-correlation of the target-plus-noise signal $x'(t)$ with the transmitted waveform $x(t)$, evaluated at lag $T_M - t$. Thus, the matched filter implements a correlator, with the transmitted waveform as the reference signal.

It is useful to determine the maximum value of SNR achieved by the matched filter. Using $H(\Omega) = \alpha X^*(\Omega) \exp(-j\Omega T_M)$ in Eq. (4.6)

$$\chi = \frac{\left| \frac{1}{2\pi} \int_{-\infty}^{\infty} X(\Omega) \left[\alpha X^*(\Omega) e^{-j\Omega T_M} \right] e^{j\Omega T_M} \, d\Omega \right|^2}{\frac{N_0}{4\pi} \int_{-\infty}^{\infty} \left| \alpha X^*(\Omega) e^{-j\Omega T_M} \right|^2 \, d\Omega}$$

$$= \frac{\left| \frac{1}{2\pi} \alpha \int_{-\infty}^{\infty} |X(\Omega)|^2 \, d\Omega \right|^2}{|\alpha|^2 \frac{N_0}{4\pi} \int_{-\infty}^{\infty} |X(\Omega)|^2 \, d\Omega}$$

$$= \frac{1}{\pi N_0} \int_{-\infty}^{\infty} |X(\Omega)|^2 \, d\Omega \qquad (4.11)$$

The energy in the signal $x(t)$ is

$$E = \int_{-\infty}^{\infty} |x(t)|^2 \, dt = \frac{1}{2\pi} \int_{-\infty}^{\infty} |X(\Omega)|^2 \, d\Omega \qquad (4.12)$$

where the second step follows from Parseval's relation. Using Eq. (4.12) in Eq. (4.11) gives

$$\chi = \frac{1}{\pi N_0} \int_{-\infty}^{\infty} |X(\Omega)|^2 \, d\Omega = \frac{E}{N_0/2} = \frac{2E}{N_0} \qquad (4.13)$$

Equation (4.13) states the remarkable result that the maximum achievable SNR depends only on the energy of the waveform, and not on other details such as its modulation. Two waveforms having the same energy will produce the same maximum SNR, provided each is processed through its own matched filter.

Two more properties of the matched filter output merit mention. The peak signal component at the matched filter output is given by Eq. (4.10) with $t = T_M$

$$y(T_M) = \int x(s) \, \alpha x^*(s) \, ds = \alpha E \qquad (4.14)$$

Also, the duration of the signal component of the matched filter output is exactly 2τ seconds, since it is the convolution of the τ-second pulse with the τ-second matched filter impulse response.

The previous results can be generalized to develop a filter that maximizes output *signal-to-interference ratio* (SIR) when the interference power spectrum is not white. In radar, this is useful for example in cases where the dominant interference is clutter, which generally has a colored power spectrum. The result can be expressed as a two-stage filtering operation. The first stage is a *whitening filter* that converts the interference power spectrum to a flat spectrum (and also modifies the signal spectrum in the process); the second stage is then a conventional matched filter as described earlier, but designed for the now-modified signal spectrum. Details are given by Kay (1998).

4.2.2 Matched filter for the simple pulse

To illustrate the previous ideas, consider a simple pulse of duration τ:

$$x(t) = \begin{cases} 1 & 0 \le t \le \tau \\ 0 & \text{otherwise} \end{cases} \qquad (4.15)$$

The corresponding matched filter impulse response is

$$h(t) = \alpha x^*(T_M - t)$$
$$= \begin{cases} \alpha & T_M - \tau \le t \le T_M \\ 0 & \text{otherwise} \end{cases} \qquad (4.16)$$

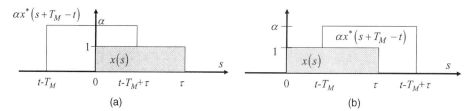

Figure 4.2 Convolution of simple pulse and its matched filter. (a) $T_M - \tau \leq t \leq T_M$. (b) $T_M \leq t \leq T_M + \tau$.

where $T_M > \tau$ for causality. Because $x(t)$ is a much simpler function than its Fourier transform (a sinc function), it is easier to work with the correlation interpretation of Eq. (4.10) to compute the output. Figure 4.2 illustrates the two terms in the integrand, helping to establish the regions of integration. Part *a* of the figure shows that

$$y(t) = \begin{cases} 0 & t \leq T_M - \tau \\ \int_0^{t-T_M+\tau}(1)(\alpha)\,ds & T_M - \tau \leq t \leq T_M \end{cases} \quad (4.17)$$

while part *b* is useful in identifying the next two regions

$$y(t) = \begin{cases} \int_{t-T_M}^{\tau}(1)(\alpha)\,ds & T_M \leq t \leq T_M + \tau \\ 0 & t \geq T_M + \tau \end{cases} \quad (4.18)$$

The result is

$$y(t) = \begin{cases} \alpha t - (T_M - \tau) & T_M \leq t \leq T_M + \tau \\ \alpha[(T_M - \tau) - t] & t \geq T_M + \tau \\ 0 & \text{otherwise} \end{cases} \quad (4.19)$$

This result is illustrated in Fig. 4.3. The matched filter output is a triangle function of duration 2τ seconds, with its peak, as expected, at $t = T_M$. The peak value is $\alpha\tau$; since the energy of the unit amplitude pulse is just τ, the peak value equals αE, as predicted.

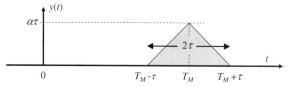

Figure 4.3 Matched filter output for a simple pulse.

The noise power at the output of the matched filter is

$$n_p = \frac{1}{2\pi} \frac{N_0}{2} \int_{-\infty}^{\infty} |H(\Omega)|^2 \, d\Omega$$

$$= \frac{N_0}{2} \int_{-\infty}^{\infty} |h(t)|^2 \, dt \quad \text{(Parsevals' relation)}$$

$$= \frac{N_0}{2} |\alpha|^2 \tau \tag{4.20}$$

The SNR is therefore

$$\chi = \frac{|\alpha\tau|^2}{(N_0/2)|\alpha|^2\tau} = \frac{2\tau}{N_0} = \frac{2E}{N_0} \tag{4.21}$$

consistent with Eq. (4.13).

4.2.3 All-range matched filtering

The matched filter was designed to maximize the output SNR at a particular time instant T_M. This raises several questions. How should T_M be chosen, and how can the range of a target be related to the resulting output? What happens if the received signal contains echoes from multiple targets at different ranges?

Start by choosing $T_M = \tau$, the minimum value that results in a causal matched filter. Now suppose the input to the matched filter is the echo from a target at an unknown range R_0, corresponding to a time delay $t_0 = 2R_0/c$. The signal component of the output of the matched filter will be

$$y(t) = \int_{-\infty}^{\infty} x(s - t_0) \alpha x^*(s + \tau - t) \, ds \tag{4.22}$$

This is just the correlation of the received, delayed echo and the matched filter impulse response. The output waveform will again be a triangle, with its peak at correlation lag zero. This occurs when $s - t_0 = s + \tau - t$, or $t = t_0 + \tau$. The matched filter output will appear as in Fig. 4.4. The peak will occur at time $t_{\text{peak}} = t_0 + \tau$, corresponding to the actual delay to the target plus the delay of the causal matched filter. The target range can be easily determined from observation of the matched filter output as $R_0 = c(t_{\text{peak}} - \tau)/2$.

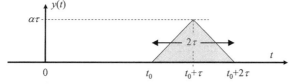

Figure 4.4 Output of the matched filter for a target at range $R_0 = ct_0/2$.

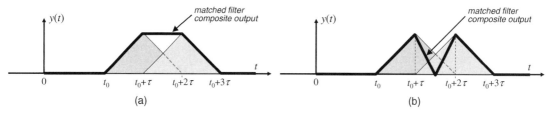

Figure 4.5 Composite matched filter response due to two scatterers separated by $c\tau/2$ meters. (a) Target responses in phase. (b) Target responses 180° out of phase.

This discussion shows that the matched filter parameter T_M can be chosen arbitrarily (typically as $T_M = \tau$). Once T_M is known, the range of a target can be determined by detecting the time at which a peak occurs at the matched filter output, subtracting T_M to get the delay to the target and back, and converting to units of range. Thus, a single choice of T_M allows detection of targets at all ranges. One simply samples the matched filter output at a series of fast-time sample instants t_k; if a peak occurs at time t_k, it corresponds to a target at range $c(t_k - T_M)/2$. If the received signal contains echoes from multiple targets at different ranges, by superposition the matched filter output will contain multiple copies of the single-pulse triangle response, one centered at the time delay (plus filter delay) of each of the various targets.

4.2.4 Range resolution of the matched filter

By determining the range separation that would result in nonoverlapping echoes, it was shown in Chap. 1 that the range resolution achieved by a simple pulse of duration τ seconds is $c\tau/2$ meters. When a matched filter is used, the output due to each scatterer is now 2τ seconds long, but is also triangular rather than rectangular in shape. Does the longer matched filter output result in a larger value of range resolution?

Before considering this question, it is useful to recall that the demodulated echo from a scatterer at range R_0 meters has not only a delay of $t_0 = 2R_0/c$ seconds, but also an overall phase shift of $\exp[j(4\pi/\lambda_t)R_0]$ radians.[†] A change of only $\lambda/4$ in range will cause a change of 180° in phase. Two overlapping target responses may therefore add either constructively or destructively in phase, and small deviations in their spacing can result in large changes in the composite response. Consider two targets at ranges $ct_0/2$ and $ct_0/2 + c\tau/2$, and assume τ is such that the two matched filter responses add in phase. Then the composite response at the matched filter output is as shown in Fig. 4.5a. The composite response is a flat-topped trapezoid. Clearly, if the separation between the two scatterers increases, a dip will begin to develop in the composite response, even when the separation is such that they remain in phase. If the

---

[†]This phase shift term was absorbed into the effective reflectivity ρ' in Chap. 2.

separation decreases, the in-phase response will still be a trapezoid, but with a higher peak and a shorter flat region as the responses overlap more. Because any increase in separation will result in a dip between the two responses, the separation of $c\tau/2$ meters is considered to be the range resolution of the matched filter output. Thus, using a matched filter does not degrade the range resolution. To reinforce this further, recall that the definition of the Rayleigh resolution is the peak-to-first null distance. Inspection of Fig. 4.3 shows that $c\tau/2$ is also the Rayleigh resolution of the simple pulse matched filter output.

Scatterers that are closer together than the Rayleigh resolution may still be resolved if the spacing is such that the individual responses add out of phase. Figure 4.5b illustrates the case for conditions that are otherwise the same as in Fig. 4.5a. Destructive interference in the region of overlap causes a deep null in the composite response. However, this null is very sensitive to the fine spacing of the scatterers and cannot be relied on to resolve two targets.

4.3 Matched Filtering of Moving Targets

Suppose a simple pulse is transmitted, $x(t) = 1, 0 \le t \le \tau$, and it echoes from a target moving toward the radar with a radial velocity of v meters per second. After demodulation, the received waveform (ignoring the overall time delay) will be $x'(t) = x(t)\exp(j\Omega_D t)$, with $\Omega_D = 4\pi v/\lambda$. Because the echo is different from $x(t)$, a filter matched to $x(t)$ will *not* be matched to $x'(t)$. If the target velocity is known, the matched filter for $x'(t)$ can be constructed

$$h(t) = \alpha x'^*(-t) = \alpha x^*(-t)e^{+j\Omega_D t} \tag{4.23}$$

The frequency response of this matched filter is

$$H(\Omega) = \alpha \int_{-\infty}^{\infty} x^*(-t)e^{+j\Omega_D t}e^{-j\Omega t} dt, \qquad t' = -t$$

$$= \alpha \left[\int_{-\infty}^{\infty} x(t')e^{-j(\Omega-\Omega_D)t'} dt' \right]^*$$

$$= \alpha X^*(\Omega - \Omega_D) \tag{4.24}$$

Thus, the matched filter for $x'(t)$ can be obtained by simply shifting the center frequency of the matched filter for $x(t)$ to the expected Doppler shift.

A more interesting situation occurs when the velocity is not known in advance, so that the receiver filter is not matched to the target Doppler shift. More generally, suppose the filter is matched to some Doppler shift Ω_i radians per second, but the actual Doppler shift of the echo is Ω_D. Choosing $T_M = 0$ for simplicity, the matched filter output will be zero for $|t| > \tau$. For $0 \le t \le \tau$ the response is

$$y(t) = \alpha \int_t^\tau \exp(j\Omega_D s)\exp[-j\Omega_i(s-t)]\, ds \tag{4.25}$$

If the filter is in fact matched to the actual Doppler shift, $\Omega_i = \Omega_D$, the output becomes

$$y(t) = \alpha \exp(j\Omega_D t) \int_t^\tau (1)\,ds$$
$$= \alpha e^{j\Omega_D t}(\tau - t) \qquad 0 \le t \le \tau \qquad (4.26)$$

The analysis is similar for negative t, $-\tau \le t \le 0$. The complete result is

$$y(t) = \begin{cases} \alpha e^{j\Omega_D t}(\tau - |t|) & -\tau \le t \le \tau \\ 0 & \text{otherwise} \end{cases} \qquad (4.27)$$

Thus, $|y(t)|$ is the usual triangular function, peaking as expected at $t = 0$.

If there is a Doppler mismatch, $\Omega_i \ne \Omega_D$, the response at the expected peak time $t = 0$ is

$$y(t)|_{t=0} = \alpha \int_0^\tau \exp(j\Omega_D s)\exp(-j\Omega_i s)\,ds$$
$$= \alpha \int_0^\tau \exp[+j(\Omega_D - \Omega_i)s]\,ds$$
$$= \frac{\alpha}{j(\Omega_D - \Omega_i)} \exp(\Omega_D - \Omega_i)s \Big|_0^\tau \qquad (4.28)$$

Defining $\Omega_{\text{diff}} \equiv \Omega_D - \Omega_i$

$$|y(0)| = \left| \frac{2\alpha \sin(\Omega_{\text{diff}} \frac{\tau}{2})}{\Omega_{\text{diff}}} \right| \qquad (4.29)$$

Equation (4.29) is plotted in Fig. 4.6. The first zero of this sinc function occurs at $F_{\text{diff}} = 1/\tau$ hertz.[†] Relatively small Doppler mismatches ($F_{\text{diff}} \ll 1/\tau$) will cause only slight reductions in the matched filter output peak amplitude. Large mismatches, however, can cause very substantial reductions.

The effect of Doppler mismatch can be either good or bad. If targets are moving and the velocities are unknown, mismatch will cause reductions in observed peaks and, if severe enough, may prevent detection. The signal processor must either estimate the target Doppler so that the matched filter can be adjusted, or construct matched filters for a number of different possible Doppler frequencies and observe the output of each to search for targets. On the other hand, if the goal is to be selective in responding only to targets of a particular Doppler shift, it is desirable to have a matched filter that suppresses targets at other Doppler shifts.

From Fig. 4.6, it is clear that the Rayleigh resolution of the Doppler mismatch response is $1/\tau$ hertz. The resolution in velocity is therefore $\lambda/2\tau$ meter per

---

[†]Note that a frequency component of $1/\tau$ hertz goes through exactly one full cycle during a pulse of duration τ seconds.

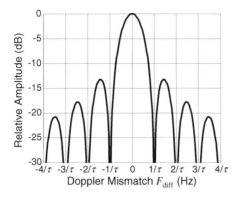

Figure 4.6 Effect of Doppler mismatch on matched filter response at expected peak time.

second. For typical pulse lengths, these are fairly large values. For example, a 10 μs pulse would exhibit a Rayleigh resolution in Doppler of 100 kHz, or in velocity at X band (10 GHz) of 1500 m/s. If finer Doppler resolution is desired, a very long pulse may be needed. For example, velocity resolution of 1 m/s at X band requires a 15 ms pulse. The range resolution is then 2250 km. This conflict between good range resolution and good Doppler resolution can be resolved using a pulse burst waveform, which will be addressed in Sec. 4.5.

4.4 The Ambiguity Function

4.4.1 Definition and properties of the ambiguity function

In the preceding sections, the matched filter response for the simple pulse waveform has been analyzed to show its behavior both in time and in response to Doppler mismatches. The *ambiguity function* (AF) is an analytical tool for waveform design and analysis that succinctly characterizes the behavior of a waveform paired with its matched filter. The AF is useful for examining resolution, side lobe behavior, and ambiguities in both range and Doppler for a given waveform, as well as phenomena such as range-Doppler coupling (introduced in Sec. 4.6.2).

Consider the output of a matched filter for a waveform $x(t)$ when the input is a Doppler-shifted response $x(t)\exp(j2\pi F_D t)$. Also assume that the filter has unit gain ($\alpha = 1$) and is designed to peak at $T_M = 0$; this merely means that the time axis at the filter output is relative to the expected peak output time for the range of the target. The filter output will be

$$y(t; F_D) = \int_{-\infty}^{\infty} x(s)\exp(j2\pi F_D s) x^*(s-t)\, ds$$
$$\equiv \hat{A}(t, F_D) \qquad (4.30)$$

which is defined as the *complex ambiguity function* $\hat{A}(t, F_D)$. An equivalent definition can be given in terms of the signal spectrum by applying basic Fourier transform properties

$$\hat{A}(t, F_D) = \int_{-\infty}^{\infty} X^*(F)X(F - F_D)\exp(j2\pi Ft)\,dF \qquad (4.31)$$

The *ambiguity function* is defined as the magnitude of $\hat{A}(t, F_D)$†

$$A(t, F_D) \equiv |\hat{A}(t, F_D)| \qquad (4.32)$$

It is a function of two variables: the time delay relative to the expected matched filter peak output, and the mismatch between that Doppler shift for which the filter was designed, and that which is actually received. For example, the AF evaluated at time $t = 0$ corresponds to the output of the actual matched filter at time $t = 2R_0/c + \tau$ for a target at range R_0. The particular form of the AF is determined entirely by the complex waveform $x(t)$.

Three properties of the ambiguity function are of immediate interest. The first states that if the waveform has energy E, then

$$|A(t, F_D)| \leq |A(0, 0)| = E \qquad (4.33)$$

Thus, when the filter is matched in both range and Doppler the response will be maximum. If the filter is not matched then the response will be less than the maximum. The second property states that total area under any ambiguity function is constant and is given by

$$\int_{-\infty}^{\infty}\int_{-\infty}^{\infty} |A(t, F_D)|^2\,dt\,dF_D = E^2 \qquad (4.34)$$

This conservation of energy statement implies that, in the design of waveforms, one cannot remove energy from one portion of the ambiguity surface without placing it somewhere else; it can only be moved around on the ambiguity surface. The third property is a symmetry relation

$$A(t, F_D) = A(-t, -F_D) \qquad (4.35)$$

In order to prove the first property, start with the square of Eq. (4.32)

$$|A(t, F_D)|^2 = \left|\int_{-\infty}^{\infty} x(s)x^*(s - t)\exp(j2\pi F_D s)\,ds\right|^2 \qquad (4.36)$$

---

†Some authors define the term "ambiguity function" as $|\hat{A}(t, F_D)|^2$ or as $\hat{A}(t, F_D)$ itself. Also, some authors define the ambiguity function as $\left|\int_{-\infty}^{\infty} x(s)\exp(j2\pi F_D s)x^*(s + t)\,ds\right|$ instead of $\left|\int_{-\infty}^{\infty} x(s)\exp(j2\pi F_D s)x^*(s - t)\,ds\right|$. The definition used here is consistent with that given in the book by Rihaczek (1996).

Applying the Schwartz inequality to Eq. (4.36) yields

$$|A(t, F_D)|^2 \leq \int_{-\infty}^{\infty} |x(s)|^2 \, ds \int_{-\infty}^{\infty} |x^*(s - t) \exp(j 2\pi F_D s)|^2 \, ds$$

$$= \int_{-\infty}^{\infty} |x(s)|^2 \, ds \int_{-\infty}^{\infty} |x^*(s - t)|^2 \, ds \quad (4.37)$$

Each integral is just the energy E in $x(t)$, so that

$$|A(t, F_D)|^2 \leq E^2 \quad (4.38)$$

The equality holds only if $x(s) = x(s - t)$ for all s, which occurs if and only if $t = F_D = 0$. Making these substitutions in Eq. (4.38) gives the equality in Eq. (4.33).

The proof of the second property starts by defining the complex conjugate of the complex ambiguity function, where

$$\hat{A}^*(t, F_D) = \int_{-\infty}^{\infty} x^*(s)x(s - t) \exp(-j 2\pi F_D s) \, ds$$

$$= \int_{-\infty}^{\infty} X(F)X^*(F - F_D) \exp(-j 2\pi F t) \, dF \quad (4.39)$$

The squared magnitude of the ambiguity function can then be written as

$$|A(t, F_D)|^2 = \hat{A}(t, F_D)\hat{A}^*(t, F_D)$$

$$= \int_{-\infty}^{\infty} \int_{-\infty}^{\infty} x(s)x^*(s - t)X(F)X^*(F - F_D) \exp[j 2\pi(F_D s - F t)] \, ds \, dF \quad (4.40)$$

The total energy in the ambiguity surface is

$$\int_{-\infty}^{\infty} \int_{-\infty}^{\infty} |A(t, F_D)|^2 \, dt \, dF_D$$

$$= \frac{1}{2\pi} \int_{-\infty}^{\infty} \int_{-\infty}^{\infty} \int_{-\infty}^{\infty} \int_{-\infty}^{\infty} x(s)x^*(s - t)X(F)X^*(F - F_D)$$

$$\times \exp[j 2\pi(F_D s - F t)] \, ds \, dF \, dt \, dF_D \quad (4.41)$$

Isolating those terms integrated over t and F_D yields the following two relationships

$$\int_{-\infty}^{\infty} x^*(s - t) \exp(-j 2\pi F t) \, dt = \exp(-j 2\pi F s) X^*(F) \quad (4.42)$$

and

$$\int_{-\infty}^{\infty} X^*(F - F_D) \exp(j 2\pi F_D \tau) \, dF_D = \exp(j 2\pi F s) x^*(s) \quad (4.43)$$

Substituting these into Eq. (4.41) yields

$$\int_{-\infty}^{\infty}\int_{-\infty}^{\infty} |A(t,F_D)|^2\,dt\,dF_D$$

$$= (1/2\pi) \int_{-\infty}^{\infty}\int_{-\infty}^{\infty} x(s)X^*(F)X(F)x^*(s)\,ds\,dF$$

$$= \int_{-\infty}^{\infty} |x(s)|^2\,ds\; 1/2\pi \int_{-\infty}^{\infty} |X(F)|^2\,dF \qquad (4.44)$$

The first integral on the right hand side of Eq. (4.44) is just the energy E of the pulse measured in the time domain; the second is, by Parseval's theorem, also the energy. Thus

$$\int_{-\infty}^{\infty}\int_{-\infty}^{\infty} |A(t,F_D)|^2\,dt\,dF_D = E^2 \qquad (4.45)$$

The symmetry property can be proved by substituting $-t$ and $-F_D$ for t and F_D, respectively, in the definition in Eq. (4.30)

$$\hat{A}(-t,-F_D) = \int_{-\infty}^{\infty} x(s)\exp(-j2\pi F_D s)x^*(s+t)\,ds \qquad (4.46)$$

Now define the change of variables $s' = s + t$

$$\hat{A}(-t,-F_D) = \int_{-\infty}^{\infty} x(s'-t)\exp(-j2\pi F_D(s'-t))x^*(s')\,ds'$$

$$= \int_{-\infty}^{\infty} x(s'-t)\exp(-j2\pi F_D s')x^*(s')\,ds'$$

$$= \exp(j2\pi F_D t)\hat{A}^*(t,F_D) \qquad (4.47)$$

Since $A(t,F_D) \equiv |\hat{A}(t,F_D)|$, Eq. (4.35) follows immediately.

It is reasonable to ask what would be an ideal ambiguity function. The answer varies depending on the intent of the system design, but a commonly cited goal is the "thumbtack" ambiguity function of Fig. 4.7, which features a single central peak, with the remaining energy spread uniformly throughout the delay-Doppler plane. The narrow central peak implies good resolution in both range and Doppler. The lack of any secondary peak implies that there will be no range or Doppler ambiguities. The uniform plateau suggests low and uniform side lobes, minimizing target masking effects. All of these features are beneficial for a system designed to make high-resolution measurements of targets in range and Doppler, or to perform radar imaging. On the other hand, a waveform intended to be used for target search might be preferred to be more tolerant of Doppler mismatch, so the Doppler shift of targets whose velocity is not yet known does not prevent their detection due to a weak response at the matched

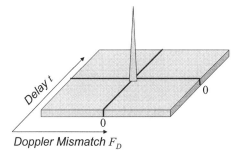

Figure 4.7 "Thumbtack" ambiguity function.

filter output. Thus, what is "ideal" in the way of an ambiguity function depends on the use to which the waveform will be put.

4.4.2 Ambiguity function of the simple pulse

As a first example of an ambiguity function, consider a simple pulse, centered on the origin and normalized to have unit energy ($E = 1$) for convenience

$$x(t) = \frac{1}{\sqrt{\tau}} \qquad \frac{-\tau}{2} \leq t \leq \frac{\tau}{2} \qquad (4.48)$$

Applying Eq. (4.30) gives, for $t > 0$

$$\hat{A}(t, F_D) = \int_{-\tau/2+t}^{\tau/2} \frac{1}{\tau} \exp(j 2\pi F_D s)\, ds$$

$$= \frac{\exp\left[j 2\pi F_D \tau/2\right] - \exp\left[j 2\pi F_D \left(-\tau/2 + t\right)\right]}{\tau j 2\pi F_D}$$

$$= \frac{1}{\tau j 2\pi F_D} e^{j 2\pi F_D t/2} \left\{ \exp\left[j 2\pi F_D \left(\frac{\tau}{2} - \frac{t}{2}\right)\right] \right.$$

$$\left. - \exp\left[-j 2\pi F_D \left(\frac{\tau}{2} - \frac{t}{2}\right)\right] \right\} \qquad (4.49)$$

The ambiguity function for $t > 0$ is the magnitude of Eq. (4.49)

$$A(t, F_D) = |\hat{A}(t, F_D)| = \left| \frac{\sin(\pi F_D (\tau - t))}{\tau \pi F_D} \right| \qquad 0 \leq t \leq \tau \qquad (4.50)$$

Repeating the derivation for $t < 0$ gives a similar result, but with the quantity $(\tau - t)$ replaced by $(\tau + t)$. The complete AF of the simple pulse is therefore

$$A(t, F_D) = \left| \frac{\sin[\pi F_D (\tau - |t|)]}{\tau \pi F_D} \right| \qquad -\tau \leq t \leq \tau \qquad (4.51)$$

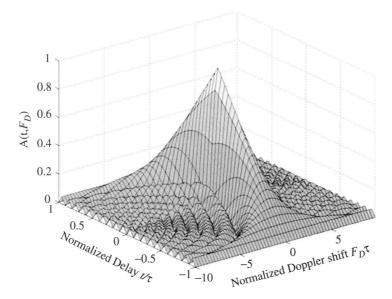

Figure 4.8 Ambiguity function of a unit-energy simple pulse of length τ.

Equation (4.51) is plotted in Fig. 4.8 in a three-dimensional surface plot, and in Fig. 4.9 as a contour plot, which is often easier to interpret and is therefore used in most cases in the remainder of this chapter. The AF for a simple pulse is a triangular ridge oriented along the delay axis. Doppler mismatches on the order of $1/\tau$ hertz or more drastically reduce and spread the matched filter output peak, as was shown previously.

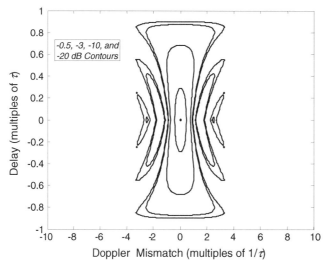

Figure 4.9 Contour plot of the simple pulse ambiguity function of Fig. 4.8.

The zero-Doppler response $A(t, 0)$ gives the matched filter output when there is no Doppler mismatch. Setting $F_D = 0$ in Eq. (4.51) and using L'Hospital's rule to resolve the indeterminate form gives

$$A(t, 0) = \left| \frac{\pi(\tau - |t|) \cos[\pi F_D(\tau - |t|)]}{\tau \pi} \right|_{F_D=0}$$

$$= \frac{\tau - |t|}{\tau} \qquad -\tau \le t \le \tau \quad (4.52)$$

Similarly, the zero-delay cut $A(0, F_D)$ gives the output of the matched filter at the expected peak time $t = 0$. Using $t = 0$ in Eq. (4.51) immediately gives

$$A(0, F_D) = \left| \frac{\sin(\pi F_D \tau)}{\tau \pi F_D} \right| \quad (4.53)$$

Equations (4.52) and (4.53) are the expected triangle and sinc functions derived previously. They are illustrated in Fig. 4.10.

A Doppler mismatch not only reduces the peak amplitude, but if severe enough, completely alters the shape of the range response of the matched filter. Figure 4.11 shows the effect of varying degrees of Doppler mismatch on the matched filter range response. These curves should be compared to Fig. 4.10a. A mismatch of $0.31/\tau$ hertz results in a reduction of about 16 percent in the peak amplitude, but the peak remains at the correct time delay. A larger shift, for example $0.94/\tau$, not only reduces the maximum output amplitude by 65 percent but eliminates the central peak altogether. By the time the mismatch is several times $1/\tau$, the response becomes completely unstructured. Note that a mismatch of n/τ hertz means that there will be n cycles of the Doppler frequency during the pulse duration τ. Also recall that for typical pulse lengths, $1/\tau$ is a large Doppler shift, so that the simple pulse still ranks as a relatively

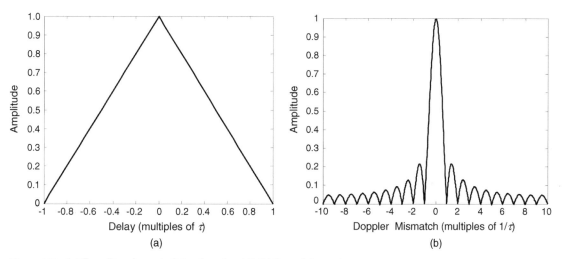

Figure 4.10 (a) Zero-Doppler cut of simple pulse AF. (b) Zero-delay cut.

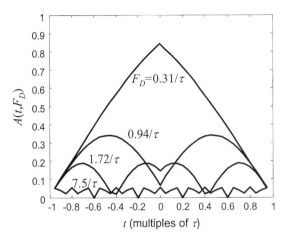

Figure 4.11 Effect of Doppler mismatch on the range response of the matched filter for the simple pulse.

Doppler-tolerant waveform. For instance, if $\tau = 10$ μs, a Doppler shift of $0.31/\tau$ is 31 kHz, corresponding to a velocity of 465 m/s, or 1040 mi/h. Even with this very large Doppler mismatch, the simple pulse matched filter output retains its basic shape, correct peak location, and suffers only the 16 percent (1.5 dB) amplitude loss.

4.5 The Pulse Burst Waveform

The flip side of the Doppler tolerance of the simple pulse described in the preceding example is that the Doppler resolution of the simple pulse is very poor. If the designer wants the radar system to respond to targets only at certain velocities, and reject targets at nearby velocities, the simple pulse is not adequate as a waveform. As observed in Chap. 1, better frequency resolution requires a longer observation time. The *pulse burst waveform* is one way to meet this requirement. It is defined as

$$x(t) = \sum_{m=0}^{M-1} x_p(t - mT) \qquad (4.54)$$

where $x_p(t)$ = single pulse of length τ
M = number of pulses in the burst, and
T = pulse repetition interval

While the constituent pulse $x_p(t)$ can be any single-pulse waveform, for the moment only the simple pulse will be considered. Figure 4.12 illustrates this waveform. The solid line forming the envelope of the sinusoidal pulses is the actual baseband waveform $x(t)$. The train of RF pulses that results when it is impressed upon a carrier is denoted, as usual, as $\tilde{x}(t)$. The total duration MT is the *coherent processing interval*(CPI), also sometimes called a *dwell*.

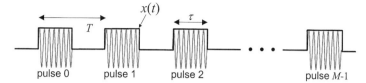

Figure 4.12 Pulse burst waveform and the resulting train of RF pulses.

4.5.1 Matched filter for the pulse burst waveform

The matched filter for the pulse burst is (with $\alpha = 1$ and $T_M = 0$)

$$h(t) = x^*(-t) = \sum_{m=0}^{M-1} x_p^*(-t - mT) \qquad (4.55)$$

and the matched filter output, given an echo from a range corresponding to a time delay t_0, is therefore

$$y(t) = \int_{-\infty}^{\infty} \left\{ \sum_{m=0}^{M-1} x_p[s - t_0 - mT] \right\} \left\{ \sum_{n=0}^{M-1} x_p^*[s - t - nT] \right\} ds$$

$$= \sum_{m=0}^{M-1} \sum_{n=0}^{M-1} \int_{-\infty}^{\infty} x_p(s - t_0 - mT) x_p^*(s - t - nT) \, ds \qquad (4.56)$$

The inner integral is the matched filter output for the constituent simple pulse. For the remainder of this section, let $t_0 = 0$ for simplicity; the results can be adjusted for any other delay t_0 by shift invariance. Denoting the simple pulse matched filter output when $T_M = 0$ as $s_p(t)$, Eq. (4.56) becomes

$$y(t) = \sum_{m=0}^{M-1} \sum_{n=0}^{M-1} s_p[-t - (m-n)T] = \sum_{m=0}^{M-1} \sum_{n=0}^{M-1} s_p^*[t - (n-m)T] \qquad (4.57)$$

where the symmetry of $s_p(t)$ has been used in the last step. Equation (4.57) states that the matched filter output is a superposition of shifted copies of $s_p(t)$. The double summation can be simplified by noting that all terms that have the same value of $(n-m)$ are identical and can be combined. For instance, there are M combinations of m and n such that $m - n = 0$, namely, all those where $m = n$. There are $M - 1$ cases where $m - n = +1$ (and another $M - 1$ cases where $m - n = -1$). Thus

$$y(t) = \sum_{m=-(M-1)}^{M-1} (M - |m|) s_p^*(t - mT) \qquad (4.58)$$

The matched filter output for the pulse burst waveform is simply a sum of scaled and shifted replicas of the output of the filter matched to a single constituent pulse.

Since the constituent pulse $x_p(t)$ is of duration τ, $s_p(t)$ is of duration 2τ. If $T > 2\tau$ as is usually the case, none of the replicas of $s_p(t)$ overlap one another. Figure 4.13 illustrates a pulse burst waveform and the corresponding matched filter output for this case and $M = 3$. The peak output is readily obtained from Eq. (4.58) with $t = 0$

$$y(0) = \sum_{m=-(M-1)}^{M-1} (M - |m|) s_p^*(-mT) = \sum_{m=-(M-1)}^{M-1} (M - |m|) s_p(mT)$$
$$= M s_p(0) = M E_p = E \tag{4.59}$$

where the second line relies on having $T > \tau$ so that $s_p(mT) = 0$ for $|m| > 0$, a somewhat looser condition than assumed in Fig. 4.13. In this equation, E_p is the energy in the single pulse $x_p(t)$, while E is the energy in the entire M-pulse waveform. Note that the peak response is M times that achieved with a single pulse of the same amplitude; this will improve signal-to-noise ratio and aid detection.

4.5.2 Pulse-by-pulse processing

The structure of Eq. (4.58) suggests that it is not necessary to construct an explicit matched filter for the entire pulse burst waveform $x(t)$, but rather that the matched filter can be implemented by filtering the data from each individual pulse with the single-pulse matched filter, and then combining those outputs. This process, called *pulse-by-pulse* processing, provides a much more convenient implementation.

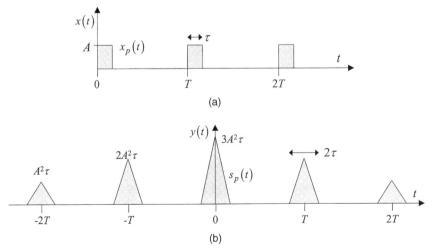

Figure 4.13 (a) Pulse burst waveform, $M = 3$. (b) Matched filter output.

Define the matched filter impulse response for the individual pulse in the burst, assuming $T_M = 0$ for simplicity

$$h_p(t) = x_p^*(-t) \qquad (4.60)$$

The output from this filter for the mth transmitted pulse, assuming a target at some delay t_l, is

$$\begin{aligned} y_m(t) &= x_p(t - t_l - mT) * h_p(t) \\ &= s_p^*(t - t_l - mT) = s_p(-t + t_l + mT) \qquad 0 \le m \le M-1 \end{aligned} \qquad (4.61)$$

Now assume that the echo from the individual pulse matched filter for the first pulse (pulse $m = 0$) is sampled at $t = t_l$; that value will be $y_0(0) = s_p(0)$. Now sample the filter response to each succeeding pulse at the same delay after its transmission (i.e., sample the same range bin for each pulse). Thus, the filter output for pulse m is sampled at $t = t_l + mT$, giving $y_m(t_l + mT) = s_p(0)$ again.

If the sample taken at time t after pulse transmission is associated with range bin l, the M samples so obtained form a discrete, constant-valued sequence $y[l, m] = s_p(0)$, $0 \le m \le M - 1$. The discrete-time causal matched filter in the slow-time (m) dimension for such a sequence is $h[m] = \alpha y^*[M - 1 - m]$; with $\alpha = 1/s_p(0)$, $h[m] = 1$, $0 \le m \le M - 1$. The output of this discrete-time matched filter is

$$z[m] = \sum_{r=0}^{M-1} y[l, r] h[m - r]$$

$$= \begin{cases} \displaystyle\sum_{r=0}^{m} y[l, r](1) & 0 \le m \le M-1 \\ \displaystyle\sum_{r=m-M+1}^{M-1} y[l, r](1) & M-1 \le m \le 2(M-1) \end{cases} \qquad (4.62)$$

The peak output will occur when the two functions in the summand completely overlap, which requires $m = M - 1$; then

$$z[M-1] = \sum_{r=0}^{M-1} y[l, r] = M s_p(0) = M E_p = E \qquad (4.63)$$

Equation (4.63) indicates that in pulse-by-pulse processing, matched filtering of the slow-time sequence from a given range bin reduces to *integrating* the slow-time samples in each range bin, and the resulting peak output is identical to that obtained with a whole-waveform continuous matched filter of Eq. (4.55). Figure 4.14 illustrates the row of slow-time samples that are integrated (after matched filtering of the single pulse in fast time) to complete the matched filtering process for the pulse burst. This operation is performed independently for each range bin.

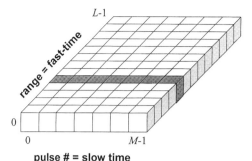

Figure 4.14 Slow-time sequence to be integrated for matched filtering of a pulse burst waveform.

4.5.3 Range ambiguity

Evaluating the pulse burst matched filter output at $t = 0$ gave the peak output for a target at the range under consideration. However, suppose the data instead contain echoes from a target an additional T seconds of delay, corresponding to $cT/2$ meters, further away. Compared to a target at the range for which the matched filter output is being evaluated, the received waveform will be unchanged except for a delay of T seconds and a reduced amplitude according to the range equation. The amplitude reduction is not pertinent to the discussion and is ignored. By shift invariance, the matched filter output will also be delayed by T seconds

$$y(t) = \sum_{m=-(M-1)}^{M-1} (M - |m|) s_p^*[t - (m+1)T] \tag{4.64}$$

Now when the matched filter output is evaluated at $t = 0$, the result is

$$y(0) = \sum_{m=-(M-1)}^{M-1} (M - |m|) s_p^*[-(m+1)T] \tag{4.65}$$

In this expression (and continuing to assume $T > \tau$), only the $m = -1$ term is nonzero, so that

$$y(0) = (M - 1) s_p(0) = (M - 1) E_p \tag{4.66}$$

This equation shows that the output at the sample time is reduced from ME_p to $(M - 1)E_p$. The situation is illustrated in Fig. 4.15. While a local peak of the matched filter output is sampled, the global peak is missed because the filter is "tuned" for the wrong delay. The result, while not zero, is a reduced-amplitude sample, reducing SNR. Equivalently, the echo appears in only $M - 1$ of the M slow-time samples integrated, because it returns after the sampling window following transmission of the first pulse, but before the sampling window following the second pulse.

The presence of a significant peak when the filter output is sampled at a delay that differs from the actual delay by T seconds will make it appear to the

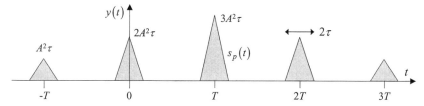

Figure 4.15 Pulse burst matched filter output when input is delayed by one PRI.

processor as though there is a target located at the range being tested, when in fact it is $cT/2$ meters away. This phenomenon is called a *range ambiguity*, and is a characteristic of pulse burst waveforms. It is not readily apparent if a peak at the matched filter output is due to a target at the apparent range, or at that range plus or minus a multiple of $cT/2$ meters.

4.5.4 Doppler response of the pulse burst waveform

To consider the effect of a Doppler mismatch on the pulse burst waveform and its matched filter, consider a target moving toward the radar at velocity v meter per second, so that its range is $R_0 - vt$ meters at time t. Again assume that the "stop-and-hop" approximation is valid, and also assume that the target motion does not exceed one range cell over the CPI, that is, $MvT < c\tau/2$; this assures that all echoes from a given target appear in the same range bin over the course of a CPI. The demodulated echoes will have a phase shift of $-(4\pi/\lambda)R(t) = -(4\pi/\lambda)(R_0 - vt)$. Adopting the pulse-by-pulse processing viewpoint and absorbing the phase component $\exp(-j4\pi R_0/\lambda)$ due to the nominal range R_0 into the overall gain, the individual matched filtered outputs for each pulse become

$$y_m(t) = x_p(t - mT) * h_p(t)$$
$$= e^{j(4\pi v/\lambda)mT} s_p(-t + mT) \qquad 0 \le m \le M - 1 \qquad (4.67)$$

The corresponding slow-time sequence is

$$y[l, m] = y_m(mT) = e^{j(4\pi v/\lambda)mT} s_p(0) \qquad 0 \le m \le M - 1$$
$$= e^{j\omega_D m} E_p \quad (\omega_D = 4\pi vT/\lambda) \qquad (4.68)$$

Integrating the slow-time samples gives

$$\sum_{m=0}^{M-1} y[l, m] = E_p \sum_{m=0}^{M-1} e^{j\omega_D m} \equiv Y[l, \omega_D)$$
$$= E_p \frac{\sin[\omega_D M/2]}{\sin[\omega_D/2]} \exp\left[-j\left(\frac{M-1}{2}\right)\omega_D\right] \qquad (4.69)$$

Equation (4.69) gives the system response to the pulse burst waveform in an arbitrary range bin l as a function of the normalized Doppler mismatch ω_D.

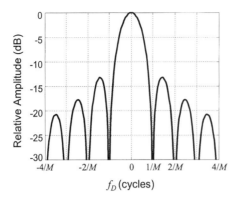

Figure 4.16 Central portion of the Doppler mismatch response of the slow-time signal using a pulse burst waveform.

This is the familiar asinc function. Figure 4.16 shows the central portion of the magnitude of this function. The zeros occur at intervals of $1/M$ cycles in normalized frequency; thus, the Rayleigh resolution in Doppler is $1/M$ cycles, or $1/MT$ hertz. MT is the duration of the entire pulse burst waveform; thus, the Doppler resolution is determined by the duration of the entire waveform instead of the duration of a single pulse. In this manner, the pulse burst waveform achieves much better Doppler resolution than a single pulse of the same duration, while maintaining the same range resolution. The cost, of course, is the time required to transmit and receive M pulses instead of one, and the computational load of integrating M samples in each range bin.

Integrating the slow-time samples of the pulse burst echo corresponds to implementing a matched filter in slow time for a signal with zero Doppler shift; in this case the expected slow-time signal is simply a constant. A matched filter for a Doppler-shifted pulse burst can be implemented by continuing to use the single-pulse matched filter in fast time, and then constructing the appropriate slow-time matched filter for the signal expected for a given Doppler shift.

Suppose the normalized Doppler shift of interest is ω_D radians. The expected slow-time signal is then of the form $A\exp(j\omega_D m)$, so after conjugation and time-reversal the slow-time matched filter coefficients will also be of the form $h[m] = \exp(+j\omega_D m)$. The matched filter peak output occurs when the impulse response and data sequence are fully overlapped, giving

$$Y[l, w; w_D] = \sum_{m=0}^{M-1} e^{-j\omega_D m} y[l, m]$$

$$= E_p \sum_{m=0}^{M-1} e^{-j\omega_D m} e^{j\omega m} = E_p \sum_{m=0}^{M-1} e^{-j(\omega-\omega_D)m}$$

$$= E_p \frac{\sin[(\omega-\omega_D)M/2]}{\sin[(\omega-\omega_D)/2]} \exp\left[-j\left(\frac{M-1}{2}\right)(\omega-\omega_D)\right] \quad (4.70)$$

which is identical to Eq. (4.69) except that the peak of the asinc function has been shifted to $\omega = \omega_D$ radians.

Note that the first line of Eq. (4.70) is simply the *discrete time Fourier transform* (DTFT) of the slow-time data sequence. Thus, a matched filter for a pulse burst waveform and a Doppler shift of ω_D radians can be implemented with a single-pulse matched filter in fast time and a DTFT in slow time, evaluated at ω_D. If ω_D is a *discrete Fourier transform* (DFT) frequency, i.e., of the form $2\pi k/K$ for some integers k and K, the slow-time matched filter can be implemented with a DFT calculation. It follows that a K-point DFT of the data $y[l, m]$ in the slow-time dimension simultaneously computes the output of K matched filters, one at each of the DFT frequencies. These frequencies correspond to Doppler shifts of $F_k = k/KT$ hertz or radial velocities $v_k = \lambda/2KT$ meter per second, $k = 0, \ldots, K - 1$. The *fast Fourier transform* (FFT) algorithm then allows very efficient search of the data for targets at various Doppler shifts by simply applying an FFT to each slow-time row of the data matrix.

4.5.5 Ambiguity function for the pulse burst waveform

Inserting the definition of the pulse burst waveform of Eq. (4.54) into the definition of the complex ambiguity function of Eq. (4.30) gives

$$\hat{A}(t, F_D) = \int_{-\infty}^{\infty} \left(\sum_{m=0}^{M-1} x_p(s - mT) \right) \left(\sum_{n=0}^{M-1} x_p^*(s - t - nT) \right) e^{j2\pi F_D s} \, ds$$

$$= \sum_{m=0}^{M-1} \sum_{n=0}^{M-1} \int_{-\infty}^{\infty} x_p(s - mT) x_p^*(s - t - nT) e^{j2\pi F_D s} \, ds \qquad (4.71)$$

Substitute $s' = s - mT$

$$\hat{A}(t, F_D) = \sum_{m=0}^{M-1} e^{j2\pi F_D mT} \sum_{n=0}^{M-1} \int_{-\infty}^{\infty} x_p(s') x_p^*(s' - t - nT + mT) e^{j2\pi F_D s'} ds' \qquad (4.72)$$

If the complex ambiguity function of the single, simple pulse $x_p(t)$ is denoted as $\hat{A}_p(t, F_D)$, the integral in Eq. (4.72) is $\hat{A}_p(t + (n - m)T, F_D)$. Thus

$$\hat{A}(t, F_D) = \sum_{m=0}^{M-1} e^{j2\pi F_D mT} \sum_{n=0}^{M-1} \hat{A}_p(t - (m - n)T, F_D) \qquad (4.73)$$

The double sum in Eq. (4.73) is somewhat difficult to deal with. Obviously, all combinations of m and n having the same difference $m - n$ result in the same summand in the second sum, but the dependence of the exponential term on m only prevents straightforward combining of all such terms. Defining $n' = m - n$, it can be shown by simply enumerating all of the combinations that the double summation of some function $f[m, n]$ can be written (Levanon, 1988; Rihaczek, 1996)

$$\sum_{m=0}^{M-1}\sum_{n=0}^{M-1} f[m,n] = \sum_{n'=-(M-1)}^{0}\sum_{m=0}^{M-|n'|-1} f[m, m-n'] + \sum_{n'=1}^{M-1}\sum_{n=0}^{M-|n'|-1} f[n+n', n] \quad (4.74)$$

Applying the decomposition of Eq. (4.74) to Eq. (4.73) gives

$$\hat{A}(t, F_D) = \sum_{n'=-(M-1)}^{0} \hat{A}_p(t - n'T, F_D) \sum_{m=0}^{M-|n'|-1} e^{j2\pi F_D mT}$$
$$+ \sum_{n'=1}^{M-1} e^{j2\pi F_D n'T} \hat{A}_p(t - n'T, F_D) \sum_{n=0}^{M-|n'|-1} e^{j2\pi F_D nT} \quad (4.75)$$

The geometric series that appears in both halves of the right hand side of this equation sums to

$$\sum_{m=0}^{M-|n'|-1} e^{j2\pi F_D mT} = \exp[j\pi F_D(M - |n'| - 1)T] \frac{\sin(\pi F_D(M - |n'|)T)}{\sin(\pi F_D T)} \quad (4.76)$$

Using this result in Eq. (4.75) and combining the two remaining sums over n' into one (and renaming the index of summation as m) gives

$$\hat{A}(t, F_D) = \sum_{m=-(M-1)}^{M-1} \hat{A}_p(t - mT, F_D) e^{j\pi F_D(M-1+m)T} \frac{\sin(\pi F_D(M - |m|)T)}{\sin(\pi F_D T)}$$
$$(4.77)$$

Equation (4.77) expresses the complex ambiguity function of the coherent pulse train in terms of the complex ambiguity function of its constituent simple pulses and the PRI.

Recall that the support in the delay axis of $\hat{A}_p(t, F_D)$ is $|t| \leq \tau$. If $T > 2\tau$, which is almost always the case, then the replications of $\hat{A}_p$ in Eq. (4.77) will not overlap, and the magnitude of the sum of the terms as m varies will be equal to the sum of the magnitude of the individual terms. The ambiguity function of the pulse burst waveform can then be written as

$$A(t, F_D) = \sum_{m=-(M-1)}^{M-1} \hat{A}_p(t - mT, F_D) \left| \frac{\sin[\pi F_D(M - |m|)T]}{\sin(\pi F_D T)} \right| \quad (T > 2\tau)$$
$$(4.78)$$

To understand this ambiguity function, it is convenient to first look at the zero Doppler and zero delay responses. The zero Doppler response is obtained by setting $F_D = 0$ in Eq. (4.78) and recalling that $A_p(t, 0) = 1 - |t|/\tau$.

$$A(t, 0) = \begin{cases} \sum_{m=-(M-1)}^{M-1} (M - |m|) \left(1 - \frac{|t - mT|}{\tau}\right) & |t - mT| < \tau \\ 0 & \text{elsewhere} \end{cases} \quad (4.79)$$

Equation (4.79) describes the triangular output of the single-pulse matched filter, repeated every T seconds and weighted by an overall triangular function $M - |m|$. Figure 4.17 illustrates this function for the case $M = 5$ and $T = 4\tau$. The ambiguity function has been normalized by the signal energy E so that it has a maximum value of 1.0. Note that, as with any waveform, the maximum occurs at $t = 0$ and the duration is twice the total waveform duration ($2MT$ in this case). The local peaks every T seconds represent the range ambiguities discussed previously in Sec. 4.5.3 and illustrated in Fig. 4.15.

The zero delay cut is obtained by setting $t = 0$ in Eq. (4.78) and recalling that $A_p(0, F_D) = \sin(\pi F_D \tau)/\pi F_D \tau$ (assuming a unit energy simple pulse)

$$A(0, F_D) = \left|\frac{\sin(\pi F_D \tau)}{\pi F_D \tau}\right| \left|\frac{\sin(\pi F_D MT)}{\sin(\pi F_D T)}\right| \quad (4.80)$$

The response is an asinc function with a first zero at $F_D = 1/MT$ hertz, repeating with a period of $1/T$ hertz. This basic behavior is weighted by a more slowly varying true sinc function with its first zero at $1/\tau$ hertz. This structure is evident in Fig. 4.18, which shows a portion of the zero delay cut for the same case with $M = 5$ and $T = 4\tau$.

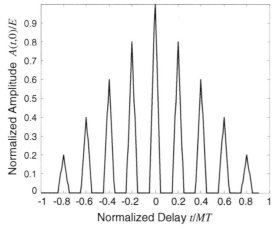

Figure 4.17 Zero-Doppler cut of the ambiguity function of a pulse burst. $M = 5$ pulses, $T = 4\tau$.

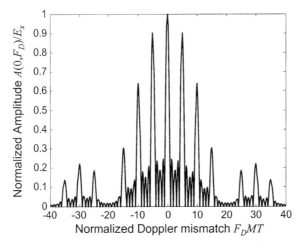

Figure 4.18 Zero-delay cut of the pulse burst ambiguity function with $M = 5$ and $T = 4\tau$.

Figure 4.19 is a contour plot of a portion of the complete ambiguity function for this waveform. Note the broadening of the response peaks in Doppler when sampling at the range-ambiguous delays such as 4τ and 8τ. (corresponding to ± 0.2 and ± 0.4 on the normalized delay scale of the contour plot). This phenomenon, caused by the $(M - |m|)$ term in the asinc term of Eq. (4.78), reflects the fact that at these range-ambiguous delays, fewer than M pulses are contributing to the matched filter local output peak. The reduced observation time

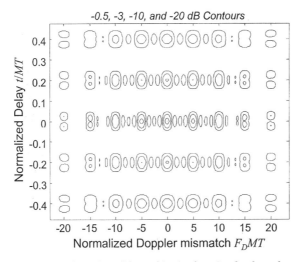

Figure 4.19 A portion of the ambiguity function for the pulse burst waveform with $M = 5$ and $T = 4\tau$. Positive frequencies only shown.

results in reduced Doppler resolution. This plot also illustrates the breakup of the well-defined peaks in delay when the Doppler mismatch reaches $1/\tau$ hertz (corresponding to 20 on the normalized Doppler scale of the plot).

Figure 4.20 is a notional illustration of the central peak of the pulse burst ambiguity function and the first ambiguities in Doppler and range. This figure summarizes how the various waveform parameters determine the resolution and ambiguities in range and Doppler. The individual pulse length τ is chosen to achieve the desired range resolution ($c\tau/2$ meters). The pulse repetition interval T sets the ambiguity interval in both range ($cT/2$ meters) and Doppler ($1/T$ hertz). Finally, once the PRI is chosen, the number of pulses in the burst determines the Doppler resolution ($1/MT$ hertz).

4.5.6 Relation of slow-time spectrum to ambiguity function

It would seem that the DTFT $Y[l, \omega_D)$ of the slow-time sequence $y[l, m]$ should be related to the variation of the complex ambiguity function $\hat{A}(t, F_D)$ in Doppler. The slow-time sequence $y[l, m]$ is obtained by sampling the output of the simple pulse matched filter output $s_p(t)$ at the same delay after transmission on each pulse. Assuming the target motion across the CPI is small compared to the range resolution (i.e., the target moves only a small fraction of a range bin during the CPI), then the amplitude of the sample taken on each pulse will be the same. This amplitude will be the maximum value $s_p(0)$ if the sampling time exactly corresponds to the target range; if the sampling time differs from that corresponding to the target range by Δt seconds, the measured amplitude on each pulse will be $s_p(\Delta t)$. Thus, the slow-time sequence in a given range bin will have constant amplitude, but the ambiguity function of the waveform will determine that amplitude.

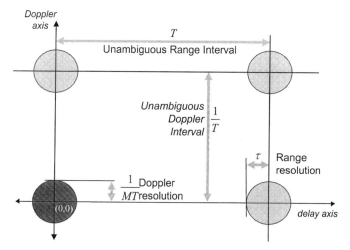

Figure 4.20 Relationship between pulse burst waveform parameters and range and Doppler resolution and ambiguities.

If there is relative motion between the radar and target, there will be a sample-to sample increase in the phase of the slow-time samples of the form $4\pi m v T/\lambda$. If the filter output is being sampled near the target range, all M pulses will contribute and the DTFT will have the $\sin(\pi F_D M T)/\sin(\pi F_D T)$ form seen in Eq. (4.80). The $\sin(\pi F_D \tau)/\pi F_D \tau$ term due to the individual pulse shape will not be observed in the DTFT; rather, this term will weight the overall amplitude of the DTFT. Finally, if the matched filter outputs are sampled at a delay corresponding to a range ambiguity, not all of them will have a signal peak present, and the Doppler resolution will degrade in $Y[l, \omega_D)$ in the same manner it did in $\hat{A}(t, F_D)$.

4.6 Frequency-Modulated Pulse Compression Waveforms

A simple pulse has only two parameters, its amplitude A and its duration τ. The range resolution $c\tau/2$ is directly proportional to τ; better resolution requires a shorter pulse. Most modern radars operate with the transmitter in saturation. That is, any time the pulse is on, its amplitude is kept at the maximum value of A; amplitude modulation (other than on/off switching) is not used. The energy in the pulse is then $A^2\tau$. This mode of operation maximizes the pulse energy, which is then also directly proportional to τ. As will be seen in Chap. 6, increasing pulse energy improves detection performance. Thus, improving resolution requires a shorter pulse, while improving detection performance requires a longer pulse. The two metrics are coupled, and in an unfortunate way, because there is effectively only one free parameter in the design of the simple pulse waveform.

Pulse compression waveforms decouple energy and resolution. Recall that a simple pulse has a Rayleigh bandwidth $\beta = 1/\tau$ hertz, and a Rayleigh resolution in time at the matched filter output of τ seconds. Thus, the *time-bandwidth product* (BT product) of the simple pulse is $\tau(1/\tau) = 1$. A pulse compression waveform, in contrast, has a bandwidth β that is much greater than $1/\tau$. Equivalently, it has a duration τ much greater than that of a simple pulse with the same bandwidth, $\tau \gg 1/\beta$. Either condition is equivalent to stating that a pulse compression waveform has a BT product much greater than one.

Pulse compression waveforms are obtained by adding frequency or phase modulation to a simple pulse. There are a vast number of pulse compression waveforms in the literature. In this text, only the most commonly used types will be described. These include linear frequency modulation, biphase codes, and certain polyphase codes. Nonlinear FM will also be briefly introduced. Many other waveforms are described in the book by Levanon and Mozeson (2004).

4.6.1 Linear frequency modulation

A linear frequency modulated waveform is defined by

$$x(t) = \cos\left(\pi \frac{\beta}{\tau} t^2\right) \qquad 0 \le t \le \tau \qquad (4.81)$$

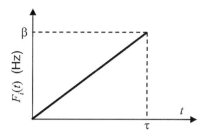

Figure 4.21 Instantaneous frequency of an LFM pulse.

The complex equivalent, which will be used here, is

$$x(t) = e^{j\pi\beta t^2/\tau} = e^{j\theta(t)} \qquad 0 \le t \le \tau \qquad (4.82)$$

The instantaneous frequency of this waveform is the derivative of the phase function

$$F_i(t) = \frac{1}{2\pi}\frac{d\theta(t)}{dt} = \frac{\beta}{\tau}t \qquad (4.83)$$

This function is shown in Fig. 4.21, assuming $\beta > 0$. Clearly $F_i(t)$ sweeps linearly across a total bandwidth of β hertz during the τ-second pulse duration. The corresponding waveform $x(t)$ (Eq. (4.81), or the real part of Eq. (4.82)) is shown in Fig. 4.22 for the case where $\beta\tau = 50$. The LFM waveform is often called a *chirp* waveform in analogy to the sound of an acoustic sinusoid with a linearly changing frequency. When β is positive, the pulse is an *upchirp*; if β is negative, it is a *downchirp*. The BT product of the LFM pulse is simply $\beta\tau$; thus $\beta\tau \gg 1$ if the LFM pulse is to qualify as a pulse compression waveform.

Figure 4.23 shows the magnitude spectrum of the LFM waveform for a relatively low BT product case ($\beta\tau = 10$), and again for a higher BT product case

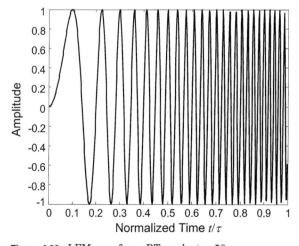

Figure 4.22 LFM waveform, BT product = 50.

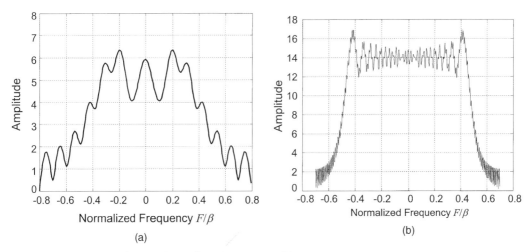

Figure 4.23 Spectrum of an LFM waveform. (a) $\beta\tau = 10$. (b) $\beta\tau = 100$.

($\beta\tau = 100$). For low BT, the spectrum is relatively poorly defined. As the BT product increases, the spectrum takes on a more rectangular shape. This seems intuitively reasonable: because the sweep is linear, the waveform spreads its energy uniformly across the spectrum.

Figure 4.24 shows the output of the matched filter for the same two chirp waveforms. The dotted line superimposed on the output waveform is the output of a matched filter for a simple pulse of the same duration. As always, the total

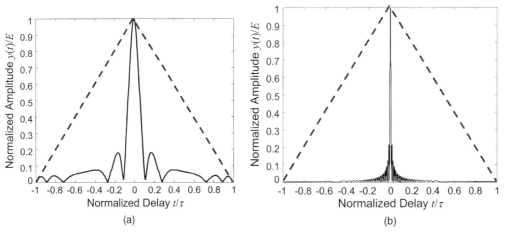

Figure 4.24 Output of matched filter for an LFM waveform. (a) $\beta\tau = 10$. (b) $\beta\tau = 100$. The dotted line is the output of a matched filter for a simple pulse of the same duration.

duration of the matched filter output is 2τ seconds. Note that in both cases, the LFM waveform results in a matched filter output with a Rayleigh resolution much narrower than τ. In fact, the Rayleigh resolution is very nearly $1/\beta$ in each case (this will be confirmed shortly), an improvement over the simple pulse by a factor of the time-bandwidth product $\beta\tau$. Also note that, unlike the simple pulse case, the matched filter output for the LFM pulse exhibits a side lobe structure. Figure 4.25 expands the central portion of Fig. 4.24b, showing the distinctly sinc-like main lobe and first few side lobes. This should not be surprising: the waveform spectrum $X(F)$ (Fig. 4.23b) is approximately a rectangle of width β hertz. Consequently, the spectrum of the matched filter output, $|X(F)|^2$, will also be approximately a rectangle of width β. The time-domain output of the matched filter is therefore expected to be approximately a sinc function with a Rayleigh resolution of $1/\beta$ seconds.

To summarize, the LFM waveform has enabled separate control of pulse energy (through its duration) and range resolution (through the pulse bandwidth). The possibility of pulse compression was created by the use of matched filters. The output of the matched filter is not a replica of the transmitted waveform $x(t)$, but of its autocorrelation function $s_x(t)$. Thus, if a waveform can be designed that has a long duration but a narrowly concentrated autocorrelation, both good range resolution and good energy can be obtained simultaneously. This in turn is accomplished by modulating a long pulse to spread its bandwidth beyond the usual $1/\tau$. Since the spectrum of the autocorrelation function is just the squared magnitude of the waveform spectrum, a spectrum spread over β hertz will tend to produce a filter output with most of its energy concentrated in a main lobe of about $1/\beta$ seconds duration. The linear FM pulse is the first example of such a waveform, but phase coded waveforms will provide more examples of this approach.

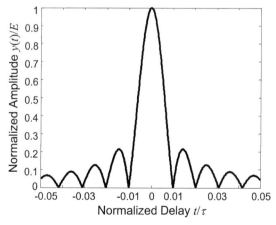

Figure 4.25 Expanded view of central portion of Fig. 4.24b.

4.6.2 The Principle of stationary phase

The Fourier transform of Eq. (4.82), derived by Rihaczek (1996), is a relatively complicated result involving the sine integral Si(F). A very useful and much simpler approximation can be derived using the *principle of stationary phase* (PSP), an advanced technique in Fourier analysis. The PSP is useful for approximate evaluation of integrals with highly oscillatory integrands; thus, it applies particularly well to Fourier transforms. Write $x(t)$ in amplitude and phase form, $x(t) = A(t)\exp[j\theta(t)]$, and consider its Fourier transform

$$X(\Omega) = \int_{-\infty}^{+\infty} \underbrace{A(t)e^{j\theta(t)}}_{x(t)} e^{-j\Omega t}\, dt \qquad (4.84)$$

Define the phase $\phi(t, \Omega)$ of the Fourier integral as the combination of the signal phase and the Fourier kernel phase

$$X(\Omega) = \int_{-\infty}^{+\infty} A(t)e^{j\theta(t)}e^{-j\Omega t}\, dt = \int_{-\infty}^{+\infty} A(t)e^{j[\theta(t)-\Omega t]}\, dt$$

$$\equiv \int_{-\infty}^{+\infty} A(t)e^{j\phi(t,\Omega)}\, dt \qquad (4.85)$$

Of course, the exact Fourier transform is known for many signals having relatively simple phase functions $\theta(t)$. The PSP is most useful when the signal phase function, and thus the total integral phase $\phi(t, \Omega)$, is continuous but nonlinear or otherwise complicated.

Define a *stationary point* of the integrand as a value of $t = t_0$ such that the first time derivative of the integral phase $\phi'(t_0, \Omega) = 0$. Then the PSP approximation to the spectrum is (Born and Wolf, 1959; Papoulis, 1962; Raney, 1992)

$$X(\Omega) \approx \sqrt{\frac{-\pi}{2\phi''(t_0, \Omega)}}\, e^{-j\frac{\pi}{4}} x(t_0) e^{j\phi(t_0, \Omega)} \qquad (4.86)$$

where $\phi''(t_0, \Omega)$ is the second time derivative of $\phi(t, \Omega)$ evaluated at $t = t_0$. If there are multiple stationary points, the spectrum is the sum of such terms for each stationary point. Equation (4.86) states that the magnitude of the spectrum at a given frequency Ω is proportional to the amplitude of the signal envelope at the time that the stationary point occurs and, more importantly, is inversely proportional to the square root of the rate of change of the frequency $\phi''(t_0, \Omega)$ at that time. The PSP also implies that only the stationary points significantly influence $X(\Omega)$.

The PSP can be applied to estimate the spectrum of the LFM waveform. The waveform is defined as

$$x(t) = A(t)e^{j\alpha t^2} \qquad A(t) = \begin{cases} 1 & -\tau/2 \le t \le +\tau/2 \\ 0 & \text{otherwise} \end{cases} \qquad \alpha \equiv \pi\frac{\beta}{\tau} \qquad (4.87)$$

Thus

$$X(\Omega) = \int_{-\infty}^{+\infty} x(t)e^{-j\Omega t}\, dt = \int_{-\infty}^{+\infty} A(t)e^{j(\alpha t^2 - \Omega t)}\, d\Omega \qquad (4.88)$$

The integrand phase and its derivatives are then

$$\phi(t,\Omega) = \alpha t^2 - \Omega$$
$$\phi'(t,\Omega) = 2\alpha t - \Omega \qquad (4.89)$$
$$\phi''(t,\Omega) = 2\alpha$$

The stationary points are found by setting $\phi'(t,\Omega) = 0$ and solving for t. In this case, there is only one stationary point, given by

$$0 = \phi'(t_0,\Omega) \Rightarrow t_0 = \frac{\Omega}{2\alpha} \qquad (4.90)$$

Inserting Eq. (4.90) into Eq. (4.86) gives

$$\begin{aligned}
X(\Omega) &\approx \sqrt{\frac{-\pi}{2\phi''(t_0,\Omega)}} e^{-j\pi/4} A(t_0) e^{j\phi(t_0,\Omega)} \\
&= \sqrt{\frac{-\pi}{2(2\alpha)}} e^{-j\pi/4} A\left(\frac{\Omega}{2\alpha}\right) e^{j\left[\alpha(\Omega/2\alpha)^2 - \Omega(\Omega/2\alpha)\right]} \\
&= j\sqrt{\frac{\pi}{4\alpha}} e^{-j\pi/4} A\left(\frac{\Omega}{2\alpha}\right) e^{-j\Omega^2/4\alpha}
\end{aligned} \qquad (4.91)$$

Recalling the finite support of the signal envelope $A(t)$, the term $A(\Omega/2\alpha)$ becomes (using $\alpha = \pi\beta/\tau$)

$$A\left(\frac{\Omega}{2\alpha}\right) = \begin{cases} 1 & -\tau/2 \leq \dfrac{\Omega}{2\alpha} \leq +\tau/2 \\ 0 & \text{otherwise} \end{cases} \Rightarrow$$

$$A\left(\frac{\Omega}{2\alpha}\right) = \begin{cases} 1 & -2\pi\left(\dfrac{\beta}{2}\right) \leq \Omega \leq +2\pi\left(\dfrac{\beta}{2}\right) \\ 0 & \text{otherwise} \end{cases} \qquad (4.92)$$

Thus, the final result is

$$X(\Omega) \approx j\sqrt{\frac{\pi}{4\alpha}} e^{-j\pi/4} e^{-j\Omega^2/4\alpha} \qquad -2\pi\left(\frac{\beta}{2}\right) \leq \Omega \leq +2\pi\left(\frac{\beta}{2}\right) \qquad (4.93)$$

Figure 4.26 compares this approximation with the exact spectrum when $\beta\tau = 100$. Equation (4.93) shows that $|X(\Omega)|$ is constant over the range $\pm\beta/2$ hertz, and is zero outside of this range. This is both intuitively satisfying, since this is exactly the range over which the instantaneous frequency of the LFM pulse sweeps, and consistent with the increasingly rectangular shape of the exact

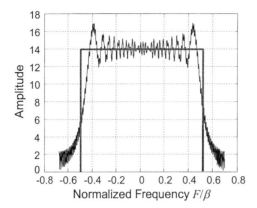

Figure 4.26 Comparison of actual spectrum and PSP approximation for an LFM pulse with $\beta\tau = 100$.

spectrum observed in Fig. 4.23 as the BT product increases. The PSP result also gives an estimate of the phase of the spectrum which, like the phase of the waveform $x(t)$, is seen to be quadratic.

4.6.3 Ambiguity function of the LFM waveform

The ambiguity function of an LFM pulse can be obtained by direct calculation, similar to the simple pulse, but with a good deal more tedium. An easier way is to introduce the "chirp property" of the ambiguity function and then apply it to the LFM case. Suppose that a waveform $x(t)$ has an ambiguity function $A(t, F_D)$. Create a modified waveform $x'(t)$ by modulating $x(t)$ with a linear FM complex chirp and compute its complex ambiguity function

$$x'(t) \equiv x(t)e^{j\pi\beta t^2/\tau}$$

$$\hat{A}'(t, F_D) = \int_{-\infty}^{\infty} x'(s)x'^*(s-t)e^{j2\pi F_D s}\,ds$$

$$= \int_{-\infty}^{\infty} x(s)e^{j\pi\beta s^2/\tau}x^*(s-t)e^{-j\pi\beta(s-t)^2/\tau}e^{j2\pi F_D s}\,ds$$

$$= e^{-j\pi\beta t^2/\tau}\int_{-\infty}^{\infty} x(s)x^*(s-t)e^{j2\pi(F_D+\beta t/\tau)s}\,ds$$

$$= e^{-j\pi\beta t^2/\tau}\hat{A}\left(t, F_D + \frac{\beta}{\tau}t\right) \tag{4.94}$$

Taking the magnitude of $\hat{A}'(t, F_D)$ gives the ambiguity function of the chirp signal in terms of the ambiguity function of the original signal without the chirp

$$A'(t, F_D) = A\left(t, F_D + \frac{\beta}{\tau}t\right) \tag{4.95}$$

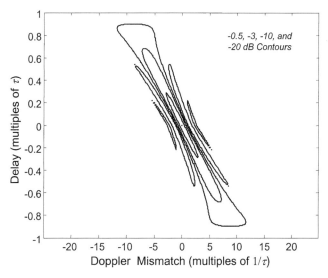

Figure 4.27 Contour plot of the ambiguity function of an LFM waveform with $\beta\tau = 10$.

Equation (4.95) states that adding a chirp modulation to a signal skews its ambiguity function in the delay-Doppler plane. Applying this property to the simple pulse AF (Eq. (4.51)) gives the AF of the LFM waveform

$$A(t, F_D) = \left| \frac{\sin\left(\pi\left(F_D + \beta t/\tau\right)(\tau - |t|)\right)}{\tau\pi\left(F_D + \beta t/\tau\right)} \right| \qquad -\tau \leq t \leq \tau \qquad (4.96)$$

Figure 4.27 is a contour plot of the AF of an LFM pulse of duration $\tau = 10$ μs and swept bandwidth $\beta = 1$ MHz; thus, the BT product is 10. The AF retains the triangular ridge of the simple pulse, but is now skewed in the delay-Doppler plane as predicted by Eq. (4.95).

The zero-Doppler cut of the LFM ambiguity function, which is just the matched filter output when there is no Doppler mismatch, is

$$A(t, 0) = \left| \frac{\sin\left[\pi\beta t\left(1 - |t|/\tau\right)\right]}{\pi\beta t} \right| \qquad -\tau \leq t \leq \tau \qquad (4.97)$$

This function was illustrated for BT products of both 10 and 100 in Fig. 4.24. The Rayleigh resolution of the LFM pulse is obtained by the examination of Eq. (4.97). Clearly the peak of $A(t, 0)$ occurs at $t = 0$. The first zero occurs when the argument of the numerator equals π, which occurs when $\beta t(1 - |t|/\tau) = 1$. For positive t, this becomes

$$\beta t - \frac{\beta t^2}{\tau} = 1 \Rightarrow t^2 - \tau t + \tau/\beta = 0 \qquad (4.98)$$

The roots of this equation are $t = (\tau \pm \sqrt{\tau^2 - 4\tau/\beta})/2 = \tau(1 \pm \sqrt{1 - 4/\beta\tau})/2$. Since the argument of the square root in the last expression is less than one, taking the negative sign gives the positive root closest to zero, and thus the Rayleigh resolution in time. This result can be simplified, however, with the following series expansion of the square root

$$\sqrt{1-x} = 1 - \frac{x}{2} - \frac{x^2}{8} - \cdots \approx 1 - \frac{x}{2} \quad (x \ll 1) \Rightarrow$$
$$t \approx \frac{\tau}{2}\left[1 - \left(1 - \frac{2}{\beta\tau}\right)\right] \approx \frac{1}{\beta} \quad (\beta\tau \gg 1) \quad (4.99)$$

Thus, the Rayleigh resolution in time is approximately $1/\beta$ seconds, corresponding to a Rayleigh range resolution ΔR of

$$\Delta R = \frac{c}{2\beta} \text{ m} \quad (4.100)$$

The zero-delay response is

$$A(0, F_D) = \left|\frac{\sin(\pi F_D \tau)}{\pi F_D \tau}\right| \quad (4.101)$$

which is simply a standard sinc function. Thus, the Doppler resolution of the LFM pulse is the same as that of a simple pulse, namely

$$\Delta F_D = \frac{1}{\tau} \text{ Hz} \quad (4.102)$$

Equation (4.101) shows that, like the simple pulse, the Doppler resolution of an LFM pulse is inversely proportional to the pulse length. Furthermore, the energy in the LFM pulse is still $A^2\tau$, directly proportional to the pulse length. Equation (4.100) shows that, *un*like the simple pulse, the range resolution is inversely proportional to the swept bandwidth. The LFM waveform has two parameters, bandwidth and duration, which can now be used to independently control pulse energy and range resolution. The pulse length is chosen (along with the pulse amplitude A) to set the desired energy, while the swept bandwidth is chosen to obtain the desired range resolution.

The expression $c/2\beta$ for range resolution is quite general. For instance, the Rayleigh bandwidth of a simple pulse is $\beta = 1/\tau$ hertz; using this in $c/2\beta$ gives $\Delta R = c\tau/2$ as before. While bandwidth and pulse length are directly related in the simple pulse, modulation of the LFM waveform has decoupled them. If $\beta\tau > 1$ for the LFM pulse, the range resolution will be better than that of a simple pulse of the same duration by the factor $\beta\tau$. Alternatively, the range resolution of a simple pulse of length τ can be matched by an LFM pulse that is longer (and thus higher energy, given the same transmitted power) by the factor $\beta\tau$.

4.6.4 Range-Doppler coupling

The skew in the ambiguity function for the LFM pulse gives rise to an interesting phenomenon. Consider the AF of Eq. (4.96). The peak of this sinc-like function will occur when

$$F_D + \frac{\beta}{\tau} t = 0 \Rightarrow t = -\frac{\tau F_D}{\beta} \tag{4.103}$$

That is, when there is a Doppler mismatch, the peak of the matched filter output will not occur at $t = 0$ as desired. Instead, the peak output will be shifted by an amount proportional to the Doppler shift. Because the target range will be estimated based on the time of occurrence of this peak, a Doppler mismatch will induce an error in measuring range. The corresponding range error will be

$$\delta R = -\frac{c \tau F_D}{2 \beta} \tag{4.104}$$

The amplitude of the peak will also be reduced by the factor $(1 - |t|/\tau) = (1 - F_D/\beta)$. Figure 4.28 illustrates the skewed ridge of the LFM ambiguity function and the relationship between Doppler shift and range measurement error.

While an incorrect range measurement is certainly undesirable, range-Doppler coupling is a useful phenomenon in some systems. A simple pulse with duration τ will have a Doppler resolution of $1/\tau$ hertz; targets with greater Doppler mismatches will produce a greatly attenuated output from the matched filter and will likely go undetected. An LFM pulse of the same duration will still produce a significant output peak for a much broader range of Doppler shifts, even though the peak will be mislocated in range. Nonetheless, the target will be more likely to be detected. Thus, the LFM waveform is more *Doppler tolerant* than the simple pulse. This makes it a good choice for surveillance applications. A relatively large range of Doppler shifts can be searched with an LFM pulse. The range error can be negated by repeating the measurements with an LFM pulse of the opposite slope, e.g., an upchirp followed by a downchirp. In this case the sign of the range error will be reversed. Averaging the two measurements will give the true range, and will also allow determination of the Doppler shift.

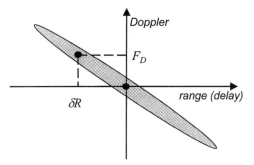

Figure 4.28 Illustration of the effect of range-Doppler coupling on apparent target range.

198 Chapter Four

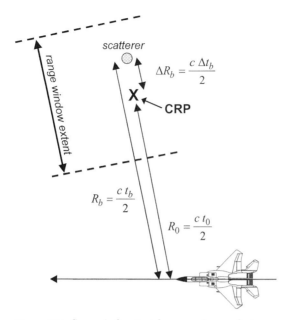

Figure 4.29 Scenario for stretch processing analysis.

4.6.5 Stretch processing

LFM waveforms are often the waveform of choice for exceptionally wideband radar systems, where the swept bandwidth β may be hundreds of megahertz or even on the order of 1 GHz. Digital processing can be difficult to implement in such systems because the high instantaneous bandwidth of the waveform requires equally high sampling rates in the A/D converter. It is difficult to obtain high quality A/D converters at these rates with wordlengths longer than perhaps eight bits with current technology. In addition, the sheer number of samples generated can be stressing for the signal processor.

Stretch processing is a specialized technique for matched filtering of wideband LFM waveforms. It is also called *deramp processing, deramp on receive, dechirp*, and *one-pass processing*. It is essentially the same as the processing used with linear *frequency-modulated continuous wave* (FMCW) radar. Stretch processing is most appropriate for applications seeking very high range resolution over relatively short range intervals (called *range windows*).

Figure 4.29 shows the scenario for analyzing stretch processing. The *central reference point* (CRP) is in the middle of the range window of interest at a range of R_0 meters, corresponding to a time delay of t_0 seconds. Consider a scatterer at range R_b and time delay t_b. The problem will be analyzed in terms of differential range or delay relative to the CRP, denoted as ΔR_b and Δt_b. The transmitted waveform is the LFM pulse of Eq. (4.82). The echo from the scatterer, with the carrier frequency included, is therefore

$$\bar{x}(t) = \zeta \exp\left[j\pi\frac{\beta}{\tau}(t-t_b)^2\right]\exp(j\Omega t) \quad t_b = t_0 + \Delta t_b \quad 0 \le t - t_b \le \tau \quad (4.105)$$

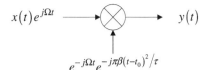

Figure 4.30 Complex equivalent receiver for stretch processor.

where ζ is proportional to the scatterer reflectivity. This echo is processed with the modified coherent receiver (in complex equivalent form) shown in Fig. 4.30. The unique aspect of this stretch receiver is the reference oscillator. It contains a conventional term $\exp(-j\Omega t)$ to remove the carrier. However, it also contains a replica of the transmitted chirp, referenced to the time delay t_0 corresponding to the CRP.

The output $y(t)$ will be, using $t_b = t_0 + \Delta t_b$

$$y(t) = \zeta \exp\left[-j2\pi \frac{\beta}{\tau}\Delta t_b(t - t_0)\right] \exp\left[j\pi \frac{\beta}{\tau}(\Delta t_b)^2\right] \quad t_0 + \Delta t_b \leq t \leq t_0 + t_b + \tau$$

(4.106)

The phase term that is quadratic in Δt_b is a complex constant. In the synthetic aperture imaging community, this is called the *residual video phase* (RVP). The other complex exponential is linear in t and therefore represents a constant-frequency complex sinusoid. By inspection, the sinusoid frequency is $F_b = -\beta \Delta t_b / \tau$ hertz. F_b is proportional to Δt_b and thus to the range of the scatterer relative to the CRP. The differential range can be obtained from the mixer output frequency as

$$\Delta R_b = \frac{-c F_b \tau}{2\beta}$$

(4.107)

Heuristically, the scatterer produces a constant frequency tone at the output of the stretch receiver because the receiver not only removes the carrier from the LFM echo, but also combines it in a mixer with a replica of the LFM with a delay corresponding to the CRP. As usual, the mixer produces sum and difference frequencies. The sum frequency is removed by a lowpass filter. (This LPF is not needed in the complex representation and is therefore not shown in Fig. 4.30, but would be required in a real implementation.) The difference frequency is the difference between the instantaneous frequency of the LFM echo and the LFM reference. Since both have the same sweep rate, this difference frequency is a constant.

If there are several scatterers distributed in range, the stretch receiver output is simply the superposition of several terms of the form of Eq. (4.106)

$$y(t) = \sum_i \zeta_i \exp\left[-j2\pi \frac{\beta}{\tau}\Delta t_{b_i}(t - t_0)\right] \exp\left[j\pi \frac{\beta}{\tau}(\Delta t_{b_i})^2\right]$$

(4.108)

Thus the output of the stretch receiver contains a different "beat frequency" tone for each scatterer. Spectral analysis of $y(t)$ can identify the ranges and amplitudes of the scatterers present in the composite echo.

It is desirable that the reference LFM chirp completely overlap the echo from a scatterer anywhere within the range window. Suppose the range window is $R_w = cT_w/2$ meters long. The leading edge of the echo from a scatterer at the nearest range, $R_0 - R_w/2$, will arrive $t_0 - T_w/2$ seconds after transmission. The trailing edge of the echo from the scatterer at the far limit of the range window, $R_0 + R_w/2$, will arrive $t_0 + T_w/2 + \tau$ seconds after transmission. Thus, data from the range window have a total duration of $T_w + \tau$ seconds. To ensure complete overlap of the reference chirp with echoes from any part of the range window, the reference chirp must be $T_w + \tau$ seconds long, and thus must also sweep over $(1 + T_w/\tau)\beta$ hertz.

The bandwidth of the receiver output can be obtained by considering the difference in beat frequencies for scatterers at the near and far edges of the range window. This gives

$$F_{b_{\text{near}}} - F_{b_{\text{far}}} = \left[-\frac{\beta}{\tau}\left(-\frac{T_w}{2}\right)\right] - \left[-\frac{\beta}{\tau}\left(+\frac{T_w}{2}\right)\right]$$
$$= \frac{T_w}{\tau}\beta \qquad (4.109)$$

If $T_w < \tau$, the bandwidth at the receiver output is less than the original signal bandwidth β. The mixer output can then be sampled with slower A/D converters, and the number of range samples needed to represent the range window data is reduced. Thus, the stretch technique is most effective for systems performing high range resolution analysis over limited range windows. Also note that while the digital processing rates have been greatly reduced, the analog receiver hardware up through the LFM mixer must be capable of handling the full instantaneous signal bandwidth.

As an example, consider a 100 µs pulse with a swept bandwidth of 750 MHz, giving a BT product of 75,000. Suppose the desired range window is $R_w = 1.5$ km, corresponding to a sampling window of $T_w = 10$ µs. In a conventional receiver, the sampling rate will be 750 Msamples per second. Data from scatterers over the extent of the range window will extend over $T_w + \tau$ seconds, requiring (750 MHz)(10 µs + 100 µs) = 82,500 samples to represent the range window. In contrast, the bandwidth at the output of the stretch demodulator will be $(T_w/\tau)\beta = 75$ MHz. The sampled time interval remains the same, so only 8250 samples are required. Restricting the analysis to a range window one-tenth the length of the pulse and using the stretch technique has resulted in a factor of 10 reduction in both the sampling rate required and the number of samples to be digitally processed.

Stretch processing of linear FM waveforms preserves both the resolution and the range-Doppler coupling properties of conventionally processed LFM. Consider the output of the stretch mixer for a scatterer at differential range Δt_b from the central reference point. This signal will be a complex sinusoid at a frequency $F_b = -\beta \Delta t_b/\tau$ hertz, observed for a duration of τ seconds. In the absence of windowing, the Fourier transform of this signal will be a sinc function with its peak at F_b and a Rayleigh resolution of $1/\tau$ hertz. The processor will be able to resolve scatterers whose beat frequencies are at least $\Delta F_b = 1/\tau$ hertz apart. The time-delay spacing that gives this frequency separation satisfies

$$\frac{1}{\tau} = \left|\frac{\beta}{\tau}\Delta t_b\right| \Rightarrow \Delta t_b = \frac{1}{\beta} \tag{4.110}$$

The corresponding range separation is then the usual result

$$\Delta R_b = \frac{c}{2}\Delta t_b = \frac{c}{2\beta} \tag{4.111}$$

To consider the effect of Doppler shift on the stretch processor, replace $x(t)$ in Eq. (4.105) with

$$\tilde{x}(t) = \varsigma_i \exp\left[j\pi\frac{\beta}{\tau}(t-t_b)^2\right] \exp(j 2\pi F_D t) \exp(j\Omega t) \quad |t-t_b| \leq \tau \tag{4.112}$$

Repeating the previous analysis, Eq. (4.106) becomes

$$y(t) = \varsigma \exp\left[-j 2\pi\left(\frac{\beta}{\tau}\Delta t_b - F_D(t-t_0)\right)\right] \exp\left[j\pi\frac{\beta}{\tau}(\Delta t_b)^2\right] \tag{4.113}$$

Equation (4.113) shows that the effect of a Doppler shift is to decrease the beat frequency F_b by F_D hertz. Since beat frequency is mapped to differential range by the stretch processor according to $\Delta R_b = -cF_b\tau/2\beta$, this implies a measured range shift of

$$\Delta R = -\frac{c\tau}{2\beta}F_D \quad \text{Hz} \tag{4.114}$$

which is the same range-Doppler coupling relationship obtained previously.

4.7 Range Side Lobe Control for FM Waveforms

It was seen in the previous section that the output of the LFM matched filter exhibits side lobes in range (equivalently, delay). These are a consequence of the approximately rectangular LFM matched filter output spectrum, which produces a sinc-like range response. The first range side lobe is, therefore, approximately 13 dB below the output peak for moderate-to-high BT products, and about -15 dB for small BT products. Side lobes this large are unacceptable in many systems that will encounter multiple targets in range due to *target masking*. This phenomenon is shown in Fig. 4.31a, where the smaller target is barely visible above the side lobes of the stronger target. The smaller target could not be reliably detected in this scenario. If the side lobes could be reduced, this masking effect could be greatly reduced, as shown in part *b* of the figure.

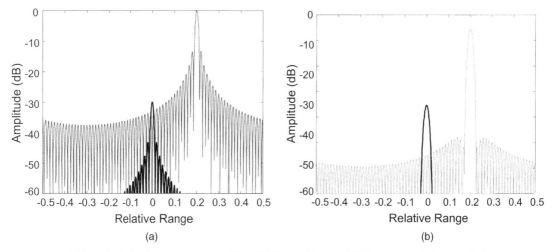

Figure 4.31 Effect of windowing on target masking. (*a*) No windowing. (*b*) Hamming window applied.

For a simple pulse, the matched filter output is a triangle function, which exhibits no side lobes. Thus, side lobe reduction is not an issue for that waveform and it will not be further discussed. For the LFM pulse, there are two basic approaches to delay side lobe reduction: shaping the receiver frequency response, and shaping the waveform spectrum.

4.7.1 Matched filter frequency response shaping

Consider receiver frequency response shaping first. Recall from *finite impulse response* (FIR) digital filter design that, to reduce side lobes of the frequency response of a digital filter, a window function is applied in the time domain to the impulse response. The goal here is to reduce side lobes in range, corresponding to the time domain, so one approach is to window the receiver frequency response in the frequency domain.

The matched filter frequency response is $H(F) = X^*(F)$, and at least for larger BT products, is therefore approximately rectangular. A modified frequency response $H'(F)$ can be obtained by multiplying $H(F)$ by a window function $w(F)$[†]

$$H'(F) = w(F)H(F) = w(F)X^*(F) \qquad (4.115)$$

Figure 4.32*a* shows a Hamming window function overlaid on the matched filter frequency response for an LFM waveform with $\beta\tau = 100$. $H'(F)$ is the product of these two functions. The resulting impulse response $h'(t)$ is shown in Fig. 4.32*b*.

[†]The lower case w is used for the frequency-domain window function $w(F)$ to emphasize that the multiplying function is the window function itself (e.g., a Hamming window) rather than its Fourier transform.

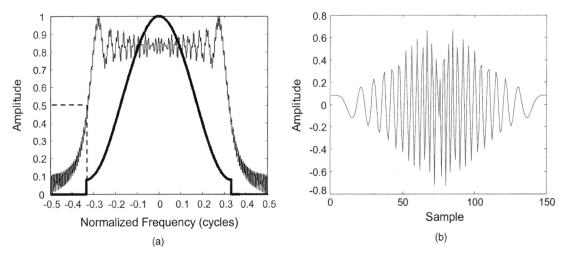

Figure 4.32 Hamming weighting of the LFM receiver frequency response. (a) Hamming window overlaid on matched filter frequency response. (b) Resulting filter impulse response $h'(t)$.

Since the matched frequency response does not have a perfectly sharp cutoff frequency, there is some uncertainty as to where in frequency to place the window cutoff. In the example, the support of the window equals the instantaneous frequency cutoff, which is 0.36 cycles on the normalized frequency scale for this particular LFM waveform. However, this choice cuts off some of the waveform energy in the side lobes, increasing the mismatched filtering losses. A case could be made for a narrower support, so that the window is applied only over the relatively flat portion of the spectrum. This would provide range side lobes more closely matching those expected for the chosen window, but would reduce the effective bandwidth, further degrading the range resolution. A case could also be made for increasing the support to maximize the output energy, but this choice might increase the range side lobes by "wasting" some of the window shape on the skirts of the LFM spectrum.

Figure 4.32b shows the matched filter impulse response for this example. Note that it suggests a windowed version of the matched filter impulse response, which would be just a conjugated, time-reversed duplicate of the transmitted LFM waveform. However, closer inspection (and comparison to Fig. 4.22) reveals differences in detail. It is important to realize that the window is applied to the matched filter frequency response, not to its impulse response.

Finally, since $H'(F) \neq X^*(F)$, the modified receiver is not matched to the transmitted LFM, and therefore the output peak and SNR will be reduced from their maximum values. This effect is evident in Fig. 4.31b, where the peak of the dominant target response is several dB lower than the unwindowed case in part a of the figure. The losses in output peak amplitude and SNR can be estimated from the window function $w(F)$. In practice, a discrete window $w[k]$ will be applied to a discrete-frequency version of $H(F)$, $H[k]$.

The loss in the peak signal output from the matched filter, denoted as LPG, is approximately

$$LPG \approx \frac{1}{K^2} \left| \sum_{k=0}^{K-1} w[k] \right|^2 \qquad (4.116)$$

where K is the window length. The loss in SNR at the matched filter output is called the *processing loss* and is approximately

$$PL \approx \frac{\left| \sum_{k=0}^{K-1} w[k] \right|^2}{K \sum_{k=0}^{K-1} |w[k]|^2} \qquad (4.117)$$

For a Hamming window, LPG is approximately 4 dB while PL is approximately 1.4 dB. Both are weak functions of K and are slightly larger for small K. These formulas are approximate due to the imperfect cutoff of the LFM spectrum and the variability in the cutoff of the window in frequency. A derivation of these formulas is deferred to Chap. 5, where they will arise again in the context of Doppler processing and where the results will be exact.

4.7.2 Waveform spectrum shaping

The principal limitation of the receiver weighting approach to range side lobe control is that the resulting filter is not matched to the transmitted waveform. An alternative approach is to design a modified pulse compression waveform whose matched filter output inherently has lower side lobes than the standard LFM. The waveform should be designed to have a spectrum shaped like that of a window function with the desired side lobe behavior. Such a waveform would combine the maximized SNR of a truly matched filter with low side lobes. There are two common ways to shape the spectrum. Both start with the idea that the LFM spectrum's relatively square shape is the result of a linear sweep rate combined with a constant pulse amplitude, resulting in a fairly uniform distribution of the signal energy across the spectral bandwidth. The spectral energy could be reduced at the edges, giving a "window-shaped" spectrum, by reducing the signal amplitude at the pulse edges while maintaining a constant sweep rate, by using a faster sweep rate at the edges with a constant pulse amplitude so as to spend less time in each spectral interval near the band edges, or both. The technique using variable sweep rates is referred to as *nonlinear FM* (NLFM).

The amplitude modulation technique implies operating the power amplifier at less than full power over the pulse length. This requires more complicated transmitter control but, more importantly, results in a pulse with less than the maximum possible energy for the given pulse length. This technique is not discussed further in this book; see the book by Levanon and Mozeson (2004) for more information.

Two methods that have been proposed for NLFM waveform design are the stationary phase method, and empirical techniques. The principle of stationary phase technique is used to design a phase function from a prototype spectral amplitude function; the instantaneous frequency function is obtained from the phase. Both are discussed in the book by Levanon and Mozeson (2004). Examples of using this technique for deriving NLFM waveforms from common window functions such as Hamming or Taylor functions are described in the chapter by Keel (2005). One empirically-developed design gives the instantaneous frequency function as (Prince, 1979)

$$F_i(t) = \frac{t}{\tau}\left(\beta_L + \beta_C \frac{1}{\sqrt{1 - 4t^2/\tau^2}}\right) \qquad |t| \leq \frac{\tau}{2} \qquad (4.118)$$

The term $\beta_L t/\tau$ represents a linear FM component, while the term involving β_C is designed to approximate a Chebyshev-shaped (constant side lobe level) spectrum. Since $F_i(t) = (1/2\pi)(d\theta(t)/dt)$, integrating and scaling this instantaneous frequency function gives the required phase modulation

$$\theta(t) = \frac{\pi\beta_L}{\tau}t^2 - \frac{\pi\beta_C\tau}{2}\sqrt{1 - 4t^2/\tau^2} \qquad |t| \leq \frac{\tau}{2} \qquad (4.119)$$

Figure 4.33 illustrates the behavior of this nonlinear FM for the case where $\beta_L\tau = 50$ and $\beta_C\tau = 20$, and the waveform is sampled at 10 times the bandwidth of the linear term (that is, $T_s = 1/10\beta_L$). The instantaneous frequency (part a of the figure) is nearly linear in the center of the pulse, but sweeps much more rapidly near the pulse edges; this reduces the spectral density at the pulse edge, resulting in the spectrum shown in part c, which has a window-like tapered shape instead of the usual nearly-square LFM spectrum. The resulting matched filter output, shown in part d, has most of its side lobes between -48 and -51 dB, with the first side lobe at -29 dB. In contrast, Fig. 4.34 illustrates the spectrum of the same waveform with β_C set to zero. This results in a linear FM waveform with the usual nearly-square spectrum. These two figures are on the same normalized frequency scale. Comparing the spectra of these two waveforms illustrates how the nonlinear term has spread and tapered the LFM spectrum to lower the matched filter side lobes. The LFM matched filter output has a peak side lobe of -13.5 dB, decaying approximately as $1/F$ at higher frequencies. The Rayleigh resolution in time of the NLFM waveform is approximately $0.8/\beta_L$, less than the $1/\beta_L$ value observed for the LFM case but greater than $1/(\beta_L + \beta_C)$.

In addition to the more difficult phase control required, the major drawback of nonlinear FM pulses is their Doppler intolerance. Figure 4.35 shows the matched filter output for the same waveform as shown earlier when a Doppler mismatch of $7/\tau$ hertz is present. While the general side lobe level remains

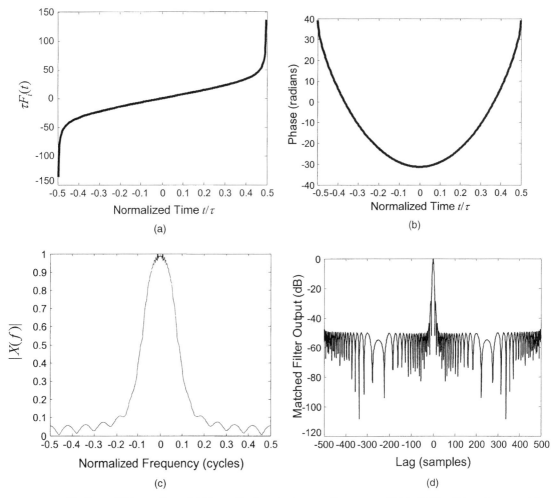

Figure 4.33 Nonlinear FM waveform. (a) Normalized instantaneous frequency $\tau F_i(t)$. (b) Resulting phase modulation function. (c) Magnitude of Fourier spectrum. (d) Magnitude of matched filter output.

largely unchanged, the main lobe is seriously degraded, exhibiting both range-Doppler coupling (a shift of the peak) and severe spreading and ambiguity caused by very high near-in side lobes. The major advantage of NLFM over linear FM with receiver weighting is that the receiver filter for the NLFM waveform is a matched filter, so that there is no reduction of the matched filter output peak.

4.8 The Stepped Frequency Waveform

The LFM waveform increases resolution well beyond that of a simple pulse by sweeping the instantaneous frequency over the desired range β within the

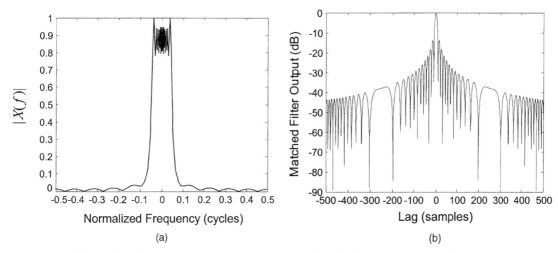

Figure 4.34 FM waveform having same linear component as that of Fig. 4.33, but no nonlinear component. (a) Magnitude of Fourier spectrum. (b) Magnitude of matched filter output.

pulse. This technique is very effective and very common, but does have some drawbacks in some systems, particularly those using very large bandwidths, on the order of hundreds of megahertz or more. First, the transmitter hardware must be capable of generating the LFM sweep. Second, all of the analog components must be able to support an instantaneous bandwidth of β hertz without introducing distortion. Even if stretch processing is used, the same is true of

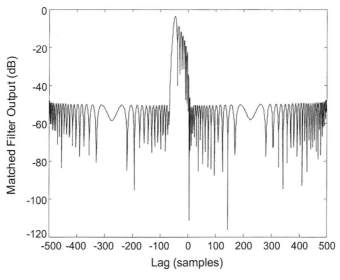

Figure 4.35 Output of NLFM matched filter when $F_D = 7/\tau$ Hz.

the receiver components up to and including the dechirp mixer and reference oscillator.

A second issue arises in systems using phase-steered array antennas. Recall from Chap. 1 that the antenna pattern of a phased array antenna is determined primarily by the array factor

$$E(\theta) = E_0 \sum_{n=0}^{N-1} a_n e^{j(2\pi/\lambda)nd \sin\theta} \qquad (4.120)$$

where d is the element spacing and the $\{a_n\}$ are the complex weights on each subarray output. The antenna is steered to a particular look direction θ_0 by setting the steering weights a_n according to

$$a_n = |a_n|e^{-j(2\pi/\lambda)nd \sin\theta_0} \qquad (4.121)$$

The magnitudes of the weights are chosen to provide the desired side lobe level. $E(\theta)$ will exhibit a peak at $\theta = \theta_0$; for example, if $|a_n| \equiv 1$, $E(\theta)$ will be an asinc function with its peak at θ_0. Note that the phases of the required weights $\{a_n\}$ are a function of the wavelength λ. However, if an LFM pulse is transmitted, the effective wavelength changes during the pulse sweep. If the system is wideband, this wavelength change will be significant and the value of θ at which $E(\theta)$ peaks will change as well. That is, the antenna look direction will actually change during the LFM sweep. This undesired frequency steering effect is an additional source of SNR loss.

Stepped frequency waveforms are an alternative technique for obtaining a large bandwidth, and thus fine range resolution without requiring intrapulse frequency modulation. A stepped frequency waveform is a pulse burst waveform. Each pulse in the burst is a simple, constant-frequency pulse; however, the RF frequency is changed during each pulse. The most common stepped frequency waveform employs a linear frequency stepping pattern, where the RF frequency of each pulse is increased by ΔF hertz from the preceding pulse. Factoring out the starting RF frequency gives the following baseband waveform

$$x(t) = \sum_{m=0}^{M-1} x_p(t - mT)e^{j2\pi m\Delta F(t-mT)} \qquad (4.122)$$

Figure 4.36 illustrates the linearly stepped frequency waveform.

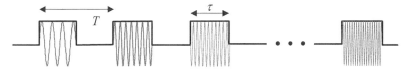

Figure 4.36 Linearly frequency-stepped waveform.

Because only simple pulses are used for each constituent pulse, the instantaneous bandwidth capability of the transmitter and receiver need be only on the order of $1/\tau$ hertz. The total bandwidth of the waveform as a whole is $M\Delta F$. When used with a phased array antenna, the time between pulses can be used to reset the phase shifters to update the $\{a_n\}$ sequence to maintain a constant steering direction θ_0 as the effective wavelength changes from pulse to pulse. The major disadvantages of this waveform are that it requires a pulse-to-pulse tunable RF frequency, along with corresponding tunable receiver center frequency, and that M PRIs are required to collect data over the desired bandwidth instead of just one.

The pulse-by-pulse processing viewpoint applied to the constant-frequency pulse burst waveform can be applied again to analyze the matched filter response for the stepped frequency waveform. Suppose the radar is stationary, and a stationary target is located at a range corresponding to a delay $t_l + \delta t$, where δt represents an incremental delay around the nominal t_l. Individual pulses are processed through the simple pulse matched filter as before, producing the output waveforms (assuming $T_M = 0$)

$$y_m(t) = s_p^*[t - (t_l + \delta t) - mT]e^{j2\pi m \Delta F [t - (t_l + \delta t) - mT]} \qquad (4.123)$$

This output is then sampled at $t = t_l - mT$ (that is, t_l seconds after the current pulse was transmitted), corresponding to range $R_l = ct_l/2$. The resulting sample becomes the lth range bin sample for the current pulse

$$\begin{aligned} y[l, m] &= y_m(t_l - mT) \\ &= s_p^*(\delta t) e^{j2\pi m \Delta F \delta t} \end{aligned} \qquad (4.124)$$

Equation (4.124) shows that the slow-time sequence at a fixed range bin l, when using a linearly stepped frequency waveform, is a discrete time sinusoid. The frequency is proportional to the displacement of the scatterer from the nominal range bin location of R_l meters. The amplitude of the sequence is weighted by the triangular simple pulse matched filter response $s_p(t)$, which has a duration of $\pm\tau$ around $t = t_l$.

Following the earlier discussion of pulse-by-pulse processing for the conventional pulse burst waveform, the slow-time matched filter impulse response for a target located at the nominal delay $t_l + \delta t$ is $h[m] = \exp(-j2\pi m\Delta F \delta t)$. Thus, the matched filter impulse response is different for every value of δt. Consider a discrete-time Fourier transform of the slow-time data

$$\begin{aligned} Y[l, \omega] &= \sum_{m=0}^{M-1} y[l, m]e^{-j\omega m} = \sum_{m=0}^{M-1} s_p^*(\delta t) e^{j2\pi m \Delta F \delta t} e^{-j\omega m} \\ &= s_p^*(\delta t) \sum_{m=0}^{M-1} e^{-j(\omega - 2\pi \Delta F \delta t)m} \end{aligned} \qquad (4.125)$$

The summation will yield an asinc function having its peak at $\omega = 2\pi\Delta F\,\delta t$. Thus, the peak of the DTFT of the slow-time data in a fixed range bin with a linearly stepped frequency waveform provides a measure of the delay of the scatterer relative to the nominal delay t_l. Specifically, if the peak of the DTFT is at $\omega = \omega_p$, the scatterer is at a differential delay

$$\delta t = \frac{\omega_p}{2\pi\Delta F} \tag{4.126}$$

Note also that the DTFT evaluated at $2\pi\Delta F\,\delta t$ is the matched filter for the slow-time sequence, so that the data samples are integrated in phase

$$Y[l,\omega] = \sum_{m=0}^{M-1} s_p^*(\delta t) e^{j2\pi m\Delta F\,\delta t} e^{-j\omega m}$$

$$= s_p^*(\delta t)\sum_{m=0}^{M-1}(1) = M s_p^*(\delta t) \tag{4.127}$$

However, the ambiguity function of the individual pulses reduces the amplitude of the slow-time samples by $|s_p(\delta t)|$.

Since the DTFT gives the matched filter output peak at an appropriate value of ω, it follows that applying a K-point DFT to the slow-time sequence implements K filters, each matched to a different differential range δt. Thus, the DFT of the slow-time data within a single range bin for a stepped frequency waveform is a map of echo amplitude versus differential range.

The DTFT of an M-point sinusoid has a Rayleigh frequency resolution of $\Delta f = 1/M$ cycles. The corresponding time resolution is $\delta t = 1/M\Delta F$; the range resolution is therefore

$$\Delta R = \frac{c}{2M\Delta F} = \frac{c}{2\beta} \tag{4.128}$$

where β is the total stepped bandwidth $M\Delta F$. Thus, the linearly stepped frequency waveform provides the same resolution as a single pulse of bandwidth β. If a K-point DFT is used to process the slow-time data, the DFT output will provide range measurements at intervals of

$$\Delta R = \frac{c}{2K\Delta F} = \frac{M}{K}\Delta R \tag{4.129}$$

Since $K \geq M$ normally, the DFT output provides echo amplitude samples at intervals equal to or less than the range resolution. This high-resolution reflectivity map is often called a *range profile*.

The total bandwidth β of the stepped frequency waveform is determined by the desired range resolution. It can be realized by various combinations of the number of frequency steps M and the step size ΔF. To determine how to choose these parameters, note that the DTFT of the slow-time data is periodic in ω with period 2π radians. Because the DTFT peak is at $\omega = 2\pi\Delta F\,\delta t$, the response

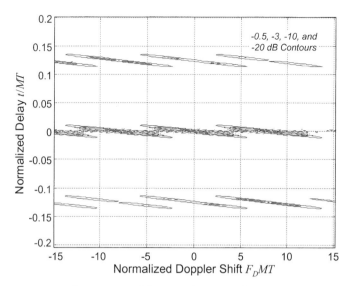

Figure 4.37 Contour plot of the central portion of the ambiguity function of a pulse burst waveform. $M = 8$, $T = 10\tau$, and $\Delta F = 0.8/\tau$.

is periodic in δt with period $1/\Delta F$. This periodicity establishes the required range bin spacing; M is then chosen to span the bandwidth required to provide the desired range resolution. The DFT range profile effectively breaks each relatively large range bin ($c/2\Delta F$ meters) into K high-resolution range bins ($c/2\beta$ meters).

Details of the Doppler response and ambiguity function of the linearly stepped frequency waveform are available in the book by Levanon and Mozeson (2004). A small central portion of the ambiguity function is shown in Fig. 4.37 for the case $M = 8$ pulses, PRI $T = 10\tau$, and a frequency step size of $\Delta F = 0.8/\tau$. The AF displays both the skewed response typical of a linear FM modulation, and the range and Doppler ambiguities typical of pulse burst waveforms. Ambiguities in delay (range) are evident at intervals of T seconds, corresponding to $1/8 = 0.125$ on the normalized scale of the figure. The width of the main response along the zero Doppler axis, though too small to see clearly in Fig. 4.37, is in fact 2τ seconds, corresponding to $\pm 1/80 = \pm 0.0125$ on the scale shown. The first zero in Doppler of the main ridge occurs at $1/MT$ hertz, corresponding to 1 on the normalized Doppler scale.

4.9 Phase-Modulated Pulse Compression Waveforms

The second major class of pulse compression waveforms are referred to as *phase coded* waveforms. A phase coded waveform has a constant RF frequency, but an absolute phase that is switched between one of N fixed values at regular intervals within the pulse length. Such a pulse can be modeled as a collection of N contiguous subpulses $x_n(t)$ of duration τ_c, each with the same frequency but a (possibly) different phase

$$x(t) = \sum_{n=0}^{N-1} x_n(t - n\tau_c)$$

$$x_n(t) = \begin{cases} \exp(j\phi_n) & 0 \leq t \leq \tau_c \\ 0 & \text{elsewhere} \end{cases} \quad (4.130)$$

Clearly, the total pulse length is $\tau = N\tau_c$. Individual subpulses are often referred to as *chips*. Phase coded waveforms are divided into *biphase codes* and *polyphase* codes. A biphase code has only two possible choices for the phase state ϕ_n, typically 0 and π; a polyphase code has more than two phase states. There are several common subcategories of each. Figure 4.1c was an example of a biphase-coded waveform.

4.9.1 Biphase codes

The most important biphase codes in radar are the *Barker codes*. Barker codes are a specific set of biphase sequences that attain an $N:1$ ratio of the peak to the highest side lobe at the output of the matched filter. A low-frequency Barker coded waveform for $N = 13$ is shown in Fig. 4.38. The phase switches are visible at $t = 5\tau_c, 7\tau_c, 9\tau_c, 10\tau_c, 11\tau_c$, and $12\tau_c$. Because there are only two phase states, the waveform is often represented by a diagram such as the one shown in Fig. 4.39, using either "+" and "−" symbols as shown, or +1 and −1 symbols. Note that biphase codes do not necessarily change phase state at every subpulse transition.

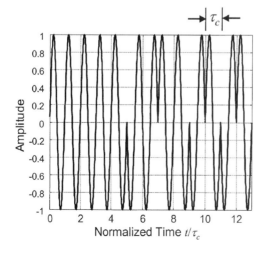

Figure 4.38 Barker coded waveform, $N = 13$.

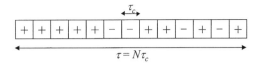

Figure 4.39 Binary sequence describing the Barker code of Fig. 4.38.

Recall that pulse compression waveforms have a bandwidth $\beta \gg 1/\tau$. Because phase coded waveforms are constant frequency, it may not be obvious that their spectrum is spread. However, the discontinuities caused by the phase transitions do spread the signal spectrum. As an example, Fig. 4.40 shows the effect of a single phase switch of 180° on the spectrum of a constant-frequency waveform. While the effect depends on the point in the pulse at which the switch occurs, clearly it significantly spreads the signal energy in frequency. Multiple phase transitions increase this effect: Fig. 4.41 compares the spectra of the 13-bit Barker coded waveform with that of a simple pulse of the same duration. The main lobe of the Barker waveform is about 13 times as wide as that of the simple pulse. In addition, the side lobes of the Barker waveform spectrum decay much more slowly than those of the simple pulse.

One of the major disadvantages of Barker codes is that there are not very many of them. Barker codes have been found only for N up to 13. All of the Barker codes are listed in Table 4.1; more than one code exists for some lengths, while none exist for others below $N = 13$. The table also lists the peak side lobe, which is simply $20 \log_{10}(1/N)$.

The matched filter output for a phase coded pulse is derived in detail by Levanon and Mozeson (2004); the result is now summarized. Denote the sequence of amplitudes of the individual pulse chips $x_n(t)$ of Eq. (4.130) as $\{A_n\} = \{\exp[j\phi_n]\}$, and express the time variable t in terms of the chip duration

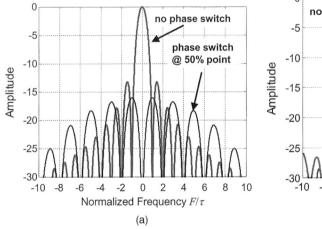
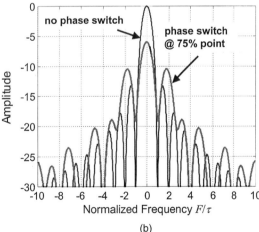

Figure 4.40 Effect of a single 180° phase switch on the spectrum of a constant-frequency pulse. (a) Phase switch occurs at $t = \tau/2$. (b) Phase switch occurs at $t = 3\tau/4$.

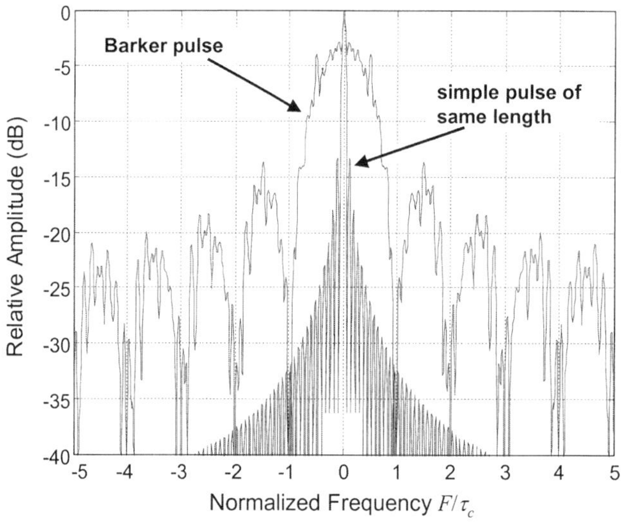

Figure 4.41 Spectra of a 13-bit Barker coded pulse and a simple pulse of the same length.

τ_c and an offset η, as $t = k\tau_c + \eta, 0 \le \eta < \tau_c$. The matched filter output, which is just the autocorrelation of $x(t)$, is

$$y(t) = s_x(t) = y(k\tau_c + \eta) = \left(1 - \frac{\eta}{\tau_c}\right) s_A[k] + \frac{\eta}{\tau_c} s_A[k+1] \tag{4.131}$$

where $s_A[k]$ is the discrete autocorrelation of the complex amplitude sequence $\{A_n\}$. Equation (4.131) describes a function that takes on the value $s_A[k]$ at $t = k\tau_c$, and is linearly interpolated (in the complex plane) between adjacent samples $s_A[k]$ and $s_A[k-1]$. Thus, the matched filter output can be determined by computing the autocorrelation of the amplitude sequence and interpolating between those values.

TABLE 4.1 Barker Codes

	Code sequence		
N	+/− Format	Octal	PSL, dB
2	+ −	2	−6.0
2	+ +	3	−6.0
3	+ + −	6	−9.5
3	+ − +	5	−9.5
4	+ + − +	15	−12.0
4	+ + + −	16	−12.0
5	+ + + − +	35	−14.0
7	+ + + − − + −	162	−16.9
11	+ + + − − − + − − + −	3422	−20.8
13	+ + + + + − − + + − + − +	17465	−22.3

As an example, consider the Barker code with $N = 13$. Representing the code sequence of Table 4.1 as the sequence $\{A_n\} = \{1,1,1,1,1,0,0,1,1,0,1,0,1\}$ gives the autocorrelation sequence $\{1,0,1,0,1,0,1,0,1,0,1,0,13,0,1,0,1,0,1,0,1,0,1,0,1\}$. Figure 4.42 illustrates the resulting autocorrelation function obtained by interpolating between the discrete autocorrelation samples. All phase codes have a peak autocorrelation value of N at time $t = 0$, where N is the code length. Barker codes have the additional property that all side lobes have a value of 1, and further that the discrete autocorrelation sequence side lobes always follow an alternating pattern of zeros and ones. The side lobe level relative to the peak is therefore $-20 \log_{10} N$, and the Rayleigh resolution is τ_c seconds in time or $c\tau_c/2$ meters in range.

Barker codes have two major disadvantages. The first, as noted earlier, is that no Barker codes of length longer than $N = 13$ are known; this limits the degree of side lobe suppression possible. The second is that they are very Doppler intolerant. A Doppler phase rotation of 360° across the full pulse is more than sufficient to completely break up the structure of the matched filter output, as seen in the 13-bit Barker ambiguity function contour plot of Fig. 4.43. As a consequence, it is common to design Barker coded waveforms to limit the Doppler phase rotation to one-quarter cycle or less, which requires that the maximum expected Doppler shift and target velocity satisfy

$$F_{D_{\max}} \tau < \frac{1}{4} \Rightarrow v_{\max} < \frac{\lambda}{8\tau} \tag{4.132}$$

This constraint limits the Doppler mismatch loss to 1 dB or less.

The limited number and length of Barker codes has led to various techniques for constructing longer biphase codes with good side lobe properties. *Combined* or *nested* Barker codes form a longer code as the Kronecker product of two

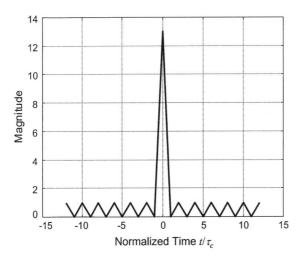

Figure 4.42 Matched filter output for a 13-bit Barker code.

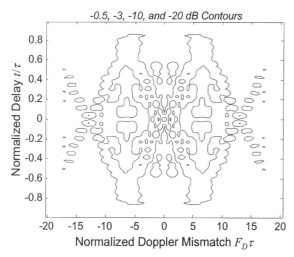

Figure 4.43 Contour plot of ambiguity function of a 13-bit Barker code.

shorter Barker codes. If an N-bit Barker code sequence is denoted as B_N, an MN-bit code can be constructed as $B_N \otimes B_M$. The Kronecker product is simply the B_M code repeated N times, with each repetition multiplied by the corresponding element of the B_N code. For example, a 20-bit code can be constructed as the product $B_4 \otimes B_5$

$$B_4 \otimes B_5 = \{1\ 1\ 1\ -1\} \otimes \{1\ 1\ 1\ -1\ 1\}$$
$$= (1)\{1\ 1\ 1\ -1\ 1\} + (1)\{1\ 1\ 1\ -1\ 1\} + (1)\{1\ 1\ 1\ -1\ 1\} + \cdots$$
$$\cdots + (-1)\{1\ 1\ 1\ -1\ 1\} + (1)\{1\ 1\ 1\ -1\ 1\}$$
$$= \{1\ 1\ 1\ -1\ 1\ 1\ 1\ 1\ -1\ 1\ 1\ 1\ 1\ -1\ 1\ -1\ -1\ -1\ -1\ 1\ 1\ -1\ 1\ 1\ 1\ -1\ 1\}$$

(4.133)

These codes have a peak side lobe higher than 1. For example, the autocorrelation of the code of Eq. (4.133) is shown in Fig. 4.44. Notice that the magnitude of the peak side lobes is 5, so that the peak side lobe level compared to the autocorrelation peak is only 1/4 instead of the 1/20 that would be obtained if a 20-bit Barker code existed.

Another technique uses pseudorandom noise sequences to generate much longer biphase codes. Pseudorandom sequences have length $N = 2^P - 1$ for some P and generally exhibit range side lobes on the order of $-10\log_{10}(N)$. For example, the matched filter output for a typical $N = 1023\,(P = 10)$ code, shown in Fig. 4.45, has peak side lobes just above -30 dB.

While the Barker codes are the only biphase codes with a peak side lobe value of 1 that are known to exist, one can seek longer codes with *minimum peak side lobe* (MPS) levels for the length of interest. These *MPS codes* are found by

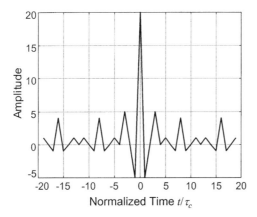

Figure 4.44 Autocorrelation of combined $B_4 \otimes B_5$ code.

exhaustive search techniques, taking advantage of certain properties of biphase code autocorrelations to prune the search somewhat. Cohen et al. (1990) give MPS codes up to $N = 48$; Coxson and Russo (2004) extend the list to $N = 70$. The peak side lobe for MPS codes of length $N \leq 28$ is 2, for $29 \leq N \leq 48$ and $N = 51$ it is 3, and for $N = 50$ and $51 \leq N \leq 69$ it is 4. Table 4.2 lists one sample code for the longest code length in each of these side lobe level regimes; sample codes for other lengths are available in the references. It is evident that the side lobe level in dB improves only very slowly as the code length increases.

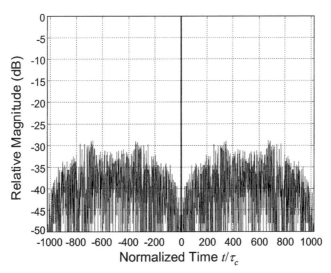

Figure 4.45 Matched filter output for a 1023-bit pseudorandom biphase code.

TABLE 4.2 Sample Minimum Peak Side Lobe Biphase Codes

Code length N	Sample code (hexadecimal)	Peak side lobe level	PSL, dB
13 (Barker)	1F35	1	−22.3
28	8F1112D	2	−22.9
51	0E3F88C89524B	3	−24.6
70	1A1133B4E3093EDD57E	4	−24.9

4.9.2 Polyphase codes

Biphase codes, as noted previously, have poor Doppler tolerance. They also suffer from precompression bandlimiting effects. As was shown in Fig. 4.41, the spectrum of a typical biphase code not only exhibits the desired main lobe spreading, but also a very slow falloff of the far side lobes. This is a direct consequence of the large phase discontinuities. Practical receivers will have a noise-limiting bandpass filter that will bandlimit the biphase waveform spectrum, smoothing the phase transitions. This has the effect of mismatching the received waveform relative to the correlator, reducing the peak gain and widening the main lobe.

Polyphase codes allow arbitrary values for the chip phases ϕ_n. Compared to biphase codes, they can exhibit lower side lobe levels and greater Doppler tolerance. A number of polyphase codes are in common use. These include the Frank codes, and the P1, P2, P3, P4, and P(n, k) codes. All are related to LFM or NLFM waveforms. Numerous other polyphase codes have been proposed; a number of these are described in the book by Levanon and Mozeson (2004).

Frank codes are codes whose length is a square, $N = M^2$ for some M. The phase sequence for a Frank code is given by

$$\phi_n = \phi(Mp + q) = \frac{2\pi}{M} pq \quad p = 0, 1, 2, \ldots, M-1 \quad q = 0, 1, 2, \ldots, M-1 \tag{4.134}$$

As an example, if $M = 4$ so $N = 16$, the sequence of phases becomes

$$\phi_n = \left\{ \underbrace{0\ 0\ 0\ 0}_{p=0}\ \underbrace{0\ \frac{\pi}{2}\ \pi\ \frac{3\pi}{2}}_{p=1}\ \underbrace{0\ \pi\ 0\ \pi}_{p=2}\ \underbrace{0\ \frac{3\pi}{2}\ \pi\ \frac{\pi}{2}}_{p=3} \right\} \tag{4.135}$$

Figure 4.46 shows the matched filter output for the case $N = 16$. Note that while the main lobe has a local minimum at $t = \tau_c$, it does not go to zero at that point as the Barker code autocorrelations do. However, the largest side lobe is 1, similar to the Barker codes. Figure 4.47 shows the ambiguity function in contour plot form. Note that the main ridge is skewed in the delay-Doppler plane, similar to the range-Doppler coupling of an LFM ambiguity function.

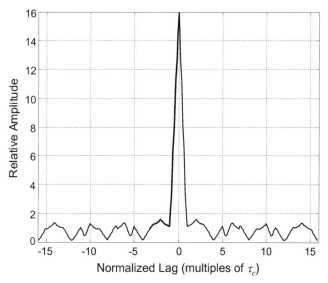

Figure 4.46 Matched filter output for a 16-bit Frank code.

The *P3 and P4 codes* of length N are given, respectively, by

$$\text{P3:} \quad \phi_n = \begin{cases} \dfrac{\pi}{N} n^2 & n = 0, 2, \ldots, N-1 \quad N \text{ odd} \\ \dfrac{\pi}{N} n(n+1) & n = 0, 1, 2, \ldots, N-1 \quad N \text{ even} \end{cases} \quad (4.136)$$

$$\text{P4:} \quad \phi_n = \dfrac{\pi}{N} n^2 - \pi n \quad n = 0, 1, 2, \ldots, N-1$$

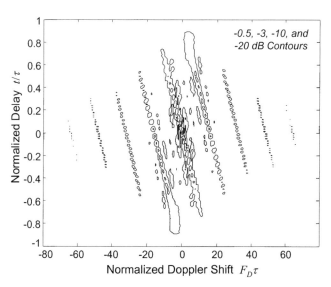

Figure 4.47 Contour plot of 16-bit Frank code ambiguity function.

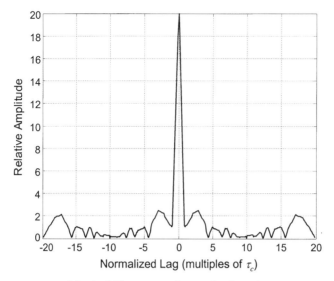

Figure 4.48 Matched filter output for a 20-bit P3 code.

Unlike the Frank code, these codes can be generated for any length N. Figure 4.48 shows the matched filter output for the $N = 20$ P3 code, while Fig. 4.49 shows the corresponding ambiguity function. Again, range-Doppler coupling is evident.

The Frank, P3, and P4 codes all are based on quadratic phase progressions, as is evident from Eqs. (4.134) and (4.136), and are therefore related to LFM

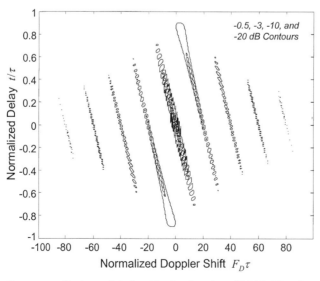

Figure 4.49 Contour plot of ambiguity function of 20-bit P3 code.

waveforms. Figure 4.50 shows the (unwrapped) phase progression of these three codes for the case $N = 16$. The P3 and P4 codes are truly quadratic, the difference being whether the minimum phase "slope" occurs at the beginning (P3) or the middle (P4) of the waveform. The smallest phase increments, and thus the minimum discontinuities in the actual RF waveform, occur where the phase slope is least. The Frank code uses a piecewise linear approximation to a quadratic phase progression. The phase increment is constant for M bits at a time, and then increases for the next M bits. This can be viewed as a phase code approximation to a stepped-frequency waveform having M steps and M bits per step (Lewis and Kretschmer, 1986).

Bandlimiting of the phase-coded waveform prior to matched filtering results in an increase in main lobe width, but a decrease in peak side lobe level, in codes that have the smallest phase increments in the middle of the codes (Lewis and Kretschmer, 1986; Levanon and Mozeson, 2004). Codes with the largest phase increments near the end exhibit the opposite behavior. Thus, of the three codes shown, the P4 will show the greatest tolerance to precompression bandlimiting in the sense of maintaining or improving its side lobe level at the matched filter output.

Just as phase codes can be designed based on linear frequency modulation waveforms, they can also be designed based on nonlinear frequency modulation waveforms. A class of codes based on NLFM waveforms, designed using the principle of stationary phase technique mentioned earlier, is given in the paper by Felhauer (1994). No closed form expression is known for these $P(n, k)$ *codes*; they must be found numerically. Typical results are very similar to those for the empirical NLFM waveforms described earlier. The effect of Doppler mismatch is similar to that observed in Fig. 4.35. This is an improvement over conventional polyphase codes, which are prone to exhibiting significantly increased side lobes near the ends of the code and, in many cases, large spurious peaks well above the general side lobe level. $P(n, k)$ codes also exhibit better tolerance

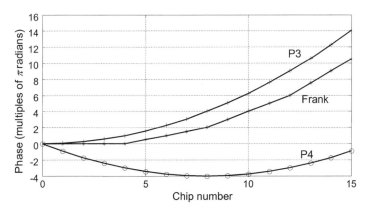

Figure 4.50 Unwrapped phase sequences of 16-bit Frank, P3, and P4 codes.

to precompression bandlimiting than do codes based on linear FM, since their spectra are already shaped by the basic NLFM design approach. Their chief disadvantage is the difficulty of their design.

4.10 Costas Frequency Codes

Costas waveforms are a class of pulse compression waveforms having aspects of both phase-coded and stepped frequency pulse burst waveforms (Costas, 1984). A Costas waveform is similar to a polyphase waveform in that it is a single pulse waveform, divided into N subpulses. It is similar to the linearly stepped frequency waveform in that, rather than maintaining a constant frequency and altering the phase of each subpulse, it alters the subpulse frequencies, stepping through a set of N frequencies that differ by ΔF hertz. Unlike the stepped frequency pulse burst, however, the Costas waveform does not step through the frequencies in linear order. The Costas pulse can be expressed as

$$x(t) = \sum_{n=0}^{N-1} x_n(t - n\tau_c)$$

$$x_n(t) = \begin{cases} \exp[jc[n](\Delta F)t] & 0 \leq t \leq \tau_c \\ 0 & \text{elsewhere} \end{cases} \quad (4.137)$$

where the sequence $c[n]$ denotes the ordering of the stepped frequencies.

Figure 4.51 shows the frequency sequence for a typical low-order Costas waveform. With proper design of the frequency step sequence, the Costas waveform can be designed to have a more thumbtack-like ambiguity function than the linearly stepped waveform. Figure 4.52 illustrates the ambiguity function of a Costas waveform with $N = 15$; the frequency step sequence was $c[n] = \{1,7,8,11,3,13,9,14,12,6,5,2,10,0,4\}$. Note the generally low and relatively uniform side lobe structure throughout the delay-Doppler plane. The construction

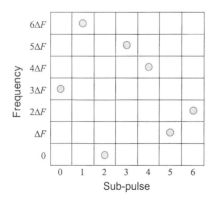

Figure 4.51 Frequency sequence for Costas waveform with $N = 7$.

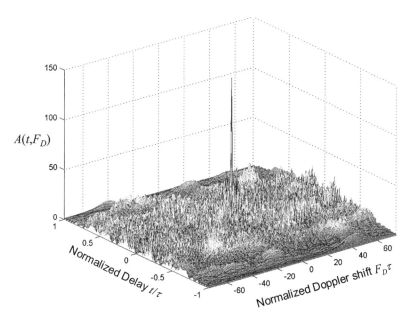

Figure 4.52 Ambiguity function for a Costas waveform with $N = 15$.

and properties of Costas waveform are discussed, and more examples given, in the book by Levanon and Mozeson (2004).

References

Born, M., and E. Wolf, *Principles of Optics*, Pergamon Press, London, 1959.

Cohen, M. N., M. R. Fox, and J. M. Baden, "Minimum Peak Side lobe Pulse Compression Codes," *Proceedings of the IEEE International Radar Conference*, pp. 633–638, Washington, DC, 1990.

Cook, C. E., and M. Bernfeld, *Radar Signals: An Introduction to Theory and Application*. Artech House, London, 1993.

Costas, J. P., "A Study of a Class of Detection Waveforms Having Nearly Ideal Range-Doppler Ambiguity Properties," *Proceedings of the IEEE*, vol. 72(8), pp. 996–1009, Aug. 1984.

Coxson, G., and J. Russo, "Efficient Exhaustive Search for Optimal-Peak Side lobe Binary Codes," *Proceedings of the IEEE Radar Conference*, Philadelphia, PA, April 2004.

Felhauer, T., "Design and Analysis of new P(n, k) Polyphase Pulse Compression Codes," *IEEE Transactions on Aerospace and Electronic Systems*, vol. 30(3), pp. 865–874, July 1994.

Harris, F. J., "On the Use of Windows for Harmonic Analysis with the Discrete Fourier Transform," *Proceeding of the IEEE*, vol. 66, no. 1, pp. 51–83, Jan. 1978.

Kay, S. M., *Fundamentals of Statistical Signal Processing, Vol. II: Detection Theory*. Prentice Hall, Upper Saddle River, NJ, 1998.

Keel, B. M., "Pulse Compression Waveforms," Chap. 18 in W. A. Holm and J. A Scheer, (eds.), *Principles of Modern Radar*. SciTech Publishing, Raleigh, NC, to appear 2006.

Levanon, N., and E. Mozeson, *Radar Signals*. Wiley, New York, 2004.

Lewis, B. L., F. K. Kretshcmer, Jr., and W. W. Shelton, *Aspects of Radar Signal Processing*. Artech House, Canton, MA, 1986.

Nathanson, F. E., (with J. P. Reilly and M. N. Cohen), *Radar Design Principles*, 2d ed. McGraw Hill, New York, 1991.

Oppenheim, A. V., and R. W. Schafer, *Discrete-Time Signal Processing*, 2d ed. Prentice Hall, Englewood Cliffs, NJ, 1999.

Papoulis, A., *The Fourier Integral and Its Applications*. McGraw-Hill, New York, 1962.

Peebles, P. Z., Jr., *Radar Principles*. Wiley, New York, 1998.

Price, R., "Chebyshev Low Pulse Compression Side lobes Via a Nonlinear FM," URSI National Radio Science meeting, Seattle, WA, June 18, 1979.

Raney, R. K., "A New and Fundamental Fourier Transform Pair," *Proceedings of the IEEE* 12th International Geoscience & Remote Sensing Symposium (IGARSS '92), pp. 106–107, 26–29 May, 1992.

Rihaczek, A. W., *Principles of High-Resolution Radar*. Artech House, Boston, MA, 1996.

Chapter 5

Doppler Processing

Doppler processing is the term applied to filtering or spectral analysis of the signal received from a fixed range over a period of time corresponding to several pulses. In general, the spectrum of the slow-time signal from a single range bin consists of noise, clutter, and one or more target signals. Figure 5.1 shows a notional generic Doppler spectrum as observed from a stationary radar. As discussed in Chap. 1, this spectrum is periodic with a period equal to the *pulse repetition frequency* (PRF), so only the "principal period" from $-PRF/2$ to $+PRF/2$ is shown. Receiver noise is spread uniformly throughout the spectrum. Clutter occupies a portion of the spectrum. The portion of the spectrum where clutter is the dominant interference is often termed the *clutter region*. The portion where noise is the dominant interference is called the *clear region*; note that the clear region is clear of clutter but not of all interference. Sometimes a *skirt region* is defined at the transition between the clutter and clear regions; in the skirt region, both noise and clutter are significant interference sources. Stationary targets will appear at zero Doppler shift, while moving targets can occur anywhere in the spectrum, as appropriate to their radial velocity relative to the radar.

In many situations, the relative amplitudes of the clutter, target, and noise signals are as shown: the target returns are above the noise floor (signal-to-noise ratio $\gg 1$), but below the clutter (signal-to-clutter ratio $\ll 1$). In this case, targets cannot be detected reliably based on amplitude in the slow-time domain alone. Doppler processing is therefore used to separate the target and clutter signals in the frequency domain. The clutter can be explicitly filtered out, leaving the target return(s) as the strongest signal present; or the spectrum can be computed explicitly so that targets outside of the clutter region can be located by finding frequency components that significantly exceed the noise floor.

In this chapter the two major classes of Doppler processing, *moving target indication* (MTI) and *pulse Doppler* processing, are described. MTI refers to the

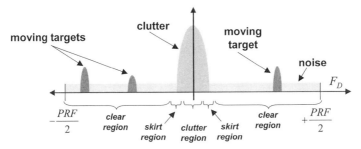

Figure 5.1 The principal period of a notional generic Doppler spectrum for a stationary radar, containing noise, clutter, and target components.

case where the slow-time signal is processed entirely in the time domain. Pulse Doppler processing refers to the case where the signal is processed in the frequency domain.† As will be seen, MTI processing produces limited information at very low computational cost; pulse Doppler processing requires more computation but produces more information and greater *signal-to-interference ratio* (SIR) improvement. Only coherent Doppler processing using digital implementations is considered, since this is the approach taken in most modern radars. Alternative systems using noncoherent Doppler processing and implementations based on analog technologies are described in the books by Eaves and Reddy (1988), Skolnik (1990; 2001), Nathanson (1991), and Schleher (1991), among many others.

5.1 Alternate Forms of the Doppler Spectrum

Before proceeding further, note that the notional Doppler spectrum of Fig. 5.1 represents a very simple case. While it is realistic for some scenarios, the Doppler spectrum for a given range bin can be greatly complicated by factors such as a moving radar platform or Doppler ambiguities caused by aliasing of the target Doppler signatures. Figure 5.2 illustrates the effect of radar platform motion, typically for a radar on an aircraft. As described in Chap. 3, the entire Doppler spectrum is shifted by the nominal radar-to-ground Doppler shift of $F_D = 2v \cos \psi / \lambda$ hertz while the main lobe is widened by $B_D = 2v\theta \sin \psi / \lambda$ hertz, where ψ is the angle between the radar line-of-sight vector and the platform velocity vector, and θ is the antenna beamwidth in radians.

In addition to this broadened *main lobe clutter* (MLC), *side lobe clutter* (SLC) and an *altitude line* (AL) are now evident as well. Side lobe clutter is clutter from echoes resulting from energy radiated and received through the radar side lobes. It is thus weaker than the main lobe clutter. Since side lobes and backlobes exist in all directions, it is possible to receive echoes from clutter directly in the direction of flight of the aircraft as well as directly behind it.

---

†Skolnik (2001) distinguishes MTI and pulse Doppler by defining pulse Doppler as a system that uses a PRF high enough to avoid blind speeds (see Sec. 5.2.4). In this text the two terms are differentiated based on the type of processing used and the information obtained.

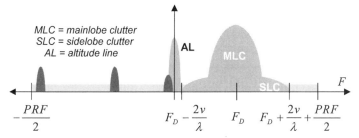

Figure 5.2 Notional Doppler spectrum for moving radar platform.

Any clutter present in these directions has a radial velocity equal to the full velocity v of the platform if in the direction of flight, or the negative of the velocity if behind the platform. Clutter at other angles can occur at any radial velocity in between $+v$ and $-v$. The altitude line results from the echo from energy transmitted through the radar side lobes straight down to the ground and back. Because this direction is normal to the velocity vector (assuming level flight), the radial velocity toward the ground is zero and the altitude line shows up at a Doppler frequency of zero, regardless of the platform velocity or antenna look direction. Although it is transmitted and received through the radar side lobes, the altitude line nonetheless tends to be relatively strong due to the relatively short vertical range and the high reflectivity of most clutter at normal incidence. Side lobe clutter does not appear in the stationary radar case of Fig. 5.1 because the side lobe echoes occur at zero Doppler and are thus part of the main lobe clutter. Similarly, the altitude line always occurs at zero Doppler and is therefore not visible until platform motion moves the main lobe clutter to a different portion of the Doppler spectrum.

Note that the target that was at the highest positive Doppler has wrapped around to the negative Doppler axis, and that clutter is now the dominant interference over a greater portion of the frequency spectrum. In many systems, the nominal Doppler shift F_D would be estimated and the slow-time data preprocessed by modulation with the sequence $\exp(j2\pi F_D m)$ to recenter the spectrum with the clutter at zero Doppler.

Figure 5.3 illustrates the effect on the generic spectrum of a PRF that is too low to represent the range of Doppler shifts present in the data. (This figure

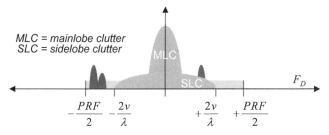

Figure 5.3 Notional generic spectrum for radar with reduced PRF.

assumes that the data have been preprocessed to recenter the clutter spectrum.) The Doppler spectrum is therefore aliased or, in more common radar terminology, the data are said to be ambiguous in Doppler. The clutter is now the dominant interference over most of the Doppler spectrum. Furthermore, one of the targets is now aliased into the clutter region. The other two still compete only with noise; they are still in the now-shrunken clear region of the spectrum.

These two examples are only a brief indication of the complications that can result in modeling the Doppler spectrum. It is important to realize that not all of the features discussed previously appear in the same range bin. For example, the altitude line will generally be evident at a near-in range bin due to the relatively short range to the ground. Main lobe clutter will be spread over multiple range bins further out, depending on the look direction of the radar. Side lobe clutter will be spread across many range bins. Thus, the clutter characteristics vary significantly with range. Thorough description of the variations in Doppler spectra for moving radar platforms and various PRF regimes are given by Morris and Harkness (1996) and Stimson (1998). Nonetheless, the simple spectrum of Fig. 5.1 has all the features needed to introduce MTI and pulse Doppler processing.

5.2 Moving Target Indication (MTI)

Figure 5.4 illustrates a two-dimensional data matrix formed from the coherently demodulated baseband returns from a series of M pulses. This matrix corresponds to one two-dimensional horizontal plane from the radar datacube of Fig. 3.1. Thus, a similar matrix exists for each phase center in the antenna system. In a single-aperture system, or at the point in an array where

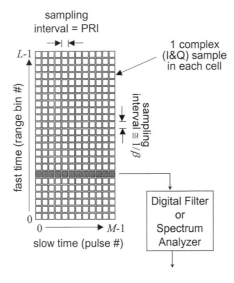

Figure 5.4 Notional two-dimensional data matrix. Each cell is one complex number.

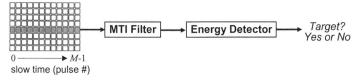

Figure 5.5 MTI filtering and detection process.

the data from multiple phase centers have been combined, there is only a single two-dimensional data matrix as shown.[†]

The samples in each column are successive samples of the returns from a single pulse, i.e., successive range bins. Each element of a column is one complex number, representing the real and imaginary (I and Q) components of one range bin. Consequently, each row represents a series of measurements from the same range bin over successive pulses. The sampling rate in the fast time or range dimension (vertical in Fig. 5.4) is at least equal to the transmitted pulse bandwidth, and therefore is on the order of hundreds of kilohertz to tens or even hundreds of megahertz. The slow-time or pulse number dimension (horizontal in Fig. 5.4) is sampled at the pulse repetition interval of the radar. Thus the sampling rate in this dimension is the PRF and is therefore on the order of ones to tens, and sometimes hundreds of kilohertz. As indicated by the shading, Doppler filtering operates on rows of this matrix.

MTI processing applies a linear filter to the slow-time data sequence in order to suppress the clutter component. Figure 5.5 illustrates the process. The type of filtering needed can be understood by considering Fig. 5.6. In this figure, it is assumed that knowledge of the platform motion and scenario geometry has been used to center the clutter spectrum at zero Doppler frequency. Clearly, some form of highpass filter is needed to attenuate the clutter without filtering out moving targets in the clear portions of the Doppler spectrum.

The output of the highpass MTI filter will be a new slow-time signal containing components due to noise and, possibly, one or more targets. This signal is passed to a detector. If the amplitude of the filtered signal exceeds the detector threshold (i.e., its energy is too great to likely be the result of noise alone), a target will be declared. Note that in MTI processing, the presence or absence of a moving target is the only information obtained. The filtering process of Fig. 5.6 does not provide any estimate of the Doppler frequency at which the target energy causing the detection occurred, or even of its sign; thus, it "indicates" the presence of a moving target, but does not determine whether the target is approaching or receding, or at what radial velocity. Furthermore, it provides no indication of the number of moving targets present. If multiple moving targets

---

[†]Not all digital processors necessarily form a data matrix similar to Fig. 5.4 explicitly. MTI processors in particular can be implemented more simply. However, the data matrix is used explicitly in many other processors, and is useful for illustration of Doppler filtering concepts.

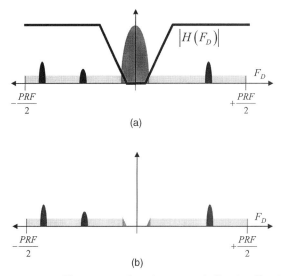

Figure 5.6 The concept of moving target indication filtering. (a) Doppler spectrum and MTI filter frequency response. (b) Doppler spectrum after filtering.

are present in the slow-time signal from a particular range, the result will still be only a "target present" decision from the detector. On the other hand, MTI processing is very simple and computationally undemanding. Despite its simplicity, a well-designed MTI can improve the SCR by several to perhaps 20 or more decibels in some clutter conditions.

5.2.1 Pulse cancellers

The major MTI design decision is the choice of the particular MTI filter to be used. MTI filters are typically low order, simple designs. Indeed, some of the most common MTI filters are based on very simple heuristic design approaches. Suppose a fixed radar illuminates a moving target surrounded by perfectly stationary clutter. The clutter component of the echo signal from each pulse would be identical, while the phase of the moving target component would vary due to the changing range. Subtracting the echoes from successive pairs of pulses would cancel the clutter components completely. The target signal would not cancel in general due to the phase changes.

This observation motivates the two-pulse MTI canceller, also referred to as the single or first-order canceller. Figure 5.7a illustrates the flowgraph of a two-pulse canceller. The input data are a sequence of baseband complex (I and Q) data samples from the same range bin over successive pulses, forming a discrete-time sequence $y[m]$ with an effective sampling interval T equal to the pulse repetition interval. The discrete time transfer function (also called the *system function*) of this linear finite impulse response (FIR, also called *tapped*

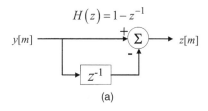

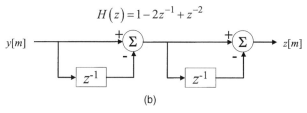

Figure 5.7 Flowgraphs and transfer functions of basic MTI cancellers. (a) Two-pulse canceller. (b) Three-pulse canceller.

delay line or *nonrecursive*) filter is simply $H(z) = 1 - z^{-1}$. The frequency response as a function of analog frequency F in hertz is obtained by setting $z = e^{j2\pi FT}$

$$\begin{aligned} H(F) &= (1 - z^{-1})|_{z=e^{j2\pi FT}} \\ &= 1 - e^{-j2\pi FT} \\ &= e^{-j\pi FT}(e^{+j\pi FT} - e^{-j\pi FT}) \\ &= 2je^{-j\pi FT}\sin(\pi FT) \end{aligned} \quad (5.1)$$

It is common to work with normalized frequency $f = FT$ cycles or with the radian equivalent, $\omega = \Omega T = 2\pi FT$ radians. For example, in terms of normalized radian frequency, the frequency response of the two-pulse canceller is

$$H(\omega) = 2je^{-j\omega/2}\sin(\omega/2) \quad (5.2)$$

Note that as F ranges from $-1/2T$ to $+1/2T$, f ranges from -0.5 to $+0.5$ and ω from $-\pi$ to $+\pi$.

Figure 5.8a plots the magnitude of the frequency response of the two-pulse canceller. Note that the filter does indeed have a null at zero frequency to suppress the clutter energy. Spectral components representing moving targets may either be partially attenuated or amplified, depending on their precise location on the Doppler frequency axis. Also note that, like all discrete-time filters, the frequency response is periodic with a period of 1 in the normalized cyclical frequency variable f, corresponding to a period of 2π in the normalized frequency variable ω or a period of $1/T = PRF$ in actual frequency in hertz. The shaded area highlights the principal period from $-PRF/2$ to $+PRF/2$; this is all that is normally plotted. Considering only this frequency range, it is clear that the frequency response is highpass in nature. The implications of the periodicity will be considered in Sec. 5.2.4.

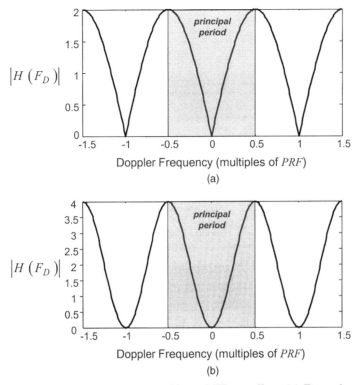

Figure 5.8 Frequency response of basic MTI cancellers. (*a*) Two-pulse canceller. (*b*) Three-pulse canceller.

The two-pulse canceller is a very simple filter; its implementation requires no multiplications and only one subtraction per output sample. As Fig. 5.8*a* shows, however, it is a poor approximation to an ideal highpass filter for clutter suppression. The next traditional step up in MTI filtering is the three-pulse (second-order or double) canceller, obtained by cascading two two-pulse cancellers. The flowgraph and frequency response are shown in Fig. 5.7*b* and Fig. 5.8*b*. The three-pulse canceller clearly improves the null breadth in the vicinity of zero Doppler, but it does not improve the consistency of response to moving targets at various Doppler shifts away from zero Doppler. It requires only two subtractions per output sample.

Despite their simplicity, the two-and three-pulse cancellers can be very effective against clutter with moderate-to-high pulse-to-pulse correlation. Figure 5.9 shows a simulated clutter sequence, formed by passing a white noise sequence through a filter with a Gaussian power spectrum having a standard deviation of $\sigma_f = 0.05$ on a normalized frequency scale; i.e., the standard deviation is 5 percent of the full spectrum width. Also shown are the outputs from the two- and three-pulse cancellers. The power in the two-pulse canceller output has been reduced by 13.4 dB relative to the unfiltered clutter sequence. For the three-pulse canceller, the reduction is 21.9 dB.

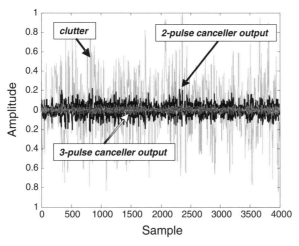

Figure 5.9 Clutter cancellation using two-and three-pulse cancellers. (See text for details.)

The idea of cascading two-pulse canceller sections to obtain higher order filters can be extended to the N-pulse canceller, obtained by cascading $N-1$ two-pulse canceller sections. The transfer function of the N-pulse canceller is therefore

$$H_N(z) = (1 - z^{-1})^{N-1} \qquad (5.3)$$

The corresponding impulse response coefficients of the filter are given by the binomial series

$$h_N[m] = \binom{N-1}{m} = (-1)^m \frac{(N-1)!}{m!\,(N-1-m)!} \qquad m = 0, \ldots, N-1 \qquad (5.4)$$

Other types of digital highpass filters could also be designed for MTI filtering. For example, an FIR highpass filter could be designed using standard digital filter design techniques such as the window method or the Parks-McClellan algorithm (Oppenheim and Schafer, 1999). To be suitable as an MTI filter, the FIR filter frequency response should have a zero at $F = 0$. In terms of the four recognized classes of FIR digital filters (see Oppenheim and Schafer, 1999, Sec. 5.7), the MTI filter can be either type I (even order with symmetric impulse response) or type IV (odd order with antisymmetric impulse response). The transfer functions of type IV filters always have a zero at $z = 1$, so that the frequency response is zero at $f = 0$, ideal for an MTI filter. The two-pulse canceller (which has an *order* of 1) is an example of a type IV filter. Type I filters do not necessarily have a zero at $f = 0$, but can be made to have one by requiring that the sum of the impulse response coefficients $h_N[m]$ equal zero. The three-pulse canceller is an example of a type I filter that has been designed to be suitable as an MTI filter.

Type II and III filters are unsuitable because they always have a zero at $z = -1$, corresponding to a frequency response null at a normalized frequency of $f = 0.5$; this creates extra undesirable blind speeds (see Sec. 5.2.4). Alternatively, *infinite impulse response* (IIR) highpass filters could be designed. Many operational radar systems, however, use two- or three-pulse cancellers for the primary MTI filtering due to their computational simplicity.

5.2.2 Vector formulation of the matched filter

The N-pulse cancellers described previously can be remarkably effective and have been widely used. Nonetheless, they are motivated by heuristic ideas. Can a more effective pulse canceller be designed? Since the goal of MTI filtering is to improve the signal-to-clutter ratio, it should be possible to apply the matched filter concept to this problem. To do so for discrete-time signals, the matched filter must first be formulated for the vector notation introduced in Chap. 1.

Consider a complex signal column vector $\mathbf{y}_m = [y[m]\ y[m-1] \cdots y[m-N+1]]'$ and a filter weight vector $\mathbf{h} = [h[0] \cdots h[N-1]]'$. The apostrophe represents matrix transpose, so that $\mathbf{y}_m$ and $\mathbf{h}$ are column vectors. For simplicity $\mathbf{y}_m$ will be written as simply $\mathbf{y}$. A single output sample z of the filter is given by $z = \mathbf{h}'\mathbf{y}$. The power in the output sample is given by

$$\text{Power in } z = |z|^2 = z^*z' = \mathbf{h}^H \mathbf{y}^* \mathbf{y}' \mathbf{h} \tag{5.5}$$

where $(\cdot)^H$ represents the Hermitian transpose.

The matched filter is obtained by finding the filter coefficient vector $\mathbf{h}$ that maximizes the signal-to-interference ratio of the filtered data. Denote the desired target signal vector by $\mathbf{t}$ and the interference vector by $\mathbf{w}$, so that $\mathbf{y} = \mathbf{t} + \mathbf{w}$. The filtered signal and interference are, respectively, $\mathbf{h}'\mathbf{t}$ and $\mathbf{h}'\mathbf{w}$. The power in the signal component is therefore $\mathbf{h}^H \mathbf{t}^* \mathbf{t}' \mathbf{h}$, and in the noise component it is $\mathbf{h}^H \mathbf{w}^* \mathbf{w}' \mathbf{h}$. Because the noise power is a random variable, its expected value is used to get meaningful results. Denote the expected value of $\mathbf{w}^*\mathbf{w}'$ as the interference covariance matrix $\mathbf{S_I}$

$$\mathbf{S_I} \equiv E\{\mathbf{w}^*\mathbf{w}'\} \tag{5.6}$$

It follows also that $\mathbf{S_I'} = \mathbf{S_I^*}$, $\mathbf{S_I} = \mathbf{S_I^H}$, and $(\mathbf{S_I^{-1}})' = (\mathbf{S_I^{-1}})^*$. With this definition the signal to interference ratio becomes

$$SIR = \frac{\mathbf{h}^H \mathbf{t}^* \mathbf{t}' \mathbf{h}}{\mathbf{h}^H \mathbf{S_I} \mathbf{h}} \tag{5.7}$$

The filter $\mathbf{h}$ that maximizes Eq. (5.7) is found using the Schwarz inequality, which in a form suitable for vector-matrix manipulations is

$$|\mathbf{p}^H \mathbf{q}|^2 \leq \|\mathbf{p}\|^2 \|\mathbf{q}\|^2 \tag{5.8}$$

with equality if and only if $\mathbf{p} = k\mathbf{q}$ for some scalar constant k. To apply Eq. (5.8), first note that the matrix $\mathbf{S_I}$ will be positive definite, so that it can be factored into the form $\mathbf{S_I} = \mathbf{A}^H\mathbf{A}$ for some matrix $\mathbf{A}$; that is, $\mathbf{A}$ is the "square root" of $\mathbf{S_I}$ in some sense. Define $\mathbf{p} = \mathbf{Ah}$ and $\mathbf{q} = (\mathbf{A}^H)^{-1}\mathbf{t}^*$. This choice is contrived so that $\mathbf{p}^H\mathbf{q} = \mathbf{h}^H\mathbf{t}^*$ and therefore $|\mathbf{p}^H\mathbf{q}|^2 = \mathbf{h}^H\mathbf{t}^*\mathbf{t}'\mathbf{h}$, which is the numerator of Eq. (5.7). The Schwarz inequality then gives

$$\mathbf{h}^H\mathbf{t}^*\mathbf{t}'\mathbf{h} \leq \|\mathbf{Ah}\|^2 \|(\mathbf{A}^H)^{-1}\mathbf{t}^*\|^2$$
$$= (\mathbf{h}^H\mathbf{S_I}\mathbf{h})(\mathbf{t}'\mathbf{S_I}^{-1}\mathbf{t}^*) \tag{5.9}$$

Rearranging Eq. (5.9) to isolate the SIR of Eq. (5.7) shows that

$$SIR \leq \mathbf{t}'\mathbf{S_I}^{-1}\mathbf{t}^* \tag{5.10}$$

with equality only when $\mathbf{p} = k\mathbf{q}$. The optimal weight vector therefore satisfies $\mathbf{Ah}_{\text{opt}} = k(\mathbf{A}^H)^{-1}\mathbf{t}^*$, or, with $k = 1$

$$\mathbf{h}_{\text{opt}} = \mathbf{S_I}^{-1}\mathbf{t}^* \tag{5.11}$$

This result is of great importance and will be used in several places throughout this text. The filtered data become, using $(\mathbf{S_I}^{-1})' = (\mathbf{S_I}^{-1})^*$

$$z = \mathbf{h}'_{\text{opt}}\mathbf{y} = \mathbf{t}^H\left(\mathbf{S_I}^{-1}\right)^*\mathbf{y} \tag{5.12}$$

5.2.3 Matched filters for clutter suppression

The results of the previous section can now be applied to design order $(N-1)$ MTI filters that are more optimal than the N-pulse canceller. Equation (5.11) shows that the linear filter that optimizes detection performance in the presence of additive interference is the FIR matched filter, and that the coefficients of the filter are given by the matrix equation

$$\mathbf{h} \equiv \begin{bmatrix} h[0] \\ \vdots \\ h[N-1] \end{bmatrix} = \mathbf{S_I}^{-1}\mathbf{t}^* \tag{5.13}$$

where $\mathbf{h}$ = column vector of filter coefficients
$\mathbf{S_I}$ = covariance matrix of the interference
$\mathbf{t}^*$ = column vector representing the desired target signal to which the filter is matched, and the asterisk denotes the complex conjugate

For a simple example, consider the first-order (length $N = 2$) matched filter. Assume the interference $w[m]$ consists of the sum of zero mean stationary white noise $n[m]$ of power (variance) σ_n^2 and zero mean stationary colored clutter $c[m]$ of power σ_c^2

$$w[m] = n[m] + c[m]$$
$$\mathbf{w} = [w[m] \quad w[m-1]]' \tag{5.14}$$

The clutter exhibits a correlation from one pulse to the next given by $\mathbf{E}\{c[m]c^*[m+1]\} = \mathbf{E}\{c^*[m]c[m-1]\} = s_c[1] = \sigma_c^2 \rho_c[1]$; the first normalized autocorrelation coefficient $\rho_c[1]$ is denoted as simply ρ for simplicity. The noise and clutter are uncorrelated with one another. The interference covariance matrix $\mathbf{S_I}$ will now be derived in detail.

The definition of $\mathbf{S_I}$ is

$$\mathbf{S_I} = \mathbf{E}\{\mathbf{w}^*\mathbf{w}'\} \equiv \begin{bmatrix} s_{11} & s_{12} \\ s_{21} & s_{22} \end{bmatrix} \quad (5.15)$$

Consider the s_{11} element. Using the fact that the noise and clutter are uncorrelated and that each is zero mean, it can be quickly concluded that

$$s_{11} = \mathbf{E}\{(n[m]+c[m])^*(n[m]+c[m])\}$$
$$= \sigma_n^2 + \sigma_c^2 \quad (5.16)$$

Next consider the s_{12} element

$$s_{12} = \mathbf{E}\{(n[m]+c[m])^*(n[m-1]+c[m-1])\}$$
$$= \mathbf{E}\{n^*[m]n[m-1]\} + \mathbf{E}\{n^*[m]c[m-1]\}$$
$$+ \mathbf{E}\{c^*[m]n[m-1]\} + \mathbf{E}\{c^*[m]c[m-1]\} \quad (5.17)$$

Again, the noise and clutter are zero mean and uncorrelated, and the noise is white, so that the expected values of the first three terms are zero. However, the last term becomes the pulse-to-pulse correlation of the clutter, so that

$$s_{12} = 0 + 0 + 0 + \mathbf{E}\{c^*[m]c[m-1]\}$$
$$= \rho\sigma_c^2 \quad (5.18)$$

It is easy to see that $s_{22} = s_{11}$ and that $s_{21} = s_{12}^*$. Thus

$$\mathbf{S_I} = \begin{bmatrix} \sigma_c^2 + \sigma_n^2 & \rho\sigma_c^2 \\ \rho^*\sigma_c^2 & \sigma_c^2 + \sigma_n^2 \end{bmatrix} \quad (5.19)$$

so that

$$\mathbf{S_I}^{-1} = \frac{1}{(\sigma_c^2+\sigma_n^2)^2 - |\rho|^2\sigma_c^4} \begin{bmatrix} \sigma_c^2+\sigma_n^2 & -\rho\sigma_c^2 \\ -\rho^*\sigma_c^2 & \sigma_c^2+\sigma_n^2 \end{bmatrix}$$
$$\equiv k \begin{bmatrix} \sigma_c^2+\sigma_n^2 & -\rho\sigma_c^2 \\ -\rho^*\sigma_c^2 & \sigma_c^2+\sigma_n^2 \end{bmatrix} \quad (5.20)$$

where k absorbs constants resulting from the matrix inversion.

To finish computing $\mathbf{h}$, a model is needed for the assumed target signal $\mathbf{t}$. For a target moving at a constant radial velocity, the expected target signal is just a discrete complex sinusoid at the appropriate Doppler frequency F_D. Following the discussion in Chaps. 2 and 4, assume the waveform is a train of simple pulses transmitted at times $t = mT$, $m = 0, \ldots, M - 1$. The individual transmitted pulses are thus of the form $\exp[j2\pi F_t(t - mT)]$, where F_t is the RF transmit frequency. If the target is at a nominal range R_0 and is moving toward the radar at a radial velocity of v meters per second, the received pulses are of the form

$$\tilde{y}(t) = \exp\left[j2\pi F_t\left(t - mT - \frac{2}{c}(R_0 - vmT)\right)\right]$$

Demodulation and sampling removes the carrier term $\exp[j2\pi F_t(t - mT)]$. The remaining phase terms over the sequence of pulses forms the slow-time discrete complex sinusoid

$$\begin{aligned} y[m] &= \exp\left[-j2\pi F_t \frac{2(R_0 - vmT)}{c}\right] \\ &= \exp\left(-j2\pi F_t \frac{2R_0}{c}\right) \exp\left(+j2\pi F_t \frac{2vmT}{c}\right) \\ &= \exp(j\phi)\exp(+j2\pi F_D mT) \quad m = 0, \ldots, M - 1 \end{aligned} \quad (5.21)$$

where F_D is the usual Doppler shift and ϕ is the phase shift corresponding to the range to the target on the first ($m = 0$) pulse.

Only N samples at a time of $y[m]$ are of interest in analyzing an N-pulse canceller. Assuming $N \leq M$ and recalling the results of Chap. 1 on vector representation of linear filtering, the series of N samples ending at $m = m_0$, $\{y[m_0], y[m_0 - 1], \ldots, y[m_0 - N + 1]\}$ can be represented in vector form as

$$\begin{aligned} \mathbf{t} &= A\left[e^{j\phi}e^{j2\pi F_D m_0 T} \quad e^{j\phi}e^{j2\pi F_D(m_0-1)T} \quad \cdots \quad e^{j\phi}e^{j2\pi F_D(m_0-N+1)T}\right]' \\ &= \hat{A}\begin{bmatrix}1 & e^{-j2\pi F_D T} & \cdots & e^{-j2\pi F_D(N-1)T}\end{bmatrix}' \end{aligned} \quad (5.22)$$

where the phase terms due to the initial range R_0 and the delay to the first sample of interest, m_0, have been absorbed into $\hat{A}$ and the signal has been renamed $\mathbf{t}$ to emphasize that it is only the target component. For the specific case $M = 2$, this becomes

$$\mathbf{t} = \hat{A}\begin{bmatrix}1 & e^{-j2\pi F_D T}\end{bmatrix}' \quad (5.23)$$

In practice, the target velocity and therefore Doppler shift are unknown; the target could be anywhere in the Doppler spectrum. The Doppler shift F_D is therefore modeled as a random variable with a uniform probability density function over $[-PRF/2, +PRF/2)$ and the expected value of $\mathbf{t}$ is computed. The expected value of the constant 1 is, of course, 1. The expected value of the second component of $\mathbf{t}$ is

$$\mathbf{E}\{e^{-j2\pi F_D T}\} = \frac{1}{PRF}\int_{-PRF/2}^{+PRF/2} e^{-j2\pi F_D T}\, dF_D = \frac{1}{PRF}\int_{-PRF/2}^{+PRF/2} e^{-j2\pi F_D/PRF}\, dF_D = 0 \quad (5.24)$$

The signal model then becomes simply

$$\mathbf{t} = \hat{A}[1 \quad 0]' \tag{5.25}$$

Finally, combining Eqs. (5.20) and (5.25) in Eq. (5.13) gives the coefficients of the optimum two-pulse filter

$$\mathbf{h} = \hat{k}\left[\sigma_c^2 + \sigma_n^2 \quad -\rho^*\sigma_c^2\right]' \tag{5.26}$$

where $\hat{k}$ absorbs all the scale factors.

To interpret this result, consider the case where the clutter is the dominant interference and is highly correlated from one pulse to the next. Then σ_n^2 is negligible compared to σ_c^2, and ρ is close to one. Absorbing σ_c^2 into $\hat{k}$, the matched filter coefficients are then 1 and approximately -1; i.e., nearly the same as the two-pulse canceller. Despite its simplicity, the two-pulse canceller is therefore nearly a first-order matched filter for MTI processing when the clutter-to-noise ratio is high and the successive clutter pulses are highly correlated. In the limit of very high clutter-to-noise ratio and perfectly correlated clutter, the two-pulse canceller is exactly the first-order matched MTI filter.

The vector matched filter derivation of the optimum two-pulse MTI filter given previously can be extended in a straightforward way to higher order MTI filters. As the order increases, the corresponding N-pulse canceller becomes a poorer approximation of the matched filter (Schleher, 1991).

It is interesting to consider the form of the optimum filter when the dominant interference is noise rather than clutter, that is, $\sigma_c^2 \ll \sigma_n^2$. In this case the optimum first-order MTI filter of Eq. (5.26) reduces to

$$\mathbf{h} \approx \tilde{k}[1 \quad 0]' \tag{5.27}$$

Equation (5.27) states that, in the presence of completely *un*correlated interference and with no knowledge of the target velocity, the filter impulse response reduces to a single impulse, meaning that the filter does nothing. In the clutter-dominated case, the filter combined the two slow-time samples because, even though constructive interference of the target could not be guaranteed, the high correlation of the clutter did guarantee that the clutter signal would be suppressed. "On average," the overall effect was beneficial. In this noise-dominated case, there is still no guarantee that the target signal will be reinforced, and in addition there is now no guarantee that the noise will be suppressed. The filter therefore does not combine the two data samples at all.

The previous analysis assumes that the target Doppler shift is unknown and therefore considers all target Doppler frequencies equally likely. It is also possible to extend the analysis and match the MTI filter to a specific Doppler shift, or to the case where the target Doppler extends only over a portion of the Doppler spectrum. These alternative assumptions manifest themselves as alternate models for the desired signal vector $\mathbf{t}$. The second case is treated in the book by Schleher (1991); here the case of a known Doppler shift for the target and a two-pulse canceller is considered. The interference and signal models are exactly the same as given earlier, except that now the target Doppler shift in $\mathbf{t}$

is not a random variable but a specific, fixed value. Therefore, it is not necessary to take an expected value of **t**. The filter coefficient vector is

$$\mathbf{h} = \mathbf{S_I}^{-1}\mathbf{t}^* = k\hat{A}^* \begin{bmatrix} \sigma_c^2 + \sigma_n^2 & -\rho\sigma_c^2 \\ -\rho^*\sigma_c^2 & \sigma_c^2 + \sigma_n^2 \end{bmatrix} \begin{bmatrix} 1 \\ e^{+j2\pi F_D T} \end{bmatrix}$$

$$= k' \begin{bmatrix} (\sigma_c^2 + \sigma_n^2) - \rho\sigma_c^2 e^{+j2\pi F_D T} \\ (\sigma_c^2 + \sigma_n^2) e^{+j2\pi F_D T} - \rho^*\sigma_c^2 \end{bmatrix} \quad (5.28)$$

While this result is easy to implement if the interference statistics are known, it is difficult to interpret. However, in the noise-limited case ($\sigma_n^2 \gg \sigma_c^2$), Eq. (5.28) reduces to

$$\mathbf{h} = k''[1 \quad e^{+j2\pi F_D T}]' \quad (5.29)$$

Equation (5.29) shows that in this case the optimum filter adds the two target samples together with a phase correction to the second so that they add in phase. In other words, the filter performs a coherent integration of the two target samples.

5.2.4 Blind speeds and staggered PRFs

The frequency response of all discrete-time filters is periodic, repeating with a period of one in the normalized cyclical frequency, corresponding to a period of $PRF = 1/T$ hertz of Doppler shift. Figure 5.8 illustrated this for the two- and three-pulse cancellers. Since MTI filters are designed to have a null at zero frequency, they will also have nulls at Doppler frequencies that are multiples of the pulse repetition frequency. Consequently, a target moving with a radial velocity that results in a Doppler shift equal to a multiple of the PRF will be suppressed by the MTI filter. Velocities that result in these unfortunate Doppler shifts are called *blind speeds* because the target return will be suppressed; the system is "blind" to such targets. From a digital signal processing point of view, blind speeds represent target velocities that will be aliased to zero frequency.

For a given PRF, the unambiguous range is

$$R_{ua} = \frac{c}{2PRF} \quad (5.30)$$

The first blind speed is simply the first nonzero null of the digital MTI filter, which occurs at $F = PRF$ hertz, converted to units of velocity

$$v_{\text{blind}} = \frac{\lambda PRF}{2} = \frac{c PRF}{2f_0} \quad (5.31)$$

As the PRF is increased for a given RF frequency, the unambiguous range decreases and the first blind speed increases. Figure 5.10 shows the unambiguous range-Doppler coverage regions that are possible. For example, each point on

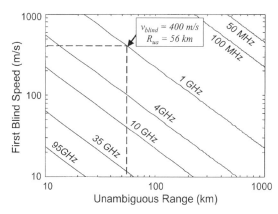

Figure 5.10 Ambiguity-free range-Doppler coverage regions.

the line marked "1 GHz" represents a combination of R_{ua} and v_{blind} that is achievable for some PRF. The dotted lines mark one example, corresponding to 400 m/s for the first blind speed and a 56-km unambiguous range. Equation (5.31) can be used to see that these values correspond to $PRF = 2667$ pulses per second.

Blind speeds could be avoided by choosing the PRF high enough so that the first blind speed exceeds any actual velocity likely to be observed for targets of interest. Unfortunately, higher PRFs correspond to shorter unambiguous ranges. It is frequently not feasible to operate at a PRF that allows unambiguous coverage of both the range and Doppler intervals of interest. For example, suppose a designer requires unambiguous range of 100 km and 100 m/s of unambiguous Doppler coverage. Figure 5.10 shows that the maximum RF must be approximately 2 GHz (the precise answer is 2.25 GHz). If the radar is required to be at X band (10 GHz), the combination of 100 km unambiguous range coverage and 100 m/s unambiguous velocity coverage is not obtainable and some ambiguity must be accepted in range, Doppler, or both.

The use of *staggered PRFs* is an alternative approach that raises the first blind speed significantly without degrading unambiguous range (Levanon, 1988; Schleher, 1991). PRF staggering can be performed on either a *pulse-to-pulse* or *dwell-to-dwell* (also called *block-to-block*) basis. The latter case is common in airborne pulse Doppler radars and is discussed in Sec. 5.3.6. Pulse-to-pulse stagger varies the pulse repetition interval, or equivalently the PRF, from one pulse to the next within a single dwell. As will be seen, this has the advantage of achieving increased Doppler coverage with a single dwell. One disadvantage is that the data are now a nonuniformly sampled sequence, making it more difficult to apply coherent Doppler filtering to the data and complicating analysis. Another is that ambiguous main lobe clutter can cause large pulse-to-pulse amplitude changes as the PRF varies, since the range of the second-time-around clutter that folds into each range cell will change as the PRF changes.

Consequently, pulse-to-pulse PRF stagger is generally used only in low PRF modes, where there are no range ambiguities.

Consider a system using P staggered PRFs $\{PRF_0, PRF_1, \ldots, PRF_{P-1}\}$. The corresponding set of pulse repetition intervals is $\{T_p\} = \{1/PRF_p\}$. Each of the PRFs can be expressed as an integer multiple of the greatest common divisor of the set

$$PRF_p = k_p \gcd(PRF_0, \ldots, PRF_{P-1})$$
$$\equiv k_p F_g \tag{5.32}$$

where F_g is the greatest common divisor of the set of PRFs. The set of integers $\{k_p\}$ is called the *staggers*,[†] and the ratio $k_m{:}k_p$ of any two of them is called a *stagger ratio*. For a given PRF, any MTI filter will exhibit blind speeds at all integer multiples of the PRF. Consequently, the first true blind Doppler frequency of a system using staggered PRFs will be the lowest frequency that is blind at all of the PRFs, i.e., the least common multiple of the set

$$F_b = \text{lcm}\,(PRF_0, \ldots, PRF_{P-1})$$
$$= F_g \,\text{lcm}\,(k_0, \ldots, k_{P-1}) \tag{5.33}$$

A complete cycle through the set of PRFs takes a total period T_{tot} equal to the sum of each of the staggered PRIs

$$T_{\text{tot}} = \sum_{p=0}^{P-1} T_p = \sum_{p=0}^{P-1} \frac{1}{PRF_p} = \sum_{p=0}^{P-1} \frac{1}{k_p F_g}$$
$$= \frac{1}{F_g}\left(\sum_{p=0}^{P-1} \frac{1}{k_p}\right) \equiv T_g \left(\sum_{p=0}^{P-1} \frac{1}{k_p}\right) \tag{5.34}$$

It is of interest to determine how much the blind speed of the staggered system is increased relative to that of an unstaggered system with the same average PRI. The average PRI is

$$T_{\text{avg}} = \frac{T_{\text{tot}}}{P} = \frac{T_g}{P} \sum_{p=0}^{P-1} \frac{1}{k_p} \tag{5.35}$$

The blind speed Doppler frequency that would be observed in an unstaggered waveform with this average PRI is

$$F_{us} = \frac{1}{T_{\text{avg}}} \tag{5.36}$$

[†]Some authors work in terms of the PRIs instead of the PRFs, and use the term "staggers" to refer to the ratio of the $\{PRI_n\}$.

Using Eqs. (5.33), (5.35), and (5.36) and noting that $F_g T_g = 1$ gives an expression for the first blind speed of the staggered PRF system in terms of the staggers $\{k_p\}$ and the blind speed of the unstaggered system with the PRF corresponding to the average PRI

$$\frac{F_b}{F_{us}} = \frac{1}{P}\text{lcm}(k_0, \ldots, k_{P-1}) \left(\sum_{p=0}^{P-1} \frac{1}{k_p} \right) \qquad (5.37)$$

For example, a two-PRF system with a stagger ratio of 3:4 would have a first blind speed 3.5 times that of a system using a fixed PRI equal to the average of the two individual PRIs. If a third PRF is added to give the set of staggers {3, 4, 5}, the first blind speed will be 15.67 times that of the comparable unstaggered system.

If a pure sinusoid $A \exp(j\Omega t)$ is input to a *linear time-invariant* (LTI) system, the output will be another pure sinusoid at the same frequency, but with possibly different amplitude and phase $B \exp(j\Omega t + \phi)$. However, if a pure sinusoid is sampled at nonuniform time intervals, the resulting series of samples, if interpreted as a conventional discrete time sequence, will *not* be equivalent to a uniformly sampled pure sinusoid at the appropriate frequency, so that the sampled signal will contain multiple frequency components. Any subsequent processing, even though itself LTI, will still result in an output spectrum containing multiple frequency components. Thus, a system utilizing nonuniform time sampling is not LTI, and the frequency response of a pulse-to-pulse staggered system cannot be determined using conventional transform techniques. Instead, an approach based on first principles can be used to explicitly compute the frequency response of a two-pulse canceller by determining its response to a pure complex sinusoid of arbitrary frequency and initial phase for the MTI filter structure of interest. Repeating for each possible sinusoid frequency, the frequency response can be determined point by point (Roy and Lowenschuss, 1970; Levanon, 1988; Schleher, 1991).

Consider the analog input $y(t) = A \exp(2\pi F_0 t + \phi_0)$ for some arbitrary starting phase ϕ_0. Define a sequence of sampling times $\{t_n\}$ in a P-stagger system as follows

$$t_n = t_0 + \sum_{p=0}^{n-1} T_{((p))_P} \qquad n = 0, \ldots, \infty \qquad (5.38)$$

where t_0 is an arbitrary starting time that is considered to be in the distant past. The notation $(())_P$ denotes evaluation of the argument modulo P. Recall that the $\{T_p\}$ are the sampling time *increments*, not the absolute sampling times. Also note that $t_n - t_{n-1} = T_{((n))_P}$. Form a discrete time sequence $x[m]$ by sampling $x(t)$ as follows

$$y[m] = y(t_m) \qquad (5.39)$$

Now consider the two-pulse canceller network of Fig. 5.7a.† Using Eqs. (5.38) and (5.39), the output $z[m] = y[m] - y[m-1]$ can be written explicitly as

$$\begin{aligned} z[m] &= A\exp[j(2\pi F_0 t_m + \phi_0)] - A\exp[j(2\pi F_0 t_{m-1} + \phi_0)] \\ &= Ae^{j[\pi F_0(t_m + t_{m-1}) + \phi_0]}\{\exp[+j\pi F_0(t_m - t_{m-1})] - \exp[-j\pi F_0(t_m - t_{m-1})]\} \\ &= Ae^{j[\pi F_0(t_m + t_{m-1}) + \phi_0]}\{\exp[+j\pi F_0 T_{((m))_P}] - \exp[-j\pi F_0 T_{((m))_P}]\} \\ &= 2jAe^{[j(\pi F_0(t_m + t_{m-1}) + \phi_0)]}\sin(\pi F_0 T_{((m))_P}) \end{aligned} \quad (5.40)$$

The "power spectrum" (magnitude-squared of the MTI filter frequency response) can be defined as the ratio of the power of the filtered sequence to the input sequence. The power of each input sample is $|y[m]|^2 = |A|^2$. The power of the output samples, $|z[m]|^2$ depends on the index m due to the varying PRIs. The output power is therefore computed as the average over one cycle of the staggered PRIs. Since the sum is over only one cycle of the sampling time increments, the $(())_P$ notation can be dropped, and the expression for the two-pulse canceller filter power spectrum becomes

$$|H_{2,P}(F_0)|^2 = \frac{\frac{1}{P}\sum_{m=0}^{P-1}|y[m]|^2}{|x[m]|^2} = \frac{4A^2\sum_{m=0}^{P-1}\sin^2(\pi F_0 T_m)}{PA^2} \quad (5.41)$$

where the notation $H_{N,P}(F)$ indicates the frequency response of an N-pulse canceller using P staggers. Finally, generalizing the specific F_0 to an arbitrary frequency F and renaming the summation index gives the squared magnitude of the frequency response of the two-pulse canceller with staggered PRIs

$$|H_{2,P}(F)|^2 = \frac{4}{P}\sum_{p=0}^{P-1}\sin^2(\pi F T_p) = \frac{4}{P}\sum_{p=0}^{P-1}\sin^2(\pi F/PRF_p) \quad (5.42)$$

The response of more general MTI filters can be obtained using a similar approach.

The actual frequency in hertz rather than normalized frequency is used in Eq. (5.42) because the nonuniform sampling rate invalidates the usual definition of normalized frequency. Figure 5.11 compares the frequency response of a two-pulse canceller using two ($P = 2$) PRFs versus conventional single-PRF operation. The staggered case uses PRFs of 750 and 1000 pulses per second; thus $F_g = 250$ Hz and the set of staggers k_p is {3, 4}. The first blind speed occurs at the least common multiple of 750 and 1000 Hz, which is 3000 Hz. The average PRF is 875 Hz, but F_{us}, which is the reciprocal of the average PRI $T_{\text{avg}} = 1.167$ ms, is 857.14 Hz. The conventional unstaggered response collected

---

†It is important to realize that the number of staggers used is independent of the canceller order selected. A two-pulse canceller can be implemented with two staggers or 10 staggers.

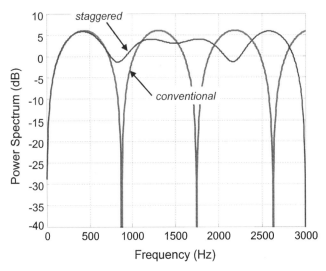

Figure 5.11 Comparison of two-pulse canceller frequency response with unstaggered waveform and 3:4 staggered waveform.

with the PRI = T_{avg} shows blind speeds at Doppler shifts equal to integer multiples of 857.14 Hz. Thus, staggering the PRF has increased the blind speed by a factor of 3.5 (=3000/857.14), consistent with Eq. (5.37). It is straightforward to show that the minimum unambiguous range in this case is reduced from the unambiguous range corresponding to T_{avg} by a factor of 6/7, or 14 percent. Figure 5.12 illustrates the frequency response when a third stagger is added to create the set {3, 4, 5}. The first true blind speed (corresponding to $F_D = 15$ kHz) occurs at 15.67 times the blind speed that would have been obtained using a fixed average PRI (corresponding to 957.5 Hz). In this case, the unambiguous range is reduced by a factor of 45/47, or 8 percent. Use of more than two PRIs in the stagger sequence and careful selection of their ratio can result in overall MTI frequency responses with less variability than shown in this simple example (Hsiao and Kretschmer, 1973; Prinsen, 1973). Another design approach uses randomized PRIs to extend the blind speed (Vergara-Domingues, 1993).

5.2.5 MTI figures of merit

The goal of MTI filtering is to suppress clutter. In doing so, it also attenuates or amplifies the target return, depending on the particular target Doppler shift. The change in signal and clutter power then affects the probabilities of detection and false alarm achievable in the system in a manner dependent on the particular design of the detection system.

There are three principal MTI filtering figures of merit in use. *Clutter attenuation* measures only the reduction in clutter power at the output of the MTI filter as compared to the input, but is simplest to compute. *Improvement*

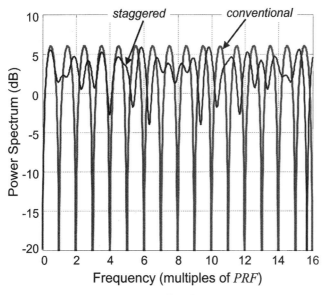

Figure 5.12 Frequency response of a pulse-to-pulse stagger system using a two-pulse canceller and three staggers with stagger ratios of 3:4:5.

factor quantifies the increase in signal-to-clutter ratio due to MTI filtering; as such, it accounts for the effect of the filter on the target as well as on the clutter. *Subclutter visibility* is a more complex measure that also takes into account the detection and false alarm probabilities and the detector characteristic. Because of its complexity, it is less often used. In this chapter, attention is concentrated on clutter attenuation CA and improvement factor I.

There are several ways to approach the calculation of the improvement factor. These include frequency domain approaches using clutter power spectra and MTI filter transfer functions, autocorrelation functions of the input and output of the MTI filter, and the vector method. Each will be illustrated in turn, starting with the frequency domain approach, which is perhaps the most intuitive.

Clutter attenuation is simply the ratio of the clutter power at the input of the MTI filter to the clutter power at the output

$$CA = \frac{\sigma_{ci}^2}{\sigma_{co}^2} = \frac{\int_{-PRF/2}^{PRF/2} S_c(F)\,dF}{\int_{-PRF/2}^{PRF/2} S_c(F)|H(F)|^2\,dF} \quad (5.43)$$

where σ_{ci}^2 and σ_{co}^2 = clutter power at the filter input and output, respectively
$S_c(F)$ = sampled clutter power spectrum expressed in terms of analog frequencies
$H(F)$ = discrete-time MTI filter frequency response

Since the MTI filter presumably reduces the clutter power, the clutter attenuation will be greater than one. In fact, clutter attenuation can be 20 dB or more in favorable conditions. However, it also depends on the clutter itself through $S_c(F)$. The shape of the clutter power spectrum and its spread in hertz are determined by the physical phenomenology and are not under the radar engineer's control. The percentage of the digital spectrum width to which a given $S_c(F)$ is mapped depends on the PRF and therefore is determined by the system design. Nonetheless, a change in clutter power spectrum due to changing terrain or weather conditions will alter the achieved clutter cancellation.

Improvement factor I is defined formally as the signal-to-clutter ratio at the filter output divided by the signal-to-clutter ratio at the filter input, averaged over all target radial velocities of interest (IEEE, 1982). Considering for the moment only a specific target Doppler shift, the improvement factor can be written in the form (Levanon, 1988)

$$I = \frac{(S/C)_{\text{out}}}{(S/C)_{\text{in}}} = \left(\frac{S_{\text{out}}}{S_{\text{in}}}\right)\left(\frac{C_{\text{in}}}{C_{\text{out}}}\right) = G \cdot CA \qquad (5.44)$$

where G is the *signal gain*. Figure 5.8 makes clear that the effect of the MTI filter on the target signal is a strong function of the target Doppler shift. Thus, G is a function of target velocity, while clutter attenuation CA is not.

To reduce I to a single number instead of a function of target Doppler, the definition calls for averaging uniformly over all target Doppler shifts "of interest." If a target is known to be at a specific velocity, the improvement factor can be obtained by simply evaluating Eq. (5.44) at the known target Doppler. It is more common to assume the target velocity is unknown a priori and use the average target gain over all possible Doppler shifts, which is just

$$G = \frac{1}{PRF}\int_{-PRF/2}^{PRF/2} |H(F)|^2\, dF \qquad (5.45)$$

Expressions for improvement factor equivalent to Eqs. (5.43) to (5.45) can be developed in terms of the autocorrelation function of the clutter and the MTI filter impulse response (Levanon, 1988; Nathanson, 1991). For low order filters such as two- or three-pulse cancellers, and clutter power spectra with either measured or analytically derivable autocorrelation functions, the resulting equations can be easier to evaluate than the frequency domain versions.

As an example of the autocorrelation approach, consider the output of the two-pulse canceller when the input is just clutter; this is $c'[m] = c[m] - c[m-1]$. The expected value of the filter output power is

$$\mathbf{E}[|c'[m]|^2] = \mathbf{E}[|c[m]|^2 - 2\text{Re}\{c[m]\,c^*[m-1]\} + |c[m-1]|^2] \qquad (5.46)$$

where Re{} denotes the real part of the argument.† Assuming that $c[m]$ is stationary

$$\mathbf{E}[|c'[m]|^2] = 2s_c[0] - 2\mathrm{Re}\{s_c[1]\} \tag{5.47}$$

where
$$s_c[k] \equiv \mathbf{E}[c[m]\,c^*[m+k]] \tag{5.48}$$

is the autocorrelation function of $c[m]$. Note that $s_c[k] = s_c^*[-k]$. Also define the normalized autocorrelation function

$$\rho_c[k] \equiv \frac{1}{s_c[0]} s_c[k] \tag{5.49}$$

The clutter attenuation component of the improvement factor in Eq. (5.44) can now be written as

$$CA = \frac{\mathbf{E}\{|c[m]|^2\}}{\mathbf{E}\{|c'[m]|^2\}} = \frac{s_c[0]}{2(s_c[0] - \mathrm{Re}\{s_c[1]\})}$$

$$= \frac{1}{2(1 - \mathrm{Re}\{\rho_c[1]\})} \tag{5.50}$$

Using Eq. (5.45) for the average gain with $H(F) = 2\sin(\pi F/PRF)$ gives $G = 2$; thus the improvement factor for the two-pulse canceller is

$$I = CA \cdot G = \frac{1}{1 - \mathrm{Re}\{\rho_c[1]\}} \tag{5.51}$$

A similar analysis can be used to derive the improvement factor for a three-pulse canceller; it is

$$I = CA \cdot G = \frac{1}{1 - \frac{4}{3}\mathrm{Re}\{\rho_c[1]\} + \frac{1}{3}\mathrm{Re}\{\rho_c[2]\}} \tag{5.52}$$

To see how these formulas are used, consider the case where the clutter spectrum is Gaussian with variance (in normalized radian frequency units) σ_ω^2, that is, $S_c(\omega) = k\exp(-\omega^2/\sigma_\omega^2)$. Assuming $\sigma_\omega \ll \pi$ so that the continuous-time Fourier transform pair for Gaussian functions can be used to a good approximation, the normalized autocorrelation function for $c[m]$ at lag k is

$$\rho_c[k] \approx e^{-(\sigma_\omega k)^2/2} \tag{5.53}$$

---

†Since the power spectrum is real-valued, the autocorrelation function must be Hermitian symmetric. It can therefore be complex valued. However, clutter is usually modeled by a power spectrum that is also an even function of frequency (i.e., symmetric about $F = 0$), for example a Gaussian clutter spectrum with a zero mean. With this extra constraint, the autocorrelation function must be real-valued also, so the Re{} operator can be dropped in Eq. (5.46) and other equations in this section.

TABLE 5.1 Improvement Factor for Gaussian Clutter Power Spectrum

Standard deviation of clutter power spectrum, Hz	Improvement factor, dB	
	Two-pulse canceller	Three-pulse canceller
$PRF/3$	0.5	0.7
$PRF/10$	7.5	12.5
$PRF/20$	13.2	21.7
$PRF/100$	24	51

Using Eq. (5.53) in Eqs. (5.51) and (5.52) gives the improvement factor for a Gaussian clutter spectrum with a two- or three-pulse canceller:

$$I = \begin{cases} \dfrac{1}{1 - e^{-\sigma_\omega^2/2}} & \text{(two-pulse canceller)} \\ \dfrac{1}{1 - \frac{4}{3}e^{-\sigma_\omega^2/2} + \frac{1}{3}e^{-2\sigma_\omega^2}} & \text{(three-pulse canceller)} \end{cases} \quad (5.54)$$

Table 5.1 shows the improvement factor predicted for two- and three-pulse cancellers for the case of a Gaussian clutter power spectrum of various spectral widths using Eq. (5.54). If the clutter spectrum is narrow compared to the PRF, the improvement factor can be 20 dB or more even for the simple two-pulse canceller. If the clutter spectrum is wide, much of the clutter power will be in the passband of the MTI highpass filter, and the improvement factor will be small.

The third approach for computing the improvement factor uses the vector analysis techniques employed in determining the matched filter for MTI. For comparison with the autocorrelation analysis given previously, consider the case where $\sigma_n^2 = 0$ (clutter only) and $\mathbf{h} = [1 \ -1]'$ (two-pulse canceller). Improvement factor is the ratio of the signal-to-interference ratio at the filter output to the ratio at the filter input. While the optimum MTI filter was derived by averaging over possible target Doppler frequencies, in evaluating the improvement factor it is assumed that any specific target has a specific Doppler frequency. The improvement factor is calculated for that specific target Doppler frequency and then averaged over allowable Doppler frequencies. Since SIR at the input does not depend on Doppler frequency, it is sufficient to do the averaging on the output SIR.

Consider therefore the signal vector given by Eq. (5.23) and the clutter covariance matrix given by Eq. (5.19) (with $\sigma_n^2 = 0$). The input SIR is just $|\hat{A}|^2/\sigma_c^2$. Equation (5.7) gave an explicit expression for the output SIR. The numerator of this expression is

$$\mathbf{h}^H \mathbf{t}^* \mathbf{t}' \mathbf{h} = |\hat{A}|^2 [1 \ -1] \begin{bmatrix} 1 & e^{-j2\pi F_D T} \\ e^{+j2\pi F_D T} & 1 \end{bmatrix} \begin{bmatrix} 1 \\ -1 \end{bmatrix}$$

$$= 2|\hat{A}|^2 (1 - \text{Re}\{e^{j2\pi F_D T}\}) \quad (5.55)$$

This is the two-pulse canceller MTI filter output signal power for a target at Doppler shift F_D hertz. Averaging over all target Doppler shifts gives

$$\mathbf{E}_{F_D}\{\mathbf{h}^H \mathbf{t}^* \mathbf{t}' \mathbf{h}\} = 2|\hat{A}|^2 \tag{5.56}$$

The denominator of Eq. (5.7) is

$$\mathbf{h}^H \mathbf{S_I} \mathbf{h} = \sigma_c^2 [1 \quad -1] \begin{bmatrix} 1 & \rho_c \\ \rho_c^* & 1 \end{bmatrix} \begin{bmatrix} 1 \\ -1 \end{bmatrix}$$

$$= 2\sigma_c^2 (1 - \mathrm{Re}\{\rho_c\}) \tag{5.57}$$

Dividing Eq. (5.56) by Eq. (5.57) gives the output signal-to-interference ratio; further dividing that ratio by the input SIR gives the improvement factor for a two-pulse canceller operating against clutter only (no noise)

$$I = \frac{1}{1 - \mathrm{Re}\{\rho_c\}} \tag{5.58}$$

Since ρ_c in the matrix formulation is the same as $\rho_c[1]$, this is the same expression obtained using autocorrelation methods in Eq. (5.51).

The two-pulse canceller is matched to the case of a target modeled by the signal vector $[1\ 0]'$ and no noise. This target model is a compromise necessitated by lack of specific a priori knowledge of the target Doppler frequency, with the result that the filter is suboptimum as compared to a filter matched specifically to the actual target Doppler frequency. An interesting question is, just how much is the improvement factor reduced by this suboptimum filter design? To answer this question, it is necessary to compute the improvement factor for the optimum filter for a specific target Doppler and compare it to the improvement factor for the averaged target model. The result must still be averaged over all target Doppler frequencies. To start, return to the case where both noise and clutter are present, so that the filter weight vector is given by Eq. (5.26).

It is not necessary to explicitly compute the output SIR for the optimum filter case; Eq. (5.10) guarantees that it equals

$$\mathbf{t}' \mathbf{S_I}^{-1} \mathbf{t}^* = k|\hat{A}|^2 [1 \quad e^{-j2\pi F_D T}] \begin{bmatrix} \sigma_c^2 + \sigma_n^2 & -\rho_c \sigma_c^2 \\ -\rho_c^* \sigma_c^2 & \sigma_c^2 + \sigma_n^2 \end{bmatrix} \begin{bmatrix} 1 \\ e^{+j2\pi F_D T} \end{bmatrix} \tag{5.59}$$

Evaluating this expression and then averaging over target Doppler frequency gives

$$\mathbf{E}_{F_D}\{SIR\}_{\mathrm{opt}} = 2k|\hat{A}|^2 (\sigma_c^2 + \sigma_n^2) \tag{5.60}$$

With noise and clutter both present, the input SIR is $|\hat{A}|^2/(\sigma_c^2 + \sigma_n^2)$. Thus, the improvement factor for the optimum filter is

$$I_{\mathrm{opt}} = 2k(\sigma_c^2 + \sigma_n^2)^2$$

$$\equiv \frac{2}{1-\beta} \tag{5.61}$$

where the explicit expression for k from Eq. (5.20) has been used, and β has been defined as

$$\beta \equiv |\rho_c|^2 \left\{ \frac{\sigma_c^4}{(\sigma_c^2 + \sigma_n^2)^2} \right\} \tag{5.62}$$

For the suboptimum case, i.e., the case where $\mathbf{h}$ is chosen using the signal model $[1\ 0]'$, the output signal power and the output interference power must again be explicitly computed to get SIR at the MTI filter output. The output signal power is

$$\mathbf{h}^H \mathbf{t}^* \mathbf{t}' \mathbf{h} = k^2 |\hat{A}|^2 \begin{bmatrix} \sigma_c^2 + \sigma_n^2 & -\rho_c \sigma_c^2 \end{bmatrix} \begin{bmatrix} 1 & e^{-j2\pi F_D T} \\ e^{+j2\pi F_D T} & 1 \end{bmatrix} \begin{bmatrix} \sigma_c^2 + \sigma_n^2 \\ -\rho_c^* \sigma_c^2 \end{bmatrix} \tag{5.63}$$

Evaluating this expression and averaging over F_D gives the average output signal power, which can be put in the form

$$\mathbf{E}_{F_D}\{\mathbf{h}^H \mathbf{t}^* \mathbf{t}' \mathbf{h}\} = k^2 |\hat{A}|^2 (\sigma_c^2 + \sigma_n^2)^2 (1 + \beta) \tag{5.64}$$

Similarly, the output interference power can be found from

$$\mathbf{h}^H \mathbf{S_I} \mathbf{h} = k^2 |\hat{A}|^2 \begin{bmatrix} \sigma_c^2 + \sigma_n^2 & -\rho_c \sigma_c^2 \end{bmatrix} \begin{bmatrix} \sigma_c^2 + \sigma_n^2 & \rho_c \sigma_c^2 \\ \rho_c^* \sigma_c^2 & \sigma_c^2 + \sigma_n^2 \end{bmatrix} \begin{bmatrix} \sigma_c^2 + \sigma_n^2 \\ -\rho_c^* \sigma_c^2 \end{bmatrix}$$

$$= k^2 |\hat{A}|^2 (\sigma_c^2 + \sigma_n^2)^3 (1 - \beta) \tag{5.65}$$

Combining Eqs. (5.64) and (5.65) gives the output SIR for the suboptimum case

$$\mathbf{E}_{F_D}\{SIR\}_{\text{sub}} = \frac{|\hat{A}|^2}{(\sigma_c^2 + \sigma_n^2)} \frac{1 + \beta}{1 - \beta} \tag{5.66}$$

Dividing by the input SIR gives the improvement factor for the suboptimum case

$$I_{\text{sub}} = \frac{1 + \beta}{1 - \beta} \tag{5.67}$$

It is now finally possible to compute the loss due to the use of a suboptimum MTI filter. Using Eqs. (5.61) and (5.67)

$$I_{\text{opt}} - I_{\text{sub}} = \frac{2}{1 - \beta} - \frac{1 + \beta}{1 - \beta} \equiv 1 \tag{5.68}$$

Thus, the average improvement factor is reduced by exactly one. Note that this is on a linear, not decibel, scale.

Equation (5.68) shows that the loss due to using the suboptimum filter is negligible except in cases where the improvement factor is very poor anyway. For example, if the optimum filter improvement factor is 10 dB ($I_{\text{opt}} = 10$), the suboptimum improvement factor will be $I_{\text{sub}} = 9 = 9.54$ dB, a loss of 0.46 dB. If I_{opt} is 20 dB, the loss will be only 0.04 dB. On the other hand, if I_{opt} approaches 2 (3 dB), I_{sub} approaches one, and no SIR improvement at all is obtained.

Additional MTI metrics can be defined. Improvement factor I is an average of the improvement in signal-to-clutter ratio over one Doppler period. At some Doppler shifts, the target is above the clutter energy, while at others it is below the clutter and therefore not detectable. I does not indicate over what percentage of the Doppler spectrum a target can be detected. The concept of *MTI visibility factor* or *target visibility V* has been proposed to quantify this effect (Kretschmer, 1986). V is the percentage of the Doppler period over which the improvement factor for a target at a specific frequency is greater than or equal to the average improvement factor I.

5.2.6 Limitations to MTI performance

The basic idea of MTI processing is that repeated measurements (pulses) of a stationary target yield the same echo amplitude and phase; thus successive pulses, when subtracted from one another, should cancel. Any effect internal or external to the radar that causes the received echo from a stationary target to vary will cause imperfect cancellation, limiting the improvement factor.

Perhaps the simplest example is transmitter amplitude instability. If two transmitted pulses differ in amplitude by 10 percent (equivalent to $20 \log_{10}(1.1/1) = 0.83$ dB), the signal resulting from subtracting the two echoes from a perfectly stationary target will have an amplitude that is 10 percent that of the individual echoes. Consequently, clutter attenuation can be no better than $20 \log_{10}(1/0.1) = 20$ dB. For a two-pulse canceller with an average signal gain G of 2 (6 dB), the maximum achievable improvement factor is 26 dB.

A more realistic analysis of the limitations due to amplitude jitter can be obtained by modeling the amplitude of the mth transmitted pulse as $A[m] = k(1 + a[m])$, where $a[m]$ is a zero mean, white random process with variance σ_a^2 that represents the percentage variation in transmitted amplitude, and k is a constant. The received signal will have a complex amplitude of the form $k'(1 + a[m]) \exp(j\phi)$, where ϕ is the phase of the received slow-time sample and the constant k' absorbs all the radar range equation factors. The average power of this signal, which is the input to the pulse canceller, is

$$\mathbf{E}\{|y[m]|^2\} = k'^2 \mathbf{E}\{1 + 2a[m] + a^2[m]\} = k'^2 (1 + \sigma_a^2) \quad (5.69)$$

The expected value of the two-pulse canceller output power will be

$$\mathbf{E}\{|z[m]|^2\} = \mathbf{E}\{|(y[m] - y[m-1])|^2\} = \mathbf{E}\{|k' e^{j\phi}(a[m] - a[m-1])|^2\}$$
$$= k'^2 \mathbf{E}\{a^2[m]\} - 2\mathbf{E}\{a[m]a[m-1]\} + k'^2 \mathbf{E}\{a^2[m-1]\}$$
$$= 2k'^2 \sigma_a^2 \quad (5.70)$$

The achievable clutter cancellation is thus

$$\frac{\text{input power}}{\text{output power}} = \frac{k'^2\left(1+\sigma_a^2\right)}{2k'^2\sigma_a^2} = \frac{1+\sigma_a^2}{2\sigma_a^2} \qquad (5.71)$$

For example, an amplitude variance of 1 percent ($\sigma_a^2 = 0.01$) limits two-pulse clutter cancellation to a factor of 50.5, or 17 dB. Note that because the average target gain G of the two-pulse canceller is $G = 2$ (3 dB), the limit to the improvement factor I is $17 + 3 = 20$ dB.

Another example is phase drift in either the transmitter or receiver. This can occur, for example, due to instability in coherent local oscillators used either as part of the waveform generator on the transmit side, or in the demodulation chains on the receiver side. Consider the weighted coherent integration of M data samples $y[m]$ with a zero-mean, stationary, white phase error $\phi[m]$

$$Z = \sum_{m=0}^{M-1} a_m y[m] e^{j\phi[m]} \qquad (5.72)$$

where the $\{a_m\}$ are the integration weights. Assume that each data sample $x[m]$ is a (possibly complex) constant A. Then the power of the weighted coherent sum in the absence of phase error $\phi[m]$ is

$$|Z|^2 = |A|^2 \left|\sum_{m=0}^{M-1} a_m\right|^2 \qquad (5.73)$$

It can be shown that the integrated power, when a white Gaussian phase error is present, is (Richards, 2003)

$$|Z|^2 = |A|^2 \left(\sum_{m=1}^{M} |a_m|^2 + e^{-\sigma_\phi^2} \sum_{\substack{m=1 \\ m\neq k}}^{M} \sum_{k=1}^{M} a_m a_k^*\right) \qquad (5.74)$$

where σ_ϕ^2 is the variance of the phase noise in radians. This can be applied to the two-pulse canceller by letting $M = 2$ and $a_0 = 1, a_1 = -1$. Equation (5.74) then gives the power at the canceller output as $2|A|^2(1 - e^{-\sigma_\phi^2})$. Since the power of a clutter sample before the canceller is $|A|^2$, the limitation on two-pulse cancellation due to the phase noise becomes just $1/2(1 - e^{-\sigma_\phi^2})$.

Other sources of limitation due to radar system instabilities include instability in transmitter or oscillator frequencies, transmitter phase drift, coherent oscillator locking errors, PRI jitter, pulse width jitter, and quantization noise. Simple formulas to bound the achievable clutter attenuation due to each of these error sources, as well as some others not mentioned, are given by Skolnik (1998), Nathanson (1991), and Schleher (1991).

External to the radar, the chief factor limiting MTI improvement factor is simply the width of the clutter spectrum itself. Wider spectra put more clutter energy outside of the MTI filter null, so that less of the clutter energy is filtered out. This effect is evident in Eq. (5.54) and was illustrated numerically in Table 5.1. The effective clutter spectrum width can be increased by radar system instabilities or by measurement geometry and dynamics. For instance, a scanning antenna adds some amplitude modulation due to antenna pattern weighting to the clutter return. The power spectrum of the measured clutter is then the convolution in the frequency domain of the actual clutter power spectrum and the squared magnitude of the Fourier transform of the amplitude modulation caused by the antenna scanning. This convolution increases the observed spectral width somewhat. In some cases, the clutter power spectrum may not be centered on zero Doppler shift. A good example is rain clutter: moving weather systems will have a nonzero average Doppler representing the rate at which the rain cell is approaching or receding from the radar system. Unless this average motion is detected and compensated, the MTI filter null will not be centered on the clutter spectrum and cancellation will be poor.

The largest source of clutter offset and spreading is radar platform motion. Recall from Chap. 3 that the motion-induced clutter bandwidth is

$$\beta_D \approx \frac{2v\theta}{\lambda} \sin \psi \qquad (5.75)$$

The offset in center frequency of the clutter spectrum can be as much as a few kilohertz for fast aircraft, while the motion-induced spectral spread can be tens to a few hundreds of hertz. This clutter spreading adds to the intrinsic spread of the clutter spectrum due to internal motion, and can often be the dominant effect determining the observed clutter spectral width and therefore determining the MTI performance limits.

5.3 Pulse Doppler Processing

Pulse Doppler processing is the second major class of Doppler processing. Recall that in MTI processing, the fast time/slow time data matrix is highpass filtered in the slow-time dimension, yielding a new fast time/slow time data sequence in which the clutter components have been attenuated. Pulse Doppler processing differs in that filtering in the slow-time domain is replaced by explicit spectral analysis of the slow-time data for each range bin. Thus, the result of pulse Doppler processing is a data matrix in which the dimensions are fast time and Doppler frequency. The spectral analysis is most commonly performed by computing the *discrete Fourier transform* (DFT) of each slow-time row of the data as shown in Fig. 5.13, but other techniques can also be used (Kay, 1988). Good general discussions of pulse Doppler processing are contained in the books by Long et al. (1990), Morris and Harkness (1996), and Stimson (1998).

The interference component of each Doppler sample will vary. Consider the notional pulse Doppler spectrum shown in Fig. 5.14. Assuming that the clutter has been centered at zero Doppler, those spectral samples at or near zero will be dominated by the strong clutter signal, even though noise is also present. Spectral samples in the clear region, away from the clutter energy, have only

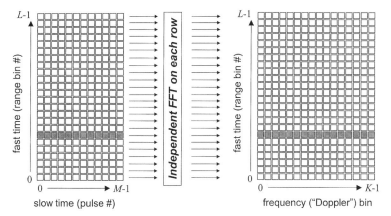

Figure 5.13 Conversion of the fast-time/slow-time data matrix to a range/Doppler matrix.

thermal noise to interfere with signal detection. Each Doppler spectrum sample is individually compared to a threshold to determine whether the signal at that range bin and Doppler frequency appears to be noise only, or noise plus a target. If the sample crosses the threshold, it not only indicates the presence of a target in that range bin, but also its approximate velocity, since the Doppler frequency bin is known. The samples that are clutter-dominated are often simply discarded on the grounds that the signal-to-interference ratio will be too low for successful detection. However, some systems use a technique called clutter mapping, discussed in Sec. 5.6.1, to attempt detection of strong targets in this clutter region.

The advantages of pulse Doppler processing are that it provides at least a coarse estimate of the radial velocity component of a moving target, including whether the target is approaching or receding, and that it provides a way to detect multiple targets, provided they are separated enough in Doppler to be resolved. The chief disadvantages are greater computational complexity of pulse Doppler processing as compared to MTI filtering and longer required dwell times due to the use of more pulses for the Doppler measurements.

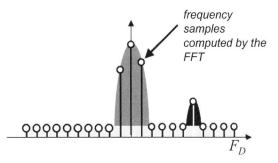

Figure 5.14 Concept of the pulse Doppler spectrum as a frequency-sampled version of an underlying discrete-time Fourier transform.

5.3.1 The discrete time Fourier transform of a moving target

To understand the behavior of pulse Doppler processing, it is useful to again consider the Fourier spectrum of an ideal, constant radial velocity, moving point target and the effects of a sampled Doppler spectrum. The issues are the same as those considered when discussing the sampling of the Doppler spectrum in Chap. 3. Consider a radar illuminating a moving target over a dwell of M pulses, and suppose a moving target is present in a particular range bin. If the target's velocity is such that the Doppler shift is F_D hertz, the slow-time received signal after quadrature demodulation is

$$y[m] = Ae^{j2\pi F_D mT} \qquad m = 0, \ldots, M-1 \qquad (5.76)$$

where T is the radar's pulse repetition interval, which is the effective sampling interval in slow time. The signal of Eq. (5.76) is the same signal considered in Chap. 3 (Eq. 3.18), except for the change from normalized frequency ω_D in radians to analog frequency F_D in hertz; they are related according to $\omega_D = 2\pi F_D T$. Equation 3.19 gave the M-pulse discrete time Fourier transform of this signal; converting to analog frequency gives

$$Y(F) = A\frac{\sin[\pi(F - F_D)MT]}{\sin[\pi(F - F_D)T]} e^{-j\pi(M-1)(F-F_D)T} \qquad F \in [-PRF/2, +PRF/2) \qquad (5.77)$$

The magnitude of this asinc function is illustrated in Fig. 5.15a for the case where $F_D = PRF/4$, $M = 20$ pulses, and $A = 1$. As would be expected, the main lobe of the response is centered at $F = F_D$ hertz. So long as $M \geq 4$, the Rayleigh (peak-to-null) main lobe bandwidth is $1/MT$ hertz; this is also the 4-dB bandwidth. The width of the mainlobe at the -3 dB points is $0.89/MT$ hertz. The first side lobe is 13.2 dB below the response peak. These main lobe width measures determine the Doppler resolution of the radar system. Note that, whichever measure is used, they are all inversely proportional to MT, which is the total elapsed time of the set of pulses used to make the spectral measurement. Thus, Doppler resolution is determined by the observation time of the measurement. Longer observation allows finer Doppler resolution.

Because of the high side lobes, it is common to use a data window to weight the slow-time data samples $y[m]$ prior to computing the DTFT. To analyze this case, replace $y[m]$ by $w[m]y[m]$ in the computation of the DTFT and again use Eq. (5.76) for the particular form of $y[m]$. After again converting from normalized frequency to hertz and recognizing that $y[m]$ is finite length, $Y(F)$ becomes

$$Y_w(F) = A\sum_{m=0}^{M-1} w[m]e^{-j2\pi(F-F_D)mT} = W(F - F_D) \qquad (5.78)$$

where the notation $Y_w(F)$ is used to emphasize that the spectrum is computed with a nontrivial window applied to the data. This is simply the Fourier transform of the window function itself, shifted to be centered on the target Doppler

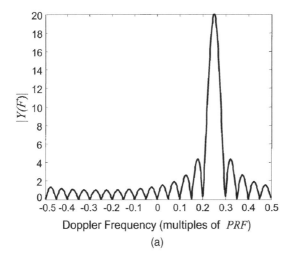

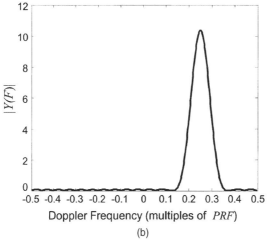

Figure 5.15 Magnitude of the discrete-time Fourier transform of an ideal moving target slow-time data sequence with $F_D = PRF/4$ and $M = 20$ pulses. (a) No window. (b) Hamming window.

frequency F_D rather than at zero. Figure 5.15b illustrates the effect of the window on the DTFT for the same data used in part a of the figure. In fact, the asinc function of Fig. 5.15a is also just the Fourier transform of the rectangular window function (equivalent to no window). Harris gives an extensive description of common window functions and their characteristics (Harris, 1978). In general, nonrectangular windows cause an increase in main lobe width, a decrease in peak amplitude, and a decrease in signal-to-noise ratio in exchange for large reductions in peak side lobe level.

It is straightforward to compute the reduction in peak amplitude and the SNR loss, given the window function $w[n]$. Consider the peak gain first.

From Eq. (5.77), the peak value of $|Y(F)|^2$ when no window is used is $A^2 M^2$. Evaluating Eq. (5.78) at $F = F_D$ gives the peak power when a window is used; this is just

$$|Y(F_D)|^2 = \left| A \sum_{m=0}^{M-1} w[m] e^{-j2\pi(0)mT} \right|^2 = A^2 \left| \sum_{m=0}^{M-1} w[m] \right|^2 \quad (5.79)$$

The ratio $|Y(F_D)|^2/|Y(F)|^2$, called the *loss in processing gain* (LPG), is

$$LPG = \frac{1}{M^2} \left| \sum_{m=0}^{M-1} w[m] \right|^2 \quad (5.80)$$

With this definition, $LPG \leq 1$, so the "loss" in dB is a negative number. Using Eq. (5.80), the LPG can be computed for any window. Values of 5 to 8 dB are common. Details depend on the specific window function, but the LPG is typically a weak function of the window length M, highest for small M and rapidly approaching an asymptotic value for large M (on the order of 100 or more).

Although the window reduces the peak amplitude of the DTFT substantially, it also reduces noise power. *Processing loss* (PL) is the reduction in SNR at the peak of the DTFT. Denoting the SNR with and without the window as χ and χ_w, respectively, it is possible to separate the effects of the window on the target and noise components of the signal

$$\frac{\chi_w}{\chi} = \frac{(S_w/N_w)}{(S/N)} = \left(\frac{S_w}{S}\right)\left(\frac{N}{N_w}\right) = LPG\left(\frac{N}{N_w}\right) \quad (5.81)$$

To determine the window's effect on the noise power, suppose $y[m]$ is a zero mean stationary white noise with variance σ^2. Then the windowed noise power is

$$\sigma_w^2 = \mathbf{E}\left\{ \left(\sum_{m=0}^{M-1} w[m]y[m]\right) \left(\sum_{l=0}^{M-1} w^*[l]y^*[l]\right) \right\}$$

$$= \mathbf{E}\left\{ \left(\sum_{m=0}^{M-1} |w[m]y[m]|^2\right) + \text{cross terms} \right\}$$

$$= \sigma^2 \sum_{m=0}^{M-1} |w[m]|^2 = N_w \quad (5.82)$$

The unwindowed noise power N can be obtained from Eq. (5.82) by setting $w[m] = 1$ for all m, giving $N = M\sigma^2$. Using Eq. (5.80) and these values for N and N_w in Eq. (5.82) gives the processing loss

$$PL = \frac{\left|\sum_{m=0}^{M-1} w[m]\right|^2}{M \sum_{m=0}^{M-1} |w[m]|^2} \quad (5.83)$$

TABLE 5.2 Properties of Some Common Data Windows

Window	Mainlobe width (relative to rectangular window)	Peak gain (dB relative to rectangular window)	Peak side lobe, dB	Signal-to-noise ratio loss, dB
Rectangular	1.0	0.0	−13	0
Hann	1.62	−6.0	−32	−1.76
Hamming	1.46	−5.4	−43	−1.35
Kaiser, $\alpha = 2.0$	1.61	−6.2	−46	−1.76
Kaiser, $\alpha = 2.5$	1.76	−8.1	−57	−2.17
Dolph-Chebyshev (50-dB equiripple)	1.49	−5.5	−50	−1.43
Dolph-Chebyshev (70-dB equiripple)	1.74	−6.9	−70	−2.10

Like the loss in peak gain, the processing loss is a weak function of M that is higher for small M but quickly approaches an asymptotic value. As an example, for the Hamming window the loss in signal-to-noise ratio is -1.75 dB for a short ($M = 8$) window, decreasing asymptotically to about -1.35 dB for long windows. Table 5.2 summarizes these four key properties of several common windows.[†] Asymptotic values for large M are given for PL and LPG. A much more extensive table, including both more metrics and many more types of windows, is given by Harris (1978).

5.3.2 Sampling the DTFT: the discrete Fourier transform

In practice, the DTFT is not computed because its frequency variable is continuous. Instead, the discrete Fourier transform is computed

$$Y[k] = \sum_{m=0}^{M-1} y[m] e^{-j2\pi mk/K} \qquad k = 0, \ldots, K-1 \qquad (5.84)$$

As discussed in Chap. 3, for a finite length data sequence, $Y[k]$ is just $Y(F)$ evaluated at $F = k/KT = k(PRF/K)$ hertz. Thus, the DFT computes K samples of the DTFT evenly spaced across one period of the DTFT.

The DFT is almost invariably computed using a *fast Fourier transform* (FFT) algorithm. Most common is the radix 2 or radix 4 Cooley-Tukey algorithm (Oppenheim and Schafer, 1999), which in addition to being fast has a number of good structural properties for implementation in either hardware or software.

[†]These data are from Harris, who uses a slightly different definition of many of the windows than is used in most data analysis and simulation packages, for reasons having to do with DFT symmetry properties. Effectively, Harris' version of an M-point shaped window (e.g., Hamming) is the first M points of the $M + 1$ point symmetrical version more commonly used. The difference is of minor consequence, particularly as M gets large. See the book by Harris (1978) for details.

However, there is no one FFT algorithm; a wide class of such algorithms exist. They differ in the data set length they are suited for; whether they optimize the number of additions, multiplications, or the sum of both; the regularity of their computational flow, how efficiently they use memory, and their quantization noise and signal scaling properties. A thorough derivation and summary of most of the algorithms in current use is given by Burrus and Parks (1985). More recently, there has been an emphasis on developing algorithms that access the data with unit stride, a feature that improves performance on many modern cache-based machines and multiprocessors (Bailey, 1990). Modern FFT software libraries such as FFTW adapt the algorithm to characteristics of the particular machine architecture on which they are hosted to maximize performance (Frigo and Johnson, 1998).

The DTFT of an ideal moving target always has the same functional form, i.e., it "looks the same." It is merely translated to the appropriate center frequency and, if a data window is used, its main lobe is widened and attenuated and the side lobes are reduced. The DFT, however, computes samples of the DTFT only at fixed Doppler frequencies. As was illustrated in Chap. 3, the appearance of a plot of the Doppler spectrum computed using a DFT can be a strong function of the relation between the actual signal frequency and the DFT frequency sample locations.

In some situations, the number of data samples available can be greater than the desired DFT size, that is, $M > K$. This occurs when there is a need to reduce the DFT size for computational reasons, or when the radar timeline permits the collection of more pulses than DTFT samples ("Doppler bins") are required and it is desirable to use the extras to improve the signal-to-noise ratio of the Doppler spectrum measurement. The data turning procedure described in detail in Chap. 3 and Fig. 3.11 allows the use of a K-point DFT while taking advantage of all of the data.

Some caution is needed in applying a data window when the data are modified by zero padding or turning. In either case a length M window should be applied to the data *before* it is either zero padded or turned. Applying a K-point window to the full length of a zero-padded sequence has the effect of multiplying the data by a truncated, asymmetric window (the portion of the actual window that overlaps the M nonzero data points), resulting in greatly increased side lobes. Applying a shortened K-point window to a turned data sequence does not increase side lobes since the full, symmetric window comes into play, but it does reduce spectral resolution and result in DFT samples that are not exactly equal to samples of the DTFT of the windowed M-point original data sequence. However, this effect is much milder than the effect of windowing errors in the zero-padding case.

Because the DFT is a sampled version of the DTFT, the peak value of the DFT obtained for a moving target signal is greatest when the Doppler frequency coincides exactly with one of the DFT sample frequencies, and decreases when the target signal is between DFT frequencies. This reduction in amplitude is called a Doppler *straddle loss*. The amount of loss depends on the particular window used. For a given signal length M, the straddle loss is always greatest for signal frequencies exactly halfway between DFT sample frequencies; and for a given frequency, the maximum straddle loss increases when the DFT size K is decreased. Usually the smallest DFT size considered is $K = M$. Thus the straddle loss can be computed by evaluating Eq. (5.84) with $y[m] = w[m]$, $K = M$, and $k = 1/2$. This is the gain at the halfway point between the $k = 0$ and $k = 1$ DFT bins; it is the same halfway between all other DFT bins as well. The computation is repeated with $k = 0$ to get the peak gain, and the ratio evaluated. To be explicit, consider the rectangular window case; then (Oppenheim and Schafer, 1999)

$$|Y[k]| = \left| \sum_{m=0}^{K-1} e^{-j2\pi mk/K} \right| = \left| \frac{\sin(\pi k)}{\sin(\pi k/K)} \right| \qquad (5.85)$$

Evaluating at $k = 1/2$ gives

$$\left| Y\left[\frac{1}{2}\right] \right| = \frac{\sin(\pi/2)}{\sin(\pi/2K)} = \frac{1}{\sin(\pi/2K)}$$
$$\approx \frac{2K}{\pi} \qquad (5.86)$$

The last step was obtained by assuming that K is large enough to allow a small angle approximation to the sine function in the denominator. $Y[0]$ is obtained either by applying L'Hospital's rule to Eq. (5.85) or computing it explicitly from Eq. (5.84); the result is $Y[0] = K$. Thus the maximum straddle loss for the DFT filterbank with no windowing (rectangular window) is

$$\text{Maximum straddle loss (rectangular window)} = \frac{2K/\pi}{K} = \frac{2}{\pi} \qquad (5.87)$$

In decibels, this is $20\log_{10}(2/\pi) = -3.92$ dB.

A similar calculation for the Hamming window (in its nonsymmetric form as defined in the paper by Harris (1978)) is carried out in the book by Levanon (1988), resulting in a smaller maximum straddle loss of 1.74 dB. Thus, while any nonrectangular window causes a reduction in peak gain, typical windows have the desirable property of having less *variability* in gain as the Doppler shift of the target varies. This effect is illustrated in Fig. 5.16, which shows the maximum DFT output amplitude as a function of the target Doppler shift for rectangular and Hamming windows for the case of $M = K = 16$. The data in this figure are obtained using the more common symmetric definition of the Hamming window. Note the general reduction in peak amplitude for the Hamming-windowed data compared to the unwindowed data. On the other hand, the *variation* in amplitude is significantly less for the windowed data

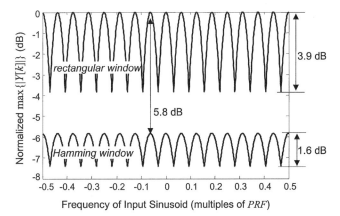

Figure 5.16 Variation of DFT output with complex sinusoid input frequency for two different data analysis windows.

(1.6 dB versus 3.9 dB); i.e., the amplitude response is more consistent.[†] This greater consistency of response is an underappreciated benefit of windowing.

5.3.3 Matched filter and filterbank interpretations of pulse Doppler processing with the DFT

Equation (5.13) defined the coefficients of the matched Doppler filter. In MTI filtering, it is assumed that the target Doppler shift is unknown. The resulting signal model of Eq. (5.22) leads to the pulse canceller as a near-optimum MTI filter for small order N. In contrast, DFT-based pulse Doppler processing attempts to separate target signals based on their particular Doppler shift. Assume that the signal is a pure complex sinusoid (ideal moving target) at a Doppler shift of F_D hertz. Based on Eq. (5.21), the model of the signal vector is then[§]

$$\mathbf{t} = \hat{A} \begin{bmatrix} 1 & e^{j2\pi F_D T} & \cdots & e^{j2\pi F_D (M-1)T} \end{bmatrix}' \tag{5.88}$$

If the interference consists only of white noise (no correlated clutter), $\mathbf{S_I}$ reduces to $\sigma_n^2 \mathbf{I}$, where $\mathbf{I}$ is the identity matrix. It follows that for an arbitrary data vector $\mathbf{x}$ the output $\mathbf{h}'\mathbf{x}$ of the matched filter becomes

$$\mathbf{h}'\mathbf{x} = \tilde{A} \sum_{m=0}^{M-1} y[m] e^{-j2\pi F_D m T} \tag{5.89}$$

[†]Comparing this figure to Table 5.2 illustrates some effects of the difference between Harris' (Harris, 1978) and more conventional definitions of the Hamming window. The symmetric version used for the figure has a peak gain loss relative to the rectangular of 5.8 dB instead of 5.4 dB, and the maximum straddle loss for the Hamming window is 1.61 dB instead of 1.74 dB.

[§]Note that, unlike the derivation of the vector matched filter, in this instance $\mathbf{t}$ is defined as a sequence of samples counting forward in time: $\mathbf{t} = [y[m], y[m+1], \ldots, y[m+M]]$; this is done for later indexing convenience.

When $F_D = k/KT = kPRF/K$ for some integer k, Eq. (5.89) is simply the K-point DFT of the data sequence $x[m]$ (to within a scale factor $\tilde{A}$). Consequently, the discrete Fourier transform is a matched filter to ideal, constant radial velocity moving target signals provided that the Doppler shift equals one of the DFT sample frequencies and the interference is white. Note that this result is very closely related to the two-pulse canceller for noise interference only considered in Sec. 5.2.3.

Since the K-point DFT computes K different outputs from each input vector, it effectively implements a bank of K matched filters at once, each tuned to a different Doppler frequency. The frequency response shape of each matched filter is just the asinc function. Specifically, denote the impulse response vector in Eq. (5.89) when $F_D = k/MT$ as $\mathbf{h}_k$. To within a scale factor

$$\mathbf{h}_k = \begin{bmatrix} 1 & e^{-j2\pi k/K} & e^{-j4\pi k/K} & \cdots & e^{-j2\pi(K-1)k/K} \end{bmatrix}' \quad (5.90)$$

The corresponding discrete time frequency response $H_k(\omega)$ is

$$H_k(\omega) = \sum_{m=0}^{K-1} h_k[m] e^{-j\omega m}$$

$$= \sum_{m=0}^{K-1} (1) e^{-j(\omega + 2\pi k/K)m} \quad (5.91)$$

This summation evaluates to the asinc function of Eq. (5.77), but shifted to a center frequency of $\omega = -2\pi k/K$, which is equivalent to $\omega = 2\pi(K-k)/K$. Thus, the kth DFT sample corresponds to filtering the data with a bandpass filter having a frequency response with an asinc function shape centered at the frequency of the $(K-k)$th DFT samples.

If the data are windowed before processing with a window function $w[m]$, Eq. (5.89) becomes (again for $F_D = k/KT$)

$$\mathbf{h}_k' \mathbf{y} = \tilde{A} \sum_{m=0}^{M-1} w[m] y[m] e^{-j2\pi mk/K} \quad (5.92)$$

The impulse response vector and frequency response are then

$$\mathbf{h}_k = \begin{bmatrix} w[0] & w[1]e^{-j2\pi k/K} & w[2]e^{-j4\pi k/K} & \cdots & w[M-1]e^{-j2\pi(M-1)k/K} \end{bmatrix}' \quad (5.93)$$

$$H_k(\omega) = \sum_{m=0}^{M-1} w[m] e^{-j(\omega + 2\pi k/K)m}$$

$$= W\left(\omega + \frac{2\pi k}{K}\right) \quad (5.94)$$

Thus, the DFT still implements a bandpass filter centered at each DFT frequency, but the filter frequency response shape now becomes that of the window function.

The relation between the DFT and a bank of filters can be made more explicit. Consider a slow-time signal $y[m]$ obtained with a long series of pulses and an

M-point window function $w[m]$. The window function can be slid along the data sequence to select a portion of the data for spectral analysis as shown in Fig. 5.17. The DTFT of the resulting sequence $w[m-p]y[p]$ is, in terms of analog frequency F

$$\begin{aligned}
Y_m(F) &= \sum_{p=-\infty}^{\infty} w[m-p]\,y[p]e^{-j2\pi FpT} \\
&= e^{-j2\pi FmT} \sum_{p=-\infty}^{\infty} w[m-p]\,e^{-j2\pi F(p-m)T}\,y[p] \\
&= e^{-j2\pi FmT}\{(w[p]\,e^{+j2\pi FpT}) * y[p]\}_{p=m}
\end{aligned} \qquad (5.95)$$

Equation (5.95) shows that, aside from a phase factor, the DTFT at a particular frequency is equivalent to the convolution of the input sequence and a modulated window function, evaluated at time m. Furthermore, if $W(F)$ is the discrete time Fourier transform of $w[m]$ (converted to an analog frequency scale), the DTFT of $w[m]\,e^{+j2\pi F_D mT}$ is $W(F+F_D)$, which is simply the Fourier transform of the window shifted so that it is centered at Doppler frequency $-F_D$ hertz. This means that measuring the DTFT at a frequency F_D is equivalent to passing the signal through a bandpass filter centered at $-F_D$ and having a passband shape equal to the Fourier transform of the window function. Since the DFT evaluates the DTFT at K distinct frequencies at once, it follows that pulse Doppler spectral analysis using the DFT is equivalent to passing the data through a bank of bandpass filters.

Of course, it is possible to build a literal bank of bandpass filters, each one perhaps individually designed, and some systems are constructed in this way. For example, the zero-Doppler filter in the filterbank can be optimized to match the expected clutter spectrum, or even made adaptive to account for changing clutter conditions. Most commonly, however, the DFT is used for Doppler spectrum analysis. This places several restrictions on the effective filterbank design. There will be K filters in the bank, where K is the DFT size; the filter center frequencies will be equally spaced, equal to the DFT sample frequencies; and all

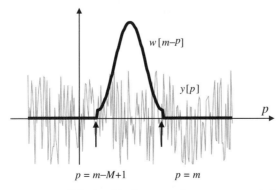

Figure 5.17 Relationship between data sequence $x[m]$ and M-point sliding analysis window $w[m]$.

the passband filter frequency response shapes will be identical, differing only in center frequency. The advantages to this approach are simplicity and speed with reasonable flexibility. The DFT provides a simple and computationally efficient implementation of the filterbank: the number of filters can be changed by simply changing the DFT size; the filter shape can be changed by simply choosing a different window; and the filter optimizes the output signal-to-noise ratio for targets coinciding with a DFT filter center frequency in a noise-limited interference environment.

5.3.4 Fine Doppler estimation

Peaks in the DFT output that are sufficiently above the noise level to cross an appropriate detection threshold are interpreted as responses to moving targets, i.e., as samples of the peak of an asinc component of the form of Eq. (5.77). As has been emphasized, there is no guarantee that a DFT sample will fall exactly on the asinc function peak. Consequently, the amplitude of the DFT sample giving rise to a detection and its frequency are only approximations to the actual amplitude and frequency of the asinc peak. In particular, the estimated Doppler frequency of the peak can be off by as much as one-half Doppler bin, equal to $PRF/2K$ hertz.

If the DFT size K is significantly larger than the number of pulses (data sequence length) M, several DFT samples will be taken on the asinc main lobe, and the largest may well be a good estimate of the amplitude and frequency of the asinc peak. Frequently, however, $K = M$ and sometimes, with the use of data turning, it is even true that $K < M$. In these cases, the Doppler samples are far apart and a half-bin error may be intolerable. One way to improve the estimate of the true Doppler frequency F_D is to interpolate the DFT in the vicinity of the detected peak.

The most obvious way to interpolate the DFT is to zero pad the data and compute a larger DFT. This approach is computationally expensive and interpolates all of the spectrum. If finer sampling is needed only over a small portion of the spectrum, the zero padding approach is inefficient.

From a digital signal processing theoretical point of view, the correct interpolation involves using all of the DFT data samples and an asinc interpolation kernel (effectively, computing a larger DFT). To understand this technique, consider computing the DTFT at an arbitrary value of ω, using only the available DFT samples. This can be done by computing the inverse DFT to recover the original time-domain data, and then computing the DTFT from those samples

$$Y(\omega) = \sum_{m=0}^{M-1} y[m] e^{-j\omega m}$$

$$= \sum_{m=0}^{M-1} \left(\frac{1}{K} \sum_{k=0}^{K-1} Y[k] e^{+j2\pi mk/K} \right) e^{-j\omega m}$$

$$= \frac{1}{K} \sum_{k=0}^{K-1} Y[k] \left\{ \sum_{m=0}^{M-1} \exp\left[-jm\left(\omega - \frac{2\pi k}{K}\right)\right] \right\} \quad (5.96)$$

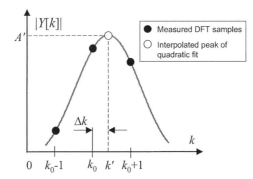

Figure 5.18 Refining the estimated target amplitude and Doppler shift by quadratic interpolation around the DFT peak.

The term in parentheses is the interpolating kernel. It can be expressed in closed form as

$$\sum_{m=0}^{M-1} \exp\left[-jm\left(\omega - \frac{2\pi k}{K}\right)\right]$$

$$= \exp\left[-j\left(\omega - \frac{2\pi k}{K}\right)(M-1)\bigg/2\right] \frac{\sin\left[\left(\omega - \frac{2\pi k}{K}\right)M\bigg/2\right]}{\sin\left[\left(\omega - \frac{2\pi k}{K}\right)\bigg/2\right]}$$

$$\equiv Q_{M,K}(\omega, k) \qquad (5.97)$$

Combining these gives

$$Y(\omega) = \frac{1}{K} \sum_{k=0}^{K-1} Y[k]\, Q_{M,K}(\omega, k) \qquad (5.98)$$

Equations (5.97) and (5.98) can be used to compute the DTFT at any single value of ω from the DFT samples. Thus, it can be applied to interpolate the values of the DFT over localized regions with any desired sample spacing. However, it remains relatively computationally expensive.

A simpler but very serviceable technique for interpolating local peaks is illustrated in Fig. 5.18. For each detected peak in the magnitude of the DFT output, a second-order polynomial (a parabola) is fit through that peak and the two adjacent magnitude data samples. Once the parabola coefficients are known, the amplitude and frequency of its peak are easily found by differentiating the formula for the parabola and setting the result to zero.

To develop this technique, assume that the DFT $Y[k]$ is a function of a *continuous* frequency index k, since the goal is to estimate a peak location assumed to be between actual sample locations. In the vicinity of the DFT peak at k_0, assume that $|Y[k]|$ is of the form

$$|Y[k]| = a_0 + a_1 k + a_2 k^2 \qquad (5.99)$$

Now consider the three measurements $|Y[k_0 - 1]|$, $|Y[k_0]|$, and $|Y[k_0 + 1]|$. If Eq. (5.99) is to be applied, these measurements must satisfy the system of equations

$$\begin{bmatrix} |Y[k_0 - 1]| \\ |Y[k_0]| \\ |Y[k_0 + 1]| \end{bmatrix} = \begin{bmatrix} 1 & k_0 - 1 & (k_0 - 1)^2 \\ 1 & k_0 & k_0^2 \\ 1 & k_0 + 1 & (k_0 + 1)^2 \end{bmatrix} \begin{bmatrix} a_0 \\ a_1 \\ a_2 \end{bmatrix} \quad (5.100)$$

The matrix of coefficients in Eq. (5.100) has the structure of a *Vandermonde matrix* (Hildebrand, 1974); its determinant is

$$\begin{vmatrix} 1 & k_0 - 1 & (k_0 - 1)^2 \\ 1 & k_0 & k_0^2 \\ 1 & k_0 + 1 & (k_0 + 1)^2 \end{vmatrix} = [k_0 - (k_0 - 1)][(k_0 + 1) - k_0][(k_0 + 1) - (k_0 - 1)] = 2$$

(5.101)

Because the determinant is nonzero, a unique solution to this system of equations is guaranteed to exist. The coefficients of the interpolating polynomial can be obtained by solving the matrix Eq. (5.100) or by a variety of other interpolation analysis methods, all of which must produce the same answer since it is unique. The final interpolation equations described here are adapted from a polynomial solution obtained by Lagrangian methods and tabulated in the book by Abramowitz and Stegun (1972).

Write the peak location k' in the form $k' = k_0 + \Delta k$ (see Fig. 5.18); thus Δk is the location of the interpolated peak relative to the index of the central sample of the three DFT samples being used for the estimate. Then the second-order polynomial passing through the three samples can be expressed as

$$|Y[k_0 + \Delta k]| = \frac{1}{2}\{(\Delta k - 1)\Delta k \, |Y[k_0 - 1]| - 2(\Delta k - 1)(\Delta k + 1)\, |Y[k_0]|$$
$$+ (\Delta k + 1)\Delta k \, |Y[k_0 + 1]|\} \quad (5.102)$$

Differentiating this equation with respect to Δk, setting the result to zero, and solving for Δk gives the estimated location of the parabola peak relative to k_0 as

$$\Delta k = \frac{-\frac{1}{2}\{|Y[k_0 + 1]| - |Y[k_0 - 1]|\}}{|Y[k_0 - 1]| - 2|Y[k_0]| + |Y[k_0 + 1]|} \quad (5.103)$$

The amplitude of the estimated peak $A' = |Y[k_0 + \Delta k]|$ is found by computing Δk and then using that result in Eq. (5.102); the result is

$$|Y[k_0 + \Delta k]| = \frac{1}{2}\{(\Delta k - 1)\Delta k \, |Y[k_0 - 1]| - 2(\Delta k - 1)(\Delta k + 1)\, |Y[k_0]|$$
$$+ (\Delta k + 1)\Delta k \, |Y[k_0 + 1]|\} \quad (5.104)$$

Note that the formula for Δk (i.e., for displacement of the estimated peak from the central sample) behaves in intuitively satisfying ways. If the first and third DFT magnitude samples are equal, $\Delta k = 0$; the middle sample is the estimated peak. If the second and third samples are equal, $\Delta k = 1/2$, indicating the estimated peak is halfway between the two samples; a similar result applies if the first and second DFT magnitude samples are equal.

This interpolation technique is ineffective when the width of the presumed main lobe response that is to be interpolated is so narrow that the apparent peak and its two neighbors are not on the same lobe of the response. This occurs when the spectrum is sampled at the Nyquist rate in Doppler, i.e., the DFT size K equals the number of data samples M; no window is applied to the data; and the data frequency does not happen to fall on a DFT frequency sample (the very situation in which interpolation is most needed). Figure 5.19 illustrates this case. The data are $M = 20$ samples of a 1.35 kHz complex sinusoid sampled at 5 kHz. The spectrum was computed using a $K = 20$ point DFT. Without interpolation, the largest DFT magnitude sample has a value of 15.15. This apparent peak amplitude is in error by 24.3 percent from the true DTFT peak value of 20. The DFT samples occur every $5000/20 = 250$ Hz, so the apparent peak occurs at a frequency of 1500 Hz, an error of 150 Hz.

If the interpolation procedure is applied to these data, poor results will be achieved because the assumption that the three points are on an approximately parabolic curve segment is not valid. For these particular data, the interpolation technique will estimate the "true" frequency and amplitude of the spectral peak as 1295.4 Hz and 15.41, respectively. The amplitude estimate is improved only slightly, to a 23 percent error. The frequency error is reduced significantly, to 54.6 Hz, but is still large.

This problem can be avoided by ensuring that the sample set is dense enough to guarantee that the three samples are all on the main lobe. One way to do

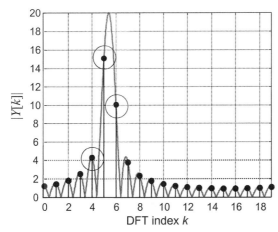

Figure 5.19 Ideal Doppler spectrum due to a moving target, sampled at the Nyquist rate.

this is to oversample in Doppler, i.e., choose $K > M$. Another is to window the data. For most common windows, the expansion of the main lobe that results is sufficient to guarantee that the apparent peak samples and its two neighbors fall on the same lobe, so that the basic assumption of a parabolic segment is more valid. Figure 5.20 illustrates this effect by applying a Hamming window to the same data used for Fig. 5.19 and again applying a 20 point DFT. Note that the peak DTFT amplitude is now 10.34 due to the effect of the Hamming window. Applying the quadratic interpolation to this spectrum gives an estimated spectral peak frequency and amplitude of 1336.6 Hz and 9.676, respectively, errors of only 6.4 percent in amplitude and 13.4 Hz in frequency. In general, the quadratic peak interpolation is more effective at improving the frequency estimate than the amplitude estimate. This is because the interpolating parabola has the same symmetry about its peak as the optimum asinc interpolating function, but is not a particularly close approximation to the asinc peak shape. A hybrid technique can be defined that combines attributes of the parabolic interpolation and the more exact asinc interpolation. The parabolic method is used to identify the frequency of the peak, and then Eq. (5.98) is used to estimate the amplitude. This approach improves amplitude accuracy while avoiding the need to compute Eq. (5.98) more than once.

Figure 5.21 illustrates the frequency estimation performance of the quadratic interpolator, both with and without Hamming windowing, on a sinusoidal data sequence of length $M = 30$. Part a of the figure illustrates the minimally sampled case $K = M$. The interpolated frequency estimates are best when the actual frequency is either very close to a sample frequency, or exactly half way between two sample frequencies. If no window is used, the worst case error of 0.23 bins occurs when the actual frequency is 0.35 bins away from a sample frequency; a Hamming window reduces this maximum error to 0.07 bins at an

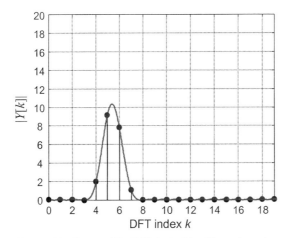

Figure 5.20 Same as Fig. 5.19, but with a Hamming window applied to the data.

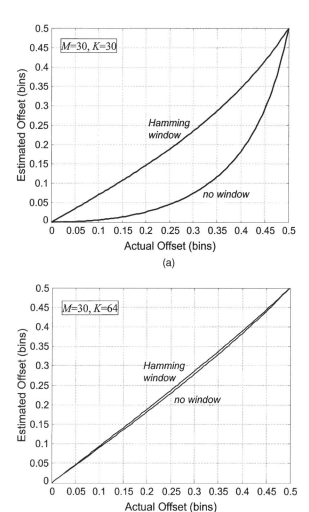

Figure 5.21 Frequency estimation performance of quadratic interpolator. $M = 30$. (a) Minimally sampled case ($K = 30$). (b) Oversampled case ($K = 64$).

offset of 0.31 bins. Figure 5.22 shows the amplitude estimation performance for the same cases. Figure 5.22a shows that the interpolator reduces the straddle loss about 0.7 to 0.8 dB, from 3.92 to 3.22 dB when no window is used, and from 1.68 to 0.82 dB when a Hamming window is used. These results confirm that the quadratic interpolator is a better frequency estimator than amplitude estimator. Though not shown, the maximum frequency estimation errors when the spectrum sampling density is slightly more than doubled, to $K = 64$, is only 0.02 bins without the window and 0.015 bins with the Hamming window,

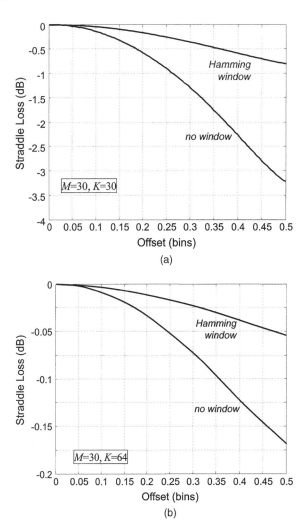

Figure 5.22 Amplitude estimation performance of quadratic interpolator. $M = 30$. (a) Minimally sampled case ($K = 30$). (b) Oversampled case ($K = 64$).

while the straddle loss is reduced to 0.2 dB without a window and only 0.05 dB with the Hamming window.

5.3.5 Modern spectral estimation in pulse Doppler processing

So far, the discrete Fourier transform, implemented with the fast Fourier transform algorithm, has been used exclusively to compute the spectral estimates needed for pulse Doppler processing. Other spectral estimators can be used. One that has been applied to radar is the *autoregressive* (AR) model, which models the actual discrete-time spectrum $Y(\omega)$ of the slow-time signal with a spectrum of the form

$$\hat{Y}(\omega) = \frac{\alpha}{1 + \sum_{p=1}^{P} a_p e^{-j\omega p}} \quad (5.105)$$

The algorithm finds the set of model coefficients $\{a_p\}$ that optimally fits $\hat{Y}(\omega)$ to $Y(\omega)$ for a given model order P. These coefficients are found by solving a set of *normal equations* (Hayes, 1996) derived from the autocorrelation of the slow-time data $y[m]$; the actual spectrum $Y(\omega)$ is not needed. Finally, the $\{a_p\}$ are used to compute an estimated spectrum according to Eq. (5.105), which can then be analyzed for target detection, pulse pair processing, or other functions.

Modeling the spectrum as shown in Eq. (5.105) is equivalent to modeling the slow-time signal $y[m]$ as the output of an IIR filter with frequency response $(1 + \sum_{p=1}^{P} a_p e^{-j\omega p})^{-1}$. The inverse filter is an FIR filter with impulse response coefficients $h[m] = a_m$; if $y[m]$ is passed through this filter, the output power spectrum will be white (provided that the actual noise power spectrum is in fact accurately modeled by Eq. (5.105)). Thus, the FIR filter designed from the model coefficients whitens the signal, removing any correlated signal components such as clutter.

Figure 5.23 illustrates the application of AR spectral estimation to design a clutter filter to enhance detection of windshear from an airborne radar (Keel, 1989). Part *a* shows the Fourier spectrum of the slow-time data from one range bin. Two peaks are evident above the noise floor. The one at zero velocity is ground clutter. The smaller peak at approximately 8 m/s is due to windblown rain. The middle plot shows the frequency response of an optimal clutter filter implemented from the $\{a_p\}$. The third plot shows the Fourier spectrum of the slow-time data after processing with the clutter filter. The ground clutter has been significantly suppressed and the weather echo is now the dominant spectral feature.

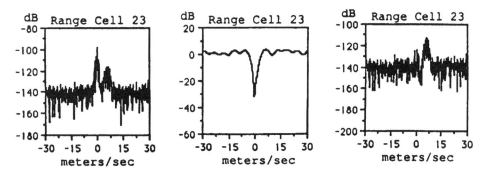

Figure 5.23 Clutter suppression and windshear detection using an autoregressive Doppler spectrum estimate. (*a*) Fourier spectrum of raw data. (*b*) Frequency response of clutter suppression filter frequency response, derived from the AR coefficients. (*c*) Fourier spectrum of filtered data. (*Figure courtesy of Dr. Byron M. Keel, GTRI.*)

5.3.6 Dwell-to-dwell stagger

Pulse Doppler processing sometimes is combined with pulse cancellers (see Sec. 5.6.1). In this case, the applicability of the concept of blind speeds is clear. If a pulse canceller is not used, there is no highpass filter, and therefore a target whose Doppler shift equals an integer multiple of the PRF will not be filtered out, as in MTI processing. However, the target energy will be indistinguishable from clutter energy, since it will alias to the dc portion of the spectrum and combine with the clutter energy. Thus, targets having Doppler shifts equal to a multiple of the PRF will still go undetected, and the corresponding target velocities are still blind speeds.

In dwell-to-dwell PRF stagger, a "dwell" (coherent processing interval) of M pulses is transmitted at a fixed PRF. A second dwell is then transmitted at a different fixed PRF. Because the blind speeds are different for each PRF used, a target that falls in a blind speed of one PRF will be visible in the others. This concept is illustrated in Fig. 5.24, which shows a notional Doppler spectrum for two different PRFs. The plots are shown on the same frequency scale. First consider the upper spectrum plot, which corresponds to data collected at PRF_1. A target whose Doppler equals PRF_1 will be aliased to zero Doppler shift, where it will be undetectable if clutter is present. If the same target scenario is measured with a lower pulse repetition frequency PRF_2 as shown in the lower half of the figure, the Doppler shift of the target no longer matches the PRF. The target energy aliases to a nonzero Doppler, where it does not compete with the clutter and is still detectable.

In some systems, as many as eight PRFs may be used. The first velocity that is completely blind at all of the PRFs is the *least common multiple* (LCM) of

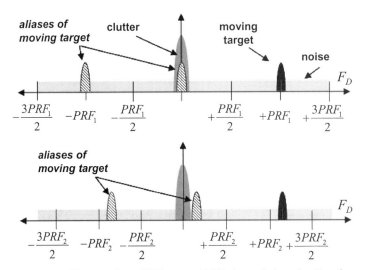

Figure 5.24 The use of two PRFs to avoid blind speeds in pulse Doppler radar. The target Doppler shift equals the PRF in the upper plot, but not in the lower plot.

the individual blind speeds, which will be much higher than any one of them alone. Target detections are accepted and passed to subsequent processing (e.g., tracking) only if they occur in some minimum fraction of the PRFs used, for example one out of two or three, or three out of eight PRFs. In medium PRF systems, particularly those using a small number of PRFs, each of the "major" PRFs used for extending the unambiguous Doppler region may be accompanied by one or two additional "minor" PRFs to resolve range ambiguities as well (see Sec. 5.5.3).

The advantages of a dwell-to-dwell stagger system are that multiple-time-around clutter can be canceled using coherent MTI and that the radar system stability, particularly in the transmitter, is not as critical as with a pulse-to-pulse stagger system (Schleher, 1991). The disadvantage is that the overall velocity response may not be very good, and that the transmission of multiple dwells consumes large amounts of the radar timeline.

5.4 Pulse Pair Processing

Pulse pair processing (PPP) is a form of Doppler processing common in meteorological radar. Unlike the MTI and pulse Doppler techniques discussed so far in this chapter, the goal of pulse pair processing is not clutter suppression to enable the detection of moving targets. In PPP, it is assumed that the spectrum of the slow-time data consists of noise and a single Doppler peak, generally not located at zero Doppler (though it could be), that is due to echo from moving weather events, typically wind-blown rain or other particulates. The goal of PPP is to estimate the radial velocity component of the wind. It is used extensively in both ground-based and airborne weather radars for storm tracking and weather forecasting. In airborne radars, it is also used for windshear detection.

Pulse pair processing assumes the radar is looking generally upward if it is ground-based, or forward if it is airborne. Consequently, it is assumed that ground clutter competing with the weather signatures is small or negligible, or has been removed by MTI filtering. The notional Doppler spectrum $S_y(F)$ assumed by PPP is shown in Fig. 5.25. It consists only of white noise and a single spectral peak due to backscatter from weather-related phenomena

$$S_y(F) = S_w(F) + S_n(F) \tag{5.106}$$

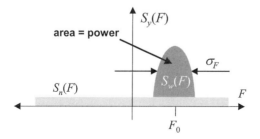

Figure 5.25 Notional slow-time power spectrum assumed in pulse pair processing.

The weather peak $S_w(F)$ is often assumed to be approximately Gaussian shaped, and is characterized by its amplitude, mean, and standard deviation. The total area of the $S_w(F)$ power spectrum component equals the power of the meteorological echo. The goal of PPP is to estimate the power, mean Doppler shift F_0, and variance (commonly called the *spectral width*) σ_F^2 of the weather component. Each of these can be estimated using either time- or frequency-domain algorithms, all of which are included under the PPP rubric.

Consider time-domain measurements first. The autocorrelation and power spectrum of the slow-time data sequence $y[m]$, $m = 0, \ldots, M-1$ obtained from M pulses sampled at a particular range bin are, respectively

$$s_y[k] \equiv \sum_{m=0}^{M-k-1} y[m]\, y^*[m+k] \tag{5.107}$$

$$S_y(\omega) = \mathbf{F}\{s_y[k]\} = |Y(\omega)|^2 \tag{5.108}$$

The power in the slow-time signal can be estimated in the time domain from the peak of the autocorrelation function

$$\hat{P}_y = s_y[0] = \sum_{m=0}^{M-1} y[m]\, y^*[m] = \sum_{m=0}^{M-1} |y[m]|^2 \tag{5.109}$$

Note that Eq. (5.109) is more accurately considered the energy in the slow-time signal; dividing by the signal duration would give a measure of power. Nonetheless, the scale factor is often ignored and the zero autocorrelation lag used directly as a power measurement.

To see how to estimate the mean frequency, ignore the noise for the moment and assume that the signal component is a pure sinusoid; note that the power spectrum of a finite segment would then be an asinc function squared. Now compute the first autocorrelation lag

$$y[m] = Ae^{j2\pi F_0 T m} \quad m = 0, \ldots, M-1$$

$$\begin{aligned}
s_y[1] &= \sum_{m=0}^{M-2} y[m] y^*[m+1] \\
&= \sum_{m=0}^{M-2} Ae^{+j2\pi F_0 T m} A^* e^{-j2\pi F_0 T (m+1)} = |A|^2 \sum_{m=0}^{M-2} e^{-j2\pi F_0 T} \\
&= |A|^2 e^{-j2\pi F_0 T} \sum_{m=0}^{M-2} (1) = |A|^2 (M-1) e^{-j2\pi F_0 T} \tag{5.110}
\end{aligned}$$

The argument of the exponential $-2\pi F_0 T$ is simply the amount of phase rotation in one sample period for a sinusoid of analog frequency F_0. The frequency can be determined from Eq. (5.110) as

$$\hat{F}_0 = \frac{-1}{2\pi T} \arg\{s_y[1]\} \tag{5.111}$$

Multiplying $\hat{F}_0$ by $\lambda/2$ converts the result into units of velocity. Although derived for the pure sinusoid, this time-domain PPP frequency estimator works well for more general signals, provided there is a single dominant frequency component with adequate signal-to-noise ratio. Note that the frequency estimate will be aliased if the Doppler frequency exceeds the PRF.

Since the complex exponential inside the summation in Eq. (5.110) does not depend on m and so could be brought out of the sum, it is not necessary to compute the full autocorrelation lag $s_y[1]$; it would suffice to simply compute $y[m]y^*[m+1]$ using only two slow-time samples. In reality, noise is present in all of the samples, and using all M available samples in the full summation averages the noise and improves the estimate quality.

To obtain a time-domain estimate of σ_F^2, assume that the Doppler power spectrum exhibits a Gaussian shape with standard deviation σ_F. Assume an estimate of F_0 is used to remove the mean Doppler component, giving a modified sequence $y'[m]$. It is convenient to start with the continuous-time equivalent $y'(t)$. $S_{y'}(F)$ will be a zero-mean Gaussian function

$$S_{y'}(F) = \frac{|A|^2}{\sqrt{2\pi}\sigma_F} e^{-F^2/2\sigma_F^2} \tag{5.112}$$

It follows that the continuous-time autocorrelation function is also Gaussian

$$s_{y'}(z) = |A|^2 e^{-2\pi^2 \sigma_F^2 z^2} \tag{5.113}$$

where the variable z represents the autocorrelation lag. If the sampling interval T is chosen sufficiently small to guarantee that $S_{y'}(1/2T) \approx 0$, the discrete time spectrum and autocorrelation will also form a Gaussian pair to a very good approximation. The sampled autocorrelation function will be

$$s_{y'}[k] = s_{y'}(z)|_{z=kT} = |A|^2 e^{-2\pi^2 \sigma_F^2 k^2 T^2} \tag{5.114}$$

and the corresponding DTFT, still in units of hertz, is

$$S_{y'}(F) = \frac{|A|^2}{\sqrt{2\pi}\sigma_F T} e^{-F^2/2\sigma_F^2} \tag{5.115}$$

Because $s_{y'}[0] = |A|^2$, the first autocorrelation lag can be written

$$s_{y'}[1] = |A|^2 e^{-\pi \sigma_F^2 T^2} = s_{y'}[0] e^{-\pi \sigma_F^2 T^2} \tag{5.116}$$

Equation (5.116) is easily solved to give an estimate for the spectrum standard deviation in terms of only $s_{y'}[0]$ and $s_{y'}[1]$

$$\hat{\sigma}_F^2 = -\frac{1}{2\pi^2 T^2} \ln \left\{ \frac{s_{y'}[1]}{s_{y'}[0]} \right\} \qquad (5.117)$$

Equations (5.109), (5.111), and (5.117) are the time-domain pulse pair processing estimators. They can be computed from only two autocorrelation lags of the slow-time data.

Equation (5.117) is sometimes simplified to avoid the natural logarithm calculation. Consider the following series expansion and approximation of $\ln(x)$

$$\ln x = \frac{x-1}{x} + \frac{1}{2}\left(\frac{x-1}{x}\right)^2 + \frac{1}{3}\left(\frac{x-1}{x}\right)^3 + \cdots$$

$$\approx \frac{x-1}{x} = 1 - \frac{1}{x} \qquad (5.118)$$

Applying Eq. (5.118) to Eq. (5.117) gives the simplified width estimator

$$\hat{\sigma}_F^2 = -\frac{1}{2\pi^2 T^2} \left\{ 1 - \frac{s_{y'}[0]}{s_{y'}[1]} \right\} \qquad (5.119)$$

The basic PPP measurements of signal power, frequency, and spectral width can also be performed in the frequency domain. The power is obtained by applying Parseval's theorem to Eq. (5.109) (Oppenheim and Schafer, 1999)

$$\hat{P}_y = \frac{1}{2\pi} \int_{-\pi}^{+\pi} S_y(\omega) \, d\omega \, \frac{1}{2\pi} \int_{-\pi}^{+\pi} |Y(\omega)|^2 \, d\omega \qquad (5.120)$$

A practical calculation uses the DFT version of Parseval's theorem with the discrete Fourier transform $Y[k]$ of $y[m]$

$$\hat{P}_y = \frac{1}{M} \sum_{k=0}^{M-1} |Y[k]|^2 \qquad (5.121)$$

There are two frequency-based methods for estimating the mean frequency of the signal. The first is a direct analog to Eq. (5.111)

$$\hat{F}_0 = \frac{-1}{2\pi T} \arg\{s_y[1]\}$$

$$= \frac{-1}{2\pi T} \arg \left\{ \frac{1}{2\pi} \int_{-\pi}^{+\pi} |Y(\omega)|^2 e^{j\omega} \, d\omega \right\} \qquad (5.122)$$

The integrand in Eq. (5.122) is real except for the $e^{j\omega}$ term. Noting that for a complex number z, $\arg\{z\} = \operatorname{atan}\{\operatorname{Im}(z)/\operatorname{Re}(z)\}$, Eq. (5.122) gives the estimator

$$\hat{F}_0 = \frac{-1}{2\pi T} \operatorname{atan}\left\{ \frac{\int_{-\pi}^{+\pi} |Y(\omega)|^2 \sin\omega\, d\omega}{\int_{-\pi}^{+\pi} |Y(\omega)|^2 \cos\omega\, d\omega} \right\} \quad (5.123)$$

The other frequency-domain frequency estimator, as well as an estimator of the spectral width, results from viewing the signal spectrum of Fig. 5.25 as a probability density function. A valid pdf must be real and nonnegative, a condition met by the power spectrum. However, a pdf must also have unit area, so the power spectrum must be normalized to ensure this is the case. By Parseval's theorem, the integral of $|Y(\omega)|^2 = 2\pi E_y$, where E_y is the energy in $y[m]$; this is the required normalization factor. For any arbitrary pdf $p_z(z)$, the mean and variance are given by

$$\bar{z} = \int_{-\infty}^{+\infty} z p_z(z)\, dz$$
$$\sigma_z^2 = \int_{-\infty}^{+\infty} (z - \bar{z})^2\, p_z(z)\, dz \quad (5.124)$$

Applying the first of these definitions to the power spectrum gives an alternative mean frequency estimator

$$\hat{F}_0 = \frac{1}{2\pi T} \int_{-\pi}^{+\pi} \omega(|Y(\omega)|^2/2\pi E_y)\, d\omega = \frac{1}{4\pi^2 T E_y} \int_{-\pi}^{+\pi} \omega|Y(\omega)|^2\, d\omega \quad (5.125)$$

Similarly, an estimator of the spectral width is

$$\hat{\sigma}_F^2 = \frac{1}{(2\pi T)^2} \int_{-\pi}^{+\pi} (\omega - \omega_0)^2 (|Y(\omega)|^2/2\pi E_y)\, d\omega$$
$$= \frac{1}{8\pi^3 T^2 E_y} \int_{-\pi}^{+\pi} (\omega - \omega_0)^2 |Y(\omega)|^2\, d\omega \quad (5.126)$$

Generally, the time-domain estimators are preferred if the signal-to-noise ratio is low or the spectral width is very narrow (Doviak and Zrnic, 1993). In the latter case, the signal is closer to the pure sinusoid assumption that motivated the time-domain estimator. In addition, the time-domain methods are more computationally efficient, because no Fourier transform calculations are required. Conversely, the frequency-domain estimators tend to provide better estimators at high SNR and large spectral widths. The frequency domain estimator also allows reduction of the noise before the estimates are calculated, reducing the estimate bias. This process, called *spectral subtraction*, is depicted in Fig. 5.26. The noise power spectrum $N(\omega)$ is estimated from a presumed clear region of the spectrum, and then simply subtracted off to form a reduced-noise power spectrum

$$S'_y(\omega) = S_y(\omega) - N(\omega) \quad (5.127)$$

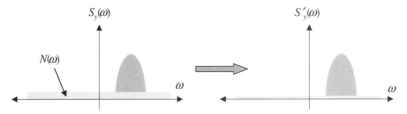

Figure 5.26 Spectral subtraction.

Because of the statistical variations of any given data set, it is possible that $S'_y(\omega)$ may have some negative values; these are usually set to zero.

Figure 5.27 shows two images from the KFFC WSR-88D NEXRAD weather radar, located in Peachtree City, Georgia, just south of Atlanta. The images were collected on March 19, 1996, and show a heavy storm in the area. While these images are much more easily viewed in color than in grayscale, some features are visible. The image on the left is a map of the power (reflectivity) estimate. Lighter grays represent areas of heavier rainfall. The image on the right is a map of the radial velocity measured by the radar, and thus of the wind speed. The large area on the left and top-left of the image with the circular inner boundary represents a range-aliased region where reliable velocity estimates are not available. The radar itself is at the center of the circle having

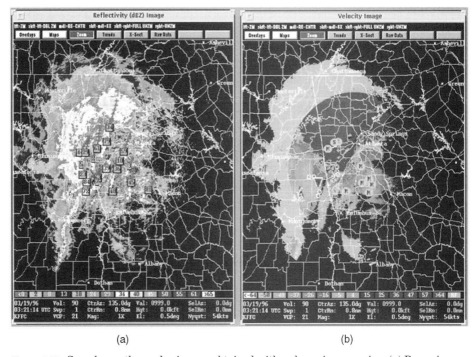

Figure 5.27 Sample weather radar images obtained with pulse pair processing. (*a*) Power image. (*b*) Velocity image. (*Images courtesy of National Severe Storms Laboratory.*)

this boundary. Inside of that radius, the black area at the top (red in the original image) represents a high velocity toward the radar, while the dark gray area inside the circular contour and to the left and bottom of the image represents high wind speeds away from the radar. Thus, the wind is blowing from the top to the bottom in this image. The various square and round markers in the power and velocity images are created by the analysis software and flag various features in the storm.

5.5 Additional Doppler Processing Issues

5.5.1 Combined MTI and pulse Doppler processing

It is not unusual to have both MTI filtering for gross clutter removal and pulse Doppler filterbanks for detailed analysis of the clutter-canceled spectrum. Since both operations are linear, the order in which they are applied would appear to make no difference to the final Doppler spectrum used for detection. However, differences in signal dynamic range can make their order significant in considering hardware effects, primarily when finite-wordlength hardware is used.

Clutter is usually the strongest component of the signal; it can be several tens of decibels above the target signals of interest. If the signal is applied to a pulse Doppler filterbank prior to MTI filtering, the side lobes of the response from the clutter around dc may swamp potential target responses at near-in velocities, masking these targets from possible detection. If the processor dynamic range is limited as well, the effect of the strong clutter signal on processor gain control may drive the target signal amplitude below the minimum detectable signal of the processor, effectively filtering out the target.

For these reasons, the MTI filter is generally placed first if both processes are used. The MTI filter will attenuate the clutter component selectively, so that the target signals become the dominant components. Subsequent finite wordlength processing will adapt the dynamic range to the targets rather than the now-suppressed clutter. In floating point processors, dynamic range is less of an issue.

5.5.2 Transient effects

All of the discussion in this chapter has assumed a steady-state scenario in the sense that the clutter spectrum is stationary and that filter transient effects have been ignored. In range-ambiguous medium and high PRF modes each received signal sample (range gate or bin) contains contributions from multiple ranges because of the multiple contributing pulses. Whenever the radar PRF changes, several pulses, known as *clutter fill pulses*, must be transmitted before a steady-state situation is achieved. For example, suppose that in steady state each range gate contains significant contributions from four pulses (four range ambiguities). Then the fourth pulse is the first one for which steady-state operation is possible. The first three pulses are clutter fill pulses, and should not be used in pulse Doppler processing. Additional pulses may be used to set the automatic gain control of the receiver and are also not used for Doppler processing.

Steady-state operation of the digital filters used for MTI processing occurs when the output value depends only on actual data input values, rather than any initial (typically zero-valued) samples used to initialize the processing. For FIR filters of order N (length $N+1$), the first N outputs are transient and are discarded in some systems. For simple single or double cancellers, these are only one or two samples. Because each slow-time sample is obtained from a different radar pulse, the N samples required to obtain steady-state filter operation represent N additional pulses above and beyond any clutter fill pulses.

5.5.3 PRF Regimes and Ambiguity Resolution

As was seen in Chap. 4, measurements made with a pulse burst waveform can be ambiguous in range, Doppler, or both. Pulse Doppler radars in particular frequently operate in scenarios that are ambiguous in one or both of the range and Doppler dimensions. Specifically, modern airborne pulse Doppler radars operate in a dizzying variety of modes having various range and Doppler span and resolution requirements. Pulse burst waveforms using a variety of constituent pulses, including simple pulses, LFM, and Barker phase codes at a minimum, are common. To meet the various mode requirements, PRFs ranging from several hundred hertz to 100 kHz or more are used.

Pulse Doppler radar operation is commonly divided into three *regimes* according to their ambiguity characteristics. Given an unambiguous range R_{ua} and unambiguous velocity v_{ua} of interest, the radar is considered to be in a *low PRF mode* if the PRF is sufficiently low to be unambiguous in range, but is ambiguous in velocity. The *high PRF mode* is the opposite: the system is ambiguous in range but not in velocity. In a *medium PRF mode*, the radar is ambiguous in both. This tradeoff is summarized in Fig. 5.28. The line plots the achievable combinations of R_{ua} and v_{ua} at 10 GHz. If the desired range and velocity coverage are 100 km and 100 m/s, respectively, the shaded area indicates a range of PRFs (in this case, 1.5 to 6.67 kHz) that will result in ambiguities in both dimensions. PRFs above 6.67 kHz will be ambiguous in range but not Doppler; those below 1.5 kHz will be ambiguous in Doppler but not range. Fortunately, techniques exist that can resolve ambiguities, although at the cost of extra measurement time and processing load.

Consider range ambiguity resolution first. Once a PRF is selected, it establishes an unambiguous range $R_{ua} = c/2PRF$. A target at an actual range $R_t > R_{ua}$ will be detected at an apparent range R_a that satisfies

$$R_t = R_a + kR_{ua} \tag{5.128}$$

for some integer k. Equivalently

$$R_a = ((R_t))_{R_{ua}} \tag{5.129}$$

where the notation $((\cdot))_x$ denotes modulo x. Normalize the range measurements to the range bin spacing ΔR, for example, $n_a = R_a/\Delta R$; then

$$n_t = n_a + kN \quad \Rightarrow \quad n_a = ((n_t))_N \tag{5.130}$$

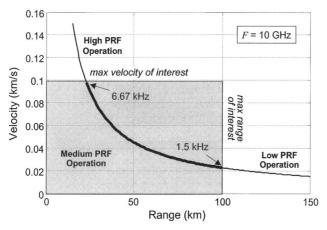

Figure 5.28 Low, medium, and high PRF regimes for a notional X-band radar.

The basic approach to resolving range ambiguities relies on multiple PRFs. Suppose that there are N_i range bins in the unambiguous range interval on PRF i; thus, $R_{ua_i} = N_i \Delta R$. Note that the unambiguous range is different for each PRF. For simplicity, assume that the range bin spacing is the same in each PRF. Then

$$n_t = n_{a_0} + k_0 N_0 = n_{a_1} + k_1 N_1 = \cdots \Rightarrow n_{a_i} = ((n_t))_{N_i} \quad (5.131)$$

The set of in Eq. (5.131) is called a set of *congruences*.

The set of congruences can be solved using the *Chinese remainder theorem* (CRT) (Trunk and Brockett, 1993). The CRT states that, given a set of r relatively prime integers $N_0, N_1, \ldots, N_{r-1}$ and the set of congruences in Eq. (5.131), there exists a unique solution (modulo $N = N_0 N_1 \cdots N_{r-1}$) for n_t given by the equations

$$n_t = k_0 \beta_0 n_{a_0} + k_1 \beta_1 n_{a_1} + \cdots + k_{r-1} \beta_{r-1} n_{a_{r-1}}$$

$$k_i = N/N_i = \prod_{j=0, j \neq i}^{r-1} N_j \quad \beta_i = ((k_i^{-1}))_{N_i} \Rightarrow ((\beta_i k_i))_{N_i} = 1 \quad (5.132)$$

To make the procedure clearer, consider the case of $r = 3$ PRFs, a common choice in airborne radar. Then n_t satisfies

$$n_t = ((\alpha_0 n_{a_0} + \alpha_1 n_{a_1} + \alpha_2 n_{a_2}))_{N_0 N_1 N_2} \quad (5.133)$$

where

$$\alpha_i = \beta_i k_i = \beta_i \prod_{\substack{j=0 \\ j \neq i}}^{2} N_j \quad (5.134)$$

(for example, $\alpha_1 = \beta_1 N_0 N_2$) and the β_I are the smallest integers such that

$$((\beta_0 \, N_1 \, N_2))_{N_0} = 1, ((\beta_1 \, N_0 \, N_2))_{N_1} = 1, ((\beta_2 \, N_0 \, N_1))_{N_2} = 1 \qquad (5.135)$$

To illustrate the procedure, suppose that the true range of a target, normalized to the range bin size, is $n_t = 19$. Further suppose that the three PRFs are chosen such that the number of range bins in the unambiguous range for each PRF are $N_0 = 11$, $N_1 = 12$, and $N_2 = 13$. On the first PRF, the target will be detected in the apparent range bin $n_{a_0} = ((19))_{11} = 8$. Similarly, $n_{a_1} = ((19))_{12} = 7$ and $n_{a_2} = ((19))_{13} = 6$. From Eq. (5.135), β_0 is the smallest integer that satisfies $((\beta_0 \cdot 12 \cdot 13))_{11} = 1$; that is, β_0 satisfies $\beta_0 \cdot 12 \cdot 13 = 156\beta_0 = 11k + 1$ for some integer k. The solution is $\beta_0 = 6$. In the same manner, it is found that $\beta_1 = 11$ and $\beta_2 = 7$. Eq. (5.134) then gives $\alpha_0 = 6 \cdot 12 \cdot 13 = 936$, $\alpha_1 = 1573$, and $\alpha_2 = 924$. Finally, Eq. (5.133) gives the estimate of the true range bin as

$$\hat{n}_t = \left((\alpha_0 n_{a_0} + \alpha_1 n_{a_1} + \alpha_2 n_{a_2}) \right)_{N_1 N_2 N_3} = 19 \qquad (5.136)$$

which is of course correct.

A serious problem with the CRT is its extreme sensitivity to errors induced by noise and range quantization. There is no guarantee that the actual range R_t will be an integer multiple of the range bin spacing ΔR as assumed previously; the target may in fact straddle range bins. In addition, noise in the measurements may cause the target to be located in an incorrect range bin. To illustrate the effect of such errors, repeat the previous example, but suppose that n_{a_2} is measured to be 7 instead of the correct value of 6. Carrying out the previous calculations will give $\hat{n}_t = 943$ instead of 19; a huge error.

Before addressing this problem directly, it is useful to introduce a *coincidence algorithm* for determining n_t. This technique is essentially a graphical implementation of the CRT (Hovanessian, 1976; Morris and Harkness, 1996). The method is best illustrated with an example. Again presume that $r = 3$ PRFs are used. Suppose that there are two targets with true ranges corresponding to range bins $n_a = 6$ and $n_b = 11$. Further suppose that PRFs are such that the number of range bins in each unambiguous range interval are $N_0 = 7$, $N_1 = 8$, and $N_2 = 9$. This means that the first target is actually unambiguous at each PRF, while the second is ambiguous at each PRF. The measured data will be

$$\begin{aligned} n_{a_0} = n_{a_1} = n_{a_2} = 6 \\ n_{b_0} = 4, \quad n_{b_1} = 3, \quad n_{b_2} = 2 \end{aligned} \qquad (5.137)$$

This measurement scenario is illustrated in Fig. 5.29.

The graphical technique proceeds by taking the pattern of detections at each PRF and replicating it as shown in Fig. 5.30. In essence, the replication implements Eq. (5.130), placing a detection at each value of $n_a + kN_0$ (and $n_b + kN_0$). These detections represent the plausible ranges for each target at each PRF. The algorithm then searches for a range bin that exhibits a detection at all three PRFs, indicating that that range bin is consistent with the

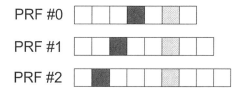

Figure 5.29 Notional measured data for illustrating coincidence algorithm for range ambiguity resolution.

measurements at all three PRFs. As shown in Fig. 5.30, this process correctly detects the true range bins $n_a = 6$ and $n_b = 11$ in this example.

The graphical interpretation suggests various methods to reduce the sensitivity of the CRT to measurement errors. In one approach, exact coincidence is not required to declare a target. Instead, a tolerance N_T is established and a detection is declared if a detection occurs in all three PRFs at some range bin $n_t \pm N_T$. Depending on the range bin size and SNR, N_T will typically be only 1 or 2 range bins. A more sophisticated version of this basic idea is described by Trunk and Kim (1994). Their method uses a systematic search of the plausible ranges to find a set composed of one plausible range from each PRF such that the set has the smallest mean-square scatter of all candidate sets.

In the last example, three PRFs proved sufficient to resolve two different range-ambiguous targets. In general, N PRFs are required to successfully disambiguate $N - 1$ targets. If the number of targets exceeds $N - 1$, *ghosts* can appear (Morris and Harkness, 1996). Ghosts are false targets resulting from false coincidences of range-ambiguous data from different targets. The problem is illustrated in Fig. 5.31, which repeats the example of Fig. 5.30 using only two of the previous three PRFs; specifically, the first and third PRFs. While targets will still be detected at the correct bins $n_a = 6$ and $n_b = 11$, a third coincidence occurs between detections from targets 1 and 2 at range bin $n_c = 20$, representing an apparent third target. Unless additional data (e.g., tracking information) is available, the signal processor has no way of recognizing that the last coincidence is among detections from different targets. Thus, the processor will declare the presence of three targets in this example, the two correct targets and one "ghost." Use of a third PRF as in Fig. 5.30 eliminates this ghost.

In a medium or high PRF mode, the radar will also suffer velocity ambiguities. This problem is identical to that of range ambiguities: given an apparent Doppler shift F_t, the actual Doppler shift must be of the form $F_t + kPRF$, for some integer k. Use of the DFT for spectral estimation results in quantization of the Doppler spectrum into Doppler bins (equivalently, velocity bins), analogous

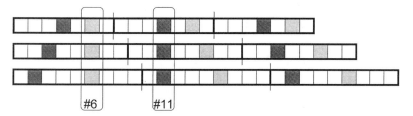

Figure 5.30 Coincidence detection of target ranges in replicated range data.

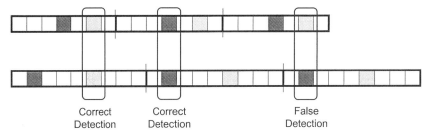

Figure 5.31 Formation of ghosts in range ambiguity resolution.

to range bins in the range dimension. The same techniques used for range disambiguation can therefore be used to resolve velocity ambiguities as well.

5.6 Clutter Mapping and the Moving Target Detector

5.6.1 Clutter mapping

All of the MTI and pulse Doppler processing discussed so far has been focused on reducing the clutter power that interferes with the signature of a moving target, thus improving the signal-to-interference ratio and ultimately the probability of detection. These techniques are not effective for targets with little or no Doppler shift, and that therefore are not separable from the clutter based on Doppler shift. *Clutter mapping* is a technique for detection of moving targets with zero or very low Doppler shift. It is typically used by ground-based scanning radars such as airport surveillance radars. It is intended for maintaining detection of targets on crossing paths, that is, passing orthogonal to the radar line of sight so that the radial velocity is zero; such targets are discarded by MTI and pulse Doppler processing. Clutter mapping can be effective if the target RCS is relatively large and the competing clutter is relatively weak, a situation depicted in Fig. 5.32.

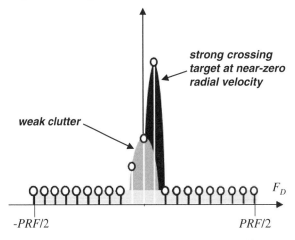

Figure 5.32 Pulse Doppler spectrum for a large RCS crossing target in weak clutter.

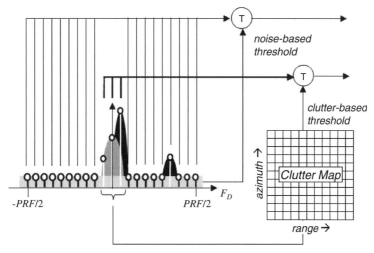

Figure 5.33 The concept of clutter mapping for detection of strong targets in clutter.

The concept of clutter mapping is shown in Fig. 5.33, which presumes that conventional pulse Doppler processing is applied to targets having Doppler shifts sufficient to separate them from the ground clutter. The output of the zero-Doppler bin is used to create a literal stored map of recent clutter echo power for each range-azimuth cell in the radar's search area. This map is updated continuously to allow for clutter variations due to weather and other environmental changes. On each scan, the received power in the nonzero Doppler bins is applied to a conventional threshold detector, using a threshold based on the noise that dominates the interference in those bins. The current zero-Doppler received power for each range-azimuth cell, instead of being discarded, is applied to a separate detector using a threshold based on the stored clutter power level for that cell. The details of threshold detection are discussed in Chap. 6; the clutter map procedure, which is a form of *constant false alarm rate* (CFAR) detection.[†]

Instead of using the zero-Doppler output of a pulse Doppler processor (typically a fast Fourier transform of the slow-time data), many clutter maps systems pass the I/Q slow-time data through a separate "zero velocity filter" as shown in Fig. 5.34. The zero velocity filter serves the opposite purpose of an MTI filter. It is a lowpass design, the output of which consists only of ground clutter and crossing target returns. The design of the zero-Doppler filter can be optimized for the clutter environment at a specific radar site, and can also be made adaptive to clutter changes, for instance due to weather in the area.

[†]CFAR techniques are introduced in Chap. 7.

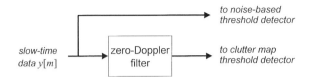

Figure 5.34 Zero-Doppler filter used to isolate low-Doppler targets and ground clutter.

5.6.2 The moving target detector

The *moving target detector* (MTD) is a term applied to the Doppler processing system used in many airport surveillance radars. The MTD combines all of the techniques discussed previously, among others, to achieve good overall moving target detection performance. A block diagram of the original MTD is shown in Fig. 5.35 (Nathanson, 1991). The upper channel begins with a standard three-pulse canceller. The clutter-cancelled output is then applied to an 8-point FFT for pulse Doppler analysis. Two PRFs are used in a block-to-block stagger to extend the unambiguous velocity region. The "frequency domain weighting" is an implementation of time-domain windowing of the data. Certain windows, including for example the Hamming, can be efficiently implemented as a convolution in the frequency domain with a 3-point kernel. The individual FFT samples are applied to a 16-range-bin CFAR threshold detector, with thresholds selected separately for each frequency bin.

To provide some detection capability for crossing targets, the lower channel uses a site-specific zero-velocity filter to isolate the echo from clutter and low-Doppler targets. The output is again applied to a clutter map threshold detector. The original MTD updated the clutter map using an 8-scan moving average, corresponding to 32 s of data history (Skolnik, 2001).

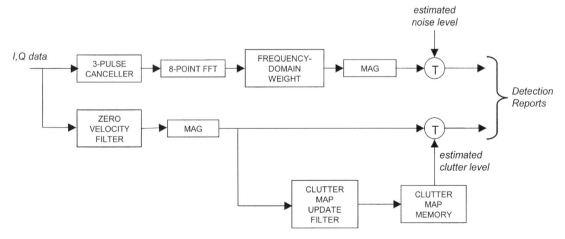

Figure 5.35 Block diagram of a complete "moving target detector" system.

The MTD design has progressed through several generations since the implementation described here. Versions used in the ASR-9 and ASR-12 airport surveillance radars are described by Taylor and Brunins (1985) and Cole et al. (1998), respectively.

5.7 MTI for Moving Platforms: Adaptive Displaced Phase Center Antenna Processing

5.7.1 The DPCA concept

MTI filtering and pulse Doppler processing provide an effective way to detect moving targets whose Doppler shift is in the clear region of the spectrum on at least one PRF. Airborne targets can generally be detected in this manner. However, slow-moving ground targets having actual Doppler shifts only slightly higher than the ground clutter will appear in the skirts of clutter spectrum at all PRFs and are therefore very difficult to detect. Recall that platform motion can substantially spread the ground clutter spectrum as described in Eq. (5.75). This spread of main lobe clutter exacerbates the problem, raising the minimum velocity at which slow-moving ground targets can be detected. This phenomenon is illustrated in Fig. 5.36, which shows the spreading of the main lobe clutter by the platform motion. Because of this spreading, clutter energy may compete directly with relatively slow-moving targets ("slow movers," typically surface targets such as vehicles on land and ships on the sea), making their detection difficult. Processing techniques intended to detect slow movers from moving platforms are referred to as *ground moving target indication* (GMTI) or *surface moving target indication* (SMTI).

Displaced phase center antenna (DPCA) processing is a technique for countering the platform-induced clutter spectral spreading. By minimizing the clutter spectral width, DPCA improves the probability of detection for slow-moving targets. It is a special case of the more general *space-time adaptive processing* (STAP) introduced in Chap. 9. The basic concept is to make the antenna appear stationary even though the platform is moving forward by electronically moving the receive aperture backward during operation. More specifically, DPCA processing attempts to compensate for aircraft motion by using multiple receive subapertures to create carefully controlled multiple virtual phase centers

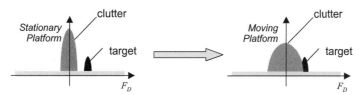

Figure 5.36 Illustration of the effect of a moving radar platform on the Doppler spectrum and the detection of "slow movers."

such that data received on one subaperture have the same virtual phase center as the data received on a *different* subaperture some time later. By properly combining pulses, staggered by this delay time, from the data streams corresponding to different receive subapertures, the illusion of data received by a *stationary* antenna can be created so that effective MTI cancellation can be achieved. References for basic DPCA are Skolnik (1980), Shaw and McAulay (1983), Staudaher (1990), and Lightstone et al. (1991).

Figure 5.37 illustrates the concept using an electronic antenna that has two subapertures. The entire antenna is used on transmission for maximum gain, so the phase center for transmission is the point **T** in the middle of the antenna. Each half of the antenna has its own receiver, so there are in effect two receive apertures, having respective phase centers **R1** and **R2** which are each Δx meters from the transmit phase center.

If the transmit phase center is located at position x_0 on the first pulse transmitted, the forward receive phase center is at $x_0 + \Delta x$ and the aft receive phase center is at $x_0 - \Delta x$, a receive phase center separation of $2\Delta x$. The effective phase center for a complete transmit-receive path is approximately halfway between the transmit and receive phase centers. Thus, for the common full-array transmit apertures and the two receive apertures, the effective two-way phase centers are at $x_0 + \Delta x/2$ and $x_0 - \Delta x/2$, a separation of Δx meters.

Now consider the motion of the platform over M_s pulses. If the pulse repetition interval is T and the platform velocity is v, the effective transmit-receive phase centers move forward by vM_sT meters in M_s PRIs. The idea of DPCA is to achieve effective MTI cancellation by combining pulses measured from the same phase center location in space. Specifically, if the **T-R1** phase center is at position

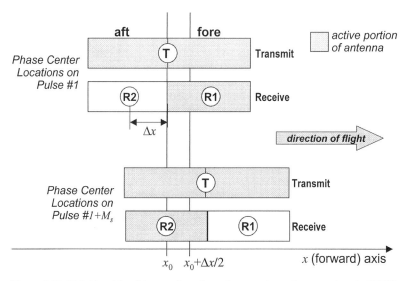

Figure 5.37 Relationship of transmit and receive aperture phase centers in DPCA processing.

$x_0 + \Delta x/2$ on the first pulse, then M_s pulses later the **T-R2** transmit phase center will be at position $x_0 - \Delta x/2 + vM_sT$. Equating these two positions gives

$$M_s = \frac{\Delta x}{2vT} \qquad (5.138)$$

M_s is the "time slip" in pulses.

The significance of the time slip given by Eq. (5.138) is that the data stream received on the aft receive aperture is geometrically equivalent to the data stream received on the forward receive aperture M_s pulses earlier. Consequently, two-pulse cancellation can be implemented by taking each sample from the **R1** data stream and subtracting the sample from the **R2** data stream taken M_s pulses later, as illustrated in Fig. 5.38. Even though these data samples were collected on different receive apertures and more than one pulse apart in time, their effective transmit-receive phase centers are the same, so they appear equivalent to successive pulses from a *stationary* antenna. The effective stationarity of the antenna then implies that the clutter spectral width is not spread by the platform motion, therefore improving the detection of slow-moving ground targets.

In general, M_s will not be integer. For example, if $\Delta x = 3$ m, $v = 200$ m/s, and $T = 2$ ms, then $M_s = 3.75$ pulses. A typical DPCA implementation will round M_s to the nearest integer for coarse alignment of the two data streams, and then use adaptive processing as described next to achieve fine clutter cancellation.

5.7.2 Adaptive DPCA

While conventional bandlimited interpolation could be used to implement fractional-PRI timing adjustments, in practice there will also be mismatches between channels that will make it impossible to achieve high cancellation ratios even if the time alignment is perfect. Adaptive processing can be combined with the basic DPCA cancellation to minimize the clutter residue at the processor output and therefore maximize the improvement factor. The following discussion of adaptive DPCA is modeled after the "suboptimum matched filter algorithm" of Shaw and McAulay (1983).

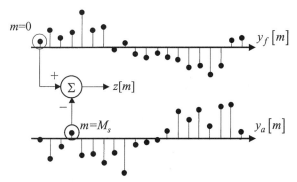

Figure 5.38 Illustration of two-pulse cancellation across two received data streams in DPCA for a time slip of approximately $M_s = 3$ PRIs.

Assume that time slip, i.e., an integer delay of one channel with respect to the other, is used to achieve coarse time alignment (to within one-half PRI) of two channels to be combined. Furthermore, divide each received signal channel into Doppler subbands using a DFT and perform the cancellation independently in each subband. This will allow the adaptive cancellation weight to be optimized separately for each subband, improving overall performance (and affording an opportunity to illustrate one example of subband-based processing). The vector analysis approach will be used to develop the adaptive filtering, so define the coarse-aligned two-channel signal vector $\mathbf{y}[l, m]$ as

$$\mathbf{y}[l, m] = \begin{bmatrix} y_f[l, m] \\ y_a[l, m + M_s] \end{bmatrix} \quad (5.139)$$

where $y_f[l, m]$ is the "fore" channel and $y_a[l, m]$ is the "aft" channel. As before, the index l is the range bin index, while m is the pulse number (slow time) index. Collect M pulses in a *coherent processing interval* (CPI) (thus $0 \leq m \leq M - 1$) and take the K-point discrete Fourier transform of the data in each range cell to get the Doppler domain data vector

$$\mathbf{Y}[l, k] = \begin{bmatrix} Y_f[l, k] \\ e^{-j 2\pi M_s k / K} Y_a[l, k] \end{bmatrix} \quad (5.140)$$

Equation (5.140) points out that the time slip adjustment adds a linear phase term to the DFT of the aft channel data. This effect on the Doppler domain data applies to both target and interference signal components.

The signals at each subaperture consist of clutter, noise, and (if present) target components. A model of the covariance matrix $\mathbf{S_I}$ of the Doppler domain interference, similar to that of Eq. (5.19), is needed. The clutter in the two channels is correlated to some degree determined by the platform motion and time slip correction. Furthermore, the clutter is not white, meaning its covariance varies as a function of the Doppler index k. The thermal noise is assumed uncorrelated between channels and is white. Therefore, $\mathbf{S_I}$ will take the form

$$\mathbf{S_I}[l, k] = \mathbf{S_I}[k] = \begin{bmatrix} \sigma_c^2[k] + \sigma_n^2 & \rho[k]\sigma_c^2[k] \\ \rho^*[k]\sigma_c^2[k] & \beta[k]\left(\sigma_c^2[k] + \sigma_n^2\right) \end{bmatrix} \quad (5.141)$$

In Eq. (5.141), the coefficient $\beta[k]$ accounts for any mismatch in the gain and frequency response or subaperture antenna patterns of the two channels. Note that the form of the covariance matrix is assumed the same in every range bin, varying only with Doppler. This assumption requires that a preprocessing gain control step compensate for the expected R^3 variation[†] in clutter power with range. Noise power does not vary with range or Doppler.

[†]Recall from Chap. 2 that the clutter power will vary as R^3 if the resolution cell is beam-limited in azimuth and pulse-limited in range. Other appropriate assumptions are used if this is not the case.

As discussed previously, the Doppler spectrum can be divided into clutter and clear regions. The clutter region is that range of the Doppler bin index k where the interference consists of both clutter and thermal noise, but with clutter dominant. The clear region is the range of k where noise is the dominant interference.

Next, a model similar to Eq. (5.23) or (5.25) for the target data in the Doppler domain is needed. A moving point target in range bin l_t is modeled by the time domain signal vector

$$\mathbf{t}[l, m] = A_t \, \delta[l - l_t] \, e^{j 2\pi F_D m T} \begin{bmatrix} \gamma_f \\ \gamma_a \end{bmatrix} \quad (5.142)$$

where γ_f and γ_a = complex scalar constants representing the unknown target phases and (possibly unequal) amplitudes in the fore and aft receive channels
A_t = target amplitude
$\delta[\cdot]$ = discrete impulse function

The target phases represented by γ_f and γ_a are determined by the absolute range and the angle of arrival, as well as the electrical lengths of the receive paths. The Doppler frequency shift is F_D hertz.

It is useful to note the relationship between the angle of arrival and the phase difference between the signals observed on different antenna subapertures. Figure 5.39 shows a wavefront impinging on two receivers (representing two antenna phase centers) separated by a distance d_{pc}. The angle of arrival of the wavefront, measured from the normal to the line connecting the two receivers, is θ_a. As shown in the figure, the additional distance the wavefront must travel after arriving at the first receiver before it reaches the second receiver is $d_{pc} \sin \theta_a$. Consequently, the phase difference in the signals received at the two sites becomes

$$\Delta \phi = \frac{2\pi}{\lambda_k} d_{pc} \sin \theta_a \quad (5.143)$$

The subscript k on the wavelength has been added to emphasize that the appropriate wavelength should be used for each Doppler bin to which Eq. (5.143) will be applied.

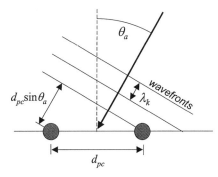

Figure 5.39 Geometry for relating angle of arrival to phase difference at two subaperture receive phase centers. The angle of arrival shown is considered negative; positive angles of arrival are clockwise from the normal to the line connecting the phase centers.

Using Eq. (5.143), γ_f and γ_a of Eq. (5.142) can be expanded as follows

$$\gamma_f = \exp\left[j\frac{2\pi}{\lambda_k}(2R + \theta_f)\right]$$
$$\gamma_a = \exp\left[j\frac{2\pi}{\lambda_k}(2R + \theta_a + \Delta\phi)\right] \quad (5.144)$$

The term $(4\pi/\lambda_k)R$ appearing in each channel is due to the two-way distance to the target, assumed to be at range R. The terms θ_f and θ_a represent the receiver phase shifts, which are different for each channel in general.

The target data model of Eqs. (5.142) and (5.144) results in the following target Doppler domain signal model

$$\mathbf{t}[l,k] = MA_t\,\delta[l - l_t]\,\delta[k - k_t]\begin{bmatrix} \gamma_f \\ e^{-j2\pi M_s k/K}\gamma_a \end{bmatrix} \quad (5.145)$$

where $k_t = (F_D KT/2\pi)$ is just the target Doppler shift converted to an equivalent DFT bin number. The two impulse functions serve to confine the response to range bin l_t and Doppler bin k_t.

As before, signal-to-interference ratio can be maximized with a matched filter that computes the scalar quantity

$$\mathbf{z}[l,k] = \left(\mathbf{S}_\mathbf{I}^{-1}[k]\mathbf{t}^*\right)'\mathbf{Y}[l,k] = \mathbf{t}^H[l,k]\left(\mathbf{S}_\mathbf{I}^{-1}[k]^*\right)\mathbf{Y}[l,k] \quad (5.146)$$

Since the target signal angle of arrival is not known a priori, a suboptimum filter is again computed by averaging the target signal vector over all values of angle of arrival, factoring out any complex constants. Also, the target location in range-Doppler space is not assumed known, so the same target model is used in each range and Doppler bin. The new target vector then becomes simply

$$\mathbf{t}[l,k] = \mathbf{t} = M\hat{A}_t\begin{bmatrix} 1 \\ 0 \end{bmatrix} \quad (5.147)$$

It was shown in Sec. 5.2.5, Eq. (5.68) that the SIR improvement factor I_{sub} with this suboptimum target assumption will be related to the optimum improvement factor I_{opt} according to $I_{\text{sub}} = I_{\text{opt}} - 1$.

The exact clutter and noise statistics are not known a priori. Consequently, $\mathbf{S}_\mathbf{I}$ also cannot be known exactly, but it can be estimated from the data. Since the clutter covariance is expected to have essentially the same form in every range bin, one way to estimate $\mathbf{S}_\mathbf{I}$ would be to compute a sample average over several range bins

$$\hat{\mathbf{S}}_\mathbf{I}[k] = \frac{1}{L_2 - L_1 + 1}\sum_{l=L_1}^{L_2}\mathbf{Y}^*[l,k]\,\mathbf{Y}'[l,k] \quad (5.148)$$

This estimate of $\hat{\mathbf{S}}_I$ then replaces the actual $\mathbf{S}_I$ in Eq. (5.146). Since the coefficients used to combine the fore and aft data streams are computed from the data itself, this is now an *adaptive* DPCA processor.

Combining Eqs. (5.140), (5.141), (5.146), and (5.147) gives the output of the DPCA system. Assuming that $\hat{\mathbf{S}}_I$ is a good approximation to $\mathbf{S}_I$ and absorbing all constants into a single constant k' gives

$$z[l,k] = k' \left\{ \beta[k] \left(1 + \frac{\sigma_n^2}{\sigma_c^2} \right) \mathbf{Y}_f[l,k] - \rho^*[k] e^{+j2\pi M_s k/K} \mathbf{Y}_a[l,k] \right\} \quad (5.149)$$

While complicated, the structure of a two-pulse canceller is clearly present in the subtraction of $\mathbf{Y}_a[l,k]$ from $\mathbf{Y}_f[l,k]$. If the interference is clutter-limited so that $\sigma_c^2 \gg \sigma_n^2$ and $\rho[k] \to 1$ so that the clutter is highly correlated across range bins, the output simplifies to

$$z[l,k] = k'\{\beta[k] \mathbf{Y}_f[l,k] - e^{+j2\pi M_s k/K} \mathbf{Y}_a[l,k]\} \quad (5.150)$$

The two-pulse canceller structure, applied to each range and Doppler bin, is clearer here. The complex exponential in the Doppler index k is equivalent to a time-domain shift of M_s samples, in accordance with the DPCA condition discussed earlier. The factor $\beta[k]$ provides a Doppler-dependent weighting factor that can be optimized to maximize cancellation in each Doppler channel.

The matched filter design assumes that $\hat{\mathbf{S}}_I$ is an estimate of the autocovariance of the interference only, i.e., it should not contain any target signal components. A practical system must take steps to ensure this is the case, perhaps by skipping range bins containing targets already in track, averaging over enough range bins to minimize any unknown target influences, prescreening the data for large amplitudes that might indicate a target, or other means. Many of the required techniques are similar to those used in constant false alarm rate detection, discussed in Chap. 7.

References

Abramowitz, M., and I. A. Stegun, "Numerical Interpolation, Differentiation, and Integration," Chap. 25 in *Handbook of Mathematical Functions with Formulas, Graphs, and Mathematical Tables*. National Bureau of Standards, U.S. Dept. of Commerce, 1972.

Bailey, D. H., "FFTs in Extended or Hierarchical Memory," *Journal of Supercomputing*, vol. 4, pp. 23–25, 1990.

Burrus, C. S., and T. W. Parks, *DFT/FFT and Convolution Algorithms*. Wiley, New York, 1985.

Cole, E. L. et al., "ASR-12: A Next Generation Solid State Air Traffic Control Radar," *Proceedings of the 1988 IEEE Radar Conference*, Dallas, TX, pp. 9–13, May 1998.

Doviak, D. S., and R. J. Zrnic, *Doppler Radar and Weather Observations*, 2d ed. Academic Press, San Diego, CA, 1993.

Eaves, J. L., and E. K. Reedy (ed.), *Principles of Modern Radar*. Van Nostrand Reinhold, New York, 1988.

Frigo, M., and S. G. Johnson, "FFTW: An Adaptive Software Architecture for the FFT," Proceedings of International Conference on *Acoustics, Speech, and Signal Processing*, vol. 3, pp. 1381–1384, 1998. Available at http://www.fftw.org.

Harris, F. J., "On the Use of Windows for Harmonic Analysis with the Discrete Fourier Transform," *Proceedings of the IEEE*, vol. 68, no. 1, pp. 51–83, Jan. 1978.

Hayes, M. H., *Statistical Digital Signal Processing and Modeling*. Wiley, New York, 1996.

Hildebrand, F. B., *Introduction to Numerical Analysis,* 2d ed. McGraw-Hill, New York, 1974.

Hovanessian, S. A., "An Algorithm for Calculation of Range in a Multiple PRF Radar," *IEEE Transactions on Aerospace and Electronic Systems*, vol. 12(2), pp. 287–290, March 1976.

Hsiao, J. K., and F. F. Krestchmer, Jr., "Design of a Staggered-p.r.f. Moving Target Indication Filter," *The Radio and Electronic Engineer*, vol. 43, no. 11, pp. 689–694, Nov. 1973.

IEEE Standard Radar Definitions, IEEE Standard 686-1982. Institute of Electrical and Electronics Engineers, New York.

Kay, S. M., *Modern Spectral Estimation*. Prentice Hall, Englewood Cliffs, NJ, 1988.

Keel, B. M., "Adaptive Clutter Rejection Filters for Airborne Doppler Weather Radar," M.S. Thesis, Clemson University, Clemson, AL, 1989.

Kretschmer, F. F., Jr., "MTI Visibility Factor," *IEEE Transactions on Aerospace and Electronic Systems*, vol. AES-22, no. 2, pp. 216–218, March 1986.

Long, W. H., D. H. Mooney, and W. A. Skillman, "Pulse Doppler Radar," Chap. 17 in M. I. Skolnik (ed.), *Radar Handbook*, 2d ed. McGraw-Hill, New York, 1990.

Morris, G. V., and L. Harkness, (eds.), *Airborne Pulse Doppler Radar*, 2d ed. Artech House, Boston, MA, 1996.

Levanon, N., *Radar Principles*. Wiley, New York, 1988.

Lightstone, L., D. Faubert, and G. Rempel, "Multipile Phase Center DPCA for Airborne Radar," *Proceedings* of the 1991 *IEEE National Radar Conference*, pp. 36–40, March 1991.

Nathanson, F. E., *Radar Design Principles,* 2d ed. McGraw-Hill, New York, 1991.

Oppenheim, A. V., and R. W. Schafer, *Discrete-Time Signal Processing*, 2d ed. Prentice Hall, Englewood Cliffs, NJ, 1999.

Prinsen, P. J. A., "Elimination of Blind Velocities of MTI Radar by Modulating the Interpulse Period," *IEEE Transactions on Aerospace and Electronic Systems*, vol. AES-9, no. 5, pp. 714–724, Sept. 1973.

Richards, M. A., "Coherent Integration Loss due to White Gaussian Phase Noise," *IEEE Signal Processing Letters*, vol. 10, no. 7, pp. 208–210, July 2003.

Richards, M. A., "Doppler Processing" Chap. 15 in W. A. Holar and J. A. Scheer (eds.), *Principles of Modern Radar*. SciTech Publishing, Raleigh, NC, to appear 2006.

Roy, R., and O. Lowenschuss, "Design of MTI Detection Filters with Nonuniform Interpulse Periods," *IEEE Transactions on Circuit Theory*, vol. CT-17, no. 4, pp. 604–612, Nov. 1970.

Schleher, D. C., *MTI and Pulse Doppler Radar*. Artech House, Boston, MA, 1991.

Shaw, G. A., and R. J. McAulay, "The Application of Multichannel Signal Processing to Clutter Suppression for a Moving Platform Radar," *IEEE Acoustics, Speech, and Signal Processing (ASSP) Spectrum Estimation Workshop II*, Nov. 10–11, Tampa, FL, 1983.

Skolnik, M. I. (ed.), *Radar Handbook*, 2d ed. McGraw-Hill, New York, 1990.

Skolnik, M. I., *Introduction to Radar Systems*, 3d ed. McGraw-Hill, New York, 2001.

Staudaher, F. M., "Airborne MTI," Chap. 16 in M. I. Skolnik (ed.), *Radar Handbook*, 2d ed. McGraw-Hill, New York, 1990.

Stimson, G. W., *Introduction to Airborne Radar*, 2d ed. SciTech Publishing, Mendham, NJ, 1998.

Taylor, J. W., Jr., and G. Brunins, "Design of a New Airport Surveillance Radar (ASR-9)," *Proceedings of the IEEE*, vol. 73, no. 2, pp. 284–289, Feb. 1985.

Trunk, G., and S. Brockett, "Range and Velocity Ambiguity Resolution," *Record* of the *1993 IEEE National Radar Conference*, pp.146–149, April 20–22, 1993.

Trunk, G., and M. W. Kim, "Ambiguity Resolution of Multiple Targets Using Pulse-Doppler Waveforms," *IEEE Transactions on Aerospace and Electronic Systems*, vol. 30(4), pp. 1130–1137, Oct. 1994.

Vergara-Dominguez, L., "Analysis of the Digital MTI Filter with Random PRI," *IEEE Proceedings, Part F*, vol. 140, no. 2, pp. 129–137, April 1993.

Chapter 6

Detection Fundamentals

As was noted in Chap. 1, the primary functions to be carried out by a radar signal processor are detection, tracking, and imaging. In this chapter, the concern is detection. In radar, this means deciding whether a given radar measurement is the result of an echo from a target, or simply represents the effects of interference. If it is decided that the measurement indicates the presence of a target, further processing is usually undertaken. This additional processing might, for instance, take the form of tracking via precise range, angle, or Doppler measurements.

Detection decisions can be applied to signals present at various stages of the radar signal processing, from raw echoes to heavily preprocessed data such as Doppler spectra or even synthetic aperture radar images. In the simplest case, each range bin (fast-time sample) for each pulse can be individually tested to decide if a target is present at the range corresponding to the range bin, and the spatial angles corresponding to the antenna pointing direction for that pulse. Since the number of range bins can be in the hundreds or even thousands, and pulse repetition frequencies can range from a few kilohertz to tens or hundreds of kilohertz, the radar can be making many thousands to literally millions of detection decisions per second.

It was seen in Chap. 2 that both the interference and the echoes from complex targets are best described by statistical signal models. Consequently, the process of deciding whether or not a measurement represents the influence of a target or only interference is a problem in statistical hypothesis testing. In this chapter, it will be shown how this basic decision strategy leads to the concept of threshold testing as the most common detection logic in radar. Performance curves will be derived for the most basic signal and interference models.

Clutter (echoes from the ground) is sometimes "interference" and sometimes the "target." If one is trying to detect a moving vehicle, ground clutter, along with noise and possibly jamming, is the interference; but if one is trying to image a region of the earth, this same terrain becomes the desired target and only noise and jamming are the interference.

An excellent concise reference for modern detection theory is given by Johnson and Dudgeon (1993, Chap. 5). When greater depth is needed, another excellent modern reference with a digital signal processing point of view is the work by Kay (1998). An important classical textbook in detection theory is the book by Van Trees (1968), while Meyer and Mayer (1973) provide a classical in-depth analysis and many detection curves specifically for radar applications.

6.1 Radar Detection as Hypothesis Testing

For any radar measurement that is to be tested for the presence of a target, one of two hypotheses can be assumed to be true

1. The measurement is the result of interference only.
2. The measurement is the combined result of interference and echoes from a target.

The first hypothesis is denoted as the *null hypothesis* H_0 and the second as H_1. The detection logic therefore must examine each radar measurement to be tested and select one of the hypotheses as "best" accounting for that measurement. If H_0 best accounts for the data, the system declares that a target was not present at the range, angle, or Doppler coordinates of that measurement; if H_1 best accounts for the data, the system declares that a target was present.[†]

Because the signals are described statistically, the decision between the two hypotheses is an exercise in statistical decision theory. A general approach to this problem is described in many texts (e.g., Kay, 1998). The analysis starts with a statistical description of the *probability density function* (pdf) that describes the measurement to be tested under each of the two hypotheses. If the sample to be tested is denoted as y, the following two pdfs are required

$$p_y(y|H_0) = \text{pdf of } y \text{ given that a target was } not \text{ present}$$

$$p_y(y|H_1) = \text{pdf of } y \text{ given that a target } was \text{ present}$$

Thus, part of the detection problem is to develop models for these two pdfs. In fact, analysis of radar performance is dependent on estimating these pdfs for the system and scenario at hand. Furthermore, a good deal of the radar system design problem is aimed at manipulating these two pdfs in order to obtain the most favorable detection performance.

[†]In some detection problems, a third hypothesis is allowed: "don't know." Most radar systems, however, force a choice between "target present" and "target absent" on each detection test.

More generally, detection will be based on N samples of data y_n forming a column vector $\mathbf{y}$

$$\mathbf{y} \equiv [y_0 \quad \ldots \quad y_{N-1}]^t \tag{6.1}$$

The N-dimensional joint pdfs $p_\mathbf{y}(\mathbf{y}|H_0)$ and $p_\mathbf{y}(\mathbf{y}|H_1)$ are then used.

Assuming the two pdfs are successfully modeled, the following probabilities of interest can be defined

Probability of Detection, P_D: The probability that a target is declared (i.e., H_1 is chosen) when a target is in fact present.

Probability of False Alarm, P_{FA}:   The probability that a target is declared (i.e., H_1 is chosen) when a target is in fact *not* present.

Probability of Miss, P_M: The probability that a target is *not* declared (i.e., H_0 is chosen) when a target is in fact present.

Note that $P_M = 1 - P_D$. Thus, P_D and P_{FA} suffice to specify all of the probabilities of interest. As the latter two definitions imply, it is important to realize that, because the problem is statistical, there will be a finite probability that the decisions will be wrong.[†]

6.1.1 The Neyman-Pearson detection rule

The next step in making a decision is to decide what the rule will be for deciding what constitutes an optimal choice between our two hypotheses. This is a rich field. The Bayes optimization criterion assigns a cost or risk to each of the four possible combinations of actual state (target present or not) and decision (select H_0 or H_1) (Johnson and Dudgeon, 1993; Kay, 1998). In radar, it is more common to use a special case of the Bayes criterion called the *Neyman-Pearson criterion*. Under this criterion, the decision process is designed to maximize the probability of detection P_D under the constraint that the probability of false alarm P_{FA} does not exceed a set constant. The achievable combinations of P_D and P_{FA} are affected by the quality of the radar system and signal processor design. However, it will be seen that for a fixed system design, increasing P_D implies increasing P_{FA} as well. The radar system designer will generally decide what rate of false alarms can be tolerated based on the implications of acting on a false alarm, which may include using radar resources to start a track on a nonexistent target, or in extreme cases even firing a weapon! Recalling that

---

[†]A fourth probability can be defined, that of choosing H_0 and thus declaring a target not present when in fact the test sample is due to interference only. This probability, equal to $1 - P_{FA}$, is not normally of direct interest.

the radar may make tens or hundreds of thousands, even millions of detection decisions per second, values of P_{FA} must generally be quite low. Values in the range of 10^{-4} to 10^{-8} are common, and yet may still lead to false alarms every few seconds or minutes. Higher-level logic implemented in downstream data processing, beyond the scope of this book, is often used to reduce the number or impact of false alarms.

Each vector of measured data values $\mathbf{y}$ can be considered to be a point in N-dimensional space. To have a complete decision rule, each point in that space (each combination of N measured data values) must be assigned to one of the two allowed decisions, H_0 ("target absent") or H_1 ("target present"). Then, when the radar measures a particular set of data values ("observation") $\mathbf{y}'$, the system declares either "target absent" or "target present" based on the preexisting assignment of $\mathbf{y}'$ to either H_0 or H_1. Denote the set of all observations $\mathbf{y}$ for which H_1 will be chosen as the region $\mathfrak{R}_1$. Note that $\mathfrak{R}_1$ is not necessarily a connected region. General expressions can now be written for the probabilities of detection and false alarm as integrals of the joint pdfs over the region $\mathfrak{R}_1$ in a N-dimensional space

$$P_D = \int_{\mathfrak{R}_1} p_{\mathbf{y}}(\mathbf{y}|H_1)\, d\mathbf{y}$$
$$P_{FA} = \int_{\mathfrak{R}_1} p_{\mathbf{y}}(\mathbf{y}|H_0)\, d\mathbf{y} \tag{6.2}$$

Because probability density functions are nonnegative, Eq. (6.2) proves a claim made earlier, namely that P_D and P_{FA} must rise or fall together. As the region $\mathfrak{R}_1$ grows to include more of the possible observations $\mathbf{y}$, either integral encompasses more of the N-dimensional space and therefore integrates more of the nonnegative pdf. The opposite is true if $\mathfrak{R}_1$ shrinks. That is, as $\mathfrak{R}_1$ grows or shrinks, so must *both* P_D and P_{FA}.[†] In order to increase detection probability, the false alarm probability must be allowed to increase as well. Loosely speaking, to achieve a good balance of performance, the points that contribute more probability mass to P_D than to P_{FA} are assigned to $\mathfrak{R}_1$. If the system can be designed so that $p_{\mathbf{y}}(\mathbf{y}|H_0)$ and $p_{\mathbf{y}}(\mathbf{y}|H_1)$ are as disjoint as possible, this task becomes easier and more effective. This point will be revisited later.

6.1.2 The likelihood ratio test

The Neyman-Pearson criterion is motivated by the goal of obtaining the best possible detection performance while guaranteeing that the false alarm probability does not exceed some tolerable value. Thus, the Neyman-Pearson decision rule is to

$$\text{choose } \mathfrak{R}_1 \text{ such that } P_D \text{ is maximized, subject to } P_{FA} \leq \alpha \tag{6.3}$$

[†]The exception occurs if points are added to or subtracted from $\mathfrak{R}_1$ for which $p_{\mathbf{y}}(\mathbf{y}|H_0)$, $p_{\mathbf{y}}(\mathbf{y}|H_1)$, or both are zero, in which case the corresponding probability is unchanged.

where α is the maximum allowable false alarm probability.[†] This optimization problem is solved by the method of Lagrange multipliers. Construct the function

$$F \equiv P_D + \lambda(P_{FA} - \alpha) \tag{6.4}$$

To find the optimum solution, maximize F and then choose λ to satisfy the constraint criterion $P_{FA} = \alpha$. Substituting Eq. (6.2) into Eq. (6.4)

$$F = \int_{\Re_1} p_{\mathbf{y}}(\mathbf{y}|H_1)\,d\mathbf{y} + \lambda \left(\int_{\Re_1} p_{\mathbf{y}}(\mathbf{y}|H_0)\,d\mathbf{y} - \alpha \right)$$

$$= -\lambda\alpha + \int_{\Re_1} \{p_{\mathbf{y}}(\mathbf{y}|H_1) + \lambda p_{\mathbf{y}}(\mathbf{y}|H_0)\}\,d\mathbf{y} \tag{6.5}$$

Remember that the design variable here is the choice of the region $\Re_1$. The first term in the second line of Eq. (6.5) does not depend on $\Re_1$, so F is maximized by maximizing the value of the integral over $\Re_1$. Since λ could be negative, the integrand can be either positive or negative, depending on the values of λ and the relative values of $p_{\mathbf{y}}(\mathbf{y}|H_0)$ and $p_{\mathbf{y}}(\mathbf{y}|H_1)$. The integral is therefore maximized by including in $\Re_1$ all the points, and only the points, in the N-dimensional space for which $p_{\mathbf{y}}(\mathbf{y}|H_1)+\lambda p_{\mathbf{y}}(\mathbf{y}|H_0) > 0$, that is, $\Re_1$ is all points $\mathbf{y}$ for which $p_{\mathbf{y}}(\mathbf{y}|H_1) > -\lambda p_{\mathbf{y}}(\mathbf{y}|H_0)$. This leads directly to the decision rule

$$\frac{p_{\mathbf{y}}(\mathbf{y}|H_1)}{p_{\mathbf{y}}(\mathbf{y}|H_0)} \underset{H_0}{\overset{H_1}{\gtrless}} -\lambda \tag{6.6}$$

Equation (6.6) is known as the *likelihood ratio test* (LRT). Although derived from the point of view of determining what values of $\mathbf{y}$ should be assigned to decision region $\Re_1$, it in fact allows one to skip over explicit determination of $\Re_1$ and gives a rule for optimally guessing, under the Neyman-Pearson criterion, whether a target is present or not based directly on the observed data $\mathbf{y}$ and a threshold $-\lambda$ (which must still be computed). This equation states that the ratio of the two pdfs, each evaluated for the particular observed data $\mathbf{y}$, should be compared to a threshold. If that "likelihood ratio" exceeds the threshold, choose hypothesis H_1, i.e., declare a target to be present. If it does not exceed the threshold, choose H_0 and declare that a target is not present. Under the Neyman-Pearson optimization criterion, the probability of a false alarm cannot exceed the original design value P_{FA}. Note again that models of $p_{\mathbf{y}}(\mathbf{y}|H_0)$ and $p_{\mathbf{y}}(\mathbf{y}|H_1)$ are required in order to carry out the LRT. Finally, realize that in computing the LRT, the data processing operations to be carried out on the observed data $\mathbf{y}$ are being specified. What exactly the required operations are depends on the particular pdfs.

The LRT test is as ubiquitous in detection theory and statistical hypothesis testing as is the Fourier transform in signal filtering and analysis. It arises as the solution to the hypothesis testing problem under several different decision

---

[†] Some subtleties that can arise if the pdfs are noncontinuous are being ignored. See the book by Johnson and Dudgeon (1993) for additional detail.

criteria, such as the Bayes minimum cost criterion, or maximization of the probability of a correct decision. Substantial additional detail is provided by Johnson and Dudgeon (1993) and Kay (1998). As a convenient and common shorthand, it is convenient to express the LRT in the following notation

$$\Lambda(\mathbf{y}) \underset{H_0}{\overset{H_1}{\gtrless}} \eta \qquad (6.7)$$

From Eq. (6.6), $\Lambda(\mathbf{y}) = p_\mathbf{y}(\mathbf{y}|H_1)/p_\mathbf{y}(\mathbf{y}|H_0)$ and $\eta = -\lambda$.

Because the decision depends only on whether the LRT exceeds the threshold or not, any monotone increasing[†] operation can be performed on both sides of Eq. (6.7) without affecting the values of observed data $\mathbf{y}$ that cause the threshold to be exceeded, and therefore without affecting the performance (P_D and P_{FA}). A well-chosen transformation can sometimes greatly simplify the computations required to actually carry out the LRT. Most common is to take the natural logarithm of both sides of Eq. (6.7) to obtain the *log likelihood ratio test*

$$\ln \Lambda(\mathbf{y}) \underset{H_0}{\overset{H_1}{\gtrless}} \ln \eta \qquad (6.8)$$

To make these procedures clearer, consider what is perhaps the simplest example, detection of the presence or absence of a constant in zero-mean Gaussian noise of variance β^2. Let $\mathbf{w}$ be a vector of i.i.d. zero mean Gaussian random variables. When the constant is absent (hypothesis H_0) the data vector $\mathbf{y} = \mathbf{w}$ follows an N-dimensional normal distribution with a scaled identity covariance matrix. When the constant is present (hypothesis H_1), $\mathbf{y} = \mathbf{m} + \mathbf{w} = m\mathbf{1}_N + \mathbf{w}$ and the distribution is simply shifted to a nonzero, positive mean[‡]

$$\begin{aligned} H_0 &: \mathbf{y} \sim N(\mathbf{0}_N, \beta^2 \mathbf{I}_N) \\ H_1 &: \mathbf{y} \sim N(m\mathbf{1}_N, \beta^2 \mathbf{I}_N) \end{aligned} \qquad (6.9)$$

where $m > 0$ and $\mathbf{0}_N$, $\mathbf{1}_N$, and $\mathbf{I}_N$ are respectively a vector of N zeros, a vector of N ones, and the identity matrix of order N. The model of the required pdfs is therefore

$$\begin{aligned} p(\mathbf{y}|H_0) &= \prod_{n=0}^{N-1} \frac{1}{\sqrt{2\pi\beta^2}} \exp\left\{-\frac{1}{2}\left(\frac{y_n}{\beta}\right)^2\right\} \\ p(\mathbf{y}|H_1) &= \prod_{n=0}^{N-1} \frac{1}{\sqrt{2\pi\beta^2}} \exp\left\{-\frac{1}{2}\left(\frac{y_n - m}{\beta}\right)^2\right\} \end{aligned} \qquad (6.10)$$

[†]A monotone decreasing operation would simply invert the sense of the threshold test.

[‡]All of the following development is fairly easily generalized for the case when m is negative or is of unknown sign. It will be seen later that radar detection generally involves working with the magnitude of the signal, thus it is sufficient to work with a positive value of m.

The likelihood ratio $\Lambda(\mathbf{y})$ and the log-likelihood ratio can be directly computed from Eq. (6.10)

$$\Lambda(\mathbf{y}) = \frac{\prod_{n=0}^{N-1} \exp\left\{-\frac{1}{2}\left(\frac{y_n-m}{\beta}\right)^2\right\}}{\prod_{n=0}^{N-1} \exp\left\{-\frac{1}{2}\left(\frac{y_n}{\beta}\right)^2\right\}} \qquad (6.11)$$

$$\ln \Lambda(\mathbf{y}) = \sum_{n=0}^{N-1} \left\{-\frac{1}{2}\left(\frac{y_n-m}{\beta}\right)^2 + \frac{1}{2}\left(\frac{y_n}{\beta}\right)^2\right\}$$

$$= \frac{1}{\beta^2} \sum_{n=0}^{N-1} m y_n - \frac{1}{2\beta^2} \sum_{n=0}^{N-1} m^2 \qquad (6.12)$$

Because of its simpler form, the log likelihood ratio will be used. Substituting Eq. (6.12) into Eq. (6.8) and rearranging gives the decision rule

$$\sum_{n=0}^{N-1} y_n \underset{H_0}{\overset{H_1}{\gtrless}} \frac{\beta^2}{m} \ln(-\lambda) + \frac{Nm}{2} \qquad (6.13)$$

Note that the right hand side of the equation consists only of constants, though not all are yet known. Equation (6.13) thus specifies that the available data samples y_n be integrated (summed) and the integrated data compared to a threshold. This integration is an example of how the LRT specifies the data processing to be performed on the measurements. Note also that Eq. (6.13) does *not* require specifically evaluating the pdfs, let alone determining what exactly is the region in N-space comprising $\Re_1$ or whether the observation $\mathbf{y}$ is in it.

In many cases of interest, the specific form of the log-likelihood ratio can be further rearranged to isolate on the left hand side of the equation only those terms explicitly including the data samples y_n, moving all other constants to the right hand side. Equation (6.13) is such a rearrangement of Eq. (6.12). The term $\sum y_n$ is called a *sufficient statistic* for this problem, and is denoted by $\Upsilon(\mathbf{y})$. The sufficient statistic, if it exists, is a function of the data $\mathbf{y}$ that has the property that the likelihood ratio (or log-likelihood ratio) can be written as a function of $\Upsilon(\mathbf{y})$, i.e., the data appear in the likelihood ratio *only* through $\Upsilon(\mathbf{y})$ (Van Trees, 1968). This means that in making a decision that is optimal under the Neyman-Pearson criterion (and the many others that lead to the LRT), knowing the sufficient statistic $\Upsilon(\mathbf{y})$ is as good as knowing the actual data $\mathbf{y}$. In particular, the decision criterion in Eq. (6.8) can be expressed as

$$\Upsilon(\mathbf{y}) \underset{H_0}{\overset{H_1}{\gtrless}} T \qquad (6.14)$$

Note that Eq. (6.13) is in the form of Eq. (6.14) with $\Upsilon(\mathbf{y}) = \sum y_n$ and $T = (\beta^2/m)\ln(-\lambda) + (Nm/2)$.

The idea of a sufficient statistic is quite rich. For example, it can be interpreted as a geometric coordinate transformation chosen to place all of the useful information in the first coordinate (Van Trees, 1968). Procedures for verifying that a statistic (a function of the data $\mathbf{y}$) is sufficient are given by Kay (1993), as is the Neyman-Fisher Factorization theorem for identifying sufficient statistics. Detailed consideration of the properties of sufficient statistics is beyond the scope of this text; the reader is referred to the previous references for greater depth.

The specific value of the threshold $\eta = -\lambda$ that will ensure that $P_{FA} = \alpha$ as desired has not yet been found. The original expression for P_{FA} was given in Eq. (6.2), but this is not very useful, since its evaluation requires the N-dimensional joint pdf of $\mathbf{y}$ and an explicit definition of the region $\mathfrak{R}_1$, which have been defined only implicitly as the points in N-space for which the LRT exceeds the still-unknown threshold. Since they are functions of the random data $\mathbf{y}$, Λ and Υ are also random variables and thus have their own probability density functions. Because of the similarity of the sufficient statistic and the log likelihood ratio in this problem, only Λ and Υ need be considered. An alternate approach to computing the LRT threshold is thus to express P_{FA} in terms of Λ or Υ, and then solve those expressions for η, or equivalently for T. The required expressions are

$$P_{FA} = \int_{\eta=-\lambda}^{+\infty} p_\Lambda(\Lambda|H_0)\,d\Lambda = \alpha \tag{6.15}$$

or
$$P_{FA} = \int_{T}^{+\infty} p_\Upsilon(\Upsilon|H_0)\,d\Upsilon = \alpha \tag{6.16}$$

As one would expect, the result depends only on the pdf of the likelihood ratio (if using Eq. (6.15)) or the sufficient statistic (if using Eq. (6.16)) when a target is not present. Given a specific model of that pdf, a specific value can be computed for η (equivalently, λ) or T.

To illustrate these results, continue the "constant in Gaussian noise" example by finding the threshold and then evaluating the performance, working with the sufficient statistic $\Upsilon(\mathbf{y})$. In this case, $\Upsilon(\mathbf{y})$ is the sum of the individual data samples y_n. Under hypothesis H_0 (no target), the samples are *independent and identically distributed* (i.i.d) $N(0, \sigma^2)$. It follows that $\Upsilon \sim N(0, N\beta^2)$. A false alarm occurs anytime $\Upsilon > T$, so

$$\alpha = P_{FA} = \int_T^{+\infty} p_\Upsilon(\Upsilon|H_0)\,d\Upsilon$$
$$= \int_T^{+\infty} \frac{1}{\sqrt{2\pi N\beta^2}} e^{\frac{-\Upsilon^2}{2N\beta^2}}\,d\Upsilon \tag{6.17}$$

Equation (6.17) is the integral of a Gaussian pdf, so the *error function* erf(x) will appear in the solution. The standard definition is (Abramowitz and Stegun, 1972)[†]

$$\text{erf}(x) \equiv \frac{2}{\sqrt{\pi}} \int_0^x e^{-t^2} dt \tag{6.18}$$

Also define the *complementary error function* erfc(x) corresponding to erf(x)

$$\text{erfc}(x) \equiv \frac{2}{\sqrt{\pi}} \int_x^{+\infty} e^{-t^2} dt = 1 - \text{erf}(x) \tag{6.19}$$

One will generally be interested in finding the value of x that results in a certain value of erf(x) or erfc(x); thus the inverse error and complementary error functions, denoted by erf$^{-1}(z)$ and erfc$^{-1}(z)$, respectively, are of interest. It follows from Eq. (6.19) that the two are related by erfc$^{-1}(z)$ = erf$^{-1}(1-z)$.[‡]

With the change of variables $t = \Upsilon/\sqrt{2N\beta^2}$, Eq. (6.17) can be written as

$$\alpha = P_{FA} = \frac{1}{\sqrt{\pi}} \int_{T/\sqrt{2N\beta^2}}^{+\infty} e^{-t^2} dt$$

$$= \frac{1}{2}\left[1 - \text{erf}\left(\frac{T}{\sqrt{2N\beta^2}}\right)\right] \tag{6.20}$$

Finally, Eq. (6.20) can be solved to obtain the threshold T in terms of the tabulated inverse error function

$$T = \sqrt{2N\beta^2}\, \text{erf}^{-1}(1 - 2P_{FA}) \tag{6.21}$$

Equations (6.20) and (6.21) show how to compute P_{FA} given T and vice-versa.

All of the information needed to carry out the LRT in its sufficient statistic form of Eq. (6.14) is now available. $\Upsilon(\mathbf{y})$ is just the sum of the data samples, while the threshold T can be computed from the number N of samples, the variance β^2 of the noise, which is assumed known, and the desired false alarm probability P_{FA}.

The performance of this detector is evaluated by constructing a *receiver operating characteristic* (ROC) curve. There are four interrelated variables of interest: P_D, P_{FA}, the noise power β^2, and the constant m whose presence is to be detected. The latter two are characteristics of the given signals, while P_{FA} is generally fixed as part of the system specifications at whatever level is deemed tolerable. Thus it is necessary only to determine P_D. The approach is identical to that used for determining P_{FA}: determine the probability density function of

---

[†]The definitions of Eqs. (6.18) and (6.19) are the same as those used in MATLAB™.

[‡]The erf$^{-1}()$ function will often be used here, even when erfc$^{-1}()$ gives a slightly more compact expression because of the wider availability of erf$^{-1}()$ functions than erfc$^{-1}()$ in MATLAB™ and similar computational software packages.

the sufficient statistic Υ under the hypothesis H_0 and integrate the area under it from the threshold to $+\infty$.

Continuing the example, note that the only change under hypothesis H_1 is that the individual data samples y_n now each have mean m, so their sum Υ has mean Nm. Thus, $\Upsilon(\mathbf{y}) \sim N(Nm, N\beta^2)$ and

$$P_D = \int_T^{+\infty} p_\Upsilon(\Upsilon|H_1)\,d\Upsilon$$

$$= \int_T^{+\infty} \frac{1}{\sqrt{2\pi N\beta^2}} e^{\frac{-(\Upsilon-Nm)^2}{2N\beta^2}}\,d\Upsilon \qquad (6.22)$$

Again applying the definition of the error function in Eq. (6.18) leads to

$$P_D = \frac{1}{2}\left[1 - \mathrm{erf}\left(\frac{T-Nm}{\sqrt{2N\beta^2}}\right)\right] \qquad (6.23)$$

Since the primary concern is the relationship between the performance metrics P_D and P_{FA}, Eq. (6.21) can be used in Eq. (6.23) to eliminate the threshold T and arrive at

$$P_D = \frac{1}{2}\left[1 - \mathrm{erf}\left\{\mathrm{erf}^{-1}(1 - 2P_{FA}) - \frac{\sqrt{N}m}{\sqrt{2\beta^2}}\right\}\right]$$

$$= \frac{1}{2}\mathrm{erfc}\left\{\mathrm{erfc}^{-1}(2P_{FA}) - \frac{\sqrt{N}m}{\sqrt{2\beta^2}}\right\} \qquad (6.24)$$

Nm is considered to be the signal component of interest (since the goal is to detect its presence) in the sufficient statistic $\Upsilon(\mathbf{y})$ (the sum of the individual data samples y_n). Since Nm is treated as a voltage, the corresponding power is $(Nm)^2$. The power of the noise component of $\Upsilon(\mathbf{y})$ is $N\beta^2$. Thus, the term $m\sqrt{N}/\beta$ is the square root of the signal-to-noise ratio χ for this problem, and Eq. (6.24) can be rewritten as

$$P_D = \frac{1}{2}\left[1 - \mathrm{erf}\left\{\mathrm{erf}^{-1}(1 - 2P_{FA}) - \sqrt{\chi/2}\right\}\right]$$

$$= \frac{1}{2}\mathrm{erfc}\left\{\mathrm{erfc}^{-1}(2P_{FA}) - \sqrt{\chi/2}\right\} \qquad (6.25)$$

Figure 6.1 illustrates how the detection and false alarm probabilities follow from the pdfs under the two hypotheses and the threshold, and how their relative values depend on the relation between the two pdfs. Two Gaussian pdfs with variance equal to one are shown. The leftmost has a zero mean, while the rightmost has a mean of 1. The left pdf is N(0, 1), while the right pdf is N(1, 1). With $N\beta^2 = 1$ and $m = 1/N$, these fit the model of Example 1.3. P_D and P_{FA} are the areas under the right and left pdfs, respectively, from the threshold

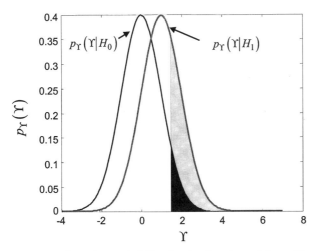

Figure 6.1 Gaussian probability density functions of the sufficient statistic when $N\sigma^2 = 1$ under hypothesis H_0 (left) and H_1 (right).

(shown as a vertical line, in this case at about $\Upsilon = 1.5$) to $+\infty$. The receiver design thus consists of adjusting the position of the threshold until the black area equals the acceptable false alarm probability. The detection probability is then the gray area (which includes the black area as well). This figure again makes it clear that P_D and P_{FA} must increase or decrease together as the threshold moves left or right. The achievable combinations of P_D and P_{FA} are determined by the degree to which the two distributions overlap.

Figure 6.2 illustrates the ROC for this problem with the SNR χ as a parameter. Part a of the figure plots the ROC using linear scales for both P_{FA} and P_D. Several features are worth noting. First, $P_{FA} = P_D$ when $\chi = 0$ (implying $m = 0$). This is to be expected since in that case, the pdf of $\Upsilon(\mathbf{y})$ is the same under either hypothesis. For a given P_{FA} and $\chi > 0$, P_D increases as the SNR increases, a result which should also be intuitively satisfying. Finally, note how abrupt the transition between near-zero and near-unity detection probabilities becomes as the SNR increases. This is a little misleading, since radars normally operate with very low values of P_{FA}; depending on the type of system, P_{FA} is typically no higher than 10^{-3} and very often is in the range of 10^{-6} to 10^{-8}. Figure 6.2b plots the same data on a logarithmic scale for P_{FA}, which better reveals the characteristics of the ROC for false alarm probabilities of interest in radar signal processing.

If the achievable combinations of P_D and P_{FA} do not meet the performance specifications, what can be done? Consideration of Fig. 6.1 suggests two answers. First, for a given P_{FA}, P_D can be increased by causing the two pdfs to move further apart when a target is present. That is, the presence of a target must cause a larger shift in the mean m of the distribution of the sufficient statistic. It was shown that m is proportional to the signal-to-noise ratio. Thus, one way to improve the detection/false alarm tradeoff is to increase the SNR.

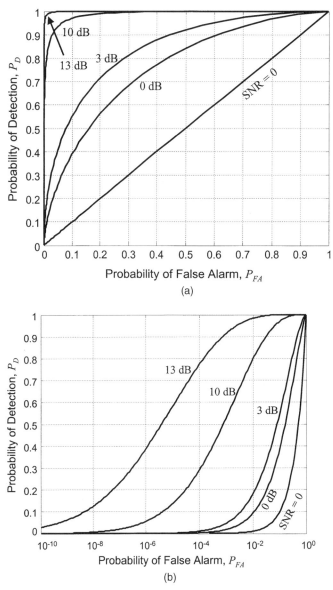

Figure 6.2 Receiver operating characteristic for the Gaussian example. (a) Displayed on a linear P_{FA} scale. (b) Displayed on a logarithmic P_{FA} scale.

This conclusion is borne out in Fig. 6.2. Figure 6.3a illustrates the effect of increased SNR on the pdfs and performance probabilities under the two hypotheses using the same threshold as Fig. 6.1.

The second way to improve the performance tradeoff is to reduce the overlap of the pdfs by reducing their variance. Reducing the noise power β^2 will reduce the variance of both pdfs, leading to the situation shown in Fig. 6.3b

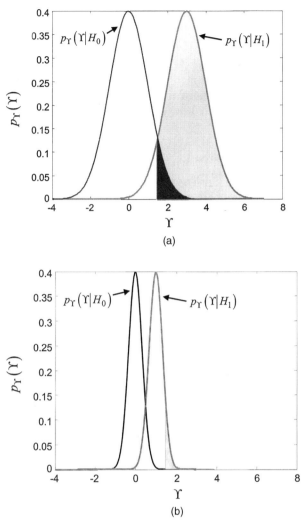

Figure 6.3 Two ways to modify the pdfs of Fig. 6.1 to improve the tradeoff between detection and false alarms. (a) Increasing the system SNR. (b) Reducing the noise power.

(where the area corresponding to P_{FA} is too small to be seen) and again improving performance. As with the first technique of increasing m, reducing β^2 again constitutes increasing the signal-to-noise ratio. Thus, consistent with Eq. (6.25), improving the tradeoff between P_D and P_{FA} requires increasing the SNR χ. This is a fundamental result that will arise repeatedly.

Radar systems are designed to achieve specified values of P_D and P_{FA} subject to various conditions, such as specified ranges, target types, interference environments, and so forth. The designer can work with antenna design, transmitter power, waveform design, and signal processing techniques, all within cost and form factor constraints. The job of the designer is therefore to develop a radar system design which ultimately results in a pair of "target absent" and "target present" pdfs at the point of detection with a small enough overlap to allow the desired P_D and P_{FA} to be achieved. If the design does not do this, the designer must redesign one or more of these elements to reduce the variance of the pdfs, shift them further apart, or both until the desired performance is obtained. Thus, a significant goal of radar system design is controlling the two pdfs analogous to those in Fig. 6.1, or equivalently, maximizing the SNR.

6.2 Threshold Detection in Coherent Systems

The Gaussian problem considered so far is useful to introduce and explain all of the major elements of Neyman-Pearson detection, such as the likelihood ratio test, probabilities of detection and false alarm, receiver operating characteristics, and the major design tradeoffs that follow. Furthermore, the problem seems "radar-like": under one hypothesis, only Gaussian noise is observed; under the other, a constant was added to the noise, which could be interpreted as the echoes from a steady target. Unfortunately, this example is not a good model for any radar detection problem due to at least three major limitations.

First, coherent radar systems that produce complex-valued measurements are of most interest. The approach, demonstrated so far only for real-valued data, must therefore be extended to the complex case; and in doing so, the complex signals and interference measurements must be modeled.

Second, there are unknown parameters. The analysis so far has implicitly assumed that such signal parameters as the noise variance and target amplitude are known, when in fact these are not known a priori but must be estimated if needed. To complicate matters further, some parameters are linked. Specifically in radar, the (unknown) echo amplitude varies with the (unknown) echo arrival time according to the appropriate version of the radar range equation. Thus, the LRT must be generalized to develop a technique that can work when some signal parameters are unknown. This extension will introduce the idea of threshold detection.

Finally, as seen in Chap. 2, there are a number of established models for radar signal phenomenology that must be incorporated. In particular, it is necessary to account for fluctuating targets, i.e., statistical variations in the amplitude of the target components of the measured data when a target is present. Furthermore, while the Gaussian pdf remains a good model for noise, in many problems the dominant interference is clutter, which may have one of the distinctly non-Gaussian pdfs discussed in Chap. 2.

The next subsections begin addressing these shortcomings by extending the LRT to coherent systems.

6.2.1 The Gaussian case for coherent receivers

An appropriate model for noise at the output of a coherent receiver was developed in Chap. 2. It was shown there that if the noise in the system prior to quadrature signal generation is a zero mean, white Gaussian process with power $\beta^2 = kT$,[†] the I and Q channels will each contain independent, identically distributed zero-mean white Gaussian processes with power $kT/2 = \beta^2/2$. That is, the noise power splits evenly but independently between the two channels. A complex noise process for which the real and imaginary parts are i.i.d. is called a *circular* random process (Dudgeon and Johnson, 1993). The expression for the joint pdf of N complex samples of the circular Gaussian random process is

$$p_\mathbf{y}(\mathbf{y}) = \frac{1}{\det\{\pi \mathbf{S_y}\}} \exp\left\{-(\mathbf{y}-\mathbf{m})^H \mathbf{S_y}^{-1}(\mathbf{y}-\mathbf{m})\right\} \qquad (6.26)$$

where $\mathbf{m}$ is the $N \times 1$ vector mean of the $N \times 1$ vector signal $\mathbf{y} = \mathbf{m} + \mathbf{w}$, $\mathbf{S_y}$ is the $N \times N$ covariance matrix of $\mathbf{y}$

$$\mathbf{S_y} = E\{\mathbf{y}\mathbf{y}^H\} - \mathbf{m}\mathbf{m}^H \qquad (6.27)$$

and H is the Hermitian (conjugate transpose) operator. In most cases the noise samples are i.i.d. so that $\mathbf{S_y} = \beta^2 \mathbf{I}_N$, which in turn means that $\det\{\pi \mathbf{S_y}\} = \pi^N \beta^{2N}$. Treatment of the case where the noise samples are not equal-variance, and the colored noise case where $\mathbf{S_y}$ is not even diagonal, is beyond the scope of this text. The reader is referred to the books by Dudgeon and Johnson (1993) and Kay (1998) for these more complex situations.

Equation (6.26) now reduces to

$$p_\mathbf{y}(\mathbf{y}) = \frac{1}{\pi^N \beta^{2N}} \exp\left\{-\frac{1}{\beta^2}(\mathbf{y}-\mathbf{m})^H(\mathbf{y}-\mathbf{m})\right\} \qquad (6.28)$$

Further simplifications occur when all of the means under H_1 are identical so that $\mathbf{m} = m\mathbf{1}_N$, where m can now be complex-valued. In this case Eq. (6.28) reduces slightly further to

$$p_\mathbf{y}(\mathbf{y}) = \frac{1}{\pi^N \beta^{2N}} \exp\left\{-\frac{1}{\beta^2}(\mathbf{y}-m\mathbf{1}_N)^H(\mathbf{y}-m\mathbf{1}_N)\right\} \qquad (6.29)$$

The LRT test for the coherent version of the previous Gaussian example can be obtained by repeating the steps in the example of Eqs. (6.22) through (6.25) using the pdf of Eq. (6.28), with $\mathbf{m} = \mathbf{0_N}$ under hypothesis H_0 and $\mathbf{m} \neq \mathbf{0_N}$ under H_1. The log likelihood ratio is

$$\ln \Lambda = \frac{1}{\beta^2}\{2\text{Re}[\mathbf{m}^H \mathbf{y}] - \mathbf{m}^H \mathbf{m}\}$$

$$= \frac{2}{\beta^2}\text{Re}\left\{\sum_{n=0}^{N-1} m^* y_n\right\} - \frac{1}{\beta^2} N|m|^2 \qquad (6.30)$$

[†]Here T refers to receiver temperature, not the detection threshold. Which meaning of T is intended should be clear from context throughout this chapter.

where the second line of Eq. (6.30) applies only to the case where the means are identical ($\mathbf{m} = m\mathbf{1}_N$).

Some interpretation of Eq. (6.30) is in order. The term $\mathbf{m}^H \mathbf{y}$ is the dot product of the complex vectors $\mathbf{m}$ and $\mathbf{y}$. As seen in Chap. 1, this dot product represents an FIR filtering operation, evaluated at the particular instant when the equivalent impulse response $\mathbf{m}^H$ and the data vector $\mathbf{y}$ completely overlap. Furthermore, since the impulse response of the filter is identical to the signal to be detected under hypothesis H_1, namely the presence of the mean $\mathbf{m}$ in the data, it is a *matched filter*, a concept discussed in detail in previous chapters. Restated, the impulse response is directly related ("matched") to the signal component whose presence one seeks to detect. In this example, the "signal" is just the vector of means, but the same reasoning applies if the elements of $\mathbf{m}$ are the samples of a modulated waveform or any other function of interest.

The second term in $\ln \Lambda$, which is the complex dot product of $\mathbf{m}$ with itself, expands to $\mathbf{m}^H \mathbf{m} = \sum_{n=0}^{N-1} |m_n|^2$, which is just the energy in $\mathbf{m}$. Denote this quantity as E. In the equal means case, this is just $E = N|m|^2$.

Finally, note the Re{} operator applied to the matched filter output $\mathbf{m}^H \mathbf{y}$. Because $\mathbf{m}$ and $\mathbf{y}$ are complex, one might be concerned that the dot product could be purely imaginary or nearly so, such that $\mathrm{Re}\{\mathbf{m}^H \mathbf{y}\} \approx 0$. The measured data $\mathbf{y}$ would then have little or no effect on the threshold test. For this example, under hypothesis H_0 $\mathbf{m} = \mathbf{0}_N$ and the Re{} operator is inconsequential. Under hypothesis H_1, each element of $\mathbf{m}$ is a complex number of the form $m_n e^{j\theta_n}$. If the target really is present, i.e., H_1 is in fact true, the elements of the measured data vector $\mathbf{y} = \mathbf{m} + \mathbf{w}$ will be of the form $m_n e^{j\theta_n} + w_n$, where w_n is a zero mean complex Gaussian noise sample. It then follows that

$$\mathbf{m}^H \mathbf{y} = \mathbf{m}^H \mathbf{m} + \mathbf{m}^H \mathbf{w}$$
$$= \sum_{n=0}^{N-1} |m_n|^2 + \sum_{n=0}^{N-1} w_n m_n e^{-j\theta_n} \qquad (6.31)$$

The first term is again the energy E in the signal $\mathbf{m}$; this is real-valued and therefore unaffected by the Re{} operator. The second term is simply weighted and integrated noise samples. While the phase of this noise component and therefore the effect of the Re{} operator is random, the sum will tend to zero as more samples are integrated.

It is evident by inspection of Eq. (6.30) that the sufficient statistic is now $\mathrm{Re}\{\mathbf{m}^H \mathbf{y}\}$. Expressing the LRT in its sufficient statistic form for the complex case

$$\Upsilon = \mathrm{Re}\{\mathbf{m}^H \mathbf{y}\} \underset{H_0}{\overset{H_1}{\gtrless}} \frac{\beta^2}{2} \ln(-\lambda) + \frac{E}{2} = T \qquad (6.32)$$

Note that if $\mathbf{m} = m\mathbf{1}_N$, the term $\mathrm{Re}\{\mathbf{m}^H \mathbf{y}\} = m \sum y_n$ and Eq. (6.32) reduces to Eq. (6.13) again. To complete consideration of the complex Gaussian case, its performance, i.e., P_D and P_{FA}, must be determined. The sufficient statistic $\Upsilon = \text{Re}\{\mathbf{m}^H \mathbf{y}\} = \text{Re}\{\sum m_n^* y_n\}$ is just a sum of Gaussian random variables, and so will also be Gaussian. To determine the performance of the coherent detector, the pdf of Υ must be determined under each hypothesis. To do so it is useful to first consider the quantity $z = \mathbf{m}^H \mathbf{y}$, which will be a complex Gaussian. First suppose hypothesis H_0 is true. In this case the $\{y_n\}$ are zero mean, and therefore so is z. Because the $\{y_n\}$ are independent, the variance of z is just the sum of the variances of the individual weighted samples

$$\text{var}(z) = \sum_{n=0}^{N-1} \text{var}(m_n^* y_n)$$

$$= \sum_{n=0}^{N-1} |m_n|^2 \beta^2 = E\beta^2 \qquad (6.33)$$

Thus, under hypothesis H_0, $z \sim N(0, E\beta^2)$. Similarly, under hypothesis H_1 $\mathbf{y} = \mathbf{m} + \mathbf{w}$ and $z \sim N(E, E\beta^2)$. Note that the mean of z is real in both cases. The power of the complex Gaussian noise splits evenly between the real and imaginary parts of z (Kay, 1998). Since $\Upsilon = \text{Re}\{\mathbf{m}^H \mathbf{y}\}$, it follows that $\Upsilon \sim N(0, E\beta^2/2)$ under H_0 and $\Upsilon \sim N(E, E\beta^2/2)$ under H_1. Following the procedure used in Example 7.2, it can be seen that

$$P_{FA} = \frac{1}{2}\left[1 - \text{erf}\left(\frac{T}{\sqrt{\beta^2 E}}\right)\right] \qquad (6.34)$$

Repeating the development of Eqs. (6.22) to (6.24) gives the probability of detection

$$P_D = \frac{1}{2}\left[1 - \text{erf}\left\{\text{erf}^{-1}(1 - 2P_{FA}) - \sqrt{\frac{E}{\beta^2}}\right\}\right]$$

$$= \frac{1}{2}\text{erfc}\left\{\text{erfc}^{-1}(2P_{FA}) - \sqrt{\frac{E}{\beta^2}}\right\} \qquad (6.35)$$

Note again that the last term in Eq. (6.35) is the square root of the energy in the "signal" $\mathbf{m}$, divided this time by the noise power β^2, i.e., the signal-to-noise ratio. Thus Eq. (6.35) can be written as

$$P_D = \frac{1}{2}[1 - \text{erf}\{\text{erf}^{-1}(1 - 2P_{FA}) - \sqrt{\chi}\}]$$

$$= \frac{1}{2}\text{erfc}\{\text{erfc}^{-1}(2P_{FA}) - \sqrt{\chi}\} \qquad (6.36)$$

Finally, in the equal means case when $\mathbf{m} = m\mathbf{1}_N$, Eq. (6.35) is similar (but not identical) to Eq. (6.24). The coherent case includes the term $\sqrt{Nm^2/\beta^2}$ instead of $\sqrt{Nm^2/2\beta^2}$ because all of the signal energy competes with only half of the noise power.

Figure 6.4 shows the receiver operating characteristic for this example. It is identical in general form to the real-valued case of Fig. 6.2, however, for a given signal-to-noise ratio the performance is better because in the coherent receiver, the signal competes with only half the noise power. For example, in this coherent case, a signal-to-noise ratio of 13 dB produces $P_D = 0.94$ at a $P_{FA} = 10^{-6}$. Figure 6.2 shows that in the real case the same χ and P_{FA} produce a P_D of just under 0.39.

6.2.2 Unknown parameters and threshold detection

In general, perfect knowledge of each of the parameters of the pdfs $p_\Upsilon(\Upsilon|H_0)$ and $p_\Upsilon(\Upsilon|H_1)$ is required to carry out the LRT, which usually means having perfect knowledge of $p_\mathbf{y}(\mathbf{y}|H_0)$ and $p_\mathbf{y}(\mathbf{y}|H_1)$. In the Gaussian example, for instance, it was assumed that the expected signal $\mathbf{y}$ is known under the various hypotheses, as well as the noise sample variance β^2. This is not the case in the real world, where the pdfs that form the likelihood ratio may depend on one or more parameters ξ that are unknown. Depending on our degree of knowledge, three cases arise

1. ξ is a random variable with a known probability density function.

2. ξ is a random variable with an unknown probability density function.

3. ξ is deterministic but unknown.

Different techniques are used to handle each of these cases. The first is the most important, because it has the greatest effect on the structure of the optimal Neyman-Pearson detector.

To illustrate the approach for handling a random parameter with a known pdf, consider yet again the complex Gaussian case. The optimal detector

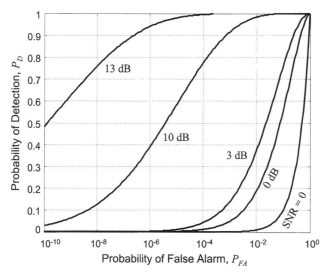

Figure 6.4 Performance of the coherent receiver on the Gaussian example.

implemented a matched filter operation $\mathbf{m}^H\mathbf{y}$, followed by the Re{} operator. The success of the matched filter structure depended on knowing exactly the constant component of $\mathbf{y} = \mathbf{m} + \mathbf{w}$ under hypothesis H_1, so that the filter coefficients could be set equal to $\mathbf{m}$ and the filter output would be real-valued. Recall that, when applied to radar, $\mathbf{y}$ under H_0 is considered to consist only of samples $\mathbf{w}$ of receiver noise, and under H_1 to consist of noisy samples $\mathbf{m} + \mathbf{w}$ of the echoes from a radar target over multiple pulses, or alternately successive fast-time samples of the waveform of one pulse echo from a target.

Claiming perfect knowledge of $\mathbf{m}$ implies knowing the range to the target very precisely, since a variation in one-way range of only $\lambda/4$ causes the received echo phase to completely reverse, i.e., to change by 180°. At microwave frequencies, this is typically only 15 to 30 cm (at L band to UHF) to a fraction of a centimeter (at millimeter wave frequencies). Because this precision is usually unrealistic, it is more reasonable to assume $\mathbf{m}$ is known only to within a phase factor $\exp(j\theta)$, where the phase angle θ is considered to be a random variable distributed uniformly over $(0, 2\pi]$ and independent of the random variables $\{m_n\}$. In other words, $\mathbf{m} = \tilde{\mathbf{m}} \exp(j\theta)$, where $\tilde{\mathbf{m}}$ is known exactly but θ is a random phase. (Note that the energy in $\tilde{\mathbf{m}}$ is the same as that in $\mathbf{m}$, that is, $\mathbf{m}^H\mathbf{m} = \tilde{\mathbf{m}}^H\tilde{\mathbf{m}}$.) This "unknown phase" assumption cannot usually be avoided in radar. What is its effect on the optimal detector and its performance?

The goal remains to carry out the LRT, so it is necessary to return to its basic definition of Eq. (6.6) and determine $p_\mathbf{y}(\mathbf{y}|H_0)$ and $p_\mathbf{y}(\mathbf{y}|H_1)$, both of which now presumably depend on θ,[†] and use the technique known as the *Bayesian approach* for random parameters with known pdfs (Kay, 1998). Specifically, compute the pdf under H_i by *separately* averaging the conditional pdfs $p_\mathbf{y}(\mathbf{y}|H_i)$ over θ

$$p_\mathbf{y}(\mathbf{y}|H_i) = \int p_\mathbf{y}(\mathbf{y}|H_i, \theta) \, p_\theta(\theta) \, d\theta \qquad i = 0, 1 \qquad (6.37)$$

The unconditional pdfs $p_\mathbf{y}(\mathbf{y}|H_i)$ are then used to define the likelihood ratio in the usual way.

As an example of the Bayesian approach for random parameters, consider again the complex Gaussian case, but now with an unknown phase in the data, $\mathbf{m} = \tilde{\mathbf{m}} \exp(j\theta)$. The conditional pdf of the observations $\mathbf{y}$ becomes, under each of the two hypotheses,

$$p_\mathbf{y}(\mathbf{y}|H_0, \theta) = \frac{1}{\pi^N \beta^{2N}} \exp\left[-\frac{1}{\beta^2} \mathbf{y}^H \mathbf{y}\right]$$

$$p_\mathbf{y}(\mathbf{y}|H_1, \theta) = \frac{1}{\pi^N \beta^{2N}} \exp\left[-\frac{1}{\beta^2} (\mathbf{y} - \tilde{\mathbf{m}} e^{j\theta})^H (\mathbf{y} - \tilde{\mathbf{m}} e^{j\theta})\right]$$

(6.38)

[†]An alternative approach called the *generalized likelihood ratio test* (GLRT), in which the unknown parameter(s) are replaced by their maximum likelihood estimates, is discussed in many detection theory texts (e.g., Kay, 1998).

Expanding the exponent in Eq. (6.38) gives

$$p_\mathbf{y}(\mathbf{y}|H_1, \theta) = \frac{1}{\pi^N \beta^{2N}} \exp\left[-\frac{1}{\beta^2}(\mathbf{y}^H \mathbf{y} - 2\mathrm{Re}\{\tilde{\mathbf{m}}^H \mathbf{y} e^{j\theta}\} + E)\right]$$

$$= \frac{1}{\pi^N \beta^{2N}} \exp\left[-\frac{1}{\beta^2}(\mathbf{y}^H \mathbf{y} - 2|\tilde{\mathbf{m}}^H \mathbf{y}|\cos\theta + E)\right] \qquad (6.39)$$

Notice that $p_\mathbf{y}(\mathbf{y}|H_0)$ does not depend on θ after all (not surprising since there is no target present in this case to present an unknown phase), so it is not necessary to apply Eq. (6.37). However, in $p_\mathbf{y}(\mathbf{y}|H_1)$ the dependence on θ is explicit. Assuming a uniform random phase and applying Eq. (6.37) under H_1 gives, after minor rearrangement

$$p_\mathbf{y}(\mathbf{y}|H_1) = \frac{1}{\pi^N \beta^{2N}} e^{-(\mathbf{y}^H \mathbf{y} + E)/\beta^2} \frac{1}{2\pi} \int_0^{2\pi} \exp\left[\frac{2}{\beta^2}|\tilde{\mathbf{m}}^H \mathbf{y}|\cos\theta\right] d\theta \qquad (6.40)$$

Equation (6.40) is a standard integral. Specifically, integral 9.6.16 in the book by Abramowitz and Stegun (1972) is

$$\frac{1}{\pi} \int_0^\pi e^{\pm z \cos\theta} d\theta = I_0(z) \qquad (6.41)$$

where $I_0(z)$ is the modified Bessel function of the first kind. Using this result and properties of the cosine function, Eq. (6.40) becomes

$$p_\mathbf{y}(\mathbf{y}|H_1) = \frac{1}{\pi^N \beta^{2N}} e^{-(\mathbf{y}^H \mathbf{y} + E)/\beta^2} I_0\left(\frac{2|\tilde{\mathbf{m}}^H \mathbf{y}|}{\beta^2}\right) \qquad (6.42)$$

The log-LRT now becomes

$$\ln \Lambda = \ln\left[I_0\left(\frac{2|\tilde{\mathbf{m}}^H \mathbf{y}|}{\beta^2}\right)\right] - \frac{E}{\beta^2} \underset{H_0}{\overset{H_1}{\gtrless}} \ln(-\lambda) \qquad (6.43)$$

or, in sufficient statistic form

$$\Upsilon = \ln\left[I_0\left(\frac{2|\tilde{\mathbf{m}}^H \mathbf{y}|}{\beta^2}\right)\right] \underset{H_0}{\overset{H_1}{\gtrless}} \ln(-\lambda) + \frac{E}{\beta^2} = T \qquad (6.44)$$

Equation (6.44) defines the signal processing required for optimum detection in the presence of an unknown phase. It calls for taking the magnitude of the matched filter output $\tilde{\mathbf{m}}^H \mathbf{y}$, passing it through the memoryless nonlinearity $\ln[I_0()]$, and comparing the result to a threshold. This result is appealing in that the matched filter is still applied to utilize the *internal* phase structure of the known signal and maximize the integration gain, but then a magnitude operation is applied because the absolute phase of the result cannot be known. Also, note that the argument of the Bessel function is the energy in the matched filter output divided by half the noise power; again, a signal-to-noise ratio.

Only half of the noise power appears because the total noise power in the complex case is split between the real and imaginary channels.

As a practical matter, it is desirable to avoid having to compute the natural logarithm and Bessel function for every threshold test, since these might occur millions of times per second in some systems. Because the function $\ln[I_0()]$ is monotonically increasing, the same detection results can be obtained by simply comparing its argument $2|\tilde{\mathbf{m}}^H\mathbf{y}|/\beta^2$ to a modified threshold. Equation (6.44) then becomes simply

$$|\tilde{\mathbf{m}}^H\mathbf{y}| \underset{H_0}{\overset{H_1}{\gtrless}} T' \qquad (6.45)$$

Figure 6.5 illustrates the optimal detector for the coherent detector with an unknown phase.

The performance of this detector will now be established. Let $z = |\tilde{\mathbf{m}}^H\mathbf{y}|$. The detection test becomes simply $z \gtrless T'$; thus the distribution of z under each of the two hypotheses is needed. As in the known phase case, under hypothesis H_0 (target absent) $\tilde{\mathbf{m}}^H\mathbf{y} \sim N(0, E\beta^2)$; thus the real and imaginary parts of $\tilde{\mathbf{m}}^H\mathbf{y}$ are independent of one another and each distributed as $N(0, E\beta^2/2)$. It was seen in Chap. 2 (or see Kay, 1998, Sec. 2.2.7) that z is Rayleigh distributed

$$p_z(z|H_0) = \begin{cases} \dfrac{2z}{E\beta^2} \exp\left(-\dfrac{z^2}{E\beta^2}\right) & z \geq 0 \\ 0 & z < 0 \end{cases} \qquad (6.46)$$

The probability of false alarm is

$$P_{FA} = \int_T^{+\infty} p_z(z|H_0)\, dz = \exp\left(\dfrac{-T'^2}{E\beta^2}\right) \qquad (6.47)$$

It is convenient to invert this equation to obtain the threshold setting in terms of P_{FA}

$$T' = \sqrt{-E\beta^2 \ln P_{FA}} \qquad (6.48)$$

Now consider hypothesis H_1, i.e., target present. In this case $\tilde{\mathbf{m}}^H\mathbf{y} \sim N(E, E\beta^2)$. Since E is real-valued, the real part $\tilde{\mathbf{m}}^H\mathbf{y}$ is distributed

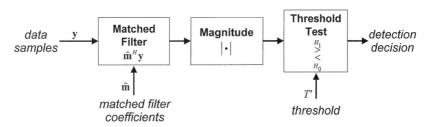

Figure 6.5 Structure of optimal detector when the absolute signal phase is unknown.

as $N(E, E\beta^2/2)$ while the imaginary part is distributed as $N(0, E\beta^2/2)$. It again follows from Chap. 2 (or see Kay, 1998, Sec. 2.2.6 or Papoulis, 1984) that the pdf of z is

$$p_z(z|H_1) = \begin{cases} \dfrac{2z}{E\beta^2} \exp\left[-\dfrac{1}{E\beta^2}(z^2 + E^2)\right] I_0\left(\dfrac{2z}{\beta^2}\right) & z \geq 0 \\ 0 & z < 0 \end{cases} \quad (6.49)$$

where $I_0(z)$ is again the modified Bessel function of the first kind. Equation (6.49) is the Rician pdf. The probability of detection is obtained by integrating it from T' to $+\infty$.

In normalized form, the required integral is

$$Q_M(\alpha, \gamma) = \int_\gamma^{+\infty} t \exp\left[-\dfrac{1}{2}(t^2 + \alpha^2)\right] I_0(\alpha t)\, dt \quad (6.50)$$

The expression $Q_M(\alpha, \gamma)$ is known as *Marcum's Q function*. It arises frequently in radar detection calculations. A closed form for this integral is not known. Algorithms for evaluating $Q_M(\alpha, \gamma)$ are compared by Cantrell and Ojha (1987). The "Communications Toolbox" optional package of MATLAB^{TM} includes a marcumq function; another MATLAB^{TM} algorithm is given by Kay (1998).

By defining a change of variables, the integral of Eq. (6.49) can be put into the form of Eq. (6.50). Specifically, choose $t = z/\sqrt{E\beta^2/2}$ and $\alpha = \sqrt{2E}/\beta$. Substituting into Eq. (6.49) and doing the integration gives

$$P_D = Q_M\left(\sqrt{\dfrac{2E}{\beta^2}}, \sqrt{\dfrac{2T'^2}{E\beta^2}}\right) \quad (6.51)$$

Finally, noting that E/β^2 is the signal-to-noise ratio χ and expressing the threshold in terms of the false alarm probability using Eq. (6.48) gives

$$P_D = Q_M\left(\sqrt{2\chi}, \sqrt{-2\ln P_{FA}}\right) \quad (6.52)$$

It is usually the case that the energy E in **m** or $\tilde{\mathbf{m}}$ is not known. Fortunately, Eq. (6.52) does not depend on E (or the noise power β^2) explicitly, but only on their ratio χ, so that it is possible to generate the ROC without this information. However, actually implementing the detector requires a specific value of the threshold T' as given in Eq. (6.48), and this does require knowledge of both E and β^2. One way to avoid this problem is to replace the matched filter coefficients $\tilde{\mathbf{m}}$ with a normalized coefficient vector $\hat{\mathbf{m}} = \tilde{\mathbf{m}}/E_{\tilde{\mathbf{m}}}$, where $E_{\tilde{\mathbf{m}}}$ is the energy in $\tilde{\mathbf{m}}$. This choice simply normalizes the gain of the matched filter to 1. The energy in this modified sequence is $\hat{E} = 1$, leading to a modified threshold

$$\hat{T} = \sqrt{-\beta^2 \ln P_{FA}} \quad (6.53)$$

The reduced matched filter gain, along with the reduced threshold, results in no change to the ROC, so that Eq. (6.52) remains valid. Setting of the threshold

$\hat{T}$ still requires knowledge of the noise power β^2; removal of this restriction is the subject of Chap. 7. The handling of unknown amplitude parameters is discussed in somewhat more detail in Sec. 6.2.4.

The performance of the envelope detector in this example is given in Fig. 6.6. The general behavior is very similar to the known phase coherent detector case of Fig. 6.4. Closer inspection, however, shows that for a given P_{FA}, the coherent detector obtains a higher P_D. To make this point clearer, Fig. 6.7 compares the detection curves for the coherent and envelope detectors (known and unknown phase, respectively), plotted two different ways. Part (a) of the figure simply repeats the 10-dB curves from the two earlier figures. At $P_{FA} = 10^{-4}$ and $\chi = 10$ dB, for example, P_D is about 0.74. This figure drops to 0.6 when the envelope detector is used.

Figure 6.7b plots the detection performance as a function of χ with P_{FA} fixed (at 10^{-6} in this example). This figure shows that, to achieve the same probability of detection, the envelope detector requires about 0.6 dB higher SNR than the coherent detector at $P_D = 0.9$, and about 0.7 dB more at $P_D = 0.5$. The extra signal-to-noise required to maintain the detection performance of the envelope detector compared to the coherent case is called an *SNR loss*. SNR losses can result from many factors; this particular one is often called the *detector loss*. It represents extra SNR that must be obtained in some way if the performance of the envelope detector is to match that of the ideal coherent detector. Increasing the SNR in turn implies one or more of many radar system changes, such as greater transmitter power, a larger antenna gain, reduced range coverage, and so forth.

The phenomenon of detector loss illustrates a very important point in detection theory: the less that is known about the signal to be detected, the higher

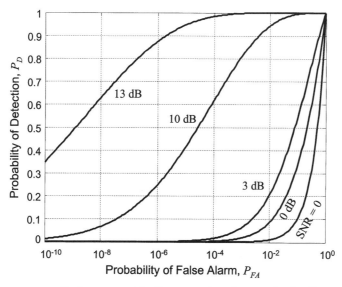

Figure 6.6 Performance of the linear envelope detector for the Gaussian example with unknown phase.

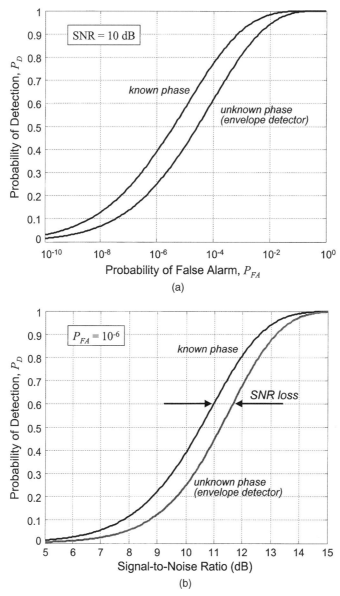

Figure 6.7 Performance difference between coherent and envelope detectors for the Gaussian example. (a) Difference in P_D for $\chi = 10$ dB. (b) Difference in P_D for $P_{FA} = 10^{-6}$.

must be the SNR to detect with a given combination of P_D and P_{FA}. In this case, not knowing the absolute phase of the signal has cost about 0.6 dB. Inconvenient though it may be, this result is intuitively satisfying: the worse the knowledge of the signal details, the worse the performance of the detector will be.

6.2.3 Linear and square-law detectors

Equation (6.44) defines the optimal Neyman-Pearson detector for the Gaussian example with an unknown phase in the data. It was shown that the $\ln[I_0(x)]$ function could be replaced by its argument x without altering the performance. In Sec. 6.3.2 a simpler detector characteristic than $\ln[I_0()]$ will again be desirable for noncoherent integration, but it will not be possible to simply substitute any monotonic increasing function. It is therefore useful to see what approximations can be made to the $\ln[I_0()]$ function.

A standard series expansion for the Bessel function holds that

$$I_0(x) = 1 + \frac{x^2}{4} + \frac{x^4}{64} + \cdots \qquad (6.54)$$

Thus for small x, $I_0(x) \approx 1 + x^2/4$. Furthermore, one series expansion of the natural logarithm has $\ln(1+z) = z - z^2/2 + z^3/3 + \cdots$. Combining these gives

$$\ln[I_0(x)] \approx \frac{x^2}{4} \qquad x \ll 1 \qquad (6.55)$$

Equation (6.55) shows that if x is small, the optimal detector is well approximated by a matched filter followed by a so-called *square law detector*, i.e., a magnitude squaring operation. The factor of four can be incorporated into the threshold in Eq. (6.44).

For large values of x, $I_0(x) \approx e^x/\sqrt{2\pi x}$, $x \gg 1$; then

$$\ln[I_0(x)] \approx x - \frac{1}{2}\ln(2\pi) - \frac{1}{2}\ln(x) \qquad (6.56)$$

The constant term on the right of Eq. (6.56) can be incorporated into the threshold in Eq. (6.44), while the linear term in x quickly dominates the logarithmic term for $x \gg 1$. This leads to the *linear detector* approximation for large x

$$\ln[I_0(x)] \approx x \qquad x \gg 1 \qquad (6.57)$$

Figure 6.8 illustrates the fit between the square law and linear approximations and the exact $\ln[I_0()]$ functions. The square law detector is an excellent fit for $x < 5$ dB, while the linear detector fits the $\ln[I_0()]$ very well for $x > 10$ dB.

Finally, note that it is easy enough to compute the squared magnitude of a complex-valued test sample as simply the sum of the squares of the real and imaginary parts. The linear magnitude requires a square root and is less computationally convenient. Section 6.5.2 presents a family of computationally simple approximations to the magnitude function.

6.2.4 Other unknown parameters

The preceding sections have shown the effect of unknown phase of the received signal on the optimal detector. However, other parameters of the received signal are also unknown in practice. The amplitude of the echo depends on all of the factors in the radar range equation, including especially the unknown target radar cross section and, at least until it is successfully detected, its range.

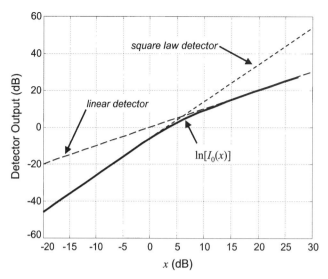

Figure 6.8 Approximation of the $\ln[I_0]$ detector characteristic by the square law detector when its argument is small, and the linear detector when its argument is large.

In addition, the target may be moving relative to the radar, so that the echo is modified by a Doppler shift.

The derivation of the magnitude-based detectors of Secs. 6.2.2 and 6.2.3 included an assumption that the received signal amplitude was known. Specifically, it was assumed that the received signal sample vector $\tilde{\mathbf{m}}$ was known, except for its absolute phase. However, as noted the absolute amplitude is also unknown in general. To determine the effect of an unknown amplitude, assume that the received signal is $A\tilde{\mathbf{m}}$, where A is an unknown but deterministic scale factor.[†] The analysis of Sec. 6.2.2 can be repeated under this assumption. The detector output under hypothesis H_0 is unchanged as would be expected, since the target echo with its unknown amplitude is not present in this case. Under hypothesis H_1, the detector output is now $\mathrm{Re}(\tilde{\mathbf{m}}^H \mathbf{y}) \sim N(A^2 E, E\beta^2/2)$. Note that the detector still considers the quantity $\tilde{\mathbf{m}}^H \mathbf{y}$ rather than $A\tilde{\mathbf{m}}^H \mathbf{y}$ because the $\tilde{\mathbf{m}}$ arises from the matched filter applied to the data and thus does not include the unknown amplitude factor A of the signal echo. Also, the quantity $E = \tilde{\mathbf{m}}^H \mathbf{m}$ is now the energy of the matched filter reference signal, while the actual signal energy becomes $A^2 E$.

The equivalent of Eq. (6.51) is now

$$P_D = Q_M\left(\sqrt{\frac{2A^2 E}{\beta^2}}, \sqrt{\frac{2T'^2}{E\beta^2}}\right) \tag{6.58}$$

[†]For instance, the earlier discussion of unknown signal energy corresponds to choosing $A = 1/E_{\tilde{\mathbf{m}}}$.

As before, the second argument of Eq. (6.58) can be written in terms of the probability of false alarm. Furthermore, because the actual signal energy is now $A^2 E$, the first argument is still $\sqrt{2\chi}$. Thus, the detection performance is still given by Eq. (6.52). The unknown echo amplitude neither requires any change in the detector structure, nor changes its performance.

Despite the unknown amplitude, the sufficient statistic was not changed. Furthermore, the probability of false alarm could be computed without knowledge of the amplitude. When both these conditions hold, the detection test is called a *uniformly most powerful* (UMP) test (Dudgeon and Johnson, 1993).

A UMP does not exist for the case where the signal delay (range) is unknown, which again is the only realistic assumption that can be made in radar. It is therefore necessary to resort to a *generalized likelihood ration test* (GLRT), in which the likelihood ratio is written as a function of the unknown signal delay Δ, and then the value of Δ that maximizes the likelihood ratio is found. Details are given by Dudgeon and Johnson (1993). The result simply requires evaluation of the matched filter over a range of delays. In practice, the matched filter is applied to the entire fast time signal; the filter output samples are simply the matched filter response for each corresponding possible target range. The maximum output is selected and compared to the threshold. If the threshold is crossed, a detection is declared. Furthermore, the value of Δ at which the maximum occurs is taken as an estimate of the target delay.

If the target is moving, an unknown Doppler shift will be imposed on the incident signal. The received echo will then be proportional not to $\tilde{\mathbf{m}}$, but to a modified signal $\tilde{\mathbf{m}}'$ where the samples of the reference signal $\tilde{\mathbf{m}}$ have been multiplied by the complex exponential sequence $\exp(j\omega_D n)$, where ω_D is the normalized Doppler shift. The required matched filter impulse response is now $\tilde{\mathbf{m}}'$; if $\tilde{\mathbf{m}}$ is replaced by $\tilde{\mathbf{m}}'$ in the derivations of Sec. 6.2.2, the same performance results as before will be obtained. Because ω_D is unknown, however, it is necessary to test for different possible Doppler shifts by conducting the detection test for multiple possible values of ω_D, similar to the procedure used to test for unknown range. If a set of K potential Doppler frequencies uniformly spaced from $-PRF/2$ to $+PRF/2$ is to be tested, the matched filter can be implemented for all K frequencies at once using the pulse Doppler processing techniques described in Chap. 5.

6.3 Threshold Detection of Radar Signals

The results of the preceding sections can now be applied to some reasonably realistic scenarios for detecting radar targets in noise. These scenarios will almost always include unknown parameters of the signal to be detected (the target), specifically, its amplitude, absolute phase, time of arrival, and Doppler shift. Detection on both a single sample of the target signal, and when multiple samples are available, is of interest. In the latter case, as discussed in Chap. 2, the target signal is often modeled as a random process, rather than a simple constant; the discussion in this chapter will be limited to the four Swerling models to illustrate both the approach and the classical, and still very useful, results obtained in these cases. Furthermore, it will be seen that the idea of pulse integration is needed in the case of multiple samples. Finally, a square-law detector will be assumed, though one important approximation that applies to linear detectors will also be introduced. Figure 6.9 represents one possible taxonomy of the most common variations on the radar detection problem. Each of these will be discussed in turn in this section, with the exception of the two diagonally hatched boxes, and of *constant-false-alarm-rate* (CFAR) techniques, which are the subject of the next chapter.

6.3.1 Coherent, noncoherent, and binary integration

The ability to detect targets is inhibited by the presence of noise and clutter. Both are modeled as random processes; the noise as uncorrelated from sample to sample, the clutter as partially correlated (including possibly uncorrelated) from sample to sample. The target is modeled as either nonfluctuating (i.e., a constant) or a random process that can be completely correlated, partially correlated, or uncorrelated from sample to sample (the Swerling models). The signal-to-interference ratio and thus the detection performance are often

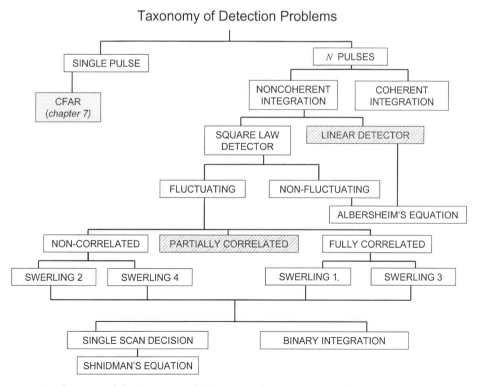

Figure 6.9 Diagram of the taxonomy of detection problems considered. (*Adapted from Levanon, 2002.*)

improved by *integrating* (adding) multiple samples of the target and interference, motivated by the idea that the interference can be "averaged out" by adding multiple samples. This idea was first discussed in Chap. 1. Thus, in general detection will be based on N samples of the target+interference. Note that care must be taken to integrate samples that represent the same range and Doppler resolution cells.

Integration may be applied to the data at three different stages in the processing chain

1. After coherent demodulation, to the baseband complex-valued (I and Q, or magnitude and phase) data. Combining complex data samples is referred to as *coherent integration*.

2. After envelope detection, to the magnitude (or squared or log magnitude) data. Combining magnitude samples after the phase information is discarded is referred to as *noncoherent integration*.

3. After threshold detection, to the target present/target absent decisions. This technique is called *binary integration*.

A system could elect to use none, one, or any combination of these techniques. Many systems use at least one integration technique, and a combination of either coherent or noncoherent with postdetection binary integration is also common. The major cost of integration is the time and energy required to obtain multiple samples of the same range, Doppler, and/or angle cell (or multiple threshold detection decisions for that cell); this is time that cannot be spent searching for targets elsewhere, or tracking already-known targets, or imaging other regions of interest. Integration also increases the signal processing computational load. Modern systems vary as to whether this is an issue: the required operations are simple, but must be performed at a very high rate in many systems.

In coherent integration, complex (magnitude and phase) data samples y_n are combined to form a new complex variable y

$$y = \sum_{n=0}^{N-1} y_n \tag{6.59}$$

As shown in Chap. 1, if the SNR of a single sample y_n is χ_1, the integrated data sample y has an SNR that is N times that of the single sample y_n, i.e., $\chi_N = N\chi_1$. That is, coherent integration attains an *integration gain* of N. Detection calculations are then based on the result for a single sample of target+noise having the improved SNR equal to χ_N. Thus, no special results are needed to analyze the case of coherent integration; one simply uses the single-sample detection results for the target and interference models of interest with the coherently integrated SNR χ_N.

In noncoherent integration, phase information is discarded. Instead, the magnitude or squared magnitude of the data samples is integrated. (Sometimes another function of the magnitude, such as the log-magnitude, is used.)

Most classical detection results have been developed for the square law detector, which bases detection on the quantity

$$z = \sum_{n=0}^{N-1} |y_n|^2 \qquad (6.60)$$

Consideration will be largely restricted to the square law detector in this section.

When coherent integration is used, detection results are obtained by using single-sample ($N = 1$) results with χ_1 replaced by the integrated χ_N. The situation for noncoherent integration is more complicated, and it will be necessary to determine the actual probability density function of the integrated variate z to compute detection results; this is done in the next subsection.

Binary integration takes place after an initial detection decision has taken place. That initial decision may be based on a single sample, or on data that have already been coherently and/or noncoherently integrated. Whatever the processing before the threshold detection, after it the result is a choice between hypothesis H_0, "target absent," and H_1, "target present." Because there are only two possible outputs of the detector each time a threshold test is made, the output is said to be binary. Multiple binary decisions can be combined in an "M out of N" decision logic in an attempt to further improve the performance. This type of integration is discussed in Sec. 6.4.

6.3.2 Nonfluctuating targets

Now consider detection based on noncoherent integration of N samples of a nonfluctuating target (sometimes called the "Swerling 0" or "Swerling 5" case) in white Gaussian noise. The amplitude and absolute phase of the target component are unknown. Thus, an individual data sample y_n is the sum of a complex constant $m = \tilde{m}\exp(j\theta)$ for some real amplitude $\tilde{m}$ and phase θ, and a white Gaussian noise sample w_n of power $\beta^2/2$ in each of the I and Q channels (total noise power β^2)

$$y_n = m + w_n \qquad (6.61)$$

Under hypothesis H_0, the target is absent and $y_n = w_n$. The pdf of $z_n = |y_n|$ is Rayleigh

$$p_{z_n}(z_n|H_0) = \begin{cases} \dfrac{2z_n}{\beta^2} e^{-z_n^2/\beta^2} & z_n \geq 0 \\ 0 & z_n < 0 \end{cases} \qquad (6.62)$$

Under hypothesis H_1, z_n is a Rician voltage density

$$p_{z_n}(z_n|H_1) = \begin{cases} \dfrac{2z_n}{\beta^2} e^{-(z_n^2+\tilde{m}^2)/\beta^2} I_0\left(\dfrac{2\tilde{m}z_n}{\beta^2}\right) & z_n \geq 0 \\ 0 & z_n < 0 \end{cases} \qquad (6.63)$$

Thus for a vector $\mathbf{z}$ of N such samples (not to be confused with the scalar sum z of samples), the joint pdfs are, for each $z_n \geq 0$

$$p_\mathbf{z}(\mathbf{z}|H_0) = \prod_{n=0}^{N-1} \frac{2z_n}{\beta^2} e^{-z_n^2/\beta^2} \tag{6.64}$$

$$p_\mathbf{z}(\mathbf{z}|H_1) = \prod_{n=0}^{N-1} \frac{2z_n}{\beta^2} e^{-(z_n^2+\tilde{m}^2)/\beta^2} I_0\left(\frac{2\tilde{m}z_n}{\beta^2}\right) \tag{6.65}$$

The LRT and log-LRT become

$$\Lambda = \prod_{n=0}^{N-1} e^{-\tilde{m}^2/\beta^2} I_0\left(\frac{2\tilde{m}z_n}{\beta^2}\right) = e^{-\tilde{m}^2/\beta^2} \prod_{n=0}^{N-1} I_0\left(\frac{2\tilde{m}z_n}{\beta^2}\right) \underset{H_0}{\overset{H_1}{\gtrless}} -\lambda \tag{6.66}$$

$$\ln \Lambda = -\frac{\tilde{m}^2}{\beta^2} + \sum_{n=0}^{N-1} \ln\left[I_0\left(\frac{2\tilde{m}z_n}{\beta^2}\right)\right] \underset{H_0}{\overset{H_1}{\gtrless}} \ln(-\lambda) \tag{6.67}$$

Incorporating the term involving the ratio of signal power and noise power on the left hand side into the threshold gives

$$\sum_{n=0}^{N-1} \ln\left[I_0\left(\frac{2\tilde{m}z_n}{\beta^2}\right)\right] \underset{H_0}{\overset{H_1}{\gtrless}} \ln(-\lambda) + \frac{\tilde{m}^2}{\beta^2} = T \tag{6.68}$$

Equation (6.68) shows that, given N noncoherent samples of a nonfluctuating target in white noise, the optimal Neyman-Pearson detection test scales each sample by the quantity $2\tilde{m}/\beta^2$, passes it through the monotonic nonlinearity $\ln[I_0()]$, and then integrates the processed samples and performs a threshold test. There are two practical problems with this equation. First, as noted earlier, it is desirable to avoid computing the function $\ln[I_0()]$ possibly millions of times per second. Second, both the target amplitude $\tilde{m}$ and the noise power β^2 must be known to perform the required scaling. The test can be simplified by using the results of Sec. 6.2.3. Applying the square law detector approximation of Eq. (6.55) to Eq. (6.68) gives the test

$$\sum_{n=0}^{N-1} \frac{\tilde{m}^2 z_n^2}{\beta^4} \underset{H_0}{\overset{H_1}{\gtrless}} T \tag{6.69}$$

Combining all constants into the threshold gives us the final noncoherent integration detection rule

$$z = \sum_{n=0}^{N-1} z_n^2 \underset{H_0}{\overset{H_1}{\gtrless}} \frac{\beta^4 T}{\tilde{m}^2} = T' \tag{6.70}$$

Equation (6.70) states that the squared magnitudes of the data samples are simply integrated and the integrated sum compared to a threshold to decide whether a target is present or not. Note that the integrated variate z is the sufficient statistic Υ for this problem.

The performance of the detector given in Eq. (6.70) must now be determined. It is convenient to scale the z_n, replacing them with the new variables $z'_n = z_n/\beta$ and thus replacing z with $z' = \sum (z'_n)^2 = z^2/\beta^2$; such a scaling does not change the performance, but merely alters the threshold value that corresponds to a particular P_D or P_{FA}. The pdf of z'_n is still either Rayleigh or Rician voltage as in Eqs. (6.62) and (6.63), but now with unit noise variance

$$p_{z'_n}(z'_n|H_0) = \begin{cases} 2z'_n e^{-z'^2_n} & z'_n \geq 0 \\ 0 & z'_n < 0 \end{cases} \tag{6.71}$$

$$p_{z'_n}(z'_n|H_1) = \begin{cases} 2z'_n e^{-(z'^2_n+\chi)} I_0(2z'_n\sqrt{\chi}) & z'_n \geq 0 \\ 0 & z'_n < 0 \end{cases} \tag{6.72}$$

where $\chi = \tilde{m}^2/\beta^2$ is the signal-to-noise ratio. Since a square law detector is being used, the pdf of $r_n = (z'_n)^2$ is needed (thus $z' = \sum r_n$); this is exponential under H_0 and a Rician power density under H_1

$$p_{r_n}(r_n|H_0) = \begin{cases} e^{-r_n} & r_n \geq 0 \\ 0 & r_n < 0 \end{cases} \tag{6.73}$$

$$p_{r_n}(r_n|H_1) = \begin{cases} e^{-(r_n+\chi)} I_0(2\sqrt{\chi r_n}) & r_n \geq 0 \\ 0 & r_n < 0 \end{cases} \tag{6.74}$$

Since z' is the sum of N scaled random variables $r_n = (z'_n)^2$, the pdf of z' is the N-fold convolution of the pdf given in Eq. (6.73) or (6.74) (Papoulis, 1984). This is most easily found using *characteristic functions* (CFs). The characteristic function $C(q)$ corresponding to a pdf $p(z)$ is given by (Papoulis, 1984)

$$C_z(q) = \int_{-\infty}^{+\infty} p_z(z) e^{jqz} \, dz \tag{6.75}$$

Note that $C_z(q)$ is just the Fourier transform of $p_z(z)$, though with the sign of the complex exponential kernel chosen opposite from the definition commonly used in electrical engineering texts. It still follows, however, that the characteristic function of the N-fold convolution of the pdfs is the product of their individual characteristic functions.

Under hypothesis H_0 the CF of r_n can be readily shown from Eqs. (6.73) and (6.75) to be

$$C_{r_n}(q) = \frac{1}{1-jq} \tag{6.76}$$

The characteristic function of z' is therefore

$$C_{z'}(q) = [C_r(q)]^N = \left(\frac{1}{1-jq}\right)^N \tag{6.77}$$

The pdf of z' is obtained by inverting its characteristic function using the inverse Fourier transform

$$p_{z'}(z'|H_0) = \frac{1}{2\pi}\int_{-\infty}^{\infty} C_{z'}(q)e^{-jqz'}\,dq \tag{6.78}$$

Using Eq. (6.77) in Eq. (6.78) and referring to any good Fourier transform table (with allowance for the reversed sign of the Fourier kernel in the definition of the characteristic function), the Erlang density (a special case of the gamma density) is obtained (Papoulis, 1984)

$$p'_z(z'|H_0) = \begin{cases} \dfrac{(z')^{N-1}}{(N-1)!}e^{-z'} & z' \geq 0 \\ 0 & z' < 0 \end{cases} \tag{6.79}$$

Note that this reduces to the exponential pdf when $N = 1$, as would be expected since in that case z' is the magnitude squared of a single sample of complex Gaussian noise.

The probability of false alarm is obtained by integrating Eq. (6.79) from the threshold value to $+\infty$. The result is (Abramowitz and Stegun, 1972)

$$P_{FA} = \int_T^{\infty} \frac{(z')^{N-1}}{(N-1)!}e^{-z'}\,dz' = 1 - I\left(\frac{T}{\sqrt{N}}, N-1\right) \tag{6.80}$$

where

$$I(u, M) = \int_0^{u\sqrt{M+1}} \frac{e^{-\tau}\tau^M}{M!}\,d\tau \tag{6.81}$$

is Pearson's form of the incomplete gamma function. For a single sample ($N = 1$), Eq. (6.80) reduces to the especially simple result

$$P_{FA} = e^{-T} \tag{6.82}$$

so that $T = -\ln P_{FA}$.

Equation (6.80) can be used to determine the probability of false alarm P_{FA} for a given threshold T or, more likely, to determine the required value of T for a desired P_{FA}. Now the probability of detection P_D corresponding to the same threshold must be determined. Start by finding the pdf of the normalized, integrated, and square-law-detected samples under hypothesis H_1. Each individual data sample r_n is Rician (Eq. (6.74)); the corresponding characteristic function is

$$C_{r_n}(q) = \frac{1}{q+1}e^{-\chi[q/(q+1)]} \tag{6.83}$$

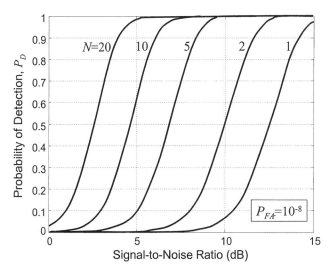

Figure 6.10 Effect of noncoherent integration on detection performance for a nonfluctuating target in complex Gaussian noise as a function of single-sample SNR.

The CF of the sum of N such samples, z', is

$$C_{z'}(q) = \left(\frac{1}{q+1}\right)^N e^{-N\chi[q/(q+1)]} \tag{6.84}$$

and the pdf of z' is[†]

$$p_{z'}(z'|H_1) = \left(\frac{z'}{N\chi}\right)^{(N-1)/2} e^{-z-N\chi} I_{N-1}(2\sqrt{N\chi z'}) \tag{6.85}$$

P_D is found by integrating Eq. (6.85). One version of the result, given by Meyer and Mayer (1973), is

$$P_D = \int_T^\infty \left(\frac{z'}{N\chi}\right)^{(N-1)/2} e^{-z-N\chi} I_{N-1}(2\sqrt{N\chi z'}) dz'$$
$$= Q_M(\sqrt{2N\chi}, \sqrt{2T}) + e^{-(T+N\chi)} \sum_{r=2}^{N} \left(\frac{T}{N\chi}\right)^{(r-1)/2} I_{r-1}(2\sqrt{N\chi T}) \tag{6.86}$$

Note that the summation term in the second line of Eq. (6.86) only contributes when $N \geq 2$. Equations (6.80) and (6.86) define the performance achievable with a noncoherent integration using a square law detector (Meyer and Mayer, 1973; DiFranco and Rubin, 1980).

Figure 6.10 shows the effect of the number of samples noncoherently integrated, N, on the receiver operating characteristic when $P_{FA} = 10^{-8}$. This figure

---

[†]Note that $I_{N-1}(x)$ is the modified Bessel function of the first kind and order $N-1$, not to be confused with the incomplete gamma function $I(u, M)$.

shows that noncoherent integration reduces the required single-sample SNR required to achieve a given P_D and P_{FA}, but not by the factor N achieved with coherent integration. For example, consider the single-sample SNR required to achieve $P_D = 0.9$. For $N = 1$, this is 14.2 dB; for $N = 10$, it drops to 6.1 dB, a reduction of 8.1 dB, but less than the 10 dB that corresponds to the factor of 10 increase in the number of pulses integrated. In Sec. 6.3.3 an estimate will be developed of this reduction in required single sample SNR, called the *noncoherent integration gain*.

6.3.3 Albersheim's equation

The performance results for the case of a nonfluctuating target in complex Gaussian noise are given by Eqs. (6.80) and (6.86). While relatively easy to implement in a modern software analysis system such as MATLAB™, these equations do not lend themselves to manual calculation. Fortunately, there does exist a simple closed-form expression relating P_D, P_{FA}, and SNR χ that can be computed by hand or with simple scientific calculators. This expression is known as *Albersheim's equation* (Albersheim, 1981; Tufts and Cann, 1983; Levanon, 1988).

Albersheim's equation is an empirical approximation to the results by Robertson (1967) for computing the single-sample SNR χ_1, required to achieve a given P_D and P_{FA}. It applies under the following conditions

- Nonfluctuating target in Gaussian (i.i.d. in I and Q) noise
- Linear (not square-law) detector
- Noncoherent integration of N samples

The estimate is given by the series of calculations

$$A = \ln\left(\frac{0.62}{P_{FA}}\right)$$
$$B = \ln\left(\frac{P_D}{1-P_D}\right) \tag{6.87}$$
$$\chi_1 = -5\log_{10} N + \left[6.2 + \left(\frac{4.54}{\sqrt{N+0.44}}\right)\right] \cdot \log_{10}(A + 0.12AB + 1.7B)\,\text{dB}$$

Note that χ_1 is in decibels, not linear power units. The error in the estimate of χ_1 is less than 0.2 dB for $10^{-7} \leq P_{FA} \leq 10^{-3}$, $0.1 \leq P_D \leq 0.9$, and $1 \leq N \leq 8096$, a very useful range of parameters. For the special case of $N = 1$, Eq. (6.87) reduces to

$$A = \ln\left(\frac{0.62}{P_{FA}}\right)$$
$$B = \ln\left(\frac{P_D}{1-P_D}\right)$$
$$\chi_1 = 10\log_{10}(A + 0.12AB + 1.7B)\,\text{dB} \tag{6.88}$$

Note that on a linear (not decibel) scale, the last line of Eq. (6.88) is just $\chi_1 = A + 0.12AB + 1.7B$.

To illustrate, suppose $P_D = 0.9$ and $P_{FA} = 10^{-6}$ are required for a nonfluctuating target in a system using a linear detector. If detection is to be based on a single sample, what is the required SNR of that sample? This is a direct application of Albersheim's equation. Compute $A = \ln(0.62 \times 10^6) = 13.34$ and $B = \ln(9) = 2.197$. Equation (6.88) then gives $\chi_1 = 13.14$ dB; on a linear scale, this is 20.59.

If $N = 100$ samples are noncoherently integrated, it should be possible to obtain the same P_D and P_{FA} with a lower single-sample SNR. To confirm this, use Eq. (6.87). The intermediate parameters A and B are unchanged. χ_1 is now reduced to -1.26 dB, a reduction of 14.4 dB. Note that in this case, the noncoherent integration gain of 14.4 dB, a factor of 27.54 on a linear scale, is much better than the $\sqrt{N}$ rule of thumb sometimes given for noncoherent integration, which would give a gain factor of only 10 for $N = 100$ samples integrated. Rather, the gain is approximately $N^{0.7}$. Albersheim's equation will be used shortly to develop an expression for estimating the noncoherent integration gain.

Albersheim's equation is useful because it requires no function more exotic than the natural logarithm and square root for its evaluation. It can thus be evaluated on virtually any scientific calculator, and is convenient to program into any programmable scientific calculator. If a somewhat larger error can be tolerated, it can also be used for square-law detector results for the nonfluctuating target, Gaussian noise case. Specifically, square law detector results are within 0.2 dB of linear detector results (Robertson, 1967; Tufts and Cann, 1983). Thus, the same equation can be used for rough calculations over the range of parameters given previously with errors not exceeding 0.4 dB.

Equations (6.87) and (6.88) provide for calculation of χ_1 given P_D, P_{FA}, and N. It is possible, however, to solve Eq. (6.87) for either P_D or P_{FA} in terms of the other and χ_1 and N, extending further the usefulness of Albersheim's equation. For instance, the following calculations show how to estimate P_D given the other factors (χ_1 is in dB)

$$\begin{aligned} A &= \ln\left(\frac{0.62}{P_{FA}}\right) \\ Z &= \frac{\chi_1 + 5\log_{10} N}{6.2 + \frac{4.54}{\sqrt{N+0.44}}} \\ B &= \frac{10^Z - A}{1.7 + 0.12A} \\ P_D &= \frac{1}{1 + e^{-B}} \end{aligned} \quad (6.89)$$

In Eq. (6.89), A and B are the same values as in Eq. (6.87), though B cannot be computed in terms of P_D, since P_D is now the unknown. A result similar to Eq. (6.89) can be derived for computing P_{FA}.

Albersheim's equation can also be used to estimate the signal-to-noise ratio gain for noncoherent integration of N samples. The noncoherent integration gain G_{nc} is the reduction in single-sample SNR required to achieve a specified P_D and P_{FA} when N samples are combined; in dB, this is given by

$$G_{nc}(N)(\text{dB}) = \chi_1|_{1\text{pulse}} - \chi_1|_{N\text{pulses}}$$
$$= 5\log_{10} N - \left[6.2 + \left(\frac{4.54}{\sqrt{N+0.44}}\right)\right] \cdot \log_{10}(A + 0.12AB + 1.7B)$$
$$+ 10\log_{10}(A + 0.12AB + 1.7B)\,\text{dB} \qquad (6.90)$$
$$= 5\log_{10} N - \left[\left(\frac{4.54}{\sqrt{N+0.44}}\right) - 3.8\right] \cdot \log_{10}(A + 0.12AB + 1.7B)\,\text{dB}$$

On a linear scale this becomes

$$G_{nc}(N) = \frac{\sqrt{N}}{k^{f(N)}} \qquad (6.91)$$

where
$$k = A + 0.12AB + 1.7B$$
$$f(N) = \left(\frac{0.454}{\sqrt{N+0.44}}\right) - 0.38 \qquad (6.92)$$

The constant k absorbs terms that are not a function of N, and the term $f(N)$ is a slowly declining function of N.

Figure 6.11 plots this estimate of G_{nc} in decibels for Albersheim's nonfluctuating, linear detector case as a function of N. Also shown are curves corresponding to $N^{0.7}$ and $N^{0.8}$. The noncoherent gain is slightly better than $N^{0.8}$ for very few samples integrated ($N = 2$ or 3), with the effective exponent on N declining slowly as N increases. G_{nc} is bracketed by $N^{0.7}$ and $N^{0.8}$ to in excess of $N = 100$ samples integrated; the gain eventually slows asymptotically to become proportional to $\sqrt{N}$ for very large N. Thus, noncoherent integration is more efficient than the $\sqrt{N}$ often attributed to it for a wide range of N, but remains less efficient than coherent integration, which achieves an integration gain of N. Nonetheless, its much simpler implementation, not requiring knowledge of the phase, means it is widely used to improve the SNR before the threshold detector.

6.3.4 Fluctuating targets

The analysis in the preceding section considered only nonfluctuating targets, often called the "Swerling 0" or "Swerling 5" case. A more realistic model allows for target fluctuations, in which the target RCS is drawn from either the exponential or chi-square pdf, and the RCS of a group of N samples follows either

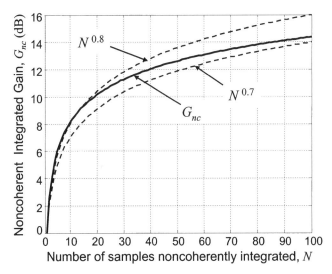

Figure 6.11 Noncoherent integration gain G_{nc} for a nonfluctuating target, as estimated using Albersheim's equation.

the pulse-to-pulse or scan-to-scan correlation model, as described in Chap. 2. Note that representing the target by one of the Swerling models 1 through 4 has no effect on the probability of false alarm, since that it is determined only by the pdf when no target is present; thus Eq. (6.80) still applies.

The strategy for determining the probability of detection depends on the Swerling model used. Figure 6.12 illustrates the approach. In all cases, the pdf of the magnitude of a single sample is Rician, so that the CF of a single square-law detected sample is still given by Eq. (6.83). However, the SNR is now a random variable because the target RCS is a random variable.

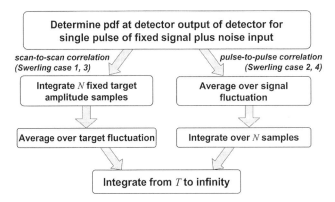

Figure 6.12 Strategy for computing P_D for Swerling target fluctuation models.

In the scan-to-scan correlation cases (Swerling models 1 and 3), the target RCS is a fixed value for all N pulses integrated to form z'. Thus, the CF of z' is the N-fold product of Eq. (6.83) with itself

$$C_{z'}(q; \chi, N) = \left(\frac{1}{q+1}\right)^N e^{-N\chi[q/(q+1)]} \qquad (6.93)$$

This is the same expression as Eq. (6.84) except that now $C_{z'}$ is written explicitly as a function of all of q, χ, and N. Next take the expected value of the CF over the target RCS (and thus the SNR)

$$\bar{C}_{z'}(q; \bar{\chi}, N) = \int_0^\infty p_\chi(\chi) C_{z'}(q; \chi, N) d\chi \qquad (6.94)$$

where $p_\chi(\chi)$ is the pdf of the SNR, and will be either exponential (Swerling 1) or chi-square (Swerling 3).

For the Swerling 1 case, the pdf of the SNR is exponential

$$p_\chi(\chi) = \frac{1}{\bar{\chi}} e^{-\chi/\bar{\chi}} \qquad (6.95)$$

where $\bar{\chi}$ is the average SNR over the N samples. Using Eqs. (6.93) and (6.95) in Eq. (6.94) gives the characteristic function, averaged over the signal fluctuations

$$\bar{C}_{z'}(q; \bar{\chi}, N) = \frac{1}{(1+q)^{N-1}[1+q(1+N\bar{\chi})]} \qquad (6.96)$$

Fourier transform tables and properties can again be applied to compute the inverse CF of Eq. (6.96), which is the pdf of z' under hypothesis H_1 for Swerling case 1

$$p_{z'}(z'|H_1) = \frac{1}{N\bar{\chi}} \left(1 + \frac{1}{N\bar{\chi}}\right)^{N-2} I\left[\frac{z'}{(1+1/N\bar{\chi})\sqrt{N-1}}, N-2\right] e^{-z'/(1+N\bar{\chi})} \qquad (6.97)$$

Integrating the pdf of Eq. (6.97) from the threshold T to $+\infty$, it is possible to arrive at the following expression for the probability of detection in the Swerling 1 case (Meyer and Mayer, 1973)

$$P_D = 1 - I\left[\frac{T}{\sqrt{N-1}}, N-2\right] + \left(1 + \frac{1}{N\bar{\chi}}\right)^{N-1}$$
$$\times I\left[\frac{T}{\left(1 + \frac{1}{N\bar{\chi}}\right)\sqrt{N-1}}, N-2\right] e^{-T/(1+N\bar{\chi})} \qquad N>1 \qquad (6.98)$$

This expression can be simplified when $P_{FA} \ll 1$ and the integrated average SNR $N\bar{\chi} > 1$; both conditions are almost always true in any scenario where target detection is to be successful. The result is

$$P_D \approx \left(1 + \frac{1}{N\bar{\chi}}\right)^{N-1} e^{-T/(1+N\bar{\chi})} \qquad (P_{FA} \ll 1, N\bar{\chi} > 1) \qquad (6.99)$$

Furthermore, Eq. (6.99) is exact when $N = 1$; in this case it reduces to

$$P_D = e^{-T/(1+\chi)} \qquad (6.100)$$

(Note that when there is only one pulse, the average SNR $\bar{\chi}$ is just the single-pulse SNR χ.) For the $N = 1$ case, Eq. (6.82) can then be used in Eq. (6.100) to write a direct relationship between P_D and P_{FA}

$$P_D = (P_{FA})^{1/(1+\chi)} \qquad (6.101)$$

In the Swerling 2 or 4 cases, the samples exhibit pulse-to-pulse correlation, meaning they are in fact uncorrelated with one another. In this case, each of the N samples integrated has a different value of SNR. Consequently, it is appropriate to average over the SNR in the single-sample CF

$$\bar{C}_r(q; \bar{\chi}) = \left(\frac{1}{q+1}\right) \int_0^\infty p_\chi(\chi) e^{-\chi\left(\frac{q}{q+1}\right)} d\chi \qquad (6.102)$$

and then perform the N-fold multiplication of the averaged single-sample CF to get the CF of the integrated data

$$\bar{C}_{z'}(q; \bar{\chi}, N) = [\bar{C}(q; \bar{\chi})]^N \qquad (6.103)$$

For the Swerling 2 case specifically, the exponential pdf can again be used for the SNR (Eq. (6.95)), applying it this time in Eq. (6.102) to arrive at

$$\bar{C}_r(q; \bar{\chi}) = \frac{1}{[1 + q(1 + \bar{\chi})]} \qquad (6.104)$$

and thus

$$\bar{C}_{z'}(q; \bar{\chi}, N) = \frac{1}{[1 + q(1 + \bar{\chi})]^N} \qquad (6.105)$$

Inverse transforming Eq. (6.105) gives the pdf of z' under hypothesis H_1 for Swerling case 2

$$p_{z'}(z'|H_1) = \frac{z'^{N-1} e^{-z'/(1+\bar{\chi})}}{(1+\bar{\chi})^N (N-1)!} \qquad (6.106)$$

Integrating Eq. (6.106) gives the probability of detection, which can be shown to be (Meyer and Mayer, 1973)

$$P_D = 1 - I\left[\frac{T}{(1+\bar{\chi})\sqrt{N}}, N-1\right]$$

$$= e^{-T/(1+\bar{\chi})} \sum_{l=0}^{N-1} \frac{1}{l!} \left(\frac{T}{1+\bar{\chi}}\right)^l \qquad (6.107)$$

The series approximation in the second line of Eq. (6.107) provides a convenient computational implementation.

Results for Swerling 3 and 4 targets can be obtained by repeating the previous analyses above for the Swerling 1 and 2 cases, but with a chi-square instead of exponential density function for the SNR

$$p_\chi(\chi) = \frac{4\chi}{\bar{\chi}^2} e^{-2\chi/\bar{\chi}} \tag{6.108}$$

Derivations of the resulting expressions for P_D can be found in the books by Meyer and Mayer (1973), DiFranco and Rubin (1980), and many other radar detection texts. Table 6.1 summarizes one form of the resulting expressions.

Figure 6.13 compares the detection performance of the four Swerling model fluctuating targets and the nonfluctuating target for $N = 10$ samples as a function of the SNR for a fixed $P_{FA} = 10^{-8}$. Assuming that the primary interest is in relatively high (>0.5) values of P_D, the upper half of the figure is of greatest interest. In this case, the nonfluctuating target is the most favorable, in the sense that it achieves a given probability of detection at the lowest SNR. The worst case (highest required SNR for a given P_D) is the Swerling case 1, which corresponds to scan-to-scan correlation and an exponential pdf of the target RCS. For instance, $P_D = 0.95$ requires $\chi \approx 6$ dB for the nonfluctuating case, but $\chi \approx 17$ dB for the Swerling 5 case, a difference of about 11 dB.

TABLE 6.1 Probability of Detection for Swerling Model Fluctuating Targets with a Square-Law Detector

Case	P_D	Comments
0 or 5	$Q_M\left(\sqrt{2N\bar{\chi}},\sqrt{2T}\right) + e^{-(T+N\bar{\chi})} \sum_{r=2}^{N} \left(\frac{T}{N\bar{\chi}}\right)^{\frac{r-1}{2}} I_{r-1}\left(2\sqrt{NT\bar{\chi}}\right)$	
1	$\left(1 + \frac{1}{N\bar{\chi}}\right)^{N-1} e^{-T/(1+N\bar{\chi})}$	Approximate for $P_{FA} \ll 1$ and $N\bar{\chi} > 1$; exact for $N = 1$
2	$1 - I\left[\frac{T}{(1+\bar{\chi})\sqrt{N}}, N-1\right]$	
3	$\left(1 + \frac{2}{N\bar{\chi}}\right)^{N-2} \left[1 + \frac{T}{1+(N\bar{\chi}/2)} - \frac{2(N-2)}{N\bar{\chi}}\right] e^{-T/(1+(N\bar{\chi}/2))}$	Approximate for $P_{FA} \ll 1$ and $N\bar{\chi}/2 > 1$; exact for $N = 1$ or 2
4	$c^N \sum_{k=0}^{N} \frac{N!}{k!(N-k)!} \left(\frac{1-c}{c}\right)^{N-k} \left\{\sum_{l=0}^{2N-1-k} \frac{e^{-cT}(cT)^l}{l!}\right\}\quad T > N(2-c)$ $1 - c^N \sum_{k=0}^{N} \frac{N!}{k!(N-k)!} \left(\frac{1-c}{c}\right)^{N-k} \sum_{l=2N-k}^{\infty} \frac{e^{-cT}(cT)^l}{l!}\quad T < N(2-c)$	$c \equiv \frac{1}{1+(\bar{\chi}/2)}$

$$P_{FA} = 1 - I\left(\frac{T}{\sqrt{N}}, N-1\right) \text{ in all cases}$$

$I(\cdot,\cdot)$ is Pearson's form of the incomplete Gamma function; $I_k(\cdot)$ is the modified Bessel function of the first kind and order k

336 Chapter Six

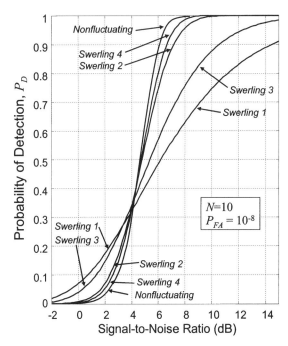

Figure 6.13 Comparison of detection performance for fluctuating (Swerling) and nonfluctuating target models using noncoherent integration of 10 pulses ($N = 10$) and a fixed probability of false alarm $P_{FA} = 10^{-8}$ with a square law detector.

At least two general conclusions can be drawn from Fig. 6.13. First, for $P_D > 0.5$, nonfluctuating targets are easier to detect than any of the Swerling cases; target fluctuations make detection more difficult, i.e., require a higher SNR for a given P_D. Second, pulse-to-pulse fluctuations (Swerling 2 and 4) aid target detectability compared to scan-to-scan detectability. For instance, a Swerling 2 target is easier to detect than the Swerling 1 that shares the same pdf for target fluctuations, and a Swerling 4 target is easier than a Swerling 3 target. Finally, note that the converse of these statements is true for detection probabilities less than 0.5.

6.3.5 Shnidman's equation

The analytic results in Table 6.1 are too complex for "back-of-the-envelope" calculations, or even for calculation on programmable calculators. Albersheim's equation provided a simple approximation for the nonfluctuating target case, but it is not applicable to fluctuating targets in general, and the Swerling models in particular. This is a serious limitation since, as seen for example in Fig. 6.13, the nonfluctuating case provides optimistic results for most parameter ranges of interest.

Fortunately, empirical approximations have also been developed for the Swerling cases. Probably the most useful result is *Shnidman's equation*

(Shnidman, 2002). Similar to Albersheim's equation, this series of equations gives the single-pulse SNR χ required to achieve a specified P_D and P_{FA} with noncoherent integration of N samples. Unlike Albersheim's equation, the results are for a square law detector; however, as noted previously, the differences are small and in any event, the square law detector is used as much or more than a linear detector.

Shnidman's "equation" is given by the following series of calculations

$$K = \begin{cases} \infty, & \text{nonfluctuating target ("Swerling 0/5")} \\ 1, & \text{Swerling 1} \\ N, & \text{Swerling 2} \\ 2, & \text{Swerling 3} \\ 2N & \text{Swerling 4} \end{cases}$$

$$\alpha = \begin{cases} 0 & N < 40 \\ \frac{1}{4} & N \geq 40 \end{cases} \tag{6.109}$$

$$\eta = \sqrt{-0.8\ln(4P_{FA}(1-P_{FA}))} + \text{sign}(P_D - 0.5)\sqrt{-0.8\ln(4P_D(1-P_D))}$$

$$X_\infty = \eta\left(\eta + 2\sqrt{\frac{N}{2} + \left(\alpha - \frac{1}{4}\right)}\right) \tag{6.110}$$

$$C_1 = \{[(17.7006P_D - 18.4496)P_D + 14.5339]P_D - 3.525\}/K$$

$$C_2 = \frac{1}{K}\left\{\exp(27.31P_D - 25.14) + (P_D - 0.8)\left[0.7\ln\left(\frac{10^{-5}}{P_{FA}}\right) + \frac{(2N-20)}{80}\right]\right\}$$

$$C_{dB} = \begin{cases} C_1 & 0.1 \leq P_D \leq 0.872 \\ C_1 + C_2 & 0.872 < P_D \leq 0.99 \end{cases}$$

$$C = 10^{C_{dB}/10} \tag{6.111}$$

$$\chi_1 = \frac{C \cdot X_\infty}{N}$$

$$\chi_1 \text{ (dB)} = 10\log_{10}(\chi_1) \tag{6.112}$$

Note that several of the equations simplify in the nonfluctuating case, $K = \infty$. Specifically, in this case $C_1 = C_2 = 0$, so that in turn $C_{dB} = 0$ and $C = 1$. The function sign(x) is $+1$ if $x > 0$ and -1 if $x < 0$.

The accuracy bounds on Shnidman's equation are somewhat looser than those specified for Albersheim's equation. Specifically, except at the extreme values of P_D for the Swerling 1 case, the error in the estimate of χ_1 is less than 0.5 dB for $0.1 \leq P_D \leq 0.99$, $10^{-9} \leq P_{FA} \leq 10^{-3}$, and $1 \leq N \leq 100$. This is a much better range for P_D than used in Albersheim's equation. The range of N is much smaller, but still large enough for almost all problems of interest. Figure 6.14 illustrates the error in the estimate of the single-pulse SNR χ_1 using Shnidman's equation for the case $P_{FA} = 10^{-6}$ and $N = 5$, and for P_D over the specified range of 0.1 to 0.99.

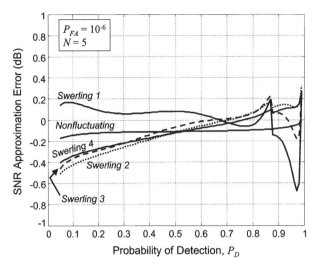

Figure 6.14 Example of error in estimating χ_1 via Shnidman's equation for $P_{FA} = 10^{-6}$ and $N = 5$.

A still more accurate approximation for the nonfluctuating and Swerling 1 cases has recently been developed (Hmam, 2003). However, it is not applicable to all of the Swerling cases, and the computations, while easy, are more extensive. Consequently, Shnidman's equation appears to be the most useful general tool currently available.

6.4 Binary Integration

Any coherent or noncoherent integration is followed finally by comparing the integrated data to a threshold. The result is a choice between two hypotheses, "target present" or "target absent," so the output is binary in the sense that it takes one of only two possible outcomes. If the entire detection process is repeated N times for a given range, Doppler, or angle cell, N binary decisions will be available. Each decision of "target present" will have some probability P_D of being correct, and a probability P_{FA} of being incorrect. To improve the reliability of the detection decision, the decision rule can require that a target be detected on some number M of the N decisions before it is finally accepted as a valid target detection. This process is called *binary integration*, "*M* of *N*" detection, or *coincidence detection* (Levanon, 1988; Skolnik, 2001).

To analyze binary integration, begin by assuming a nonfluctuating target so that the probability of detection P_D is the same for each of N threshold tests. Then the probability of *not* detecting an actual target (i.e., the probability of a miss) on one trial is $1 - P_D$. If there are N independent trials, the probability of missing the target on all N trials is $(1 - P_D)^N$. Thus, the probability of detecting the target on at least one of N trials, denoted as the *cumulative probability* P_{CD}, is

$$P_{CD} = 1 - (1 - P_D)^N \qquad (6.113)$$

Table 6.2 shows the single-trial probability of detection required to achieve $P_{CD} = 0.99$ as a function of N. Clearly, a "1 of N" decision rule achieves a high cumulative probability of detection with relatively low single-trial probabilities of detection. In other words, the "1 of N" rule increases the effective probability of detection.

The trouble with the "1 of N" rule is that it "works" for the probability of false alarm also. The probability of at least one false alarm in N trials is the cumulative probability of false alarm, P_{CFA}

$$P_{CFA} = 1 - (1 - P_{FA})^N \tag{6.114}$$

Assuming that $P_{FA} \ll 1$, Eq. (6.114) can be approximated as

$$\begin{aligned} P_{CFA} &= 1 - (1 - P_{FA})^N \\ &= 1 - \left(1 - N \cdot P_{FA} - \frac{N(N-1)}{2} P_{FA}^2 - \cdots \right) \\ &\approx 1 - (1 - N \cdot P_{FA}) = N \cdot P_{FA} \end{aligned} \tag{6.115}$$

where the binomial series expansion was used to obtain the second line of Eq. (6.115). Equation (6.115) shows that the "1 of N" rule *increases* P_{FA} by a factor of approximately N.

A binary integration rule that reduces the required single-trial P_D but increases P_{FA} does not accomplish the goal of decreasing the required single-sample SNR to achieve a specified P_D and P_{FA}. What is needed is a binary integration rule that increases P_{CD} compared to P_D, while leaving P_{CFA} equal to, or less than P_{FA}. An "M of N" strategy provides better results.

Consider the cumulative probability P_C of H successes in N trials, when the probability of success on a single trial is p; it is

$$P_C = \sum_{r=M}^{N} \binom{N}{r} p^r (1-p)^{N-r} \tag{6.116}$$

where

$$\binom{N}{r} \equiv \frac{N!}{(N-r)! r!} \tag{6.117}$$

Equation (6.116) can be applied to the cumulative probability of false alarm by letting $p = P_{FA}$, and to the probability of detection by letting $p = P_D$. In the former case, a "success" is a false alarm, i.e., the event that has a probability of p; in the latter case, a "success" is a correct detection. Consider the specific

TABLE 6.2 Single-Trial P_D Needed to Achieve $P_{CD} = 0.99$

N	1	2	4	10	20	100
P_D	0.99	0.90	0.68	0.37	0.2	0.045

example of a "2 of 4" rule, that is, $N = 4$ and $H = 2$. Using these parameters in Eq. (6.116) gives

$$P_C = \sum_{r=2}^{4} \frac{24}{(4-r)!r!} p^r (1-p)^{4-r}$$
$$= 6p^2(1-p)^2 + 4p^3(1-p) + p^4 \tag{6.118}$$

To determine the effect of this rule on the probability of false alarm, let $p = P_{FA}$. Assuming that $P_{FA} \ll 1$, Eq. (6.118) can be approximated by its first term and simplified to obtain

$$P_{CFA} \approx \binom{N}{M} P_{FA}^M = 6 P_{FA}^2 \tag{6.119}$$

Thus, the "2 of 4" rule will result in a cumulative false alarm probability that is less than the single-trial P_{FA} as desired. Because single-trial values of P_D are not necessarily very close to 1, Eq. (6.118) cannot easily be approximated in a simple form similar to Eq. (6.119). Table 6.3 shows the cumulative probability obtained using a "2 or 4" rule for various values of the single-trial probability p. The three cases above the dotted line are appropriate for considering the effect on likely single-trial probabilities of detection, while the two cases below the line are examples of the effect on likely single-trial probabilities of false alarm. This table shows that the "2 of 4" rule not only reduces the probability of false alarms, it also increase the probability of detection so long as the single-trial P_D is reasonably high.

This example illustrates the characteristics required of an "M of N" rule. For small values of p, P_C should be less than or equal to p so that the rule reduces false alarm probabilities. For larger values of p, P_C should be larger than p so that detection probabilities are increased by binary integration. To show the effect of the "M of N" rule on large and small single-trial probabilities, Fig. 6.15 plots the ratio of P_C to p for $N = 4$ and all four possible choices of M. If the ratio is greater than 1, P_C is greater than p; this should be the case for values of p appropriate to single-trial detection probabilities. Conversely, for small values of p, appropriate to false alarm probabilities, the ratio should be

TABLE 6.3 Cumulative Probability Using a "2 of 4" Rule

p	P_C
0.5	0.688
0.8	0.973
0.9	0.996
10^{-3}	5.992×10^{-6}
10^{-6}	6.0×10^{-10}

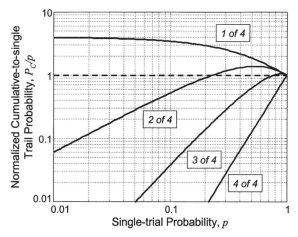

Figure 6.15 Ratio of the cumulative probability to the single-trial probability for an "M of 4" binary integration rule.

less than 1. Figure 6.15 shows that the ratio is greater than 1 for all values of p for the "1 of 4" rule, consistent with the earlier discussion. Similarly, the "4 of 4" rule results in a ratio that is always less than 1, good for false alarm reduction but bad for improving detection. The "2 of 4" and "3 of 4" rules both provide good false alarm reduction for small values of p and detection improvement for large values of p. However, the "3 of 4" rule increases probabilities only for p equal to approximately 0.75 or higher, a relatively narrow range; and the increase is very slight. The "2 of 4" rule improves detection for values of p down to approximately 0.23, still well above any likely single-trial false P_{FA}. Thus the "2 of 4" rule appears to be the best choice when $N = 4$.

Note that a nonfluctuating target has been implicitly assumed, since the single-trial P_D was assumed to be the same on each trial. The results can be extended to fluctuating targets (Weiner, 1991; Shnidman, 1998). One result of these analyses is that for a given Swerling model, P_{FA} specification, SNR, and number of samples N, there is a value of M, M_{opt}, that maximizes P_D. M_{opt} can be estimated as

$$M_{\text{opt}} = 10^b N^a \tag{6.120}$$

where the parameters a and b are given in Table 6.4 for the various Swerling models for $P_D = 0.9$ and $10^{-8} \leq P_{FA} \leq 10^{-4}$ ("Swerling 0" is the nonfluctuating case) (Shnidman, 1998).

Other binary integration strategies exist having somewhat different properties. For instance, the decision logic can require that the M hits be contiguous, rather than just any M hits out of N. See the book by Skolnik (1988) for references to the literature for these and other variations on binary integration.

TABLE 6.4 Parameters for Estimating M_{opt}

Swerling model	a	b	Range of N
0	0.8	−0.02	5–700
1	0.8	−0.02	6–500
2	0.91	−0.38	9–700
3	0.8	−0.02	6–700
4	0.873	−0.27	10–700

6.5 Useful Numerical Approximations

As has been seen, calculations of detection and false alarm probabilities, or of signal-to-noise ratios required to achieve a given P_D and P_{FA}, are plagued by difficult-to-evaluate functions; even the simplest calculations involve nontrivial functions such as error functions or Marcum's Q function, or their inverses or integrals. Thus, one is driven to tabulated results or to computer calculations for accurate estimation of radar performance probabilities. However, there do exist some numerical approximations and computational techniques that allow estimates of radar performance in some cases with nothing more complicated than a scientific calculator. Two such particularly useful techniques, Albersheim's and Shnidman's equations, were discussed in Sec. 6.3.3 and 6.3.5, respectively. In this section, a few more useful techniques are collected.

6.5.1 Approximations to the error function

For Gaussian problems, it has been seen that the error function is of critical importance in calculating probabilities. A number of approximations and bounds to erf(x) are available in the book by Abramowitz and Stegun (1972). A relatively simple approximation is given by

$$t = \frac{1}{1 + 0.4707x}$$

$$\text{erf}(x) \approx 1 - e^{-x^2}\{0.3480242t - 0.0958798t^2 + 0.7478556t^3\} \quad (6.121)$$

One pair of upper and lower bounds on erf(x) is

$$1 - \frac{2}{\sqrt{\pi}} \frac{e^{-x^2}}{x + \sqrt{x^2 + 2}} < \text{erf}(x) < 1 - \frac{2}{\sqrt{\pi}} \frac{e^{-x^2}}{x + \sqrt{x^2 + \frac{4}{\pi}}} \quad (6.122)$$

Figure 6.16 plots erf(x), the approximation of Eq. (6.121), and the bounds of Eq. (6.122). Equation (6.121) is an excellent fit over the entire range. The lower bound is fairly tight over the entire range, while the upper bound is best used only for $x > 0.8$.

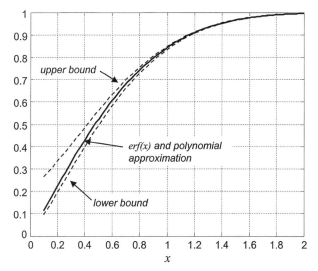

Figure 6.16 Error function erf(x) and the upper and lower bounds of Eq. (6.122).

Another calculator-friendly approximation to the error function is given by the equations (Vedder, 1988)

$$a = \sqrt{\frac{8}{\pi}} \qquad b = \frac{(4-\pi)}{3\pi}\sqrt{\frac{2}{\pi}}$$

$$\text{erf}(x) \approx \frac{2}{1+e^{-\sqrt{2}x(a+2bx^2)}} - 1 \qquad (6.123)$$

This approximation to erf(x) is not plotted because the difference is visually indistinguishable. Examination of the error in Vedder's approximation shows that it peaks at a value of approximately 6.27×10^{-4} when $x = -1.64$, and -6.27×10^{-4} when $x = +1.64$. This is accurate enough to allow the approximation to be used for all values of x in most cases.

Vedder also gave a formula for the inverse error function

$$A = \frac{a}{3b} \qquad B = \frac{\ln\left[\left(\frac{X+1}{2}\right)^{-1} - 1\right]}{2b}$$

$$\text{erf}^{-1}(X) \approx -\sqrt{2A}\sinh\left\{\frac{1}{3}\sinh^{-1}(BA^{-3/2})\right\} \qquad (6.124)$$

Again, the visual match between erf$^{-1}(X)$ and Eq. (6.124) is excellent. The error is greatest at the extremes, that is, $X \to \pm 1$; it is less than 1 percent for $|X| < 0.98$ and less than 2 percent for $|X| < 0.995$.

6.5.2 Approximations to the magnitude function

In Secs. 6.2.2 and 6.2.3 it was shown that the problem of unknown phase leads to a detector structure based on a function of the magnitude of the matched filter output. Depending on the SNR regime, the ideal $\ln[I_0(x)]$ detector characteristic is approximated by either a linear or square law detector. Computing the squared magnitude from the complex I and Q data is simple, requiring only two multiplies and one add per sample. Efficiently computing the linear magnitude is more troublesome. Filip (1976) in his paper described a family of 13 simple approximations to the magnitude function and calculated the SNR loss of each as compared to an exactly linear detector. If a complex data sample is of the form $z = I + jQ = A\exp(j\theta)$, the approximations to $R = |z|$ are all of the form

$$\hat{R} = \begin{cases} a_1 \max(|I|,|Q|) + b_1 \min(|I|,|Q|) & 0 \leq \theta \leq \theta_B \\ a_2 \max(|I|,|Q|) + b_2 \min(|I|,|Q|) & \theta_B \leq \theta \leq \pi/4 \end{cases} \quad (6.125)$$

Clearly this is a two-region approximation, with θ_B determining the breakpoint between the regions. The max() and min() functions serve to reflect an arbitrary complex number into the first octant. Different members of the family are defined by different choices of the parameters $\{a_i\}$ and $\{b_i\}$.

Filip gave 13 variations corresponding to 13 different parameter choices. Table 6.5 lists four of the 13 approximations, identifying them by the same numbers used by Filip (1976). Approximation number 3 is an example of a very simple estimate, using only one region and binary coefficients. The loss at a relatively stressing $P_D = 0.999$ and $P_{FA} = 10^{-6}$ is 0.0939 dB. Number 8 is an example of a two-region approximation with binary coefficients; the loss is reduced to 0.0245 dB. Approximation number 10 is designed to minimize the error variance while maintaining a zero mean error. It achieves a loss of 0.0629 dB. Finally, number 13 is an example of an equiripple approximation. At 0.0021 dB, it has the lowest loss of any of Filip's approximations, but also requires the most computation, possibly defeating its purpose.

TABLE 6.5 Coefficients of Approximations to the Magnitude Function

Filip's approx. No.	a_1	b_1	$\tan(\theta_B)$	a_2	b_2	SNR loss, dB at $P_D = 0.999$, $P_{FA} = 10^{-6}$
3	1	1/2	1	–	–	0.0939
8	1	1/4	1/2	3/4	3/4	0.0245
10	0.948	0.393	1	–	–	0.0629
13	0.9903	0.19698	0.414214	0.83954	0.56094	0.0021

References

Abramowitz, M., and I. A. Stegun, *Handbook of Mathematical Functions with Formulas, Graphs, and Mathematical Tables*. National Bureau of Standards, U.S. Dept. of Commerce, 1972.

Albersheim, W. J., "Closed-Form Approximation to Robertson's Detection Characteristics," *Proceedings of IEEE*, vol. 69, no. 7, p. 839, July 1981.

Brennan, L. E., and I. S. Reed, "A Recursive Method of Computing the Q Function," *IEEE Transactions on Information Theory*, vol. IT-11, no. 2, pp. 312–313, April 1965.

Cantrell, P. E., and A. K. Ojha, "Comparison of Generalized Q-function Algorithms," *IEEE Transactions on Information Theory*, vol. IT-33, no. 4, pp. 591–596, July 1987.

DiFranco, J. V., and W. L. Rubin, *Radar Detection*. Artech House, Dedham, MA, 1980.

Filip, A. E., "A Baker's Dozen Magnitude Approximations and Their Detection Statistics," IEEE *Transactions on Aerospace & Electronic Systems*, vol. AES-12, no. 1, pp. 86–89, Jan. 1976.

Hmam, H., "Approximating the SNR Value in Detection Problems," *IEEE Transactions on Aerospace and Electronic Systems*, vol. AES-39, no. 4, pp. 1446–1452, Oct. 2003.

Johnson, D. H., and D. E. Dudgeon, *Array Signal Processing*. Prentice Hall, Englewood Cliffs, NJ, 1993.

Kay, S. M., *Fundamentals of Statistical Signal Processing, Vol. I: Estimation Theory*. Prentice Hall, Upper Saddle River, NJ, 1993.

Kay, S. M., *Fundamentals of Statistical Signal Processing, Vol. II: Detection Theory*. Prentice Hall, Upper Saddle River, NJ, 1998.

Levanon, N., *Radar Principles*. Wiley, New York, 1988.

Meyer, D. P., and H. A. Mayer, *Radar Target Detection*. Academic Press, New York, 1973.

Papoulis, A., *Probability, Random Variables, and Stochastic Processes*, 2d ed. McGraw-Hill, New York, 1984.

Robertson, G. H., "Operating Characteristic for a Linear Detector of CW Signals in Narrow Band Gaussian Noise," *Bell System Technical Journal*, vol. 46, no. 4, pp. 755–774, April 1967.

Shnidman, D. A., "Radar Detection Probabilities and Their Calculation," *IEEE Transactions on Aerospace and Electronic Systems*, vol. AES-31, no. 3, pp. 928–950, July 1995.

Shnidman, D. A., "Binary Integration for Swerling Target Fluctuations," *IEEE Transactions on Aerospace and Electronic Systems*, vol. AES-34, no. 3, pp. 1043–1053, July 1998.

Shnidman, D. A., "Determination of Required SNR Values," *IEEE Transactions on Aerospace and Electronic Systems*, vol. AES-38, no. 3, pp. 1059–1064, July 2002.

Skolnik, M. I., *Introduction to Radar Systems*, 3d ed. McGraw-Hill, New York, 2001.

Tufts, D. W., and A. J. Cann, "On Albersheim's Detection Equation," *IEEE Transactions on Aerospace and Electronic Systems*, vol. AES-19, no. 4, pp. 643–646, July 1983.

Van Trees, H. L., *Detection, Estimation, and Modulation Theory, Part I: Detection, Estimation, and Linear Modulation Theory*. Wiley, New York, 1968.

Vedder, J. D., "Approximation to the Normal Probability Distribution," *NASA Tech Briefs*, p. 96, Feb. 1988.

Weiner, M. A., "Binary Integration of Fluctuating Targets," *IEEE Transactions on Aerospace and Electronic Systems*, vol. AES-27, no. 1, pp. 11–17, Jan. 1991.

Chapter 7

Constant False Alarm Rate (CFAR) Detection

Standard radar threshold detection, as discussed in Chap. 6, assumes that the interference level is known and constant. This in turn allows accurate setting of a threshold that guarantees a specified probability of false alarm. In practice, interference levels are often variable. *Constant false alarm rate* (CFAR) detection, also frequently referred to as "adaptive threshold detection" or "automatic detection," is a set of techniques designed to provide predictable detection and false alarm behavior in realistic interference scenarios.

7.1 The Effect of Unknown Interference Power on False Alarm Probability

In Chap. 6, the detection and false alarm performance of a square law detector were considered for a target in complex white Gaussian interference as a function of the Swerling target model and number of pulses noncoherently integrated. It was shown in Eq. (6.82) that for the simplest case of a single data sample ($N = 1$), the false alarm probability is

$$P_{FA} = e^{-T} \tag{7.1}$$

where T is the detector threshold. Solving Eq. (7.1) gave the threshold value required to achieve a specified probability of false alarm, namely $T = -\ln P_{FA}$. This analysis was done in terms of the normalized linear detector output $z' = z/\beta$, where β^2 is the total noise power (I and Q channels) of the interference. In terms of the unnormalized data sample z and a square law detector the appropriate threshold is

$$T = -\beta^2 \ln P_{FA} \tag{7.2}$$

and the probability of false alarm is

$$P_{FA} = e^{-T/\beta^2} \tag{7.3}$$

Note that the threshold T is proportional to the interference power, i.e., it is of the form $T = \alpha\beta^2$, and that the multiplier α is a function of the desired false alarm probability.

To tune the square law detector for a particular radar system, an acceptable value of P_{FA} must be chosen; the threshold is then computed according to Eq. (7.2). The probability of detection that will be achieved is determined by the target *signal-to-noise ratio* (SNR).

Accurate setting of the threshold requires accurate knowledge of the interference power β^2. In some systems this is known, but in many it is not. When the interference is principally receiver noise, it is possible to measure β^2 and calibrate the detector. In day-to-day operation, however, the receiver noise will vary over time due to factors such as temperature changes and component aging. Temperature compensation and periodic recalibration, if possible, can combat this problem. If the total interference power is significantly affected by external sources, the variability can be much more severe. In very low noise radar systems, a significant part of the noise power is cosmic noise. The total receiver interference then varies with the look direction and the time of day. In conventional radars, the total interference power can be affected by in-band *electromagnetic interference* (EMI). For example, UHF radars can be affected by television transmissions, while certain wireless communication services can compete with higher frequency radars, especially in urban areas. If the dominant interference is ground clutter, its power will vary radically with the type of terrain being illuminated and even the weather and seasons. For instance, open desert has a relatively low reflectivity, while refrozen snow can have a very high reflectivity. Finally, the dominant interference can be hostile electromagnetic emissions deliberately directed at the radar system (jamming). In this case, the interference power can be extremely high.

In any of these cases, the observed P_{FA} will vary from the intended value. To see how significant this variation might be, let P_{FA0} be the intended probability of false alarm when the actual interference power is the expected value of β_0^2; thus $T = -\beta_0^2 \ln P_{FA0}$. Now suppose the actual interference power is β^2. The actual P_{FA}, using Eq. (7.3) with the threshold designed assuming an interference power of β_0^2 from Eq. (7.2), will be

$$P_{FA} = \exp\left(\frac{\beta_0^2 \ln P_{FA0}}{\beta^2}\right) = \exp\left(\ln P_{FA0}^{(\beta_0^2/\beta^2)}\right)$$

$$= P_{FA0}^{(\beta_0^2/\beta^2)} \tag{7.4}$$

and the *increase* in false alarm probability will be a factor of

$$\frac{P_{FA0}}{P_{FA}} = (P_{FA0})^{[(\beta_0/\beta)^2 - 1]} \tag{7.5}$$

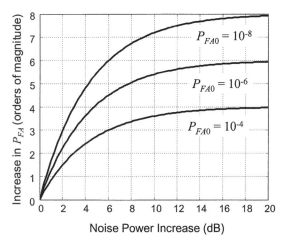

Figure 7.1 Increase in probability of false alarm for fixed threshold due to increase in noise power.

Figure 7.1 plots Eq. (7.5) for three different values of the design false alarm probability. This figure shows that even modest increases of 2 dB can cause an unintended increase in P_{FA} of 1.5 to 3 orders of magnitude, with the largest changes occurring when the desired P_{FA} is lowest. Clearly, such sensitivity to small changes in interference power or, equivalently, small errors in setting the threshold will have major impacts on radar performance.

7.2 Cell-Averaging CFAR

7.2.1 The effect of varying P_{FA}

The reason for the large increase in P_{FA} observed in Fig. 7.1 is that the threshold T was based on an incorrect value for the interference power β^2. More generally, as the interference power at the output of the radar receiver varies, the actual P_{FA} will vary widely. From a system point of view, this is highly undesirable. When the interference power rises, the number of false alarms will also rise, possibly by orders of magnitude. It might seem that the difference between a probability of false alarm of 10^{-8} and a rate of 10^{-6} may be insignificant, but consider the example of a simple radar system with a PRF of 10 kHz and 200 range bins. If each range bin is tested, this system makes $(10{,}000)(200) =$ two million detection decisions per second. With $P_{FA} = 10^{-8}$, false alarms occur, on an average, only once every 50 seconds. If P_{FA} rises to 10^{-6}, the system is confronted with an average of two false alarms every second.

How much of a concern this increase is depends on the impact of a false alarm in the overall radar system. This could range from an increased demand on radar or signal processor resources to confirm or reject the false alarm or to start unneeded tracks, to increased cluttering of an operator display, and even in extreme cases to firing of a weapon.

If the interference power drops below that assumed when calculating the threshold, the false alarm probability will drop. This may seem inconsequential or even desirable, but a reduced P_{FA} represents a threshold that is higher than necessary to achieve the system design goals. Since P_{FA} and P_D always increase or decrease together as discussed in Chap. 6, this means that the probability of detection is less than could be achieved with a correctly set threshold.

7.2.2 The cell-averaging CFAR concept

In order to obtain predictable and consistent performance, the radar system designer would usually prefer a constant false alarm rate. To achieve this, the actual interference power must be estimated from the data in real time, so that the detector threshold can be adjusted to maintain the desired P_{FA}. A detection processor that can maintain a constant P_{FA} is called a *constant false alarm rate* (CFAR) processor.

Figure 7.2 shows a generic radar detection processor. The detector shown is for a system using range-Doppler processing, but other systems might consider only a one-dimensional vector of range cells in making a decision. Still other systems might perform the detection process on a radar image, so that the individual cells are pixels in a two-dimensional image. Whatever the form of the data, the detector will test each available data sample for the presence or absence of a target. The current cell under test, denoted by x_i in Fig. 7.1, is compared against a threshold determined by the interference power. If the value of the data in the test cell exceeds the threshold, the processor declares a target present at the appropriate range and velocity (or range, or image location, as appropriate). The next cell is then tested and so forth until a target present/target absent decision has been made for all cells of interest.

To set the threshold for testing cell x_i, the interference power *in the same cell* must be known. Since it may be variable, it must be estimated from the data. The approach used in CFAR processing is based on two major assumptions:

- The neighboring cells contain interference with the same statistics as the cell under test (called *homogeneous* interference), so that they are representative of the interference that is competing with the potential target.

- The neighboring cells do *not* contain any targets; they are interference only.

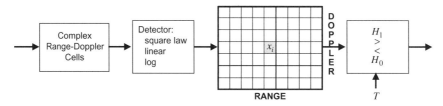

Figure 7.2 Generic detection processor.

Under these conditions, the interference statistics in the cell under test can be estimated from the measured samples in the adjoining cells.

The statistics that must be estimated are determined by the statistics needed to implement the threshold test. For Gaussian interference and linear or square law detectors, the interference will be Rayleigh or exponential distributed, respectively. In either case, the interference *probability density function* (pdf) has only one free parameter, the mean interference power. Thus, the CFAR processor must estimate the mean interference power in the cell under test by using the measured data in the adjoining cells.

For a more specific example, consider the square law case. The pdf of a cell x_i, assuming the interference is *independent identically distributed* (i.i.d.) in the I and Q signals with power $\beta^2/2$ in each (for a total power of β^2), is then

$$p_{x_i}(x_i) = \frac{1}{\beta^2} e^{-x_i/\beta^2} \tag{7.6}$$

As seen in Eq. (7.2), knowledge of β^2 is needed to set the threshold. When exact knowledge of β^2 is not available, it must be estimated.

Assume that N cells in the vicinity of the cell under test are used to estimate β^2, and that the interference in each is independent and identically distributed. The joint pdf of a vector $\mathbf{x}$ of N such samples is

$$p_{\mathbf{x}}(\mathbf{x}) = \frac{1}{\beta^{2N}} \prod_{i=1}^{N} e^{-x_i/\beta^2}$$

$$= \frac{1}{\beta^{2N}} e^{-\left(\sum_{i=1}^{N} x_i\right)/\beta^2} \tag{7.7}$$

Equation (7.7) is the likelihood function Λ for the observed data vector $\mathbf{x}$. The maximum likelihood estimate of β^2 is obtained by maximizing Eq. (7.7) with $\sum x_i$ held constant (Kay, 1993). It is equivalent and more convenient to maximize the log-likelihood function

$$\ln \Lambda = -N \ln(\beta^2) - \frac{1}{\beta^2} \left(\sum_{i=1}^{N} x_i \right) \tag{7.8}$$

Setting the derivative of Eq. (7.8) with respect to β^2 equal to zero gives

$$\frac{d(\ln \Lambda)}{d(\beta^2)} = 0 = -N \left(\frac{1}{\beta^2} \right) - \left(-\frac{1}{(\beta^2)^2} \right) \left(\sum_{i=1}^{N} x_i \right) \tag{7.9}$$

Solving Eq. (7.9) for β^2 gives the unsurprising result that the maximum likelihood estimate is just the average of the available samples

$$\widehat{\beta^2} = \frac{1}{N}\sum_{i=1}^{N} x_i \qquad (7.10)$$

The required threshold is then estimated as a multiple of the estimated interference power

$$\hat{T} = \alpha \widehat{\beta^2} \qquad (7.11)$$

Because the interference power and thus the threshold are estimated from an average of the power in the cells adjoining the test cell, this CFAR approach is referred to as *cell averaging CFAR* (CA CFAR). Because the interference power is estimated rather than known exactly, the scale factor α will not have the same value as in Eq. (7.2); it will be derived in Sec. 7.3.

7.2.3 CFAR reference windows

Equation (7.10) showed that the parameter of the exponential pdf describing the square-law detected data is estimated from an average of N adjoining data samples. Figure 7.3 shows two examples of how the samples to be averaged

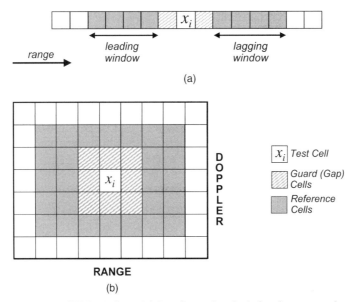

Figure 7.3 CFAR windows. (*a*) One-dimensional window for range-only processor. (*b*) Two-dimensional window for range-Doppler processor.

are selected. Figure 7.3a shows a one-dimensional data vector of range cells with the cell under test, x_i, in the middle. The data in the grey cells to either side, representing data from ranges nearer and farther from the radar than the cell under test, are averaged to estimate the noise parameter. These cells are called the *reference cells*. The cross-hatched cells immediately adjacent to the cell under test, called *guard cells*, are excluded from the average. The reason is that a target, if present, might straddle range cells. In this case, the energy in the cell adjacent to x_i would contain both interference and target energy and would therefore not be representative of the interference alone. The extra energy from the target would tend to raise the estimate of the interference parameter. For instance, in the square law detector case, the estimate of β^2 would be too high, resulting in a threshold that was too high and a lower P_{FA} and P_D than intended. If the system range resolution is such that anticipated targets could extend over multiple range cells, more than one guard cell would be skipped on each side of the cell under test. The combined reference cells, guard cells, and cell under test are referred to as the *CFAR window*.

Figure 7.3b shows a typical two-dimensional equivalent to the one-dimensional case, in this case applied to a range-Doppler matrix after detection. Both the guard region and the reference window are now two dimensional. A two-dimensional CFAR window could also be applied to synthetic aperture radar imagery, in which case the two dimensions would be simply the range and cross-range spatial dimensions. In the range-Doppler case, the cell averaging CFAR might be applied only over certain range and Doppler cells because ground clutter renders the interference nonhomogeneous in Doppler and possibly in range.

7.3 Analysis of Cell-Averaging CFAR

The intent of the adaptive calculation of the threshold is to provide a constant false alarm rate despite varying interference power levels. In this section, the detector performance is analyzed for the case of a square law detector and i.i.d. exponential-distributed (post-detection) interference to see if this goal has been achieved. The threshold computed according to Eq. (7.11) will be a random variable, and thus so will be the probability of false alarm. The detector will be considered to be CFAR if the expected value of P_{FA} does not depend on the actual value of β^2.

7.3.1 Derivation of CA CFAR threshold

Combining Eqs. (7.10) and (7.11) gives an expression for the estimated threshold

$$\hat{T} = \frac{\alpha}{N} \sum_{i=1}^{N} x_i \qquad (7.12)$$

Define $z_i = (\alpha/N)x_i$; thus $\hat{T} = \sum_{i=1}^{N} z_i$. Using standard results from probability theory and Eq. (7.6) gives the pdf of z_i

$$p_{z_i}(z_i) = \frac{N}{\alpha\beta^2} e^{-Nz_i/\alpha\beta^2} \qquad (7.13)$$

The pdf of $\hat{T}$ is the Erlang density (a special case of the gamma density) seen previously in Eq. (6.79):

$$p_{\hat{T}}(\hat{T}) = \left(\frac{N}{\alpha\beta^2}\right)^N \frac{\hat{T}^{N-1}}{(N-1)!} e^{-N\hat{T}/\alpha\beta^2} \qquad (7.14)$$

The P_{FA} observed with the estimated threshold will be $\exp(-\hat{T}/\beta^2)$. This is now also a random variable; its expected value is

$$\overline{P}_{FA} = \int_{-\infty}^{\infty} e^{-\hat{T}/\beta^2} p_{\hat{T}}(\hat{T}) d\hat{T}$$

$$= \left(\frac{N}{\alpha\beta^2}\right)^N \frac{1}{(N-1)!} \int_{-\infty}^{\infty} \hat{T}^{N-1} e^{-[(N/\alpha)+1]\hat{T}/\beta^2} d\hat{T} \qquad (7.15)$$

Completing this standard integral and performing some algebraic manipulations gives the final result

$$\overline{P}_{FA} = \left(1 + \frac{\alpha}{N}\right)^{-N} \qquad (7.16)$$

For a given desired $\overline{P}_{FA}$, the required threshold multiplier is obtained by solving Eq. (7.16) to obtain

$$\alpha = N\left(\overline{P}_{FA}^{-1/N} - 1\right) \qquad (7.17)$$

Note that $\overline{P}_{FA}$ does not depend on the actual interference power β^2, but only on the number N of neighboring cells averaged and the multiplier. Thus, the cell-averaging technique exhibits CFAR behavior.

7.3.2 Cell-averaging CFAR performance

Now that a rule for selecting the CA CFAR threshold has been determined, the detection performance can be determined. Equation (6.100) shows that for a single test cell sample of a Swerling 1 target with the threshold $\hat{T}$, $P_D = \exp[-\hat{T}/(1+\chi)]$, where χ is the signal-to-noise ratio. Since there is only a single sample, fluctuation models are irrelevant and so this result also applies for Swerling 2 targets. The expected value of P_D is obtained by averaging over the threshold

$$\overline{P}_D = \int_{-\infty}^{\infty} e^{-\hat{T}/(1+\chi)} p_{\hat{T}}(\hat{T}) d\hat{T} \qquad (7.18)$$

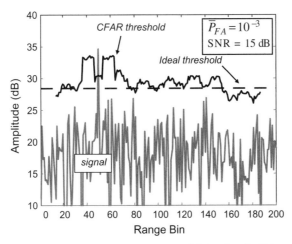

Figure 7.4 Example of cell-averaging CFAR threshold behavior.

This integral is the same general form as Eq. (7.15); the result is

$$\overline{P}_D = \left(1 + \frac{\alpha}{N(1+\chi)}\right)^{-N} \tag{7.19}$$

Note that this also does not depend on the interference power. However, this result is based on the assumption of Gaussian I/Q noise, a square law detector, Swerling 1 or 2 target, and a single test cell.

Figure 7.4 illustrates the operation of cell-averaging CFAR. The simulated data correspond to Gaussian I/Q noise with power $10\log_{10}(\beta^2) = 20$ dB. A single nonfluctuating target with a power of 35 dB is present in range bin 50; the SNR is thus $10\log_{10}(\chi) = 15$ dB. If the desired $P_{FA} = 10^{-3}$, Eq. (7.2) gives the ideal $T = 691$, equal to 28.4 dB. This threshold level is indicated on the plot. Note that the ideal threshold is a multiple of $-\ln(P_{FA}) = 6.91$ times the true interference power; equivalently, the threshold is 8.4 dB above the interference power level.

Consider a CA CFAR with leading and lagging windows of 10 cells each after 3 cell guard regions; thus $N = 20$ cells are averaged to estimate the interference power.[†] From Eq. (7.17), the multiplier α will be 8.25, placing the threshold about 9.2 dB above the *estimated* mean power. The line labeled "CFAR threshold" shows the computed threshold as the reference window slides across the data. Except in the vicinity of the target, the estimated threshold tracks the ideal threshold well, staying within 2 dB across most of the data. Note that the data exceed the CFAR threshold only at range bin 50. In this case, the CFAR detector works very well: a detection would correctly be declared when

---

[†]Unless otherwise stated, this same arrangement of lead, lag, and guard cells is used in all examples in this chapter.

the CFAR test cell is located at range bin 50, but there are no false alarms (threshold crossings) at any other range bins.

The increase in the threshold to either side of the target location is characteristic of cell-averaging CFAR. For the particular CFAR window configuration used here, when the test cell is between range bins 37 and 46, the cell containing the target will be in the leading reference window and will be included in the estimate of the interference power. The estimated power $\widehat{\beta^2}$ and, in turn, the computed threshold $\hat{T}$ will be significantly raised. This phenomenon repeats when the test cell is between bins 54 and 63, so that the target is in the lagging reference window. When a target is in the reference window, the assumption that all of the reference cells share the same interference statistics as the test cell is violated, and the estimate of the interference power is unreliable. However, when the test cell is located at the cell containing the target, the reference windows contain only noise samples and the threshold falls to an appropriate level, allowing target detection. The extent of the elevated threshold regions to either side of a target equals the extent of the leading and lagging windows. The extent of the region of normal threshold level between the two elevated regions equals the total number of guard cells plus one (for the test cell).

As the number of reference cells N becomes large, the estimate $\widehat{\beta^2}$ should converge to the true value β^2, and the average probabilities of detection and false alarm should also converge to the values obtained in Chap. 6. To see this, it is easier to work with $\ln \overline{P}_{FA}$ than with $\overline{P}_{FA}$ itself

$$\ln \overline{P}_{FA} = \ln\left\{\left(1+\frac{\alpha}{N}\right)^{-N}\right\}$$

$$= -N \ln\left(1+\frac{\alpha}{N}\right) = -N\left\{\frac{\alpha}{N} - \frac{1}{2}\left(\frac{\alpha}{N}\right)^2 + \cdots\right\} \quad (7.20)$$

Taking the limit as $N \to \infty$

$$\lim_{N\to\infty}\{\ln \overline{P}_{FA}\} = -N\left\{\frac{\alpha}{N}\right\} \Rightarrow \lim_{N\to\infty} \overline{P}_{FA} = e^{-\alpha} \quad (7.21)$$

Similarly
$$\lim_{N\to\infty}\{\overline{P}_D\} = e^{-\alpha/(1+\chi)} \quad (7.22)$$

Combining Eqs. (7.21) and (7.22) gives the relation

$$\overline{P}_D = (\overline{P}_{FA})^{1/(1+\chi)} \quad (7.23)$$

which is identical to the results obtained in Chap. 6 (see Eq. (6.101)) for the Swerling 1, $N = 1$ case with a known interference power.

All of the previous discussion has been for a square-law detector. Similar analyses can be carried out for a linear detector, however, the results are more difficult to obtain in closed form. Suppose that $\{w_i\}$ are the output of a linear detector. The threshold will be set according to

$$\hat{T} = \kappa \left(\frac{1}{N} \sum_{i=1}^{N} w_i \right) \tag{7.24}$$

Raghavan (1992) obtains a formula relating $\overline{P}_{FA}$ and $\hat{T}$ that must be solved iteratively and is numerically difficult; however, his exact results show excellent agreement with the approximation (Di Vito and Moretti, 1989):

$$\kappa = \sqrt{N(\overline{P}_{FA}^{-1/N} - 1)\left(c - (c-1)e^{1-N}\right)} \tag{7.25}$$

where $c = 4/\pi$. Notice the similarity to Eq. (7.17). The square root is a consequence of using a linear rather than square-law detector. For $N > 4$ the $(c-1)\exp(1-N)$ term is negligible, and since $c \approx 1.27$, $\kappa \approx 1.13\sqrt{\alpha}$. Furthermore, although the square law detector CA CFAR performs marginally better than the linear detector, its performance is virtually identical for parameter values of practical interest (Raghavan, 1992).

7.3.3 CFAR loss

In the example of Fig. 7.4 it was noted that for the parameters given, if the interference power were known exactly, the ideal threshold would be 8.4 dB above the mean power; but if the power had to be estimated from the data with $N = 20$, the threshold would be 9.2 dB above the estimated power level. This higher threshold compared to the interference power level is necessary to compensate for the imperfectly known interference power and guarantee the desired $\overline{P}_{FA}$. Because the threshold multiplier is increased in CFAR, the average probability of detection for a target of a given SNR will be decreased relative to the known-interference case. Alternately, to achieve a specified $\overline{P}_D$ for a given $\overline{P}_{FA}$, a higher SNR will be required than would be were the interference power known exactly. This increase in SNR required to achieve specified detection statistics when using CFAR techniques is called the *CFAR loss*.

To quantify the CFAR loss in the case of a CA CFAR, combine Eqs. (7.16) and (7.19) to eliminate the multiplier α and solve for the value of SNR required to achieve a specified combination of $\overline{P}_{FA}$ and $\overline{P}_D$. The result is a function of the number of samples averaged and is denoted by χ_N

$$\chi_N = \frac{(\overline{P}_D/\overline{P}_{FA})^{1/N} - 1}{1 - \overline{P}_D^{1/N}} \tag{7.26}$$

As $N \to \infty$, the estimate of interference power converges to the true value and so $\overline{P}_{FA}$ and $\overline{P}_D$ will converge to the values given by Eqs. (7.21) and (7.22). Similarly combining these two equations gives the value of SNR, denoted by χ_∞,

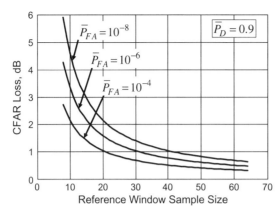

Figure 7.5 Cell-averaging CFAR loss for Swerling 1/2 target in Gaussian I/Q interference with $P_D = 0.9$.

required to achieve the specified probabilities when the interference estimate is perfect

$$\chi_\infty = \frac{\ln(\overline{P}_{FA}/\overline{P}_D)}{\ln(\overline{P}_D)} \quad (7.27)$$

The CFAR loss is then simply the ratio (Levanon, 1988; Hansen and Sawyers, 1980)

$$\text{CFAR loss} = \frac{\chi_N}{\chi_\infty} \quad (7.28)$$

Figure 7.5 plots Eq. (7.28) for a $\overline{P}_D$ of 0.9 and three values of $\overline{P}_{FA}$. The loss is greatest for lower values of $\overline{P}_{FA}$ and decreases as expected when the number of reference cells increases. For small ($N < 20$) reference windows, the CFAR loss can be several dB. High losses make values of N less than 10 unacceptable in most cases. Although not shown here, the CFAR loss also increases with increasing $\overline{P}_D$ for a given $\overline{P}_{FA}$ and N.

Although the previous results were derived for a Swerling 1 or 2 target, Nathanson (1991) cites literature stating that the CFAR loss is roughly the same for all of the Swerling target models, as well as the nonfluctuating "Swerling 0" case.

7.4 CA CFAR Limitations

The cell-averaging CFAR concept relies on two major assumptions

- Targets are isolated; specifically, targets are separated by at least the reference window size, so that no two are ever in the reference window at the same time.

- All of the reference window interference samples are independent and identically distributed, and that distribution is the same as that of the interference component in the cell containing the target; in other words, the interference is homogeneous.

While useful in many situations, either or both of these conditions are frequently violated in real-world scenarios, particularly when the dominant interference is clutter, i.e., echo from terrain, rather than thermal noise. In this section the effect on cell averaging CFAR of violating these assumptions is discussed, and then some modifications that combat these effects are described.

7.4.1 Target masking

Target masking occurs when two or more targets are present such that, when one target is in the test cell, one or more targets are located among the reference cells. Assuming that the power of the target in the reference cell exceeds that of the surrounding interference, its presence will raise the estimate of the interference power and thus of the CFAR threshold. The target(s) in the reference window can "mask" the target in the test cell because the increased threshold will cause a reduction in the probability of detection, i.e., the detection is more likely to be missed. Equivalently, a higher SNR will be required to achieve a specified $\overline{P}_D$. The elevated threshold surrounding the test cell in Fig. 7.4 illustrates a threshold increase of 3 to 4 dB when the target is in the reference window for the parameters used there.

Figure 7.6 is an example of target masking. As before, the interference level is 20 dB, the target in range bin 50 has an SNR of 15 dB, and the threshold is computed using 20 reference cells and a desired $\overline{P}_{FA}$ of 10^{-3}. However, a second target with an SNR of 20 dB in range bin 58 elevates the estimated interference power when the first target is in the test cell. This increase in

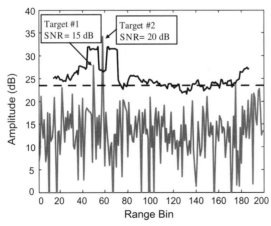

Figure 7.6 Illustration of target masking. (See text for details.)

threshold is sufficient to prevent detection of the first target in this case. On the other hand, the 15-dB target does not affect the threshold enough to prevent detection of the second, stronger target.

Precise analysis of the effect of a target in the reference cells is conceptually simple but somewhat complicated in practice. However, a relatively simple estimate that illustrates the approximate effect of an interfering target can be derived. Consider a single interfering target with power γ_i that contaminates only one of the N CFAR reference cells. The SNR of this interferer is $\chi_i = \gamma_i/\beta^2$. The expected value of the new threshold will be

$$\mathbf{E}\{\hat{T}'\} = \mathbf{E}\left\{\frac{\alpha}{N}\left(\gamma_i + \sum_{i=0}^{N-1} x_i\right)\right\}$$

$$= \frac{\alpha\gamma_i}{N} + \alpha\beta^2 = \alpha\left(1 + \frac{\chi_i}{N}\right)\beta^2 \qquad (7.29)$$

Thus, $\mathbf{E}\{\hat{T}'\}$ is again a multiple of the interference power β^2 as in Eq. (7.11) but with a multiplier α' given by

$$\alpha' \equiv \alpha\left(1 + \frac{\chi_i}{N}\right) \qquad (7.30)$$

The elevated threshold will decrease both the probability of detection and the probability of false alarm. Using Eq. (7.30) and Eq. (7.17) in Eq. (7.19) gives an expression for the new value of $\overline{P}_D$ in terms of the *original* design value of $\overline{P}_{FA}$

$$\overline{P}'_D = \left[1 + (\overline{P}_{FA}^{-1/N} - 1)\left(\frac{1 + \chi_i/N}{1 + \chi}\right)\right]^{-N} \qquad (7.31)$$

Note that if $\chi_i \to 0$ (no interfering target) or $N \to \infty$ (target influence becomes negligible), then $\overline{P}'_D \to \overline{P}_D$. Figure 7.7a illustrates this for one example, where $\overline{P}_{FA} = 10^{-3}$ and N is either 20 or 50 cells. The probability of detection without the interferer for these two cases is about 0.78 and 0.8, respectively.

Another way to characterize the effect of an interfering target is by the increase in SNR required to maintain the original value of $\overline{P}_D$. Let χ' be the value of SNR required to attain the original $\overline{P}_D$ using the elevated threshold $\hat{T}'$. Equation (7.19) expressed $\overline{P}_D$ in terms of the original value of χ and threshold multiplier α. Approximately the same relationship will determine the detection probability $\overline{P}'_D$ attained with the new threshold multiplier α' and SNR χ'. Thus, $\overline{P}'_D$ will equal $\overline{P}_D$ if

$$\frac{\alpha}{N(1+\chi)} = \frac{\alpha'}{N(1+\chi')} \qquad (7.32)$$

Using Eq. (7.30) in Eq. (7.32) leads to

$$\chi' = \left(1 + \frac{\chi_i}{N}\right)(1+\chi) - 1 \qquad (7.33)$$

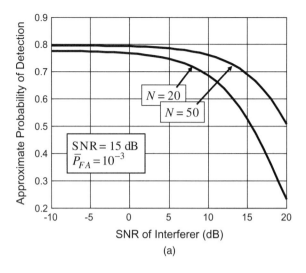

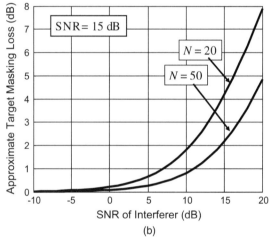

Figure 7.7 Approximate effect of interfering target on cell averaging CFAR. Threshold set for $P_{FA} = 10^{-3}$. (a) Reduction in P_D. (b) Equivalent masking loss.

Figure 7.7b plots this approximate "target masking loss" χ'/χ in decibels for the same conditions as in Fig. 7.7a.

The results given in Eqs. (7.31) and (7.33) are only approximations. A more careful analysis would mimic the basic CA CFAR analysis of Secs. 7.3.1 and 7.3.2 by finding the probability density function of the threshold in the presence of an interfering target, then using that pdf to find the expected values of P_D and P_{FA}. This approach is complicated by the fact that the interfering target changes the pdf of the cell containing it. For instance, if the interferer is nonfluctuating, the pdf of the power in its cell will be Rician, while all of the remaining cells will still be exponentially distributed. The pdf of the threshold will be a mixture of the Rician and exponential pdfs. To avoid calculating this pdf, the expected value of the threshold was used in the expressions for the case of no interfering target. This gives a simple approximation that behaves correctly in the limits of large and small interferer SNR χ_i.

Figure 7.8 illustrates a related phenomenon, self-masking by a distributed target. The interference, detector, and target characteristics are the same as in Fig. 7.4, with the exception that the physical extent of the target is now greater than a range bin, so that the target signature is spread over three

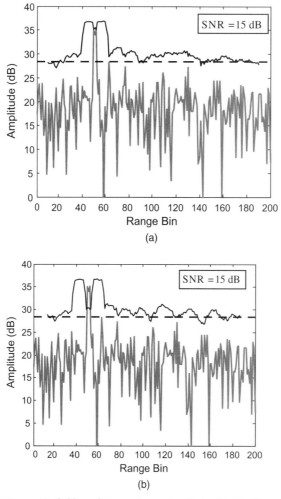

Figure 7.8 Self masking and guard cells in CA CFAR. (See text for details.) (*a*) Threshold using no guard cells. (*b*) Threshold using three guard cells to each side of the test cell.

consecutive cells. Figure 7.8a shows the effect when no guard cells are used. When one of the three target cells is the test cell, the other two contaminate the interference estimate, in this case raising the threshold just enough to prevent detection. This effect is the reason for using guard cells. Their impact is illustrated in Fig. 7.8b, where three guard cells are used to each side of the test cell. This lengthens the total CFAR window slightly, but assures that when the test cell is centered on the target, the adjoining target cells do not contaminate the interference estimate and the target is now detected.

7.4.2 Clutter edges

If the dominant interference is clutter, rather than thermal noise or jamming, the clutter can often be highly heterogeneous. The radar may illuminate a patch of terrain that is, for instance, part open field and part forested, or part land and part water. When the test cell is at or near the boundary between two clutter regions having different reflectivities, the statistics in the leading and lagging window will not be the same. Such *clutter edges* can cause both false alarms at the edge, and allow masking of targets in the lower-reflectivity region, but near the edge.

Figure 7.9 shows one of the two principal effects of heterogeneous clutter. The first 100 bins have a mean interference power of 20 dB. The clutter power rises suddenly by 10 dB, to a mean of 30 dB, in the last 100 bins, simulating a change in terrain type, perhaps from open field to a wooded area. One target is present, at bin 50. Two ideal thresholds corresponding to $P_{FA} = 10^{-3}$ are shown, one for each of the two clutter regions. The threshold estimated by the cell averaging CFAR tracks each region, but with a transition region in the range of bins 87 through 113. In this example, the clutter happens to include a high-amplitude fluctuation near the clutter edge. Because the CFAR threshold

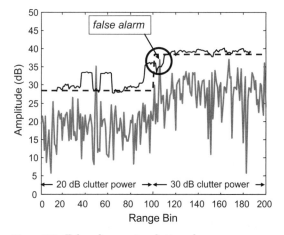

Figure 7.9 False alarms at a clutter edge.

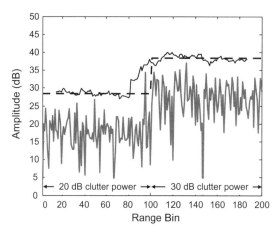

Figure 7.10 Target masking at a clutter edge. Clutter parameters are the same as in Fig. 7.9.

does not rise to the correct level for the new clutter level until several cells after the transition, the clutter spike crosses the CFAR threshold and a false alarm occurs. However, the target at bin 50 is detected normally.

Figure 7.10 illustrates the second effect of clutter edges. The clutter regions are the same as in Fig. 7.9, but the particular clutter sequence is different. A target with an SNR of 15 dB is located at bin 95, five bins away from the clutter edge. The CFAR window uses ten cells in both the leading and lagging windows, as well as three guard cells on either side of the test cell, for a total CFAR window length of 27 cells. Thus, when the test cell is centered over the target, the leading reference window is mostly filled by clutter from the high-reflectivity region, elevating the threshold above the target and causing a missed detection. (Note that this example also does not suffer a clutter-edge false alarm.) This technique has served as the basis for an operational countermeasures technique. For example, by operating close to a tree line, tanks or other vehicles can make it more difficult for radar-guided munitions to detect them.

7.5 Extensions to Cell-Averaging CFAR

The performance limitations caused by nonhomogeneous clutter and interfering targets have led to the development of numerous extensions to the cell-averaging CFAR concept, each designed to combat one or more of the deleterious effects. These techniques are often somewhat heuristically motivated, and can be difficult to analyze exhaustively due to the many variations in clutter nonhomogeneity, target and interfering target signal-to-noise ratio, CFAR window size, and CFAR detection logic.

One common CFAR extension is the *smallest-of cell-averaging CFAR* (SOCA CFAR); the method is also known as the *least of cell averaging CFAR*.

This technique is intended to combat the masking effect caused by an interfering target among the CFAR reference cells, as was shown in Fig. 7.6. In an N-cell SOCA approach, the lead and lag windows are averaged separately to create two independent estimates $\widehat{\beta_1^2}$ and $\widehat{\beta_2^2}$ of the interference mean, each based on $N/2$ reference cells. The threshold is then computed from the smaller of the two estimates in a manner similar to Eq. (7.11)

$$\hat{T} = \alpha_{SO} \min\left(\widehat{\beta_1^2}, \widehat{\beta_2^2}\right) \tag{7.34}$$

If an interfering target is present in one of the two windows, it will raise the interference power estimate in that window. Thus, the lesser of the two estimates is more likely to be representative of the true interference level and should be used to set the threshold.

Because the interference power is estimated from $N/2$ cells instead of N cells, the threshold multiplier α required for a given design value of $\overline{P}_{FA}$ will be increased. It is tempting to conclude that the threshold multiplier α_{SO} for SOCA CFAR could be calculated using Eq. (7.17), but with N replaced by $N/2$. However, a more careful analysis shows that the required multiplier is the solution of the equation (Weiss, 1982)

$$\overline{P}_{FA}/2 = \left(2 + \frac{\alpha_{SO}}{(N/2)}\right)^{-N/2} \left\{\sum_{k=0}^{\frac{N}{2}-1} \binom{\frac{N}{2}-1+k}{k} \left(2 + \frac{\alpha_{SO}}{(N/2)}\right)^{-k}\right\} \tag{7.35}$$

This equation must be solved iteratively. As an example, for $\overline{P}_{FA} = 10^{-3}$ and $N = 20, \alpha_{SO} = 11.276$. In contrast, the CA CFAR multiplier is $\alpha = 8.25$ for the same conditions. Recall that α_{SO} will be applied to an estimate of β^2 based on $N/2$ cells, while α will be applied to an estimate based on N cells.

Figure 7.11a compares the behavior of conventional CA CFAR and SOCA CFAR on simulated data containing two closely-spaced targets of SNR 15 and 20 dB, a 10-dB clutter edge, and a third 15-dB target near the clutter edge. As before, the lead and lag windows are both 10 samples (thus $N = 20$), and there are 3 guard cells to each side of the test cell. The ideal threshold shown is based on $\overline{P}_{FA} = 10^{-3}$. The threshold multipliers are $\alpha = 8.25$ for CA CFAR and $\alpha_{SO} = 11.276$ for SOCA CFAR as discussed previously.

Note that the CA CFAR masks the weaker of the two closely-spaced targets and also fails to detect the target near the clutter edge, but does not exhibit a false alarm at the clutter edge in this instance. The SOCA threshold, in contrast, easily allows detection of both targets; the half of the CFAR window contaminated by the other target is simply ignored by the SOCA logic. Similarly, the SOCA CFAR detects the target near the clutter edge, again because the half of the window containing the higher-power clutter is ignored.

Figure 7.11a also shows the principal failing of the SOCA method. Although the CA CFAR did not exhibit a false alarm at the clutter edge, the SOCA CFAR does. This is a natural consequence of the SOCA logic. As the CFAR window crosses a clutter edge, there will be a region in which the test cell is in the

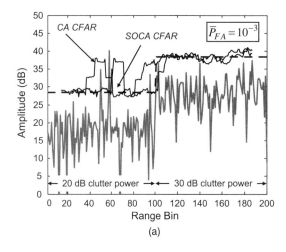

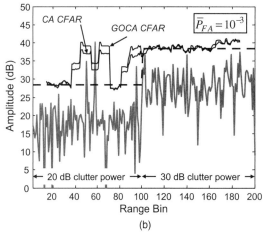

Figure 7.11 Comparison of conventional, "smallest-of," and "greatest of" cell averaging CFAR with multiple targets and a clutter edge. (*a*) CA and SOCA CFAR. (*b*) CA and GOCA CFAR.

higher interference power region, while one of the lead or lag windows is filled mostly or entirely with samples of the lower power interference. The SOCA logic ensures that the threshold is then based on the lower interference power, significantly raising the probability that the clutter in the test cell will cross the threshold.

For systems and environments in which closely-spaced targets are unlikely but the clutter is highly nonhomogeneous, clutter-edge false alarms may be of much more concern than target masking. In this case, the observations above suggest that a *greater-of cell-averaging CFAR* (GOCA CFAR) logic be used.

As with the SOCA technique, the lead and lag windows are averaged separately, but now the threshold is based on the larger of the two averages

$$\hat{T} = \alpha_{\text{GO}} \max\left(\widehat{\beta_1^2}, \widehat{\beta_2^2}\right) \tag{7.36}$$

Similar to the SOCA case, the GOCA threshold multiplier is the solution of the equation (Weiss, 1982)

$$\overline{P}_{FA}/2 = \left(1 + \frac{\alpha_{\text{GO}}}{(N/2)}\right)^{-N/2} - \left(2 + \frac{\alpha_{\text{GO}}}{(N/2)}\right)^{-N/2}$$
$$\times \left\{ \sum_{k=0}^{\frac{N}{2}-1} \binom{\frac{N}{2}-1+k}{k} \left(2 + \frac{\alpha_{\text{GO}}}{(N/2)}\right)^{-k} \right\} \tag{7.37}$$

For $N = 20$ and $\overline{P}_{FA} = 10^{-3}$, $\alpha_{\text{GO}} = 7.24$.

Figure 7.11b illustrates the performance of the GOCA CFAR logic on the same example used in Fig. 7.11a. The GOCA threshold is now equal to or higher than the CA CFAR threshold. Not surprisingly, the GOCA logic successfully avoids the false alarm at the clutter edge. However, the strong target masks the weaker target. Furthermore, the weaker target even makes detection of the stronger target more marginal, although in this case the detection is successful. The GOCA CFAR also misses the target near the clutter edge due to the masking effect of the elevated clutter. Pace and Taylor (1994) provide additional analysis of the GOCA CFAR for the case of a linear detector and several of the approximations to the magnitude function previously discussed in Chap. 6.

In the absence of clutter edges and interfering targets, the SOCA and GOCA CFARs will still estimate the threshold using only half of the reference cells. Consequently, they will exhibit an additional CFAR loss above and beyond that of the conventional CA CFAR. This additional loss is less than 0.3 dB for the GOCA CFAR over a wide range of parameters (Hansen and Sawyers, 1979). The additional loss for the SOCA CFAR is greater, especially for small values of N. It is necessary to use $N > 32$, approximately, to insure that the additional loss is less than 1 dB over a wide range of $\overline{P}_{FA}$ values (Weiss, 1982). A mitigating influence for either approach is that the use of a split window may allow a larger value of N than would normally be used in a conventional cell-averaging CFAR. The reason is that the window size in CA-CFAR is limited by concern over nonhomogeneous clutter. Since it is known that only half the window will actually be used, a larger value of N can be tolerated.

Still another way to combat the target masking problem is *censored* or *trimmed mean* CFAR (Ritcey, 1986). In these techniques, the M reference cells ($M < N$) having the highest power are discarded and the interference power is estimated from the remaining $N - M$ cells. In some versions of trimmed mean CFAR, both the highest and lowest power reference cells are discarded. Consider an example where $M = 2$. If an interfering target is present, but is confined to only one or two cells (or if two interferers are present, each confined to one cell), the censoring process will completely eliminate their elevating effect on the estimate of interference power. There will, however, be a small additional CFAR loss due to the use of only $N - M$ cells instead of N cells (Ritcey and Hines, 1989). Proper selection of M requires some knowledge of the maximum number of interferers to be expected, as well as whether they will be confined to one cell or will be distributed over multiple cells. Typically, one-quarter to one-half of the reference window cells are discarded (Nathanson, 1991). In addition, implementation of the technique requires logic to rank order the reference cell data, a significant implementation consideration at the speeds at which real-time CFAR calculations often must be done.

Many additional variations on the approaches described previously can be used. For example, censoring can be combined with any of the CA, SOCA, or GOCA techniques. A more elaborate approach attempts to examine the behavior of the interference in the lead and lag windows and then choose an appropriate CFAR algorithm. One version of these ideas computes the mean and the variance in each of the lead and lag windows. If the variance in a window exceeds a certain threshold, it is assumed that the data in that window are not homogeneous Rayleigh interference, most likely due to target contamination. A series of logical decisions then determines whether to combine the windows for a CA CFAR using the data from both windows, use CA CFAR using only one window of data, or use GOCA or SOCA CFAR (Smith and Varshney, 2000). For example, if the means differ by less than a specified threshold, and the variances in each window are less than the variance threshold, conventional CA CFAR is used to set the detection threshold. If instead the means do differ by more than that threshold, GOCA CFAR is applied. If the variance in one window exceeds the variance threshold, but the other window does not, CA CFAR based only on the low-variance window is applied. If both windows have variances exceeding the variance threshold, SOCA CFAR is applied.

Another recent attempt to develop a CFAR algorithm that provides good performance in the presence of clutter edges and target masking, while maintaining performance near to that of CA CFAR in homogeneous clutter, is the *switching CFAR* (S-CFAR) (Van Cao, 2004). In this approach, the CFAR reference window is divided into two groups, not necessarily contiguous: those cells above a threshold set as a fraction of the test cell value, and those below. If the number of cells in the low amplitude group exceeds some threshold N_t, typically set to about one half of the total number N of reference cells, all N cells are used in a cell averaging calculation. If the number of low amplitude cells is less than N_t, the threshold is set based only on the low amplitude cells. The principal advantage appears to be reduced losses compared to order statistic CFAR (OS CFAR, described in the next section) due to masking targets and somewhat improved clutter-edge performance, while avoiding the need for sorting required by OS CFAR.

Yet another approach that has been proposed to combat masking is the use of alternate detector laws (i.e., not linear or square-law). By far the most common is log CFAR, which applies conventional cell-averaging CFAR logic to the

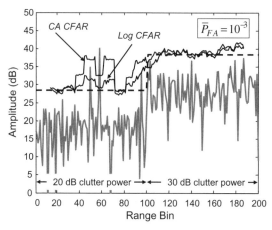

Figure 7.12 Comparison of CA CFAR and log CFAR on the same data as Fig. 7.11.

logarithm of the received power samples.[†] There appears to be no simple closed form analysis equivalent to Eqs. (7.6) through (7.11) for determining the relationship between an average of the log-detected data and the interference power β^2. However, motivated by considering the logarithm of the threshold computation for a square law detector seen in Eq. (7.11), the log CFAR threshold is computed by adding an offset to the averaged logarithmic data

$$\hat{T}_{\log} = \frac{1}{N} \sum_{i=1}^{N} 10 \log_{10}(x_i) + \alpha_{\log} \qquad (7.38)$$

In general, applying a logarithmic transformation to the data compresses its numerical dynamic range. This was an important implementation advantage in older systems built using analog or fixed-point digital hardware, but is less of a consideration with more modern processors. However, applying the CFAR calculation to the logarithmic data has the additional advantage that isolated interferers in the reference window do not have as great an influence on the numerical value of the estimated interference mean, thus reducing target masking effects. This effect is clearly shown in Fig. 7.12, which shows the same data set used previously containing closely-spaced targets with 15 and 20 dB signal-to-clutter ratio, and a 10-dB clutter edge. The two targets are easily detected. Unfortunately, log CFAR exhibits poor performance at clutter edges, in particular an increased vulnerability to false alarms at clutter edges. In this example, not only does a false alarm occur at the clutter edge, but the target near the edge is not detected.

[†]The base of the logarithm affects the specific offset needed to set the threshold but is otherwise unimportant. In Eq. (7.38) it are assumed that the log data are on a decibel scale.

No explicit expression is known for finding the required threshold offset $\alpha_{\log}$ as a function of $\overline{P}_{FA}$. Results have been obtained for $\overline{P}_D$ and $\overline{P}_{FA}$ that can be solved numerically to find suitable values of the threshold (Novak, 1980). To create Fig. 7.12, a Monte Carlo simulation was instead used to determine the required value by trial and error. The result for $N = 20$ reference cells and $\overline{P}_{FA} = 10^{-3}$ is $\alpha_{\log} = 11.85$ dB.

Hansen and Ward (1972) have estimated the CFAR loss of a log CFAR detector using Monte Carlo simulation techniques. They found that log CFAR increases the CFAR loss relative to the linear detector and that in homogenous clutter, the number of log CFAR reference window cells $N_{\log}$ required to achieve the same CFAR loss as an N-cell conventional CA CFAR using a linear detector is

$$N_{\log} = 1.65N - 0.65 \tag{7.39}$$

Thus, the use of the log detector increases the required CFAR window size by about 65 percent to avoid increasing the CFAR loss. Hansen and Ward also state that for $N > 8$, the log CFAR loss in dB is about 65 percent more than the loss for a CA CFAR with the same value of N. This statement is consistent with the data of Fig. 7.5 and Eq. (7.39) in the sense that, for a given P_{FA} and N, the CA CFAR loss in dB increases by about 65 percent if the number of cells is decreased by about 65 percent. For $N < 8$, the differential is less, and the two have the same performance for $N = 1$ (though this is not a case of practical interest).

7.6 Order Statistic CFAR

An alternative to cell-averaging CFAR is the class of *rank-based* or *order statistic* CFARs (OS CFAR). Proposed primarily for combating masking degradations, OS CFAR retains the one-dimensional or two-dimensional sliding window structure of CA CFAR, including guard cells if desired, but does away entirely with averaging of the reference window contents to explicitly estimate the interference level. Instead, OS CFAR rank orders the reference window data samples $\{x_1, x_2, \ldots, x_N\}$ to form a new sequence in ascending numerical order, denoted by $\{x_{(1)}, x_{(2)}, \ldots, x_{(N)}\}$. The kth element of the ordered list is called the kth *order statistic*. For example, the first order statistic is the minimum, the Nth order statistic is the maximum, and the $(N/2)$th order statistic is the median of the data $\{x_1, x_2, \ldots, x_N\}$. In OS CFAR, the kth order statistic is selected as representative of the interference level and a threshold is set as a multiple of this value

$$\hat{T} = \alpha_{\text{OS}} x_{(k)} \tag{7.40}$$

Note that the interference is thus estimated from only one actual data sample, instead of an average of all of the data samples. Nonetheless, the threshold in fact depends on all of the data, since all of the samples are required to determine which will be the kth largest.

It will be shown that this algorithm is in fact CFAR (i.e., does not depend on the interference power β^2), and the threshold multiplier required to achieve a specified $\overline{P}_{FA}$ will be determined. The analysis follows (Levanon 1988). To simplify the notation, consider the square-law detected output x_i normalized to its mean, $y_i = x_i/\beta^2$; this will have an exponential pdf with unit mean. The rank-ordered set of reference samples $\{y_i\}$ are denoted by $\{y_{(i)}\}$. For a given threshold T, the probability of false alarm will be

$$P_{FA}(T) = \int_T^{+\infty} e^{-y}\,dy = e^{-T} \tag{7.41}$$

The average P_{FA} will be computed as

$$\overline{P}_{FA} = \int_0^{+\infty} P_{FA}(T)p_T(T)\,dT \tag{7.42}$$

where $p_T(T)$ is the pdf of the threshold. Because T is proportional to the kth-ranked reference sample $y_{(k)}$, it is necessary to find the pdf of $y_{(k)}$. This tedious derivation can be found in the book by Levanon (1988); in terms of the probability density function $p_{y_i}(y)$ and corresponding *cumulative distribution function* (CDF) $P_{y_i}(y)$, the result is

$$p_{y_{(k)}}(y) = k\binom{N}{k} P_{y_i}^{k-1}(y)\left[1 - P_{y_i}(y)\right]^{N-k} p_{y_i}(y) \tag{7.43}$$

For i.i.d. Gaussian I/Q noise, the pdf and CDF of a single normalized reference sample y_i are

$$p_{y_{(k)}}(y) = e^{-y}$$

$$P_{y_{(k)}}(y) = \int_0^y p_{y_{(k)}}(y')\,dy' = 1 - e^{-y} \tag{7.44}$$

Using Eq. (7.44) in Eq. (7.43) gives the pdf of the kth ranked sample

$$p_{y_{(k)}}(y) = k\binom{N}{k}[e^{-y}]^{N-k+1}[1-e^{-y}]^{k-1} \tag{7.45}$$

The threshold is $\hat{T} = \alpha_{OS}\, y_{(k)}$, so the pdf of $\hat{T}$ is $p_{\hat{T}}(\hat{T}) = (1/\alpha_{OS})p_{y_{(k)}}(\hat{T}/\alpha_{OS})$

$$p_{\hat{T}}(\hat{T}) = \frac{k}{\alpha_{OS}}\binom{N}{k}\left[e^{-\hat{T}/\alpha_{OS}}\right]^{N-k+1}\left[1 - e^{-\hat{T}/\alpha_{OS}}\right]^{k-1} \tag{7.46}$$

Inserting this result into Eq. (7.42) gives

$$\overline{P}_{FA} = \int_0^{+\infty} e^{-\hat{T}}\frac{k}{\alpha_{OS}}\binom{N}{k}\left[e^{-\hat{T}/\alpha_{OS}}\right]^{N-k+1}\left[1 - e^{-\hat{T}/\alpha_{OS}}\right]^{k-1} d\hat{T}$$

$$= \frac{k}{\alpha_{OS}}\binom{N}{k}\int_0^{+\infty} e^{-(\alpha_{OS}+N-k+1)\hat{T}/\alpha_{OS}}\left[1 - e^{-\hat{T}/\alpha_{OS}}\right]^{k-1} d\hat{T} \tag{7.47}$$

With the change of variable $T' = \hat{T}/\alpha_{OS}$ this becomes the slightly more convenient form

$$\overline{P}_{FA} = k \binom{N}{k} \int_0^{+\infty} e^{-(\alpha_{OS}+N-k+1)T'} \left[1 - e^{-T'}\right]^{k-1} dT' \tag{7.48}$$

Utilizing integral 3.312(1) of the book by Gradshteyn and Ryzhik (1980) gives

$$\begin{aligned}\overline{P}_{FA} &= k \binom{N}{k} B(\alpha_{OS} + N - k + 1, k) \\ &= k \binom{N}{k} \frac{\Gamma(\alpha_{OS} + N - k + 1)\Gamma(k)}{\Gamma(\alpha_{OS} + N + 1)}\end{aligned} \tag{7.49}$$

where $B(\cdot,\cdot)$ is the beta function and in turn can be expressed in terms of the gamma function $\Gamma(\cdot)$ as shown. For integer arguments, $\Gamma(n) = (n-1)!$ and Eq. (7.49) therefore reduces for integer α_{OS} to

$$\begin{aligned}\overline{P}_{FA} &= k \frac{N!}{k!(N-k)!} \frac{(k-1)!(\alpha_{OS}+N-k)!}{(\alpha_{OS}+N)!} \\ &= \frac{N!(\alpha_{OS}+N-k)!}{(N-k)!(\alpha_{OS}+N)!} \quad (\alpha_{OS} \text{ integer})\end{aligned} \tag{7.50}$$

Figure 7.13 plots $\overline{P}_{FA}$ as a function of α_{OS} for two choices of OS windows, one with $N = 20$ and one with $N = 50$. In the first case, the $k = 15$th order statistic is chosen to set the threshold, while in the second the $k = 37$th order statistic

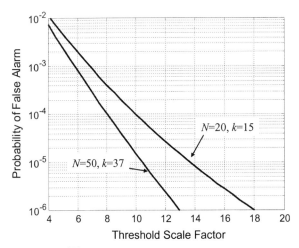

Figure 7.13 $\overline{P}_{FA}$ versus threshold scale factor α_{OS} in order statistics CFAR. The selected order statistic is chosen as approximately 0.75N.

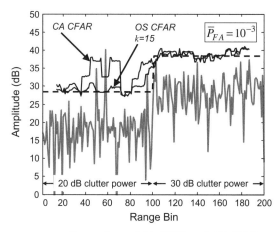

Figure 7.14 Comparison of CA CFAR and OS CFAR on the same data as Fig. 7.11.

is selected. Plots such as these can be used to determine the threshold multiplier needed to achieve a specified $\overline{P}_{FA}$ for a given OS CFAR window configuration. For example, with $N = 20$ and $k = 15$, a multiplier of $\alpha_{OS} = 6.857$ gives $\overline{P}_{FA} = 10^{-3}$.

Figure 7.14 shows the performance of the OS CFAR on the same data set considered previously using $N = 20$ again and choosing the $k = 15$th order statistic to set the threshold. The use of the ordered statistic instead of a mean estimate makes the detector almost completely insensitive to masking by closely spaced targets so long as the number of cells contaminated by interfering targets does not exceed $N - k$. In this example, both closely-spaced targets are readily detected. Although they were used in Fig. 7.14 for consistency, guard cells are less important in an OS CFAR, since the rank ordering process will not be affected by targets spreading outside of the test cell. The effect of the window length and choice of order statistic k on the behavior at clutter edges are discussed in (Rohling, 1983). If $k \leq N/2$, there will be extensive false alarms at clutter edges. Thus, k is usually chosen to satisfy $N/2 < k < N$. Typically, k is on the order of $0.75N$ (Nathanson, 1991).

To determine the OS CFAR loss, it is necessary to determine the signal-to-noise ratio required to obtain a specified $\overline{P}_D$ for a given $\overline{P}_{FA}$ and compare that result to the ideal threshold or to CA CFAR in homogeneous interference. By repeating the analysis above but using the Swerling 1 or Swerling 2 model (Rayleigh amplitude pdf) for the test cell, it can be shown that the average probability of detection is also given by Eq. (7.49), but with α_{OS} replaced by (Levanon, 1988)

$$\alpha_{OS}^D = \frac{\alpha_{OS}}{1 + \bar{\chi}} \qquad (7.51)$$

where $\bar{\chi}$ is the average SNR of the target. By varying α_{OS} or α_{OS}^D, curves of $\overline{P}_D$ versus $\overline{P}_{FA}$ can be developed and the CFAR loss measured. An alternative analysis approach is given by Blake (1988). In the absence of interfering targets, the OS-CFAR suffers a small additional loss over CA CFAR. The value depends on both k and N but is typically on the order of 0.3 to 0.5 dB. If interfering targets are present, the OS CFAR loss increases only very slowly until the number of interferers exceeds $N - k$, the number of "ignored" high-rank cells. In contrast, the loss in CA CFAR increases very rapidly in this case due to the elevated estimate of the interference power (Blake, 1988). Thus, OS CFAR losses are lower than CA CFAR losses in the presence of interferers. Additional results on the performance of OS CFAR, including the effect of noncoherent integration and its behavior in Weibull clutter, are available in the paper by (Shor and Levanon, 1991).

7.7 Additional CFAR Topics

7.7.1 Adaptive CFAR

"Adaptive CFAR" is the name given to a growing class of CFAR algorithms designed to improve performance in nonhomogeneous clutter. Generally, they dispense with the fixed CFAR window structure (usually with half of the cells in each of the lead and lag windows). Instead they typically construct a statistical test to determine if the reference cells span one clutter field or two, i.e., whether or not the reference cell data are homogeneous. If not homogeneous, the algorithms estimate not only the clutter statistics in each field, but at which cell the transition from one to the other occurs (and therefore which type of clutter is competing with the target in the cell under test).

The basic approach to adaptive CFAR for nonhomogeneous clutter was described in Finn (1986). The algorithm assumes a CFAR window of total length N cells, including the cell under test, that spans two clutter fields, i.e., two regions with different statistical parameters. The clutter cell at which the statistics change, say cell $M + 1$, is the clutter edge; however, M is not known. Initially, it is also assumed that the clutter follows the usual square-law detected, exponential distribution. The algorithm starts by setting $M = 1$ and computing the sample mean $\widehat{\beta^2_{1(1)}}$ in the first clutter regions, which is only cell 1 when $M = 1$, and the sample mean $\widehat{\beta^2_{2(1)}}$ in the second region comprising the remaining $N - 1$ cells. The notation (1) in this subscript indicates that the estimates for the two regions are for the transition point $M = 1$. This process is repeated for $M = 2, \ldots, N - 1$. Thus, a pair of sample means $\widehat{\beta^2_{1(M)}}$ and $\widehat{\beta^2_{2(M)}}$ are computed for each of the N possible transition points between the two-clutter regions.

The next step is to choose the most likely transition point M_t. The maximum likelihood estimate of this transition point is the value M_t of M that maximizes the log-likelihood function (Finn, 1986)

$$L_M = -\left\{M \ln \widehat{\beta^2_{1(M)}} - (N - M) \ln \widehat{\beta^2_{2(M)}}\right\} \quad (7.52)$$

Once M_t is identified, it is also known if the cell under test is in the first or second clutter region. Standard CA CFAR, using the appropriate mean estimate and the number of cells with which it is estimated, can then be applied. For example, if the cell under test is in the first region, the threshold would be set according to

$$\alpha = M_t \left(\overline{P}_{FA}^{-1/M_t} - 1 \right)$$
$$\hat{T} = \alpha \widehat{\beta_{1(M_t)}^2} \tag{7.53}$$

Note that this procedure is effectively SOCA CFAR when the cell under test is in the low clutter region, and GOCA CFAR when it is in the high clutter region (assuming the transition point is correctly located).

Two modifications to this basic procedure are described in (Finn, 1986). The discussion above does not allow for the possibility that the clutter is uniform. Thus, an additional likelihood test is conducted to compare L_{M_t} as computed in Eq. (7.52) with $M = M_t$, to the corresponding metric under the assumption of uniform clutter, $L_0 = -N \ln \widehat{\beta_{1(0)}^2} = -N \ln \widehat{\beta^2}$. If $L_0 > L_{M_t}$, the clutter is assumed homogeneous and conventional CA CFAR is applied.

When the two-region clutter hypothesis is accepted, the estimate of the transition point M_t can of course be incorrect. If the target is in fact in the low clutter region but is incorrectly determined to be in the high clutter region, the high clutter statistics will be used to set the threshold. There will then be an increased probability that the high clutter will mask the target. If the target is in the high clutter region but is incorrectly determined to be in the low clutter region, the low clutter statistics will be used to set the threshold, which will then be too low. While the target will have an enhanced probability of detection, the false alarm probability will rise, possibly dramatically. For this reason, the adaptive CFAR algorithm is modified to bias the decision in favor of the hypothesis that the target is in the high clutter region. While this will increase masking effects somewhat, it avoids the generally more damaging problem of large increases in the false alarm rate. Details are given by Finn (1986).

Again, many of the extensions to CA CFAR can also be applied to the adaptive CFAR. It can be applied to log normal or Weibull clutter by computing both sample means and variances in each region. The data in each region can be censored prior to estimating the statistics. Order statistic, rather than cell-averaging, rules can be used to set the threshold. Many such variations are available in the literature, as are algorithms building on the basic adaptive concept but applying different statistical estimators and decision logics.

7.7.2 Two-parameter CFAR

All of the results discussed in this chapter so far have assumed Rayleigh voltage/exponential power interference, which is the appropriate model when the primary interference is Gaussian noise, whether the source is low-level receiver noise or high-level noise jamming. Only one parameter, the mean power (β^2 in this chapter), is required to completely specify the pdf. However, as discussed in

Chap. 2, many types of clutter are best modeled by more complicated probability density functions such as the log-normal or Weibull pdf. Unlike the exponential pdf, these are *two-parameter* distributions and estimates of both the mean and variance (or a related parameter such as skewness) must be estimated in order to characterize these distributions. Specifically, a threshold control mechanism must be based on estimates of both parameters if it is to exhibit CFAR behavior.

An example of a CFAR algorithm for log-normal clutter is given in the paper by Schleher (1977). The receiver uses a log detector, so the detected samples $\{x_i\}$ are normally distributed. The CFAR structure is a conventional cell averaging approach on the log data. The threshold is computed as follows:

$$\hat{\mu} = \frac{1}{N} \sum_{i=1}^{N} x_i$$

$$\hat{s} = \sqrt{\frac{1}{N} \sum_{i=1}^{N} (x_i - \hat{\mu})^2} \qquad (7.54)$$

$$\hat{T} = \hat{\mu} + \alpha \hat{s}$$

This CFAR threshold calculation could clearly be combined with many of the embellishments discussed earlier for CA CFAR, such as SOCA or GOCA rules or censoring.

Because of the need to estimate two parameters, the CFAR loss is greater with two parameter distributions than with the Rayleigh/exponential distribution, and in fact can be very large, especially for small numbers of reference cells. For example, with $\overline{P}_D = 0.9$, $\overline{P}_{FA} = 10^{-6}$, and $N = 32$ reference cells, the CFAR loss using Eq. (7.54) in log-normal clutter is approximately 13 dB (Schleher, 1977). From Fig. 7.5, a conventional CA CFAR in Rayleigh/exponential clutter with the same detection statistics and window size has a CFAR loss of just under 1 dB.

The same calculations are used to set the CFAR threshold in Weibull clutter, though the specific values of α needed vary from the log-normal case. Two proposed Weibull detectors, the so-called log-t detector and another based on maximum likelihood estimates of the Weibull pdf parameters, have been shown to be equivalent to Eq. (7.54) (Gandhi et al., 1995).

Order statistic CFARs have also been proposed for two-parameter clutter. One example combines OS CFAR in each of the lead and lag windows with a greatest-of logic to estimate the interference mean, and then uses the single parameter Eq. (7.36) to set the threshold (Rifkin, 1994). Since the second (skewness) parameter of the pdf is not estimated implicitly or explicitly, the multiplier α must be made a function of the skewness, implying in turn that the skewness parameter must be known to correctly set the threshold. Performance results again suggest that choosing the order statistic k to be about $0.75N$ provides the best performance against interferers and uncertainty in the skewness parameter (Rifkin, 1994).

7.7.3 Clutter map CFAR

In Chap. 5, the technique of clutter mapping for detection of stationary or slowly moving targets by ground-based, fixed-site radars when the competing zero-Doppler clutter was not too strong was discussed. The threshold for each range-angle cell was computed as a multiple of the measured clutter in the same cell. The clutter measurement was obtained as a simple first-order recursive filter of the form

$$\hat{x}[n] = (1-\gamma)\hat{x}[n-1] + \gamma x[n] \qquad (7.55)$$

where $\hat{x}[n]$ is the estimate of the clutter reflectivity at time n (n usually indexes complete radar scans of an area) and $x[n]$ is the currently measured clutter sample at time n. The factor γ controls the relative weight of the current measurement versus the preceding measurements. This equation is applied separately to each range-angle cell of interest. The detection threshold is then set for each range-angle cell as

$$\hat{T}[n] = \alpha \hat{x}[n-1] \qquad (7.56)$$

Note that the threshold for testing for a target on the current scan is based on the clutter estimate from the previous scan. The current data are *not* included because, if the cell under test does contain a target, it would distort the clutter measurement and raise the threshold, creating a self-masking effect. Even with this precaution, self-masking can cause CFAR losses of several dB if slow-moving targets are present, so that they persist in a map cell for more than one scan (Lops, 1989). If the target persists in the map cell for a number of scans approaching the number of scans integrated to form the clutter map, detectability is essentially lost entirely.

The first-order difference Eq. (7.55) corresponds to an IIR filter with the impulse response

$$h[n] = \gamma(1-\gamma)^n u[n] \qquad (7.57)$$

where $u[n]$ is the unit step function. Consequently, the output of the filter can also be expressed as a convolution with $h[n]$

$$\hat{x}[n] = x[n] * h[n]$$
$$= \gamma \sum_{m=0}^{\infty} (1-\gamma)^m x[n-m] \qquad (7.58)$$

This equation shows that, similar to CA CFAR, the threshold will be based on an average of clutter measurements, but there are several important differences. In CA CFAR, clutter samples taken during the same pulse or dwell and from cells adjoining the cell under test in spatial position, Doppler, or both are used to estimate the clutter level. The clutter must be assumed spatially homogeneous so that these adjoining cells represent the interference in the cell under test accurately. The various CA CFAR extensions discussed previously are all motivated by real-world violations of this assumption. In clutter mapping, the threshold is based on clutter samples taken from the cell under test but on previous time intervals. Thus, spatial homogeneity is not required, but homogeneity in time, i.e., statistical stationarity, is. Both types of CFAR require that the clutter samples be uncorrelated (in space or Doppler for CA CFAR, in time for clutter mapping) for the analyses given here to be valid. Finally, Eq. (7.58) shows that the threshold is based on an infinite, weighted sum of previous samples rather than the finite sums of CA CFAR and its variants.

It is also possible to estimate the clutter mapping threshold using a simple average of a finite number of previous measurements. One version of the *moving target detector* (MTD) is said to have used an 8-scan average, covering about 32 seconds of data (Skolnik, 2001). In this case, the CA CFAR analyses would apply to the clutter map as well. However, for computational simplicity the recursive filter of Eq. (7.55) is most often used. Nitzberg (1986) derives the average probabilities of false alarm and detection when the threshold is computed with the recursive scheme. They are

$$\overline{P}_{FA} = \prod_{m=0}^{\infty} \{1 + \alpha\gamma(1-\gamma)^m\}^{-1} \qquad (7.59)$$

$$\overline{P}_D = \prod_{m=0}^{\infty} \left\{1 + \frac{\alpha}{1+\bar{\chi}}\gamma(1-\gamma)^m\right\}^{-1} \qquad (7.60)$$

These formulas are slow to converge in practice; a more rapidly converging variation is given by Levanon (1988). Again, α can be varied to generate curves of $\overline{P}_D$ versus $\overline{P}_{FA}$ and CFAR loss can then be determined by comparing these curves to the case where the interference is known exactly. The case $\gamma = 0$ in fact corresponds to the ideal case. For every increase of 0.2 in γ, the CFAR loss increases approximately 3 dB (Levanon, 1988). A crude approximate formula for the CFAR loss is (Taylor, 1990)

$$\text{Loss (dB)} = -5.5 \log_{10} \overline{P}_{FA} \left(1 + \frac{2}{\gamma}\right)^{-1} \qquad (7.61)$$

This formula is of limited accuracy, especially for large γ, but may be useful for rough calculations. Another implementation of the MTD used a recursive filter as described previously with a feedback coefficient of $\gamma = 7/8$ (Nathanson, 1991). A rational value of γ with a denominator that is a power of two is particularly amenable to the fixed point implementations used in early versions of the MTD, since the division by the denominator value (eight in this case) can be implemented by simply right-shifting the binary data.

Some clutter map systems combine multiple range cells at a given azimuth direction to form a single, larger map cell (Conte and Lops, 1997). This then allows the introduction of any of several standard CFAR techniques to combine the range cells and improve the map cell clutter estimate. Either cell-averaging or order statistics CFAR can be applied to the range cells to form the basic clutter power estimate. However, the basic CA or OS approaches can be extended as appropriate for the environment with any of the techniques discussed previously for conventional CFAR: censoring, guard cells, smallest-of or greatest-of (SOCA and GOCA) detectors, log detectors, and two-parameter estimation algorithms (Lops, 1996; Conte and Lops, 1997).

7.7.4 Distribution-free CFAR

The CFAR processors discussed so far assume a specific form of the pdf of the interference in order to determine the value of the threshold (equivalently, the value of the threshold multiplier α). For instance, the particular form of Eq. (7.16), and thus of the formula for α in Eq. (7.17), is a result of having assumed an exponential distribution for the square-law detected interference-only samples. If noise is the dominant interference source, this is not a significant constraint. However, for a system operating in a clutter limited environment, or in an environment where the dominant interference varies between clutter of various distributions, noise, and jamming, a threshold setting algorithm based on a particular interference pdf may produce large errors in the threshold setting when another interference pdf dominates. For this reason, threshold setting algorithms that do not depend on the particular pdf of the interference are of interest. Such techniques are called *distribution free* or *nonparametric* CFAR algorithms (DF CFAR).

Historically, DF CFAR has been based most often on a two stage "double-threshold" approach, in which the first stage threshold converts the raw data into binary detection/no detection decisions. This decision is repeated over multiple pulses or scans, and the individual detection decisions for a given resolution cell are combined using an "M out of N" rule as described in Chap. 6. The pdf of the output of the first stage is binomial, since the data are binary at that point, independent of the input pdf. Four variations on this idea are discussed in the book by Barrett (1987). Two of them, the "double threshold detector" and the "rank order detector" use conventional cell-averaging or OS CFAR in the first stage to set a specific $\overline{P}_{FA}$ and therefore require knowledge of the interference pdf to set the first stage threshold. These are therefore not truly distribution-free.

The "modified double threshold detector" replaces the deterministically computed first stage threshold with a feedback circuit that monitors $\overline{P}_{FA}$ at the first stage output and adjusts the threshold to approximate the desired value. This technique requires large amounts of data to estimate the observed $\overline{P}_{FA}$ but will work for any input distribution. In the *rank sum double quantizer*, the first stage does not actually threshold the data, but instead computes the rank of the test cell compared to the reference cells and passes the rank number, instead of a binary detection/no detection decision, to the second stage. The second stage integrates this rank number over multiple pulses or scans, producing a random variable with a distribution that depends only on the correlation between the data samples from successive pulses or scans (Barrett, 1987). A two-parameter cell-averaging CFAR computation is applied to this variable to obtain the final detection decision.

A more modern approach based on rank order ideas is given by Sarma and Tufts (2001). Consider a set of N reference samples $\{y_i\}$ and the kth order statistic for this set, $y_{(k)}$. The pdf for $y_{(k)}$ was given in Eq. (7.43). The *coverage* C is defined as the probability that a reference sample y_i is greater than $y_{(k)}$. Note that $0 \leq C \leq 1$. Since the cumulative distribution function $P_{y_i}(y)$ is the probability that y_i is *less* than some value y, it follows that

$$C = 1 - P_{y_i}(y_{(k)}) \tag{7.62}$$

and the pdf of C becomes

$$p_C(C) = k \binom{N}{k} (1-C)^{k-1} C^{N-k} \tag{7.63}$$

Furthermore, the expected value of C is

$$\mathbf{E}\{C\} = \int_0^1 C p_C(C)\, dC = \frac{N+1-k}{N+1} \tag{7.64}$$

Now consider a system that uses the kth rank order statistic of the reference cell data as the threshold value for the cell under test. This differs from a conventional OS CFAR, which computes a threshold as a multiple of the kth rank order statistic, where the multiplier α_{OS} is a function of the pdf of the interference data. If instead $y_{(k)}$ is itself the threshold, no multiplier α is required. Under these conditions, C will be the probability of false alarm. Equations (7.63) and (7.64) show that the pdf of C, and thus also its expected value $\overline{C} = \overline{P}_{FA}$, do not depend on the pdf of the raw data $\{y_i\}$. Thus, the detector is a distribution-free CFAR.

A limitation of this DF CFAR is that only certain values of $\overline{P}_{FA}$ are achievable, and they depend on the number N of reference cells. $\overline{P}_{FA}$ takes on the discrete set of values given by Eq. (7.64) as k varies from 1 to N. The minimum possible value is obtained with $k = N$ and is

$$\overline{P}_{FA_{\min}} = \frac{1}{N+1} \tag{7.65}$$

Equation (7.65) illustrates another limitation of this technique: small values of $\overline{P}_{FA_{\min}}$ require large values of N. More realistically, practical limitations on the reference window size N limit this method to relatively high values of $\overline{P}_{FA}$. For a given value of N, the design value of the probability of false alarm, say $\overline{P}_{FA_d}$, must be chosen to be greater than or equal to $\overline{P}_{FA_{\min}}$. Assuming this condition is satisfied, the rank order to be used as a threshold is the one that produces a value of $\overline{P}_{FA}$ as close to $\overline{P}_{FA_{\min}}$ as possible without exceeding it. That rank is given by

$$k = \left\lceil (N+1)(1 - \overline{P}_{FA_d}) \right\rceil \tag{7.66}$$

While the false alarm probability does not depend on the pdf of the interference, the detection probability does. For exponentially distributed interference and target (Swerling 1 case), the average probability of detection is (Sarma and Tufts, 2001)

$$\overline{P}_D = \prod_{i=0}^{k-1} \frac{N-i}{N-i+(1+\gamma)^{-1}} \qquad (7.67)$$

As with the other CFAR detectors, Eqs. (7.66) and (7.67) can be used to determine the CFAR loss of the DF CFAR. The additional loss over a CA CFAR for the Swerling 1 case is typically less than about 0.4 dB.

7.7.5 System-level control of false alarms

It has been seen in this and the preceding chapter that achieving good detection performance (high $\overline{P}_D$, low $\overline{P}_{FA}$) requires a *signal-to-interference ratio* (SIR) on the order of 15 dB or better at the point of detection. For a given target RCS, the SIR is determined in part by basic radar system design choices reflected in the radar range equation: transmitter power, antenna gain, operating frequency, and noise figure. Furthermore, the fundamental goal of many of the techniques of radar signal processing discussed in other chapters of this text is to improve the SIR before the point of detection. Examples include matched filtering, pulse compression, MTI, pulse Doppler processing, and space-time adaptive processing. Once the SIR has been maximized, the detector, whether fixed or adaptive threshold, sets the actual threshold value and thus the false alarm probability. The SIR then determines the detection probability.

In some cases the detection probability may still be lower than required. In this event, the threshold may be lowered, increasing $\overline{P}_D$ but also increasing $\overline{P}_{FA}$. Additional techniques can then be applied at other stages of the overall system processing in order to reduce $\overline{P}_{FA}$ back to an acceptable level. Several options are discussed in the book by Nathanson (1991); their applicability depends on the particular system involved. If jamming is present, a sidelobe blanker or sidelobe canceller can be used to further improve the SIR before the detector (assuming more advanced techniques such as STAP have not already been applied). After the detector, a clutter map may be used in some systems to reject false alarms due to fixed clutter discretes or known *radio frequency interference* (RFI) sources. If valid targets can be expected to extend over more than one range, azimuth, or Doppler cell, apparent detections that occupy only a single cell can be rejected as false alarms after the detector, lowering the system $\overline{P}_{FA}$. Finally, an apparent target can be tracked to make sure it recurs over multiple scans; if it does not, it is rejected as a false alarm. If it does, its kinematics can be tracked over multiple scans. If the target track violates reasonable bounds on velocity and acceleration, it can be assumed to be a false alarm, quite possibly due to the presence of electronic countermeasures, and the track can be rejected. Thus, control of the overall system false alarm rate can be spread over virtually all stages of the system.

References

Barrett, C. R., Jr., "Adaptive Thresholding and Automatic Detection," Chap. 12 in *Principles of Modern Radar*, (J. L. Eaves and E. K. Reedy, (eds.)). Van Nostrand Reinhold, New York, 1987.

Blake, S., "OS-CFAR Theory for Multiple Targets and Nonuniform Clutter," *IEEE Transactions on Aerospace and Electronic Systems*, vol. AES-24, no. 6, pp. 785–790, Nov. 1988.

Conte, E., and M. Lops, "Clutter-Map CFAR Detection for Range-Spread Targets in Non-Gaussian Clutter, Part I: System Design," *IEEE Transactions on Aerospace and Electronic Systems*, vol. AES-33, no. 2, pp. 432–442, April 1997.

Di Vito, A., and G. Moretti, "Probability of False Alarm in CA-CFAR Device Downstream From Linear-law Detector," *Electronics Letters*, vol. 25, no. 5, pp. 1692–1693, Dec. 1989.

Finn, H. M., "A CFAR Design for a Window Spanning Two Clutter Fields," *IEEE Transactions on Aerospace and Electronic Systems*, vol. AES-22, no. 2, pp. 155–169, March 1986.

Gandhi, P. P., and S. A. Kassam, "Analysis of CFAR Processors in Nonhomogeneous Background," *IEEE Transactions on Aerospace and Electronic Systems*, vol. AES-24, no. 4, pp. 427–445, July 1988.

Gandhi, P. P., E. Cardona, and L. Baker, "CFAR Signal Detection in Nonhomogeneous Weibull Clutter and Interference," *Record of the IEEE International Radar Conference*, pp. 583–588, 1995.

Gradshteyn, I.S., and I. M. Ryzhik, *Tables of Integrals, Series, and Products*, A. Jeffrey (ed.). Academic Press, New York, 1980.

Hansen, V. G., and H. R. Ward, "Detection Performance of the Cell-Averaging LOG/CFAR Receiver," *IEEE Transactions On Aerospace and Electronic Systems*, vol. AES-8, no. 5, pp. 648–652, Sept. 1972.

Hansen, V. G., and J. H. Sawyers, "Detectability Loss Due to 'Greatest Of' Selection in a Cell-Averaging CFAR," *IEEE Transactions on Aerospace and Electronic Systems*, vol. AES-16, no. 1, pp. 115–118, Jan. 1980.

Kay, S. M., *Fundamentals of Statistical Signal Processing, Vol. I: Estimation Theory*. Prentice Hall, Upper Saddle River, NJ, 1993.

Levanon, N., *Radar Principles*. Wiley, New York, 1988.

Lops, M., and M. Orsini, "Scan-by-scan averaging CFAR," *IEE Proceedings*, Part F, vol. 136, no. 6, pp. 249–253, Dec. 1989.

Lops, M., "Hybrid Clutter-map/*L*-CFAR Procedure for Clutter Rejection in Nonhomogeneous Environment," *IEE Proceedings of Radar, Sonar, and Navigation*, vol. 143, no. 4, pp. 239–245, Aug. 1996.

Nathanson, F. E., (with J. P. Reilly and M. N. Cohen), *Radar Design Principles*, 2d ed. McGraw-Hill, New York, 1991.

Nitzberg, R., "Clutter Map CFAR Analysis," *IEEE Transactions on Aerospace and Electronic Systems*, vol. AES-22, no. 4, pp. 419–421, July 1986.

Novak, L. M., "Radar Target Detection and Map-Matching Algorithm Studies," *IEEE Transactions on Aerospace and Electronic Systems*, vol. AES-16, no. 5, pp. 620–625, Sept. 1980.

Raghavan, R. S., "Analysis of CA-CFAR Processors for Linear-Law Detection," *IEEE Transactions on Aerospace and Electronic Systems*, vol. AES-28, no. 3, pp. 661–665, July 1992.

Rifkin, R., "Analysis of CFAR Performance in Weibull Clutter," *IEEE Transactions on Aerospace and Electronic Systems*, vol. AES-30, no. 2, pp. 315–329, April 1994.

Ritcey, J. A., "Performance Analysis of the Censored Mean-Level Detector," *IEEE Transactions on Aerospace and Electronic Systems*, vol. AES-22, no. 4, pp. 443–454, July 1986.

Ritcey, J. A., and J. L. Hines, "Performance of Max-Mean-Level Detector With and Without Censoring," *IEEE Transactions on Aerospace and Electronic Systems*, vol. AES-25, no. 2, pp. 213–223, March 1989.

Rohling, H., "Radar CFAR Thresholding in Clutter and Multiple Target Situations," *IEEE Transactions on Aerospace and Electronic Systems*, vol. AES-19, no. 4, pp. 608–620, July 1983.

Sarma, A., and D. W. Tufts, "Robust Adaptive Threshold for Control of False Alarms," *IEEE Signal Processing Letters*, vol. 8, no. 9, pp. 261–263, Sept. 2001.

Schleher, D. C., "Harbor Surveillance Radar Detection Performance," *IEEE Journal of Oceanic Engineering*, vol. OE-2, no. 4, pp. 318–325, Oct. 1977.

Shor, M., and N. Levanon, "Performance of Order Statistics CFAR," *IEEE Transactions on Aerospace and Electronic Systems*, vol. AES-27, no. 2, pp. 214–224, March 1991.

Smith, M. E., and P. K. Varshney, "Intelligent CFAR Processor Based on Data Variability," *IEEE Transactions on Aerospace and Electronic Systems*, vol. AES-36, no. 3, pp. 837–847, July 2000.

Taylor, J. W., "Receivers," Chap. 3 in M. Skolnik (ed.), *Radar Handbook*, 2d ed. McGraw-Hill, New York, 1990.

Van Cao, T.-T., "A CFAR Thresholding Approach Based on Test Cell Statistics." *Proceedings of the 2004 IEEE Radar Conference*, pp. 349–354, April 26–29, 2004.

Weiss, M., "Analysis of Some Modified Cell-Averaging CFAR Processors in Multiple-Target Situations," *IEEE Transactions on Aerospace and Electronic Systems*, vol. AES-18, no. 1, pp. 102–114, Jan. 1982.

Chapter 8

Introduction to Synthetic Aperture Imaging

When it was first developed, radar had two primary functions, detection and tracking. To these have been added high-resolution radar imaging in two and, more recently, three dimensions. The technique of high-resolution two-dimensional radar imaging is usually called *synthetic aperture radar* (SAR); the reason will become clear in Sec. 8.1.2. SAR is most often applied to imaging of static ground scenes; thus, the "target" in SAR operation is the ground clutter. Applications of radar imagery include cartography, land use analysis, oceanography, and numerous military applications such as reconnaissance, surveillance, battle damage assessment, ground target classification and identification, navigation, and more. SAR maps are routinely generated from both airborne and spaceborne platforms, and with resolutions ranging from several tens of meters down to a few inches.

Figure 8.1a is an example of a SAR image produced in the mid-1990s. Collected by the Sandia National Laboratories' K_u band radar, this image obtains a resolution of 3 m at ranges of tens of kilometers. Figure 8.1b is an aerial photograph of the same scene; close examination reveals many similarities as well as many significant differences in the appearance of the scene at radar and visible wavelengths. Figure 8.2 is another example, a SAR image of the national mall area of Washington, DC.

Despite the impressive quality of the SAR images, a human observer would likely prefer photographs for purposes of understanding and analyzing the scene. Referring again to Fig. 8.1, though printed here in black and white, the original photograph is in color, whereas the SAR image is monochrome since SAR measures only the scalar quantity of reflectivity.[†] The photograph

---

[†] "False color" or "pseudo color" SAR imagery is often produced by combining multiple SAR images of the same scene collected at different polarizations and/or radar frequencies.

(a)

Figure 8.1 Comparison of optical and SAR images of the Albuquerque airport. (a) K_u band (15 GHz) SAR image, 3-m resolution. (b) Aerial photograph. (Images courtesy of Sandia National Laboratories.)

has finer resolution than the SAR image. The SAR image exhibits a granular speckle, often referred to as "salt-and-pepper" noise, typical of coherent imaging systems (including holograms, for instance) but absent in the noncoherent optical image. Close inspection reveals differences in phenomenology that can confound image analysis. For example, in the bottom center of the photograph there is a large concrete pad area on which there are three rectangular buildings; the concrete appears light in color, the buildings dark. In the SAR image, the contrast is reversed. Another example is the painted stripes visible in the photograph at the ends of the runways (top center and right); these are entirely absent in the SAR image.

(b)

Figure 8.1 (*Continued*)

Why then is SAR of interest? The answer becomes apparent if the comparison of Fig. 8.1 is repeated on a cloudy night. The photograph would become a solid black, since the ground would not be visible through the clouds and the sun would not be present to illuminate even the clouds. The SAR image would be unchanged, because the SAR is an active system that provides its own illumination, and because microwaves pass through clouds and other weather with little attenuation. Thus, radar provides a means for surveillance at any time of day or night and in a much wider range of weather conditions. Figure 8.3 compares two images taken from the space shuttle of the Manhattan and Long Island areas of New York. Both were taken in April 1994 at 3:00 A.M., though on different days, from an altitude of approximately 233 km. The lower image is a SAR map formed from three bands of radar data collected by the *shuttle*

Figure 8.2 Synthetic aperture radar image of the Capitol Mall area in Washington, DC. (Image courtesy of Sandia National Laboratories.)

Figure 8.3 Comparison of optical (top) and radar (bottom) images of the Manhattan and Long Island, New York area as viewed from the space shuttle at 3:00 A.M. (Image courtesy NASA/JPL-CalTech.)

imaging radar-C (SIR-C) instrument.[†] The upper image is a photograph. The radar image shows the full outline of Long Island. The photograph only shows those portions of the island that are well illuminated at 3:00 A.M., which does not include the northern edges of the island.

The history of SAR is described briefly by Sherwin (1962) and Wiley (1985). The SAR concept was first described and demonstrated by Carl Wiley of Goodyear Aircraft in 1951. The technique discussed in his patent (Wiley, 1965) would now be categorized as *Doppler beam sharpening* (DBS). Since then, SAR has undergone several significant technology phases. The late 1950s and early 1960s developed the original concept and implementations of what is now known as *stripmap SAR*. In these pre-Moore's Law days, SAR radar data were collected on photographic film and image formation was performed using remarkably elegant optical processing systems (Cutrona, 1966; Brown, 1969; Harger, 1970; Ausherman, 1980; Elachi, 1988). The 1960s saw the development of *spotlight SAR*, generally credited to Jack Walker of the Environmental Research Institute of Michigan (ERIM) (Ausherman et al., 1984).[‡] In the 1970s, digital processing for SAR image formation was developed (Kirk, 1975), while the range-Doppler algorithm (Wu, 1982) significantly improved the attainable resolution and image size. In the 1970s, David Munson of the University of Illinois published the connection between spotlight SAR imaging and certain forms of computerized tomography ("CAT scanning") (Munson et al., 1983). This observation was the first of several that significantly expanded the capabilities of SAR algorithms by moving them beyond the range-Doppler viewpoint adopted in their early development and adapting techniques from other fields, such as tomography and seismic prospecting (Munson and Visentin, 1989; Cafforio et al., 1991).

SAR is the first of two advanced radar signal processing techniques to which this text provides an introductory overview. (The other, space-time adaptive processing, is the subject of Chap. 9.) Beginning in the mid-1990s, a number of excellent textbooks on SAR have become available (Curlander and McDonough, 1991; Carrara et al., 1995; Jakowatz et al., 1996; Soumkeh, 1999; Franceschetti and Lanari, 1999; Cumming and Wong, 2005). The reader is referred to these for in-depth discussion of SAR processing. This chapter begins with a heuristic overview of the SAR concept from two points of view: synthetic antenna apertures, and Doppler resolution. These are sufficient to derive many of the fundamental equations describing SAR resolution, coverage, and sampling requirements, and to describe the nature of the SAR data set. The signal processing required for SAR image formation is then addressed more directly, describing a basic SAR data model and introducing three of the most common algorithms for SAR image formation. Finally, the concept of the interferometric SAR approach to three-dimensional radar imaging is introduced.

[†]This image is a grayscale representation of a false color image generated by assigning red to the L band, horizontal polarization transmit/horizontal polarization receive ("HH") data; green to the L band HV data; and blue to the C band HV data.

[‡]Subsequently part of Veridian Corporation and then General Dynamics Corporation.

8.1 Introduction to SAR Fundamentals

8.1.1 Cross-range resolution in radar

To be useful, a radar map must provide adequate resolution for its intended use; this may range from tens of meters to fractions of a meter. Furthermore, this resolution should be available in both the range and cross-range dimensions, since there is no reason to prefer one over the other in most operational scenarios. Finally, the resolution should be maintained throughout the imaged scene. Sufficient range resolution for radar mapping is relatively easy to achieve using the pulse compression techniques discussed in Chap. 4. Comparable cross-range[†] resolution, however, is not possible in conventional operation, often termed *real-beam imaging*.

Figure 8.4 illustrates cross-range resolution in a real-beam, forward-looking radar. The antenna scans in azimuth angle. It has an azimuth beamwidth of θ_{az} radians; thus, at a range R_0, the width of the beam is $R_0\theta_{az}$ meters to a good approximation. The cross-range dimension is the direction orthogonal to range. As discussed in Chap. 2, the receiver output for a fixed range as a function of azimuth scan angle is the range-averaged reflectivity convolved with the two-way antenna voltage pattern. In Fig. 8.4a, the two scatterers are separated in cross-range by less than one beamwidth, so the receiver output will blur the response to the two scatterers together (see Fig. 2.19). In Fig. 8.4b, they are separated by more than the beamwidth so that the receiver output for the appropriate range bin as a function of scan angle will show two distinct peaks. By convention, the two scatterers are therefore considered just resolvable if they are separated by the width of the antenna beam. Assuming narrow azimuth beamwidths, the cross-range resolution ΔCR is well approximated by

$$\Delta CR = 2R_0 \sin\left(\frac{\theta_{az}}{2}\right) \approx R_0 \theta_{az} \tag{8.1}$$

The beamwidth θ_{az} in this equation is usually taken as the two-way 3-dB beamwidth.

As shown in Chap. 1, the beamwidth of a conventional antenna is of the form

$$\theta_{az} = k\frac{\lambda}{D_{az}} \quad \text{rad} \tag{8.2}$$

where D_{az} is the width of the antenna in the azimuth dimension. The scale factor k depends on the antenna design. It is as little as 0.89 for an ideal aperture

---

[†]"Cross-range" is the direction orthogonal to the radar range direction, and thus to the radar antenna boresight. It differs from azimuth in that azimuth specifies an angular displacement relative to the boresight, while cross-range specifies a displacement in an orthogonal Cartesian coordinate. If the radar is sidelooking and the platform is not crabbed, the cross-range direction is parallel to the platform velocity vector. Thus the cross-range dimension is sometimes referred to as the along track dimension.

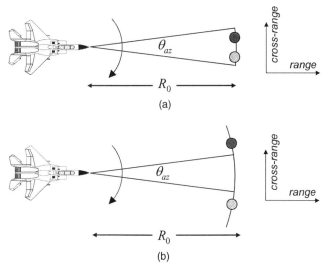

Figure 8.4 Resolution in cross-range of two scatterers at the same range. (*a*) Unresolved in cross-range. (*b*) Resolved in cross range.

antenna, but more often is on the order of 1.2 to 2.0 for practical antenna designs. In general, the lower the antenna sidelobes, the higher is k. Here it is sufficient to say that k is on the order of one and ignore it in subsequent calculations for compactness. Combining Eqs. (8.2) and (8.1) gives the cross-range resolution of a real beam radar as

$$\Delta CR_{\text{real beam}} = \frac{R\lambda}{D_{az}} \tag{8.3}$$

The resolution indicated by Eq. (8.3) is not acceptable for imaging purposes. Unlike range resolution, the cross-range resolution degrades in proportion to range instead of being constant through the image. Far more important, the cross-range resolution is too coarse for useful images. Consider some typical numbers. An airborne tactical X band radar (10 GHz) with a 1-m antenna width would achieve a cross-range resolution of 300 m at 10 km range. A satellite in low earth orbit (770 km altitude) operating in C band (5 GHz) with a 10-m antenna would exhibit a cross-range resolution of 4.6 km. These numbers are too coarse for useful imagery.

Equation (8.3) suggests that cross-range resolution can be improved by restricting the operating range, using higher frequencies, or using larger antennas. Considering the airborne example, a change of two orders of magnitude is required to improve the resolution from 300-m to the 3-m resolution of Fig. 8.1*a*. This requires a change to either a 1 THz radar frequency, limitation to only 100 m operating range, an increase in the antenna size to 100 m, or some combination of less drastic but still very large changes in these parameters.

Such large changes seem impractical. For instance, Fig. 8.5 compares the approximate relative size of a 100-m phased array antenna and a typical fighter aircraft. It seems unlikely that such a large antenna could be constructed and flown on that aircraft.

8.1.2 The synthetic aperture viewpoint

In fact, Fig. 8.5 does suggest a way in which fine cross-range resolution could be achieved. Rather than constructing a large physical phased array antenna to meet the requirements of Eq. (8.3), consider implementing only a single array element of the antenna, and then utilizing the platform motion to move that element through successive element positions to form the complete array. At each element position, a pulse is transmitted and the resulting fast-time data collected. When the element has traversed the length of the complete array, the data from each position is coherently combined in the signal processor to create the effect of a large phased array antenna with elements at each of the positions. The individual "element" can be the conventional antenna on the platform. In effect, the usual combining of phased array element signals in microwave hardware is performed instead in the signal processor. The system thus "synthesizes" a large phased array antenna aperture by operating a single element from multiple locations in space; hence the name, "synthetic aperture radar" (in some older literature, "synthetic array radar"). This process is suggested by Fig. 8.6, which shows an aircraft that has collected data at four positions along the array; data are still to be collected at several more positions. While good range resolution is obtained via pulse compression, the cross-range resolution of the fast time samples is wide and increases with range. After processing, the data have high resolution in both range and cross-range. Because data from multiple pulses are combined to form the effective high-resolution beam, the scene being imaged should not change while the data are collected, so that each pulse represents data from the same scenario. This again emphasizes that SAR is intended primarily for imaging static scenes.

The radar does not usually collect data over just a single synthetic aperture length sufficient to obtain the desired narrow beamwidth. Instead, it operates continuously during flight, producing an ongoing sequence of fast time data vectors from different positions along the flight path. The effective synthetic aperture size D_{SAR} is determined by selecting the number of spatial positions from which data will be combined to form a narrow effective beam. The amount

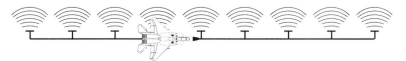

Figure 8.5 Relative size of a fighter aircraft and a phased array antenna large enough to achieve $\Delta CR = 3$ m at X band and 10 km range.

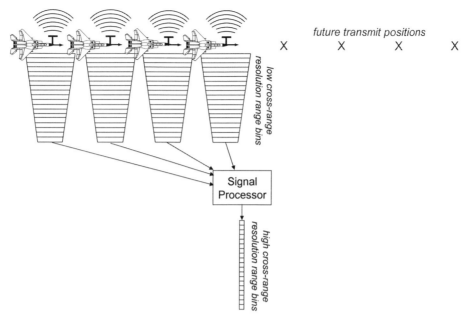

Figure 8.6 The concept of synthetic aperture operation.

of data combined is expressed in terms of the *aperture time* T_a, which is related to the synthetic aperture size by the platform velocity

$$D_{SAR} = vT_a \qquad (8.4)$$

By combining a sliding window of data covering T_a seconds or, equivalently, D_{SAR} meters, a series of narrow effective beams can be formed at successive cross-range positions. For each such position, the signal processor forms a set of range bins with high resolution in both range and cross-range centered at the cross-range position corresponding to the center of the synthetic aperture, as illustrated in Fig. 8.7.[†] This series of range traces thus forms a two-dimensional radar image of the scene.

The mode of operation implied by Fig. 8.7 is called *sidelooking stripmap SAR*. Because the synthetic aperture is formed by the forward motion of the radar platform, the array face is naturally oriented perpendicular to the flight path. Thus, the effective antenna pattern is oriented orthogonal to the velocity vector, a configuration referred to as sidelooking radar. In stripmap SAR, the physical antenna that serves as the array element is not actively scanned; it is instead locked into the sidelooking position, and its antenna pattern therefore moves across the ground as the aircraft flies forward. Conventional phase steering

---

[†]Successive synthetic apertures are usually offset by only one data collection position; thus they overlap. The two data subsets are shown nonoverlapping here for clarity.

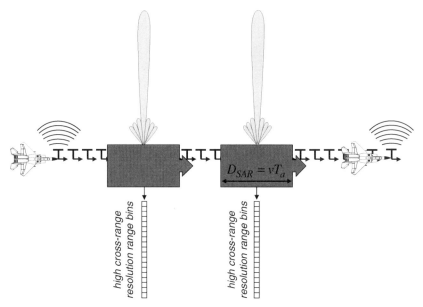

Figure 8.7 Forming multiple synthetic apertures by combining sliding subsets of continuously-collected data.

combined with mechanical steering of the physical antenna can be used to orient the effective SAR beam at an angle to the velocity vector other than 90°; this is referred to as a *squinted SAR*. For simplicity, only the sidelooking case is considered in this chapter.

In a conventional phased array antenna, all elements are present and active at once on both transmit and receive for each pulse; thus, the phase center of the antenna is in the middle of the physical structure for both transmit and receive. For a synthetic phased array, this is not the case: only one element at a time is active, so the phase center for transmit and receive moves across the face of the synthetic aperture as the data are collected. Consequently, the antenna pattern for a synthetic array differs somewhat from that of a physical array. To see this, consider the geometry of Fig. 8.8, which shows a scatterer at a range R and angle θ from the center of an array. Assuming that R is large compared to the array size, the range from the nth element is well approximated as

$$R_n = R - nd \sin\theta \tag{8.5}$$

where d is the element spacing. Suppose the signal $\exp(j\Omega t)$ is transmitted from the nth element. The received echo will be, ignoring scale factors

$$y_n(t) = \exp\left\{j\Omega\left[t - \frac{2}{c}(R - nd \sin\theta)\right]\right\} \tag{8.6}$$

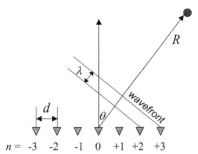

Figure 8.8 Geometry of synthetic array.

In synthetic array operation, each element is operated separately and the outputs combined. Thus, the total output is

$$y(t) = \sum_{n=-M}^{+M} y_n(t) = \sum_{n=-M}^{+M} \exp\left\{j\Omega\left[t - \frac{2}{c}(R - nd\sin\theta)\right]\right\}$$

$$= e^{j\Omega t} e^{-j4\pi R/\lambda} \sum_{n=-M}^{+M} \exp[-j\Omega(R - nd\sin\theta)]$$

$$= e^{j\Omega t} e^{-j4\pi R/\lambda} \left\{\frac{1}{2M+1} \frac{\sin[(2M+1)\Omega d \sin\theta/c]}{\sin(\Omega d \sin\theta/c)}\right\} \quad (8.7)$$

The term in brackets determines the amplitude of the received signal and therefore specifies the two-way voltage pattern of the antenna. The Rayleigh beamwidth is the angle to the first null, which occurs when the argument of the sine function equals π. Defining the total aperture size as $D_{SAR} = (2M+1)d$, this occurs when

$$\sin\theta_{SAR} \approx \theta_{SAR} = \frac{\lambda}{2D_{SAR}} \quad (8.8)$$

In comparison, the array factor of a conventional phased array, given in Eq. (1.14), gives a Rayleigh beamwidth of $\theta = \lambda/D$. Thus, the synthetic array has a beamwidth half that of a conventional array of the same aperture size.

It is now possible to determine the cross-range resolution obtained by the synthetic array. Combining Eqs. (8.4) and (8.8) gives

$$\Delta CR = R\theta_{SAR} = \frac{\lambda R}{2D_{SAR}} = \frac{\lambda R}{2vT_a} \quad (8.9)$$

Equation (8.9) is a fundamental result that relates the amount of data combined, represented through the aperture time T_a, and the SAR cross-range resolution. Solving for T_a gives the design equation

$$T_a = \frac{\lambda R}{2v\Delta CR} \quad (8.10)$$

Equation (8.10) also gives the first hint of complications in SAR signal processing: the aperture time required to obtain a constant cross-range resolution is proportional to range, implying that the required processing is different at different ranges.

Equation (8.10) also suggests that arbitrarily fine resolution can be obtained, at least in principle, by letting T_a become large. However, the maximum practical value of aperture time is limited by the physical antenna on the platform. Consider Fig. 8.9. When the aircraft is at the position on the right, the target is just entering the mainbeam of the physical antenna; when the aircraft is at the position on the left, the target is just exiting the mainbeam. For aircraft positions before or after this interval, the target is not in the physical antenna mainbeam, and the data collected outside of this interval will have no significant contribution from the target. Thus, any one scatterer contributes to the SAR data only over a maximum synthetic aperture size equal to the travel distance between the two points shown, which equals the width of the physical antenna beam at the range of interest, namely $R\theta_{az}$. The corresponding maximum effective aperture time is $R\theta_{az}/v$. Inserting this result into Eq. (8.9) and using Eq. (8.2) (with $k = 1$) gives a lower bound on sidelooking stripmap SAR cross-range resolution of

$$\Delta CR_{\min} = \frac{D_{az}}{2} \qquad (8.11)$$

Equation (8.11) is a remarkable result. It states that the lower bound on stripmap SAR resolution is independent of range, a desirable result for imaging, and more importantly, is much smaller than the real-beam resolution of Eq. (8.3). Considering the same two examples given previously, the lower bound

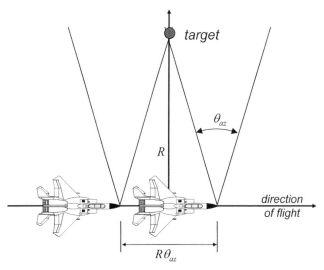

Figure 8.9 Limitation of the aperture time by the physical antenna beamwidth.

on cross-range resolution becomes 0.5 m instead of 300 m for the airborne case, and 5 m instead of 4.6 km for the spaceborne case. (In practice, this resolution will be degraded by 50 to 100 percent by the use of windows for sidelobe control in the processing.) Furthermore, the lower bound on cross-range resolution does not depend on the wavelength. Finally, note that Eq. (8.11) states that to improve the cross-range resolution, the physical antenna size should be reduced! It is rare that improved performance requires reducing the antenna size. In SAR, this occurs because a smaller physical antenna will have a broader beamwidth θ_{az}, allowing a larger maximum synthetic aperture size. Of course, shrinking the antenna size also reduces the SNR drastically and also reduces the maximum area imaging rate. Finally, note that in many cases the lower bound of Eq. (8.11) will be finer than required. In this case, the aperture time is simply limited to whatever value is required to obtain the desired resolution, as given by Eq. (8.10).

Resolutions of 1 to 10 m at ranges of 10 to 50 km, cruise velocities of 100 to 200 m/s, and frequencies from 10 to 35 GHz are typical of many operational airborne SAR radars, and thus might be considered "mainstream" parameters. Figure 8.10a is a representative plot of aperture time versus radar frequency with ΔCR as a parameter using Eq. (8.10). While low-frequency, high resolution SARs can demand very long aperture times, for most systems T_a is in the range of a few tenths of a second to one or two seconds. For spaceborne systems, frequencies are most often in the range of 1 to 5 GHz, with newer systems ranging up to 10 GHz. In *low earth orbit* (LEO), the velocity is about 7500 m/s and the range about 770 km. Figure 8.10b repeats the calculation of T_a for this scenario. Again, "mainstream" aperture times are on the order of a few tenths of a second to one or two seconds.

While the stripmap resolution lower bound of Eq. (8.11) is finer than needed in many cases, in some high resolution applications it may not be good enough, especially when the degradation due to windowing for sidelobe control is considered. The solution is to construct a longer synthetic aperture. To keep a given scatterer within the physical antenna mainbeam, it then becomes necessary to abandon the restriction that the antenna is not scanning. As the radar flies the synthetic aperture, inertial navigation data are used to actively scan the antenna so as to keep its boresight pointed at the center of a *region of interest* (ROI) on the ground. In so doing, the ability to map a continuous strip is sacrificed for a higher resolution map of the ROI. Once enough data have been collected to image the ROI, the antenna can be re-steered to image another discrete region. This mode of operation, termed *spotlight SAR*, is illustrated in Fig. 8.11. Because the radar essentially rotates around the region being imaged, spotlight SAR resolution is often expressed in terms of the rotation angle γ of the radar boresight vector as the platform traverses the synthetic aperture length. Clearly, $D_{SAR} = 2R \sin(\gamma/2) \approx R\gamma$ for small γ. Using this equivalence in Eq. (8.9) gives the alternate expression

$$\Delta CR = \frac{\lambda}{2\gamma} \qquad (8.12)$$

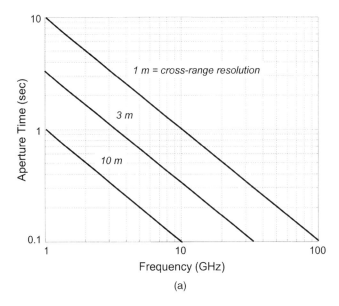

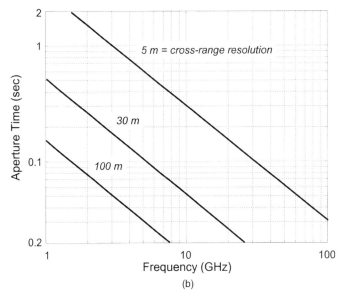

Figure 8.10 Aperture time versus cross-range resolution and radar frequency. (a) Airborne platform, $v = 150$ m/s and $R = 10$ km. (b) Spaceborne platform in low earth orbit, $v = 7500$ m/s and $R = 770$ km.

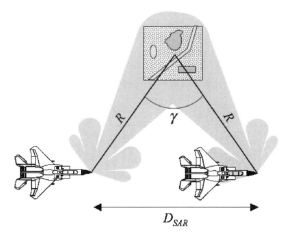

Figure 8.11 Spotlight mode SAR operation.

Figure 8.12 is a 1-m resolution spotlight SAR image of the Pentagon in Washington, DC, clearly showing remarkable details of the five rings of the building, individual trees, and the surrounding road network.

8.1.3 Doppler viewpoint

The fundamental SAR resolution Eq. (8.9) or (8.10) was derived above from a basic antenna theory point of view. It can also be derived from a Doppler processing point of view. In fact, this approach is more consistent with the original conception of SAR and serves as a better starting place for considering SAR image formation algorithms.

Consider two scatterers at range R separated in cross-range by ΔCR meters. The angular separation of the two scatterers is then

$$\Delta\theta = \frac{\Delta CR}{R} \tag{8.13}$$

In Chap. 2 it was seen that two scatterers separated in angle by θ radians around a nominal angle of $\phi = 90°$ (sidelooking) have a difference in Doppler shift of $2v\theta/\lambda$ hertz. Using this result and Eq. (8.13) gives the Doppler difference between two scatterers separated by ΔCR meters

$$\Delta F_D = \frac{2v\Delta CR}{\lambda R} \tag{8.14}$$

These two scatterers could be resolved by Doppler processing of the slow time data, provided that the Doppler resolution is no larger than ΔF_D hertz. It was seen in Chap. 1 that frequency resolution is inversely proportional to signal duration in time, $\Delta F = 1/T$. In this scenario, the signal duration is

Figure 8.12 One meter resolution spotlight SAR image of the Pentagon. (Courtesy of Sandia National Laboratories.)

the aperture time T_a. Thus, two scatterers separated in cross-range by ΔCR meters can be resolved if $T_a = 1/\Delta F_D$ or more, which gives

$$T_a = \frac{\lambda R}{2v\,\Delta CR} \tag{8.15}$$

which is identical to Eq. (8.10).

The Doppler viewpoint suggests a starting point for SAR image formation algorithms. Figure 8.13 illustrates a two-dimensional view[†] of a sidelooking stripmap SAR radar. A scatterer **P** at range R_P is located at a cross-range position x_P relative to the antenna boresight, and is within the physical antenna mainbeam. This scatterer will produce a Doppler-shifted echo. Pulse Doppler processing of the slow time data in the range bin corresponding to

---

[†]This is equivalent to viewing the three-dimensional scenario in the *slant plane* formed by the velocity vector and the vector traced out by the motion of the antenna boresight along the ground plane.

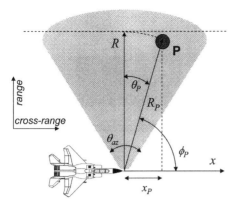

Figure 8.13 Geometry relating cross-range position to Doppler shift.

R_P will produce a peak in the Doppler spectrum at $F_{DP} = (2v/\lambda)\cos(\phi_P) = (2v/\lambda)\sin(\theta_P)$, where the angle ϕ_P of the scatterer from the forward-looking velocity vector has been used in previous chapters, but the angle θ_P is measured from sidelooking, which is more relevant here. The corresponding cross-range position x_P is approximately R_P times the angular offset from sidelooking

$$x_P = R_P \sin^{-1}\left(\frac{\lambda F_{DP}}{2v}\right) \approx \frac{\lambda R_P F_{DP}}{2v} \qquad (8.16)$$

Equation (8.16) provides a mapping of the Doppler axis to cross-range position. Note that the mapping is range-dependent, and therefore different for each range bin. Some systems use only a single mapping based on the nominal range at the center of the imaged swath.

An image is thus formed by collecting a fast-time/slow-time data set that covers the range window of interest and provides an aperture time adequate for the desired resolution. The Doppler spectrum is computed at each range of interest. The Doppler axis is then remapped using Eq. (8.16), resulting in a range/cross-range image. If the same aperture time (slow time data set size) is used in each range bin, this particular SAR imaging algorithm is called *Doppler beam sharpening* (DBS). If the aperture time is increased proportional to range so as to maintain constant cross-range resolution, it becomes a simple form of the *range-Doppler algorithm*. There are several limitations, primarily the problems of *range migration* and *quadratic phase errors*, that limit the scene size and resolution obtainable with this simple algorithm. Before discussing these, it is useful to consider some additional aspects of SAR operation.

8.1.4 SAR coverage and sampling

To determine the amount of terrain which a radar can image, the size of the image in two dimensions must be considered. In stripmap SAR, the along-track extent of the image is unlimited; so long as the radar platform continues collecting and processing data, the image is extended, rather like a scroll. What determines the range extent of the image, often called the *swath length*?

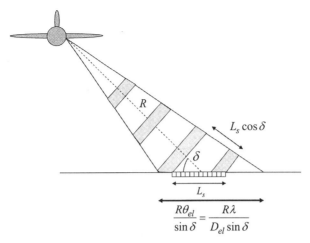

Figure 8.14 Illustration of swath length and unambiguous range.

Consider Fig. 8.14, which illustrates a sidelooking SAR scenario viewed in the along-track direction. The nominal grazing angle is δ radians. Scatterers outside of the mainbeam elevation beamwidth θ_{el} will not produce significant echoes. Thus, the swath length L_s is upper bounded by the projection of the elevation beam onto the ground plane. This is the same geometry considered in discussing beam-limited resolution cell size in Chap. 2; the maximum swath length is therefore

$$L_{s_{max}} = \frac{R\theta_{el}}{\sin\delta} = \frac{R\lambda}{D_{el}\sin\delta} \qquad (8.17)$$

The swath length can be made less than this by simply collecting fast time samples over a more limited extent. In the illustration, the actual swath length $L_s < L_{s_{max}}$ is indicated by the row of range samples within the elevation beam. Since the radar platform moves forward at v meters per second while imaging a swath L_s meters deep, the area coverage rate is

$$A = vL_s \quad \text{m}^2/\text{s} \qquad (8.18)$$

Because imaging radars are attempting to map the ground clutter scene, foldover of clutter due to range ambiguities is to be avoided. It is not necessary that the radar be unambiguous out to the full range R. Rather, the requirement is that echoes from the far edge of the swath on one pulse be received before echoes of the next pulse from the near edge of the swath. The shaded bands within the antenna beam in Fig. 8.14 indicate successive pulses in flight, with a spacing just large enough to meet this requirement. The PRI must evidently satisfy

$$T \geq \frac{2}{c}(L_s\cos\delta) \Rightarrow PRF \leq \frac{c}{2L_s\cos\delta} \qquad (8.19)$$

A second constraint on the PRF is due to the Doppler bandwidth as discussed in Chap. 2. This constraint, for the sidelooking case, is

$$PRF \geq \frac{2v\theta_{az}}{\lambda} = \frac{2v}{D_{az}} \qquad (8.20)$$

where the relation $\theta_{az} = \lambda/D_{az}$ has been used in the last step. It is interesting to note from Eq. (8.20) that, since the platform moves forward at v meters per second while transmitting pulses at a rate of at least $2v/D_{az}$ pulses per second, an equivalent condition is that a pulse be transmitted every $D_{az}/2$ meters or less of forward motion.

Typical Doppler bandwidths $2v/D_{az}$ vary from a few hundreds of hertz to a few kilohertz, while typical range swaths vary from 10 to 50 km. The combination of these constraints results in SAR PRFs of several hundred to a few kilohertz in most cases. Thus, SAR is a low PRF radar mode.

Note that the Doppler bandwidth constraint of Eq. (8.20) also potentially constrains the swath length. Specifically, combining Eqs. (8.19) and (8.20) and solving for L_s gives another bound for the stripmap swath length:

$$L_s \leq \frac{cD_{az}}{4v\cos\delta} \quad \text{m} \qquad (8.21)$$

Using this result in (8.18) gives a mapping rate constraint of

$$A \leq \frac{cD_{az}}{4\cos\delta} \quad \text{m}^2/\text{s} \qquad (8.22)$$

Note that, while good SAR resolution encourages small antenna azimuth sizes (Eq. (8.11)), high mapping coverage rates encourage larger azimuth antenna extents. Finally, if the swath length is limited by the antenna elevation beamwidth, Eqs. (8.17), (8.19), and (8.20) can be combined to derive the inequality

$$D_{az}D_{el} \geq \frac{4vR\lambda}{c\tan\delta} \qquad (8.23)$$

This result specifies a lower bound on the antenna area.

Equations (8.21) and (8.22) are often referred to as the stripmap "SAR *swath constraint* and *mapping rate* equations, while Eq. (8.23) is called the *antenna area constraint*. It is important to realize that Eqs. (8.21) and (8.22) are special cases that apply only when the PRF is set equal to or greater than the Doppler bandwidth, and the PRF becomes the limiting factor for the swath extent. If L_s is limited more strongly by the elevation beamwidth or by other design considerations, more general equations such as Eqs. (8.17) and (8.18) must be used. The area constraint equation applies only when the maximum beam-limited swath length is used. A number of variations and extensions of these ideas for other cases are described by Freeman et al. (2000).

The previous equations apply to stripmap SAR. It is straightforward to estimate the mapping rate for spotlight SAR operation as well. From Eq. (8.12), an image with resolution ΔCR requires that the antenna LOS rotate through

$\lambda/2\Delta CR$ radians. The platform must therefore travel $\lambda R/2\Delta CR$ meters; this will require $\lambda R/2v\Delta CR$ seconds. This is the time required for one spotlight SAR image. The number of images per unit time, N_{spot}, is therefore upper bounded by

$$N_{\text{spot}} \leq \frac{2v\Delta CR}{\lambda R} \quad \text{images/second}$$
$$= \frac{7200v\Delta CR}{\lambda R} \quad \text{images/hour} \qquad (8.24)$$

8.2 Stripmap SAR Data Characteristics

Up to the point where the complex image is finally passed through a detector to convert it to pixel data suitable for display, the SAR signal processing system is nominally linear. (It is *not* shift invariant, as will become clear.) Consequently, the behavior of the SAR system, including such characteristics as resolution, sidelobes, and the response to complex multi-scatterer scenes, can be understood by considering the response to a single isolated point scatterer. In the nomenclature of linear systems analysis, the data set generated by a single point scatterer is called the *point scatterer response* (PSR) or the SAR system *impulse response*. The former term is preferred because the latter term is generally associated with shift invariant systems.

8.2.1 Stripmap SAR Geometry

Figure 8.15 defines the geometry for sidelooking stripmap SAR data acquisition. The *ground plane*, which is the surface to be imaged, is defined by the x and y_g axes. The dimension y_s defines the *slant range* dimension, and the plane defined by the x and y_s axes is called the *slant plane*. Because the radar measures delay, this is the more natural plane in which to discuss imaging. Of course, real terrain varies in height. This aspect will be discussed in Sec. 8.6. For the present, it is assumed that the ground plane scene is two-dimensional.

Consider a scatterer **P** located at coordinates $(x_P, y_{gP}, 0)$ in the ground plane; it is assumed this is within the antenna beam so that a significant echo is received. In the slant plane, the coordinates of **P** are $(x_P, y_{sP}, 0)$. At time t, the phase center of the radar antenna is located at coordinates $(u = vt, 0, h)$. Note that this implies flying at constant altitude and velocity along the $+x$ direction.

The range from the radar to **P** is the Euclidean distance

$$R = R(u) = \sqrt{(u-x_P)^2 + y_{gP}^2 + h^2} = \sqrt{(u-x_P)^2 + R_P^2}$$
$$= R_P\sqrt{1 + \frac{(u-x_P)^2}{R_P^2}} \qquad (8.25)$$

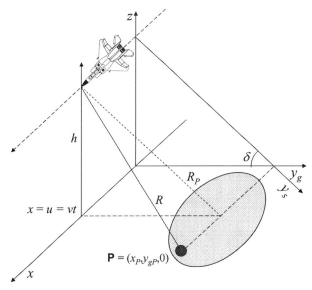

Figure 8.15 Geometry for stripmap SAR data acquisition.

Equation (8.25) shows that in a stripmap SAR, the range between the radar and an arbitrary scatterer varies hyperbolically as the radar moves along the synthetic aperture. The form of the range variation is invariant to the scatterer's along-track position x_P, in the sense that the x_P dependence of $R(u)$ involves only the position of the aircraft relative to the scatterer, $(u - x_P)$. More explicitly, if the range variation for a scatterer at x_{P1} is denoted by $R_1(u)$, the range variation for a scatterer at position $x_{P2} = x_{P1} + \Delta x$ is $R_2(u) = R_1(u - \Delta x)$. In contrast, the range variation does vary with the absolute slant or ground range of the target, R_P or y_{gP} (which appears through R_P).

In the analysis of low- to medium-resolution SAR systems it is common to further simplify Eq. (8.25) by using the series expansion of the square root $\sqrt{1+x} = 1 + \frac{1}{2}x + \frac{1}{8}x^2 + \cdots$ and keeping only the first two terms, giving

$$R(u) \approx \left[1 + \frac{(u - x_P)^2}{2R_P^2}\right] R_P = R_P + \frac{(u - x_P)^2}{2R_P}$$

$$= R_P + \frac{u^2}{2R_P} - u\frac{x_P}{R_P} + \frac{x_P^2}{2R_P} \tag{8.26}$$

This assumption requires $|u - x_P|^2 \ll R_P^2$. Since the maximum value of $|u - x_P|$ of interest is $D_{SAR}/2$, this is equivalent to stating that the synthetic aperture size is short compared to the nominal range. Equation (8.26) shows that the range from the radar to the target varies approximately quadratically as the data set is collected. Since the received phase of the target echo is shifted from the transmitted phase by an amount proportional to range, namely

$\phi = (4\pi/\lambda)R$, if follows that the absolute phase of the received echoes will also vary approximately quadratically. In Chap. 4, it was seen that a quadratic phase function corresponds to linear frequency modulation. LFM is used to obtain high resolution in fast time by spreading the waveform bandwidth. Equation (8.26) shows that the slow-time data from a given scatterer will also have an approximately quadratic phase modulation, which will spread the slow-time bandwidth and enable high cross-range resolution. In this case, however, the modulation is induced by the changing geometry due to platform motion relative to the imaged scene.

It is useful to determine the equivalent of the LFM time-bandwidth product for this cross-range chirp. For a given scatterer position x_P, the received phase will vary with aperture position u as $\phi(u) = (4\pi/\lambda)R(u)$. The instantaneous frequency corresponding to this phase modulation is obtained as usual as

$$K_{u_i} = \frac{d\phi(u)}{du} = \frac{4\pi}{\lambda R_P}(u - x_P) \quad \text{rads/m} \quad (8.27)$$

Note that, since the independent variable u is a spatial coordinate in meters, instead of time in seconds, the frequency is a spatial frequency. As u varies over the synthetic aperture length of D_{SAR} meters, the spatial frequency bandwidth will sweep over $(4\pi/\lambda R_P)D_{SAR}$ radians per meter $= 2D_{SAR}/\lambda R_P$ cycles per meter. Defining the latter quantity as the spatial bandwidth β_u, the "space-bandwidth" product of the cross-range chirp is

$$\beta_u u = \frac{2D_{SAR}^2}{\lambda R_P} \quad (8.28)$$

Just as matched filtered LFM chirps have a time resolution equal to their duration divided by their BT product, it should be possible, by proper processing, to achieve a cross-range resolution in SAR that is the synthetic aperture size divided by the space-bandwidth product. Note that dividing D_{SAR} by Eq. (8.28) does in fact give the result of Eq. (8.9).

The variation in range over the synthetic aperture is called *range migration*. It can be broken into two components, *range walk* ΔR_w and *range curvature* ΔR_c. For an aperture size $D_{SAR} = vT_a$ and sidelooking operation, assume the aperture extends from $u = -D_{SAR}/2$ to $u = +D_{SAR}/2$; that is, it is centered around $u = 0$. Depending on the relative values of x_P and D_{SAR}, the platform may or may not pass by the scatterer during the aperture time. Range walk is the difference between the range to **P** at the beginning and end of the synthetic aperture

$$\begin{aligned}\Delta R_w &= R(-D_{SAR}/2) - R(+D_{SAR}/2) \\ &= -[(-D_{SAR}/2) - (+D_{SAR}/2)]\frac{x_P}{R_P} \\ &= \frac{D_{SAR}}{R_P}x_P = \frac{vT_a}{R_P}x_P \quad (8.29)\end{aligned}$$

Note that this is just the change over the aperture time in the linear term in u of Eq. (8.26). Range curvature is the variation in the quadratic term of $R(u)$, which has its maximum at either extreme of the aperture position, $u = \pm D_{SAR}/2$, and its minimum at $u = 0$

$$\Delta R_c = R(+D_{SAR}/2) - R(0)$$
$$= \frac{D_{SAR}^2}{8R_P} = \frac{v^2 T_a^2}{8R_P} \qquad (8.30)$$

Range walk is greatest for scatterers at the edge of a scene, where x_P is largest. Both the range walk and range curvature depend on T_a and R_P. If a constant aperture time is used, range walk and curvature both decrease as $1/R_P$. However, if constant cross-range resolution is desired, as is often the case, T_a must increase proportionally to range. In this case, ΔR_w is constant over range (for a given x_P), while ΔR_c increases in proportion to R_P.

Range migration is significant only if it exceeds the range bin size. If so, echo samples from a single scatterer will start in different range bins on different pulses, complicating the process of combining them to form a high resolution image. Note that both range walk and range curvature can be computed using only parameters which are known to the radar system: ownship velocity, aperture time, and range. Thus, it is possible for the radar platform data processor to estimate the range migration parameters.

8.2.2 Stripmap SAR data set

The fast-time/slow-time data complex baseband data set used to form the SAR image is called the *phase history*. Figure 8.16 is the real part of the received data from a single point scatterer located at $x_P = 0$, $y_{gP} = 10$ km. The data corresponds to a 5 GHz radar at an altitude of 5 km. The pulse is an LFM upchirp of bandwidth $\beta = 10$ MHz and duration $\tau = 5$ μs, sampled at a fast-time rate of 30 Msamples/second. The PRF is 500 Hz and the platform velocity is 250 m/s. The aperture time is $T_a = 13.5$ s. With these parameters, the expected resolutions are 15 m in range and only 0.1 m in cross-range. While not very realistic, this example results in a data set with exaggerated curvatures for illustrating some general characteristics of stripmap SAR data.

Figure 8.16a shows the complete two-dimensional PSR. The range curvature is clearly evident. The complicated pattern of the data within the point spread response reflects the changing amplitude of the chirp pulse. Each fast-time (ground range) slice through the data set is simply an echo of the transmitted chirp. For instance, the slice of data in the thin rectangular box at azimuth +500 m is shown in part b of the figure; the upchirp is clear. All other fast time slices are identical except for their start time and initial phase offset, determined by the differing ranges to the target from each point along the radar platform's synthetic aperture.

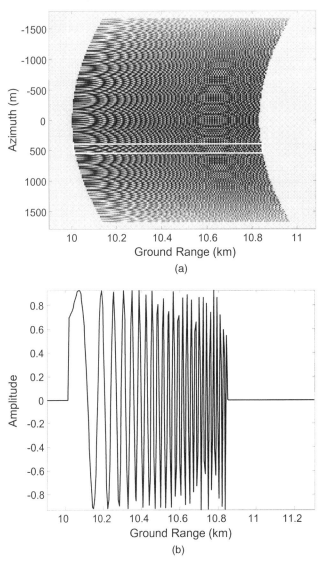

Figure 8.16 (a) Real part of the received data from a point scatterer located at $x_P = 0$, $y_{gP} = 10$ km. (b) Fast-time slice at azimuth position $u = 500$ m. (See text for radar parameters.)

Figure 8.17 shows a cross-range slice of the data in Fig. 8.16a, taken at ground range 10.4 km. Part a shows the complete cross-range slice, emphasizing the attenuation caused by the physical antenna pattern at the edges of the data set, when the point scatterer is near the edge of the antenna mainbeam. The segment inside the rectangular box is expanded in Fig. 8.17b, where it can be seen that the motion-induced phase modulation forms a chirp that passes through zero frequency at $u = x_P$ (broadside from the scatterer). Thus, the PSR is (approximately) a two-dimensional chirp function.

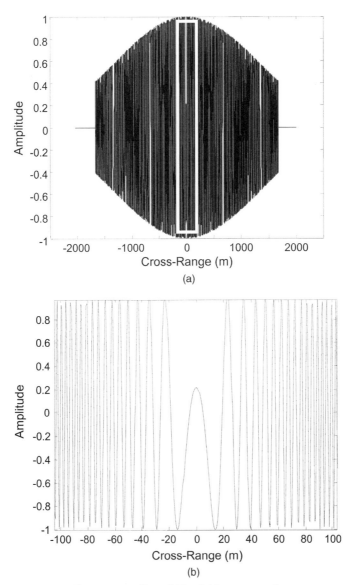

Figure 8.17 Cross range slice of Fig. 8.16a at ground range $y_{gP} = 10.4$ km. (a) Complete slice. (b) Central portion of the slice.

The size of the data set which must be processed by a stripmap SAR can be quite large. For a swath length L_s, range resolution ΔR, and pulse time-bandwidth product $\beta\tau$, the number of fast-time samples required is

$$L = \frac{L_s}{\Delta R} + (\beta\tau - 1) \tag{8.31}$$

This result assumes a fast time sampling rate equal to the bandwidth β; if oversampling is used, L increases accordingly. The number of slow time samples equals the number of pulses within the aperture time

$$M = T_a PRF$$
$$= \left(\frac{\lambda R}{2v \Delta CR}\right)\left(\frac{2v\theta_{az}}{\lambda}\right) = \frac{R\theta_{az}}{\Delta CR} \qquad (8.32)$$

Notice that this is just the number of cross-range resolution cells in the physical beamwidth $R\theta_{az}$. Equation (8.32) assumes that the PRF equals the clutter bandwidth. Again, if oversampling is used, M increases accordingly.

As an example, consider the SIR-C, which operates at L band with a swath width of 15 km from an orbital range of 250 km, range and cross-range resolution of 15 m, and a 17 μs pulse with a 10 MHz bandwidth. The antenna size is 10 m, suggesting a beamwidth θ_{az} of 0.03 rad = 1.72°. These parameters result in $L = 1170$ fast time samples per pulse, and $M = 500$ pulses in the aperture time and thus contributing to each image pixel. This large number of pulses contributing to the image of each point target differentiates SAR from the much simpler, but related, pulse Doppler processing. In pulse Doppler, a CPI is typically a few tens of pulses at most, corresponding to much shorter aperture times. Returning to the SIR-C example, $(1170)(500) = 585{,}000$ samples contribute to each point scatterer image. If properly coherently integrated, a 57.7-dB integration gain over noise can be achieved.

8.3 Stripmap SAR Image Formation Algorithms

For a general scene, the stripmap SAR data set is the superposition of a large number of weighted and shifted replicas of the PSR. Recall that the PSR is, in general, a function of range but is independent of cross-range. The goal of any SAR image formation algorithm is to compress the PSR of each scatterer into an impulse-like function with the appropriate location and amplitude, as shown notionally in Fig. 8.18.

A variety of stripmap SAR image formation algorithms exist. They vary in resolution capability, scene size capability, and computational complexity. In general, finer resolution, longer standoff range, lower radar frequency, squint geometries, and greater scene size require more elaborate and computationally expensive algorithms (Bamler, 1992; Gough and Hawkins, 1997). In this section, two basic imaging algorithms are introduced, Doppler beam sharpening and the range-Doppler algorithm. More advanced techniques such as the *range migration algorithm* (also called the $\omega - k$ algorithm) and the *chirp scaling algorithm* are described in detail by Bamler (1992), Gough and Hawkins (1997), Francaschetti and Lanari (1999), Soumekh (1999), and Cumming and Wong (2005).

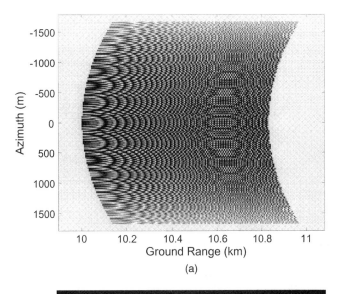

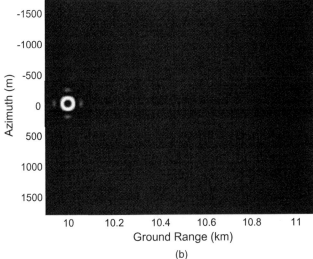

Figure 8.18 The goal of SAR image formation algorithms is to transform the two-dimensional PSR in (a) into the point target image in (b).

8.3.1 Doppler beam sharpening

Doppler beam sharpening is the original form of SAR envisioned by Wiley (1965). It uses a constant aperture time for all ranges, so that the cross-range resolution is proportional to range. It is the simplest SAR algorithm, and is suitable only for relatively coarse resolution imagery; nonetheless, it provides a substantial improvement over real beam cross-range resolution and has relatively low computational requirements.

DBS presumes that the PSR is compressed in the fast time dimension by conventional pulse compression, whether implemented by direct convolution, fast convolution, or stretch processing. Ignoring fast-time sidelobes, the fast-time output of the matched filter for a pulse transmitted from aperture position u can then be approximated as $A\exp(j\phi)\delta_D(t - 2R(u)/c)$, where $\delta_D(\cdot)$ is the continuous-time Dirac impulse function, the phase ϕ due to the round trip travel is $-(4\pi/\lambda)R(u)$, and the constant A absorbs all amplitude factors. Using Eq. (8.26), the slow time phase variation is therefore

$$\phi(u) \approx -\left(\frac{4\pi}{\lambda}\right)\left(R + \frac{u^2}{2R} - u\frac{x}{R} + \frac{x^2}{2R}\right)$$

$$\approx -\left(\frac{4\pi}{\lambda}\right)\left(R - u\frac{x}{R} + \frac{x^2}{2R}\right) \tag{8.33}$$

where the last step assumes that the quadratic (in u) phase term can be neglected; this assumption will be revisited in Sec. 8.3.2. The subscript P has been dropped on the coordinates x and R for simplicity. The instantaneous cross-range spatial frequency K_u is

$$K_u = \frac{d\phi(u)}{du} \approx \frac{4\pi}{\lambda}\frac{x}{R} \quad \text{rads/m} \tag{8.34}$$

so that

$$x = \frac{\lambda R}{4\pi}K_u \tag{8.35}$$

Equation (8.35) relates scatterer cross-range position to spatial frequency. Alternatively, x can be expressed in terms of Doppler (temporal) frequency. Using $u = vt$ in Eq. (8.33), the Doppler frequency F_D due to a scatterer at (x, R) is $(1/2\pi)(d\phi/dt)$, so that

$$x = \frac{\lambda R}{2v}F_D \tag{8.36}$$

a relation that is more natural to the DBS name. This is the same as Eq. (8.16) presented previously.

The algorithm implied by Eq. (8.36) is diagrammed in Fig. 8.19. It requires only one *fast Fourier transform* (FFT) per range bin, followed by a rescaling of the axes to express the result in range/cross-range coordinates. Typically, windowing or other sidelobe suppression techniques would be included in the processing in both dimensions.

Figure 8.20 is the result of applying a simple DBS algorithm to simulated data from a test array. The magnitude of the DBS image in dB is shown. Point scatterers were simulated at all combinations of x and R equal to -1000, 0, or $+1000$ m relative to the center of a scene [the *central reference point* (CRP)] at a nominal range of 20 km. The radar is therefore located above the image.

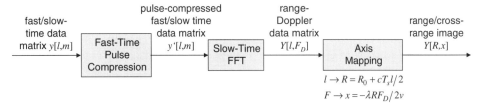

Figure 8.19 Block diagram of Doppler beam sharpening algorithm.

The image resolution is $\Delta R = \Delta CR = 50$ m. The aperture time was 40 ms. The radar frequency was 10 GHz, and an LFM pulse with $\beta = 3$ MHz and $\tau = 5$ μs ($\beta\tau = 15$) was used at a PRF of 7.5 kHz. The Doppler frequency axis was scaled to cross range position using Eq. (8.36) with R set equal to the nominal range $R_{CRP} = 20$ km. Each of the scatterers is well-focused, with clearly defined range and cross-range sidelobes. Scatterers off the center line of $x = 0$, however, are imaged at slightly further away from the radar than those on the center line because of the contribution of x to the total range. In addition, the columns of scatterers bow in slightly as range increases. This is because the spectrum in each range bin was remapped from Doppler frequency to cross-range position using a scale factor based on the same central range of 20 km.

These geometric distortions are readily corrected after the basic image formation. The cross-range displacement is corrected by simply updating the cross-range scale for display of each row with the appropriate range R for each range

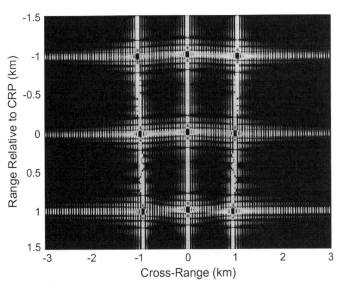

Figure 8.20 Magnitude in dB of the DBS image of a test array. (See text for parameters.)

bin, instead of using R_{CRP} for all of the range bins. The result of this operation is shown in Fig. 8.21a. The scatterers are now aligned vertically at the correct cross-range coordinates of -1, 0, or $+1$ km. However, the range shift for scatterers off the center line is still present. This can be corrected by computing in each cross range bin an actual range $R' = \sqrt{R^2 + x^2}$, and then shifting that cross-range column by the difference of $R' - R$ meters. Interpolation will be required to accommodate partial-range bin shifts. Figure 8.21b shows the result of this operation; the nine scatterers are now accurately centered on correct coordinates. The curvature of the cross-range sidelobes nicely reflects the range curvature that was present in the original DBS image and has now been corrected.

The DBS algorithm assumes that the echo samples from a given scatterer remain in the same range bin over the aperture time, so that the FFT of the slow time data will integrate all of the scatterer samples. This can be assured by requiring that the range walk ΔR_w not exceed some fraction (typically one-half) of a range bin over the aperture. Using Eq. (8.29), this condition becomes

$$\Delta R_w = \frac{D_{SAR}}{R}x = \frac{vT_a}{R}x \leq \frac{\Delta R}{2} \tag{8.37}$$

Substituting from Eq. (8.15) for T_a gives the constraint

$$x \leq \frac{\Delta R \cdot \Delta CR}{\lambda} = \frac{\Delta CR^2}{\lambda} \tag{8.38}$$

where the last step assumes "square pixels," that is, $\Delta R = \Delta CR$. Equation (8.38) is a constraint on the maximum value of x and thus the maximum scene width in Doppler beam sharpening. The total scene width of $2x$ is plotted in Fig. 8.24 in the next section.

The resolution performance of DBS is sometimes quantified in a Doppler *beam sharpening ratio* (BSR), defined as the ratio of the real-beam cross-range resolution to the DBS cross-range resolution. Using Eqs. (8.3) and (8.9), the BSR is

$$\begin{aligned} BSR &= \frac{\Delta CR_{\text{real beam}}}{\Delta CR_{\text{DBS}}} = \frac{(R\lambda/D_{az})}{(\lambda R/2vT_a)} \\ &= \frac{2vT_a}{D_{az}} \\ &= 2\frac{D_{SAR}}{D_{az}} \end{aligned} \tag{8.39}$$

where the last step used the relation $D_{SAR} = vT_a$. Thus, the BSR is just twice the ratio of the synthetic and real aperture sizes. The factor of two reflects the difference in synthetic and real array patterns, discussed in Sec. 8.1.2.

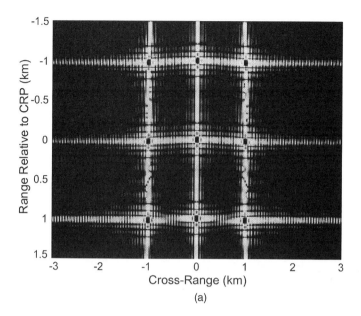

(a)

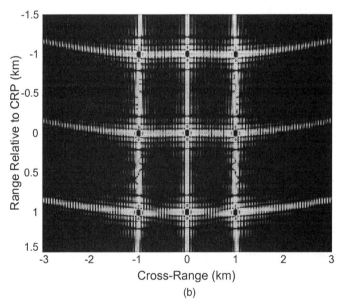

(b)

Figure 8.21 Correction of DBS geometric distortion. (*a*) Rescaling of cross-range for range variation. (*b*) Shifting of range for cross-range displacement.

8.3.2 Quadratic phase error effects

The term $u^2/2R$ was neglected in simplifying Eq. (8.33). Inclusion of this term adds a quadratic phase component to the slow time signal. Because the DBS algorithm associates a frequency component with a cross-range coordinate, the instantaneous frequency variation implied by the quadratic phase will tend to smear the cross-range response of the processor. The ideal response of the DBS processor to a point scatterer is simply the discrete time Fourier transform of a sinusoid. Figure 8.22a shows the DTFT of the signal $y[m] = \exp(j\phi_{\max}m^2/M^2)$, $-M \leq m \leq +M$, where $\phi_{\max}$ is the peak phase error. This sequence can be viewed as the product of a phase error sequence and an ideal sinusoid of frequency zero, that is, a vector of ones. Increasing the quadratic phase error attenuates and spreads the main lobe; this represents a loss of brightness and resolution, respectively, in the DBS image of a scatterer (Richards, 1993). Note that, as the maximum phase error approaches π radians, the well-defined main lobe/side lobe structure of the response is breaking down. Figure 8.22b repeats the experiment, but with a Hamming window applied to the data before the DTFT. The Hamming window greatly moderates the effects of the phase error. The gain and resolution *variations* are reduced, and the general shape of the response is better maintained. This increased robustness of response is yet another benefit of windowing data.

Figure 8.23 plots the loss in peak amplitude and the increase in 3-dB main lobe width of the DTFT as the peak quadratic phase error increases. The moderating effect of the Hamming window is again evident. These plots are for a signal length of 101 samples. The degree of degradation is a mild function of sequence length. For shorter signals, the degradations are greater; for longer signals, they are somewhat less. For the example shown, limiting the maximum quadratic phase error to $\phi_{\max} \leq \pi/2$ limits the loss in amplitude to 1 dB and the resolution increase to 7 percent, even without a window. This suggests that uncompensated quadratic phase errors can be ignored so long as they do not exceed $\pi/2$ radians over the aperture time. Tighter limits can be adopted if higher quality standards are desired.

The assumption that the u^2/R term in Eq. (8.33) can be neglected can now be revisited. This term contributes a quadratic phase term to $\phi(u)$. The maximum value of u of interest is $D_{SAR}/2 = vT_a/2$. Requiring that this term be limited to $\pi/2$ gives

$$\frac{4\pi}{\lambda}\frac{v^2 T_a^2}{4R} \leq \frac{\pi}{2} \tag{8.40}$$

Using $T_a = \lambda R/2v\Delta CR$ gives the constraint

$$\Delta CR \geq \frac{1}{2}\sqrt{\lambda R} \tag{8.41}$$

Thus, the quadratic phase term can be ignored in DBS processing so long as this constraint is observed. Figure 8.24 plots the constraint of Eq. (8.41) for four

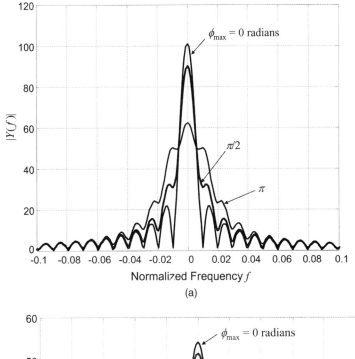

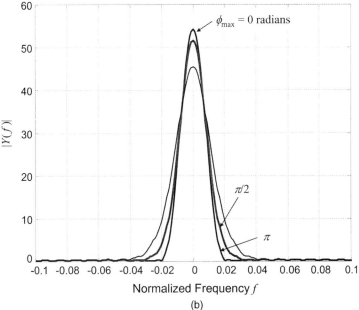

Figure 8.22 The effect of quadratic phase error on the DTFT of a 101-point sinusoid. (*a*) No window. (*b*) Hamming window.

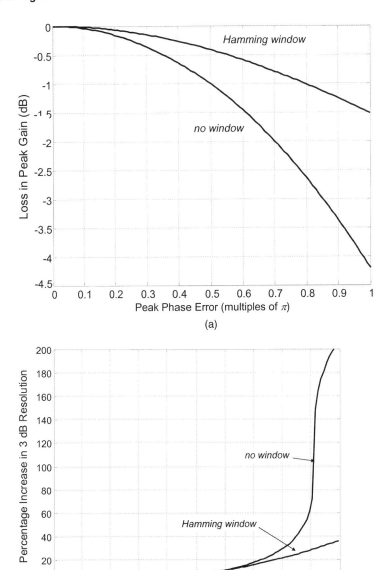

Figure 8.23 The effect of quadratic phase error on the DTFT of a 101-point sinusoid. (*a*) Reduction in peak amplitude. (*b*) Percentage increase in 3-dB resolution.

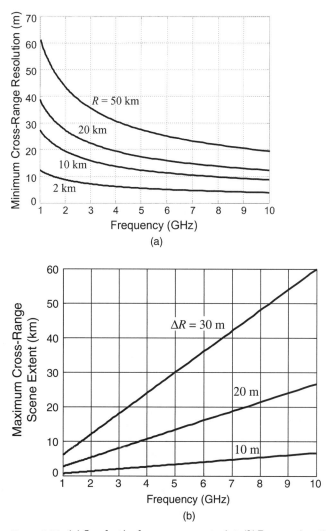

Figure 8.24 (a) Quadratic phase error constraint. (b) Range migration constraint.

values of range appropriate to airborne radars, as well as the DBS scene size constraint of Eq. (8.38). Taken together, these two constraints show that while DBS processing can achieve resolutions much better than real beam systems, it is most appropriate for imaging relatively large scenes at moderate-to-low resolution, and is best used at higher radar frequencies.

Figure 8.25a shows a portion of the DBS image for the same target array, but imaged with a resolution goal of 10 m. The standoff range is now 50 km, while the radar frequency remains 10 GHz. This scenario meets the range

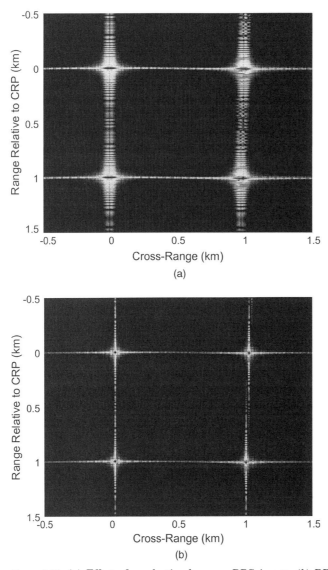

Figure 8.25 (a) Effect of quadratic phase on DBS image. (b) DBS image with azimuth dechirp.

migration constraint of Eq. (8.38) (Fig. 8.24b), but does not meet the quadratic phase error constraint of Eq. (8.41) (Fig. 8.24a). The slow time quadratic phase term represents a cross-range spatial frequency chirp. Because spatial frequency maps to cross-range position, the scatterer response is smeared in cross range. (This is the same effect illustrated in Fig. 8.22.) Range resolution is not affected.

This smearing can be corrected by compensating the data to remove the quadratic phase term. The required correction is implemented as a range-bin-dependent phase multiplication in each slow time row of the raw data $y[l, m]$

$$\begin{aligned} y'[l, m] &= y[l, m] \exp\left[j\left(\frac{4\pi}{\lambda}\right)\left(\frac{u^2}{2R}\right)\right] \\ &= y[l, m] \exp\left\{j\left(\frac{4\pi}{\lambda}\right) \frac{\left[v\left(m - \frac{M-1}{2}\right)T\right]^2}{2\left(R_0 + \frac{c(l-1)T_f}{2}\right)}\right\} \end{aligned} \quad (8.42)$$

where T_f = fast time sampling interval
T = slow time sampling interval (pulse repetition interval)
R_0 = range corresponding to the first range bin
M = number of slow-time samples

Figure 8.25b shows the effect of this compensation, called *azimuth dechirp*, on the same data. The full cross-range resolution is restored.

Because of the short aperture times required for its modest resolution, DBS is often used in an actively scanning mode, unlike higher resolution radar imaging modes (Stimson, 1998). As long as the region of interest stays within the area illuminated by the mainbeam over the aperture time, the scanning is of little consequence. In keeping with the scanning operation, DBS is often used in squint mode rather than in a sidelooking configuration. In this event, range walk becomes large and must be compensated. Additional enhancements to DBS can also be implemented. For instance, *range migration correction* can be implemented to compensate for fractional-bin range curvature, and *secondary range compression* can improve focusing of targets at large cross-range displacements from the line of sight. These extensions are discussed in the book by Schleher (1991).

8.3.3 Range-Doppler algorithms

As resolution becomes finer, standoff range shorter, or radar frequency lower, both range curvature and azimuth quadratic phase tend to become more pronounced. For example, Figure 8.26 illustrates the amount of range curvature as a function of cross-range resolution for an L band (1 GHz) radar at a standoff range $R_0 = 50$ km and velocity $v = 150$ m/s. If it is assumed that the range resolution is set approximately equal to the cross-range resolution (to obtain "square pixels"), the range curvature is completely insignificant when $\Delta CR = 50$ m, and is still only about 1/5th of a range bin when $\Delta CR = 10$ m. Augmented DBS algorithms are effective for these cases. However, when ΔCR is reduced to 3 m, the curvature rises to about five range bins, assuming $\Delta R = 3$ m also. DBS algorithms are unsuitable in this situation. To achieve well-focused higher-resolution imagery, it is necessary to develop an algorithm that can address significant range migration and quadratic phase. The *range-Doppler* (RD)

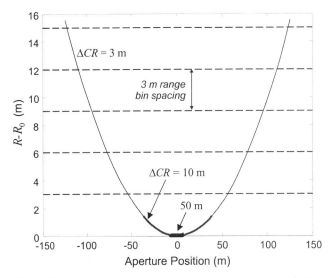

Figure 8.26 Increase in range curvature as cross-range resolution becomes finer.

algorithm is a family of algorithms in widespread use that offer this capability while maintaining separable range and cross-range processing.

The RD algorithm assumes that pulse compression is performed first. For a single point scatterer, the output of the fast-time matched filter for the pulse transmitted from aperture position u is a narrow peak at time delay (assuming that the filter delay is removed) $2R(u)$, surrounded by sidelobes. This can be approximated by an impulse at that range, $\delta_D[t - 2R(u)]$. The responses from the series of pulses follow the range migration curve of Eq. (8.25). Thus, after pulse compression, the response of the stripmap SAR system to a scatterer at (x, R) is (ignoring amplitude factors)

$$h(u, t; R) = \delta_D(t - R(u)) = \delta_D\left(t - \frac{2}{c}\sqrt{(u-x)^2 + R^2}\right) \quad (8.43)$$

The response is parameterized by R but not x because its shape is invariant to x, but does vary with R. To derive the range-Doppler algorithm, begin by expressing R in terms of differential range ΔR around a nominal standoff range R_0, similar to the analysis of stretch processing in Chap. 4

$$\begin{aligned}
R(u) &= \sqrt{(u-x)^2 + R^2} = \sqrt{(u-x)^2 + (R_0 + \delta R)^2} \\
&= \sqrt{(u-x)^2 + R_0^2 + 2R_0\delta R + \delta R^2} \\
&= \sqrt{(u-x)^2 + R_0^2 + 2R_0\delta R}
\end{aligned} \quad (8.44)$$

where the last step assumes $|\delta R| \ll R_0$, that is, the swath length is small compared to the standoff range. Define $R_x^2 = (u-x)^2 + R_0^2$; then

$$R(u) = R_x \sqrt{1 + \frac{2R_0 \delta R}{R_x^2}}$$

$$\approx R_x \left(1 + \frac{R_0 \delta R}{R_x^2}\right) = R_x + \frac{R_0 \delta R}{R_x}$$

$$= R_x + \frac{R_0 \delta R}{\sqrt{(u-x)^2 + R_0^2}}$$

$$\approx R_x + \delta R = \sqrt{(u-x)^2 + R_0^2} + \delta R \qquad (8.45)$$

where the last line assumes that $|u - x| \ll R_0$ (cross-range extent of the scene is small compared to the standoff range). Using Eq. (8.45) in Eq. (8.43) gives the form of the PSR used in the RD algorithm

$$h(u, t; R_0) = \delta_D\left[\left(t - \frac{2}{c}\delta R\right) - \frac{2}{c}\sqrt{(u-x)^2 + R_0^2}\right] \qquad (8.46)$$

Unlike the general PSR, this form is invariant to both x and R. While there is a range dependence in the form of R_0, the RD algorithm uses a single fixed value of R_0.

The image could be formed by performing a two-dimensional convolution (matched filtering) of $h(u, t; R_0)$ with the data set $y(u, t)$. However, the processing is usually done in the frequency domain by computing the two-dimensional Fourier transforms of $h(u, t; R_0)$ and $y(u, t)$, $H(K_u, \Omega; R_0)$ and $Y(K_u, \Omega)$, and forming the image as

$$f(u, t; R_0) = \mathbf{F}^{-1}\{H^*(K_u, \Omega; R_0) Y(K_u, \Omega)\}$$
$$f(x, \delta R; R_0) = f(u, ct/2; R_0) \qquad (8.47)$$

To proceed, an expression is needed for the RD SAR system transfer function $H(K_u, \Omega; R_0)$, the two-dimensional Fourier transform of $h(u, t; R_0)$

$$H(K_u, \Omega; R_0) = \int_{-\infty}^{+\infty} \left(\int_{-\infty}^{+\infty} h(u, t; R_0) e^{-j\Omega t} dt\right) e^{-jK_u u} du \qquad (8.48)$$

Note that K_u is a spatial frequency in radians per meter and Ω is a temporal frequency in radians per second. Using Eq. (8.46) and performing the Fourier transform over the fast time variable t gives the intermediate result

$$\mathcal{H}(K_u, t; R_0) = \int_{-\infty}^{+\infty} \exp\left\{-j\left[K_u u + \frac{4\pi}{\lambda}\left(\delta R - \sqrt{(u-x)^2 + R_0^2}\right)\right]\right\} du$$
$$(8.49)$$

The slow-time transform can be completed using the principle of stationary phase (Bamler, 1992; Raney, 1992); the result is

$$H(K_u, \Omega; R_0) = \sqrt{\frac{\pi c R_0}{j\Omega}} \exp\left\{+jR_0\left[\frac{2\Omega}{c} - \sqrt{\left(\frac{2\Omega}{c}\right)^2 - K_u^2}\right]\right\}$$

$$\approx A \exp\left\{+jR_0\left[\frac{2\Omega}{c} - \sqrt{\left(\frac{2\Omega}{c}\right)^2 - K_u^2}\right]\right\} \quad (8.50)$$

The last step recognizes that most SAR systems, despite their high resolution, are relatively narrowband. The fast-time frequency Ω varies over a limited range (usually less than 10 percent), so the amplitude factor is relatively constant.

The range-Doppler algorithm is diagrammed in Fig. 8.27. The major advantages of the RD algorithm are its ability to process the entire range swath (provided it is small compared to the standoff range), compensating for both range migration and quadratic phase, while utilizing the computational efficiency of the two-dimensional FFT. Since this version operates in the spatial frequency/temporal frequency domain, however, it is not clear why the technique is called the *range*-Doppler algorithm. In fact, the name range-Doppler derives from a version of the algorithm obtained by additional simplifications of Eq. (8.50).

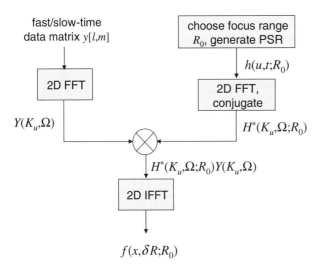

Figure 8.27 Block diagram of range-Doppler algorithm.

Using a binomial expansion and assuming $2\Omega/c \gg K_u$, the argument of the complex exponential in H can be simplified as follows

$$R_0\left(\frac{2\Omega}{c} - \sqrt{\left(\frac{2\Omega}{c}\right)^2 - K_u^2}\right) = R_0\left(\frac{2\Omega}{c} - \frac{2\Omega}{c}\sqrt{1 - \frac{K_u^2}{(2\Omega/c)^2}}\right)$$

$$\approx R_0\left[\frac{2\Omega}{c} - \frac{2\Omega}{c}\left(1 - \frac{K_u^2}{2(2\Omega/c)^2}\right)\right] = \frac{cK_u^2 R_0}{4\Omega} \quad (8.51)$$

Most radar systems, even high-resolution SAR systems, are narrowband in the sense that the waveform bandwidth $\Delta\Omega$ is much less than the radar frequency Ω_0. Expanding Ω about Ω_0 further simplifies Eq. (8.51) to

$$\frac{cK_u^2 R_0}{4\Omega} = \frac{cK_u^2 R_0}{4(\Omega_0 + \Delta\Omega)} = \frac{cK_u^2 R_0}{4\Omega_0(1 + \Delta\Omega/\Omega_0)}$$

$$\approx \frac{cK_u^2 R_0}{4\Omega_0}\left(1 - \frac{\Delta\Omega}{\Omega_0}\right) = \frac{cK_u^2 R_0}{4\Omega_0} - \frac{cK_u^2 R_0}{4\Omega_0^2}\Delta\Omega \quad (8.52)$$

In the second line, the term $(1 + \Delta\Omega/\Omega_0)^{-1}$ was expanded in a binomial series and only the first two terms retained. Finally, the simplified transfer function is

$$H(K_u, \Delta\Omega; R_0, \Omega_0) \approx A\exp\left(j\frac{cK_u^2 R_0}{4\Omega_0}\right)\exp\left(-j\frac{cK_u^2 R_0}{4\Omega_0^2}\Delta\Omega\right) \quad (8.53)$$

The approximate transfer function of Eq. (8.53) implies two distinct operations for performing the RD cross-range compression. For a given spatial frequency K_u, the second term is a linear phase in $\Delta\Omega$, corresponding to a shift in the fast time dimension of $cK_u^2 R_0/4\Omega_0^2$ seconds. This shift in the data "straightens out" the range curvature in the PSR. The other phase term compensates the slow-time quadratic phase modulation of the data, essentially performing an LFM pulse compression in slow time. Figure 8.28 is a purely notional illustration of the effect of each of these steps on the data, beginning with the raw fast time/slow time data matrix in step 1 and continuing to the compressed point spread response in step 4.

The fast time interpolation can be implemented as a frequency domain multiplication, or as an explicit shift and interpolation in the fast time domain. In the latter case, the data are operated on in the time domain in one dimension (fast time, or range) and the frequency domain in the other (spatial frequency), hence the name "range-Doppler" algorithm.

Figure 8.29 illustrates the performance of the range-Doppler algorithm on two simulated scenarios, using the full RD point spread response of Eq. (8.50). In Fig. 8.29a, the RD algorithm is applied to a collection of point scatterers

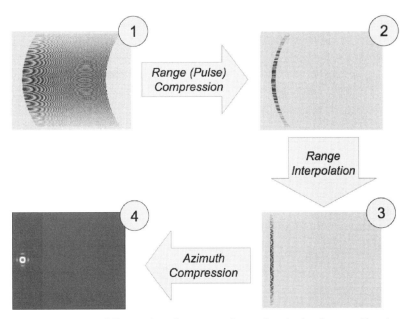

Figure 8.28 Notional illustration of sequence of operations in classic range-Doppler algorithm.

in a $50 \times 50 \text{ m}^2$ area, sensed with an X band (9.5 GHz) radar from a standoff range of 7.5 km. While the range and cross-range resolutions are only 0.5 m, this example is relatively unchallenging due to the high RF frequency and long standoff range. All five scatterers are well-focused, with no visible degradations in the scene. Figure 8.29b is a more challenging case. Although the resolution is relaxed to 1.0 m in both dimensions, the image area is larger (100 m by 100 m) and, most importantly, the RF frequency is lowered to L band (1.5 GHz). Both of these factors increase the variability of the range curvature as a function of range. Although differences in side lobe structure are evident, the scatterer at the center reference range R_0 remains well focused. However, the cross-range resolution of the other scatterers degrades as they are located further away from the scene center.

8.3.4 Depth of focus

The exact stripmap PSR of Eq. (8.43) is range-dependent. The range-Doppler algorithm linearizes the PSR about a particular range R_0. However, if applied over a deep enough swath, the variation in the PSR with R will become significant, and failure to address this variation will result in increasingly poor focusing at increasing ranges from R_0. Advanced algorithms such as the *range-migration algorithm* (Bamler, 1992; Gough and Hawkins, 1997) fully account for the variation in the PSR with range, but are not discussed in this introduction. Another approach is to break the desired swath into N subswaths centered on ranges $R_{01}, \ldots, R_{0N}$. The PSR is updated and the RD algorithm applied

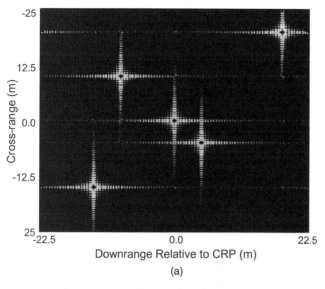

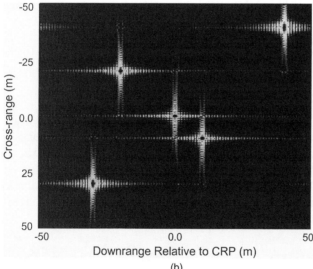

Figure 8.29 Magnitude in dB of the range-Doppler image of a test array. (See text for parameters.) (a) X band, long standoff range, small scene size. (b) L band, short standoff range, larger scene size. (Images courtesy of Dr. Gregory A. Showman, GTRI.)

independently in each subswath. The resulting images are mosaiced together to form the complete image. The subswath extent over which the RD algorithm can be applied without significant defocusing is called the *depth of focus*.

The depth of focus is determined by finding the change in range R_0 which will cause a specified change in range curvature over the aperture time.

The "tolerable" variation is often taken as $\lambda/8$, corresponding to a two-way variation of $\lambda/4$ and thus a phase change of $\pi/2$ radians. Equation (8.30) gave a formula for range curvature. Differentiating with respect to range gives the change in curvature as nominal range changes; setting that quantity equal to $\lambda/4$ gives the depth of focus as (Carrara et al., 1995)

$$DOF = \frac{\lambda R_0^2}{D_{SAR}^2} = \frac{\lambda R_0^2}{(vT_a)^2} = \frac{4(\Delta CR)^2}{\lambda} \tag{8.54}$$

Consider SAR imaging of a scene with a swath depth of 10 km. If imaged using an X band (10 GHz) radar with a cross-range resolution of 3 m, the depth of focus is 1.2 km. To use the RD algorithm, the swath should be broken into at least nine and possibly more subswaths, with the PSR updated for each. If the scene was imaged with an L band (1 GHz) radar at 1 m resolution, the depth of focus is only 13 m, requiring at least 750 subswaths to use the RD algorithm. The large number of subswaths indicates that the RD algorithm is poorly suited to this scenario, and a more advanced algorithm such as range migration should be used instead. As another example, the scenario of Fig. 8.29a has a DOF of 31.67 m, roughly equal to the approximately 37 m range spread of the scatterers, while Fig. 8.29b has a DOF of 20 m, well under the range spread of approximately 37 m in that scenario.

8.4 Spotlight SAR Data Characteristics

The spotlight SAR scenario was illustrated in Fig. 8.11. Many spotlight SAR systems utilize a linear FM waveform with stretch processing. Assume an LFM pulse that sweeps from $F_0 - \beta/2$ to $F_0 + \beta/2$ hertz in the time interval from $-\tau/2$ to $+\tau/2$ seconds. Adapting to the notation of this chapter, it was shown in Chap. 4 (Eq. 4.108) that on a pulse taken from aperture position u, an ensemble of scatterers at range $\delta R_i = c\delta t_i$ relative to a central range R_0 results in a signal component at the stretch mixer output

$$y(t) = \sum_i \tilde{\rho}_i \exp\left(-j2\pi \frac{\beta}{\tau}\delta t_i (t - t_0)\right) \exp\left(j\pi \frac{\beta}{\tau}(\delta t_i)^2\right) \tag{8.55}$$

where $t_0 = 2R_0/c$ and $\tilde{\rho}_i$ is the complex reflectivity of the echo from the ith scatterer,[†] located at range $R_0 + \delta t_i$. Extending this formula to a continuum of scatterers gives

$$y(t) = w(t) \int_{-\infty}^{+\infty} \tilde{\rho}(\delta t) \exp\left(-j2\pi \frac{\beta}{\tau}\delta t (t - t_0)\right) \exp\left(j\pi \frac{\beta}{\tau}(\delta t)^2\right) d(\delta t) \tag{8.56}$$

[†] $\tilde{\rho}_i$ includes range weighting and other range equation factors; see Sec. 2.7.

In this equation, $w(t)$ is a rectangular window function that limits the data extent to that of the reference chirp in the stretch mixer. Specifically

$$w(t) = \begin{cases} 1 & t_0 - \dfrac{L_s}{c} - \dfrac{\tau}{2} \leq t \leq t_0 + \dfrac{L_s}{c} + \dfrac{\tau}{2} \\ 0 & \text{otherwise} \end{cases} \quad (8.57)$$

Consider the Fourier transform of $\tilde{p}(\delta t)$

$$\tilde{P}(\Omega) = \int_{-\infty}^{+\infty} \tilde{p}(\delta t) \exp(-j\Omega\, \delta t)\, d(\delta t) \quad (8.58)$$

Comparing Eq. (8.58) to Eq. (8.56) shows that

$$y(t) = w(t)\tilde{P}\left(\Omega_0 + \frac{2\pi\beta}{\tau}(t - t_0)\right)$$

$$= \tilde{P}(\Omega) \qquad \Omega \in \Omega_0 \pm 2\pi\left(1 + \frac{L_s}{(c\tau/2)}\right)\frac{\beta}{2}$$

$$= \tilde{P}(F) \qquad F \in F_0 \pm \left(1 + \frac{L_s}{(c\tau/2)}\right)\frac{\beta}{2} \quad (8.59)$$

provided that the *residual video phase* (RVP) term $\exp[j\pi\beta/(\delta t)^2\tau]$ can be ignored; this condition will be reconsidered later. Assume that $L_s \ll c\tau/2$, that is, the swath length is small compared to the pulse duration. Then

$$y(t) = \tilde{P}(F) \qquad F \in F_0 \pm \frac{\beta}{2} \quad (8.60)$$

Equation (8.60) states the remarkable result that the *time* domain output of the stretch mixer traces out the Fourier transform $\tilde{P}(F)$ of the range profile $\tilde{p}(\delta t)$ over the frequency interval $[F_0 - \beta/2,\ F_0 + \beta/2]$. This is intuitively satisfying, since this is the swept bandwidth of the LFM pulse. $\tilde{P}(F)$ can be rescaled into units of spatial frequency via the transformation $F \to cK_R/4\pi$. Using $\Delta R = c/2\beta$, the mixer output range becomes $K_R \in (4\pi/\lambda_0) \pm (\pi/\Delta R)$ rad/m. Figure 8.30 is a notional illustration of the portion of the range profile spectrum $\tilde{P}(F)$ or $\tilde{P}(K_R)$ measured on a single pulse by the stretch processor.

As discussed in Sec. 2.7.3, $\tilde{p}(\delta t)$ is a transformation of the scatterers in a two-dimensional scene into the one-dimensional range profile. This concept is illustrated in Fig. 8.31. The complex reflectivities of all of the scatterers on the isorange contour corresponding to a delay of $t_0 + \delta t_0$ are integrated to form the value of the range profile at time $t_0 + \delta t_0$, $\tilde{p}(\delta t_0)$. "CRP" again marks the central reference point, corresponding to a range $R_0 = ct_0/2$. If the width of the illuminated area is much less than the nominal range, $R_0\theta_{az} \ll R_0$, the lines of integration become nearly straight. The range averaging then takes the form of a *projection* in the tomographic sense of the two-dimensional scene into a one-dimensional function.

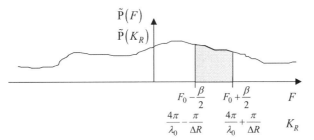

Figure 8.30 Portion of the range profile bandwidth measured by the stretch processor, in both temporal and spatial frequency units.

Consider the two coordinate systems shown in Fig. 8.32. The (p_θ, q_θ) axes are rotated by θ radians with respect to the (u, R) axes. The coordinate systems are related by the equations

$$u = p_\theta \cos\theta - q_\theta \sin\theta$$
$$R = p_\theta \sin\theta + q_\theta \cos\theta \tag{8.61}$$

A projection of a two-dimensional effective reflectivity scene $\rho'(u, R)$ into a one-dimensional cross-range-averaged reflectivity range profile $\tilde{\rho}_\theta(p_\theta)$ is defined by Dudgeon and Mersereau (1984).

$$\tilde{\rho}_\theta(p_\theta) = \int_{-\infty}^{+\infty} \rho'(p_\theta \cos\theta - q_\theta \sin\theta, \, p_\theta \sin\theta + q_\theta \cos\theta) \, dq_\theta \tag{8.62}$$

The *projection-slice theorem* of Fourier analysis then states that the one-dimensional Fourier transform of the projection is a slice of the two-dimensional

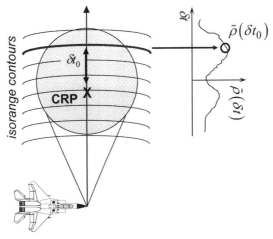

Figure 8.31 Projection of a two-dimensional scene into a one-dimensional angle-averaged range profile.

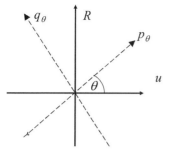

Figure 8.32 Coordinate systems for defining a two-dimensional projection.

Fourier transform of the original function $\rho'(u, R)$ (Dudgeon and Mersereau, 1984)

$$\int_{-\infty}^{+\infty} \tilde{\rho}_\theta(p_\theta) e^{-jp_\theta U} dp_\theta = P'(U\cos\theta, U\sin\theta) \tag{8.63}$$

Figure 8.33 illustrates the consequences of the LFM/stretch data acquisition and the projection-slice theorem for spotlight SAR. The scene $\rho'(u, R)$ on the left represents a patch of terrain, perhaps containing a road, building, stand of trees, and small lake. The drawing on the right represents the (unknown) Fourier transform $P'(K_u, K_R)$ of the scene in spatial frequency units. The radar views the scene from a particular angle, and the resulting stretch mixer output provides a measurement of the two-dimensional Fourier transform of the scene along the same angle, as shown on the right side of the sketch. Because of the limited bandwidth of the stretch measurement, only a portion of the total slice, highlighted in a lighter color, is actually measured (see Fig. 8.30).

As the aircraft flies along the synthetic aperture, it continues to transmit pulses while steering the antenna to remain aimed at the CRP of the region of interest. Each successive pulse therefore measures a projection of $\rho'(u, R)$ at a new angle and, therefore, a segment of a slice of $P'(K_u, K_R)$ at that same angle. Thus, over a series of pulses the system collects data over an annular region of

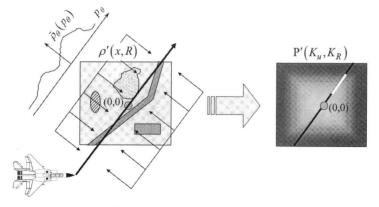

Figure 8.33 Spotlight SAR data acquisition model.

the two-dimensional spectrum of the image. This model of the spotlight SAR data set is depicted in Fig. 8.34. The extent of the annulus is $4\pi\beta/c$ radians per meter = $2\beta/c$ cycles per meter in the radial direction. The spatial range resolution is therefore $(2\beta/c)^{-1} = \Delta R$ meters, as expected. The width of the annulus at its center $K_u = 4\pi/\lambda_0$ is $(4\pi/\lambda_0)\gamma$ radians per meter, so the spatial cross-range resolution is $\lambda/2\gamma$, again as expected. Finally, note that the spectral data are on a polar, rather than rectangular, grid in (K_u, K_R). For this reason, the data set is referred to as *polar format data*.

An interesting question is the number of projections required to reconstruct the image; the answer determines the required PRF. Suppose that the final image $\rho'(u, R)$ desired is L_u by L_s meters. To avoid aliasing artifacts over a region of this size, Nyquist sampling theory requires that samples of $P'(K_u, K_R)$ be no more than $1/L_u$ cycles per meter = $2\pi/L_u$ radians per meter apart in K_u; similarly, samples should be no more than $1/L_s$ cycles per meter = $2\pi/L_s$ radians per meter apart in K_R. Consider the K_u dimension. At the center of the annulus $(K_R = 4\pi/\lambda_0)$, a sample spacing of $2\pi/L_u$ corresponds to an angular spacing between successive slices of $\lambda_0/2L_u$ radians. The number of slices required to span the total angular extent of the annulus is then

$$N_\gamma = \frac{\gamma}{(\lambda_0/2L_u)} = L_u\left(\frac{2\gamma}{\lambda_0}\right) = \frac{L_u}{\Delta CR} \tag{8.64}$$

which is simply the number of resolution cells in the cross-range extent. The linear distance traveled by the radar platform between slices, and the resulting PRI, is

$$\Delta u = R_0\frac{\lambda_0}{2L_u} \Rightarrow PRI = \frac{\lambda_0 R_0}{2v L_u} \tag{8.65}$$

Note that if L_u is set equal to its practical maximum of the illuminated beamwidth so that $R_0\theta_{az} = R_0\lambda_0/D_{az}$, the platform travel between pulses is again $D_{az}/2$.

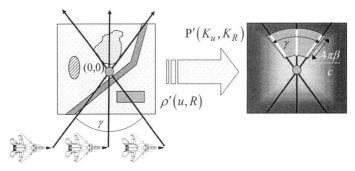

Figure 8.34 Spotlight SAR data acquisition maps an annular region in Fourier space.

The number of range samples needed (equivalent to time samples of the stretch mixer output for each pulse) is the radial extent of the annulus divided by the sample spacing in K_R, which reduces to the number of range cells in the range extent

$$N_R = \frac{(4\pi\beta/c)}{(2\pi/L_s)} = \frac{L_s}{\Delta R} \qquad (8.66)$$

Since the stretch mixer output traces out the two-dimensional Fourier transform slice, the interval $1/L_R$ between samples in K_R specifies the time Δt between samples at the mixer output, using the mapping of time to spatial frequency $K_u = (4\pi\beta/c\tau)t$

$$\Delta t = \frac{c\tau}{4\pi\beta}\frac{2\pi}{L_R} = \frac{c\tau}{2\beta L_R} \qquad (8.67)$$

It is important to remember that this spotlight data model assumes an LFM waveform and a stretch receiver. In addition, two assumptions were made in deriving the results above. The first is that the RVP term in Eq. (8.56) can be ignored. The second is that the curvature in the isorange contours can be ignored in modeling the mixer output $\tilde{\rho}(\delta t)$ as a projection $\tilde{\rho}_\theta(p_\theta)$. Both assumptions are valid provided the scene size is limited. The limitation due to isorange curvature can be shown to be

$$L_u < 2\Delta CR\sqrt{\frac{2R_0}{\lambda}} \qquad (8.68)$$

while the limitation due to RVP is

$$L_u < \frac{2\Delta CR \cdot F_0}{\sqrt{\alpha/\pi}} \qquad (8.69)$$

Both are derived by Jakowatz (1996, Appendix B). Figure 8.35 plots these two constraints for $R_0 = 10$ km and $\tau = 10$ μs. The limit on scene size due to RVP is very loose; the limit due to isorange curvature is more constraining.

8.5 The Polar Format Image Formation Algorithm for Spotlight SAR

Given the spotlight SAR data model previously, the basic image formation algorithm is fairly obvious: inverse Fourier transform the available two-dimensional spectral segment of $P'(K_u, K_R)$ to estimate the image $\rho'(u, R)$. However, a conventional *inverse discrete Fourier transform* (IDFT) algorithm assumes the spectral data are on a rectangular grid in (K_u, K_R), while the spotlight data are on a polar grid. Consequently, an extra step is required to interpolate the data from a polar to a rectangular grid before the IDFT can be applied. In practice, the data will often be windowed in both dimensions after interpolation in order to reduce side lobes in the final image. The resulting *polar format algorithm* (PFA) is diagrammed in Fig. 8.36.

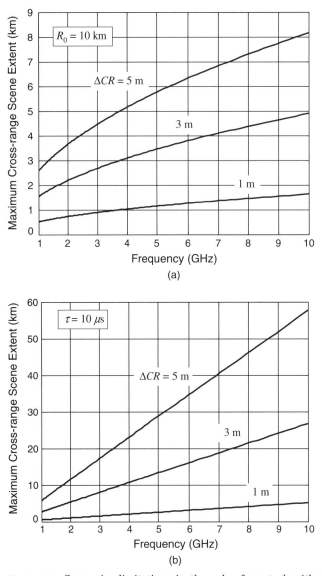

Figure 8.35 Scene size limitations in the polar format algorithm. (*a*) Limit due to isorange contour curvature. (*b*) Limit due to residual video phase.

The key step in the PFA is the polar-to-rectangular interpolation of the Fourier data. While a two-dimensional interpolation would be theoretically ideal, separable schemes using successive one-dimensional interpolations in two different dimensions are generally preferred for simplicity. Two separable approaches are commonly used. In either approach, a rectangular grid of desired (K_u, K_R) sample locations is first established based on scene size and

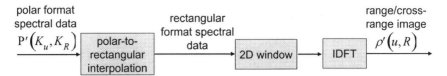

Figure 8.36 Polar format image formation algorithm for spotlight SAR.

resolution considerations. The rectangular region covered by this grid must be contained within the annular region of available data. The first approach, illustrated in Fig. 8.37a, is the more commonly used of the two. The complex polar format spectral data are interpolated along each radial to the *keystone* grid depicted in the center of the figure. After this step, the data are evenly spaced in K_R but not in K_u. A second series of one-dimensional interpolations, now in the K_u dimension, align the samples to the desired grid in K_u. Note that in principle, the keystone grid could be directly measured by the SAR system by varying the mixer output sample rate on successive pulses in a carefully controlled fashion.

The second approach, depicted in Fig. 8.37b, interpolates first along constant-radius lines to align data on the desired K_u sample locations. The intermediate grid is then interpolated in the K_R dimension to the final rectangular grid (Munson, 1985).

In the radial-keystone scheme, the interpolation is based on a separable two-dimensional sinc kernel that arises from the bandlimited nature of the data. Practical implementations window the sinc kernel in each dimension to both limit side lobes and provide a finite-length interpolating kernel. Interpolating kernels typically range from 7th to 11th order, and the total computational load

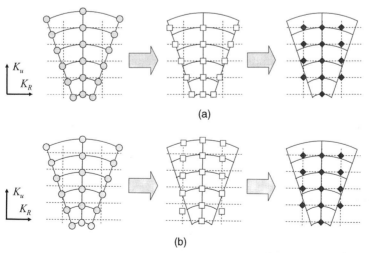

Figure 8.37 Two schemes for separable polar-to-rectangular interpolation. (*a*) Radial-keystone interpolation. (*b*) Angular-range interpolation.

for polar-to-rectangular interpolation often rivals that of the two-dimensional inverse FFT used for the final image formation. The angular-range scheme uses a periodic asinc kernel in the angular interpolation and a sinc kernel in the range interpolation. A comparison of several of these schemes is given by Munson et al. (1985). A detailed discussion of the implementation of the radial-keystone approach is given by Jakowatz et al. (1996, Chapter 3).

At higher RF frequencies, the annular region of polar format spectral data are further from the origin in (K_u, K_R) space. If the resolution is relatively large, the required bandwidth is also small, so the extent of the annulus in range and angle is small. In this case the polar grid may be nearly rectangular, in which case the expensive polar-to-rectangular interpolation can be avoided and the image formed with a simple inverse two-dimensional FFT. This case is called the *rectangular format algorithm* (RFA). It is essentially the DBS algorithm, but with an inverse FFT required in the range dimension to transform the stretch receiver output back to the fast time domain. Consequently, the constraints to DBS processing (see Fig. 8.24) apply to the RFA algorithm as well.

While the polar format algorithm is a commonly used technique in many operational spotlight SAR systems, other classes of algorithms can be applied to spotlight image formation. Application of the range migration and chirp scaling algorithms to spotlight SAR is described in the book by Carrara et al. (1995). Interest is growing in *convolution-backprojection* (CBP) methods, despite their greater computational cost, due to their increased flexibility in addressing non-rectangular data formats and avoiding interpolations, which are a principal source of image degradation in the PFA algorithm (Desai and Jenkins, 1992). Iterative reconstruction techniques (Dudgeon and Mersereau, 1984) have also been suggested.

8.6 Interferometric SAR

One of the newest developments in synthetic aperture radar is the ability to do high resolution imaging in three dimensions using interferometric techniques, commonly called IFSAR. The basic approach employs two complex SAR images of a scene formed using two displaced apertures. The apertures may be physically separate receive apertures on a single antenna structure, often with a common transmit aperture. Alternatively, IFSAR can be implemented using images collected from a conventional single aperture system on multiple passes.

8.6.1 The effect of height on a SAR image

The output of a SAR image formation algorithm is a *complex*-valued two-dimensional image: both an amplitude and phase for each pixel. Conventional two-dimensional SAR imaging discards the phase of the final image, displaying only the magnitude information. In IFSAR, the pixel phase data are retained.

Introduction to Synthetic Aperture Imaging

Because SAR image formation is a nominally linear process, the complex amplitude $f(x, y_g)$ of a pixel at ground coordinates (x, y_g) and elevation of $z = 0$ can be viewed as the product of four factors

$$f(x, y_g) = A \cdot G \cdot e^{j 4\pi R_f(x, y_g)/\lambda} \rho(x, y_g)$$
$$= A \cdot |G| \cdot |\rho(x, y_g)| \cdot \exp\left\{ j \left[\phi_G + \phi_\rho(x, y_g) - \frac{4\pi}{\lambda} R_f(x, y_g) \right] \right\} \quad (8.70)$$

In this equation, ρ is the complex reflectivity of the pixel, the complex exponential term is the phase shift due to the range to the pixel, G is the complex gain of the receiver and SAR image formation algorithm, A contains all range equation factors, and R_f is the range from the aperture phase center to the ground coordinates (x, y_g). The phase of the pixel is therefore

$$\phi_f(x, y_g) = \phi_G + \phi_\rho(x, y_g) - \frac{4\pi}{\lambda} R_f(x, y_g) \quad (8.71)$$

Now consider the two scatterers **P1** and **P2** illustrated in Fig. 8.38. Both are at ground range y_{g1}, but one is at an elevation $z = h_0$ relative to some unknown reference plane, while the other is at an elevation $z = h_0 + \Delta h$. They are observed from two distinct radar apertures at an altitude $z = h_0 + Z$ and separated horizontally by a *baseline* B. Each aperture independently transmits a radar waveform, receives the data, and forms a complex SAR image of the

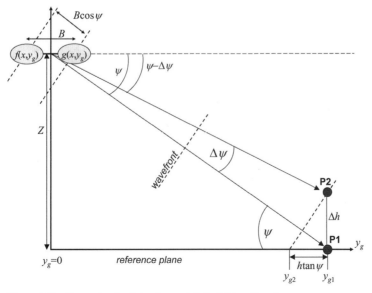

Figure 8.38 Geometry for interferometric height estimation.

scene; these images are denoted by $f(x, y_g)$ and $g(x, y_g)$.[†] The baseline should be orthogonal to the flight path; therefore the direction of the aircraft motion is into the page.[‡] The radar range is great enough that the incoming wavefront can be considered planar. If the depression angle from the middle of the baseline to **P1** is ψ, the difference in range to the two aperture phase centers is well approximated as $B \cos \psi$. The difference in received phase at the two apertures then becomes, using Eq. (8.71)

$$\phi_{fg} \equiv \phi_f - \phi_g \approx -\frac{4\pi}{\lambda} B \cos \psi \qquad (8.72)$$

Now consider scatterer **P2**, also located at ground range y_{g1} but elevated by Δh meters. Because the radar measures time delay and thus slant range, the echo from **P2** will be indistinguishable from that of a scatterer located on the ground plane at the range where the wavefront impacting **P2** also strikes the ground. As shown in Fig. 8.38, this ground coordinate is

$$y_{g2} = y_{g1} - \Delta h \tan \psi \qquad (8.73)$$

Because SAR images are two-dimensional, **P2** will be incorrectly imaged at range y_{g2}.[§] The imaging of the elevated scatterer at an incorrect range coordinate is termed *layover*, because the scatterer appears to have been shifted toward the radar. As illustrated, the layover is only in the range coordinate. In squinted operation, layover occurs in both range and cross-range; details are given by Sullivan (2000) and Jakowatz et al. (1996).

Elevating the scatterer at y_{g1} to height Δh will also change the depression angle from the center of the IFSAR baseline to the scatterer. This change can be found by differentiating Eq. (8.72) with respect to the grazing angle

$$\frac{d\phi_{fg}}{d\psi} = \frac{4\pi}{\lambda} B \sin \psi \quad \Rightarrow \quad \Delta \psi = \frac{\lambda}{4\pi B \sin \psi} \Delta \phi_{fg} \qquad (8.74)$$

This equation states that a change in the interferometric phase difference of $\Delta \phi_{fg}$ implies a change in depression angle to the scatterer of $\Delta \psi$ radians. To relate this depression angle change to an elevation change, consider Fig. 8.37 again, which shows that

$$\frac{\Delta h}{H} \approx \frac{\Delta \psi}{\tan \psi} \qquad (8.75)$$

Eliminating y_g using a series approximation of the tangent function, and assuming that $\Delta \psi$ is small, Eq. 8.75 can be reduced to

$$(Z - \Delta h) \tan \psi = Z \tan \psi - Z \Delta \psi \quad \Rightarrow \quad \Delta h = Z \cot \psi \, \Delta \psi \qquad (8.76)$$

[†]A variation that uses a common transmit aperture and two independent receive apertures is discussed in the book by Sullivan (2000).

[‡]The two apertures could also be displaced vertically, with similar results.

[§]SAR images are naturally formed in the slant plane, but are usually projected into the ground plane for display.

Finally, using Eq. (8.76) in Eq. (8.74) gives a measure of how much the interferometric phase difference for a given pixel will change if the scatterer elevation changes (Carrara et al., 1995)

$$\Delta h = \frac{\lambda Z \cot \psi}{4\pi B \sin \psi} \Delta \phi_{fg} \qquad (8.77)$$

Equation (8.77) is the basic result of IFSAR.

The phase map $\phi_{fg}(x, y_g)$ can be written as

$$\phi_{fg}(x, y_g) = \phi_{fg}(x_0, y_{g0}) + \Delta \phi_{fg}(x, y_g) \qquad (8.78)$$

Multiplying Eq. (8.78) by the scale factor in Eq. (8.77) gives

$$h(x, y_g) = \frac{\lambda Z \cot \psi}{4\pi B \sin \psi} [\phi_{fg}(x_0, y_{g0}) + \Delta \phi_{fg}(x, y_g)]$$
$$\equiv h_{\text{offset}} + \Delta h(x, y_g) \qquad (8.79)$$

Thus, given an interferometric phase difference map $\phi_{fg}(x, y_g)$, multiplying by the scale factor of Eq. (8.77) gives a height map that gives the height variations from pixel-to-pixel, but has an unknown overall offset. That is, the IFSAR technique provides a good measure of *relative* height versus spatial location.

Because the radar signal processor can only measure the phase of a signal sample as the arctangent of the ratio of its imaginary and real parts, only the *wrapped* interferometric phase $((\phi_{fg}))_{2\pi}$ can be measured, where the notation $((\cdot))_{2\pi}$ indicates arithmetic modulo 2π. Consequently, $\Delta \phi_{fg}$ is also measured modulo 2π.

The two SAR images required for IFSAR processing can be generated in one of two basic ways. In *one-pass IFSAR*, the approach described previously, the radar platform has two displaced receive apertures so that the data for both images are collected on a single pass. This approach has the advantages of operational simplicity, much greater ease of aligning the trajectories of the two apertures, and no decorrelation of the scene between passes. *Two-pass IFSAR* uses a conventional single-receiver SAR system and flies two independent passes past the scene to be imaged. This approach requires careful alignment of the two flight paths, which can be difficult to achieve in airborne systems but is very feasible for orbiting radars, and can be implemented with existing conventional single-receiver SAR sensors.

8.6.2 IFSAR processing steps

Formation of an IFSAR image involves the following major steps

- Formation of the two individual SAR images, $f[l, m]$ and $g[l, m]$
- *Registration* of the two images
- Formation of the wrapped interferometric phase map $((\phi_{fg}[l, m]))_{2\pi}$
- Smoothing of $((\phi_{fg}[l, m]))_{2\pi}$ to reduce phase noise

- Two-dimensional *phase unwrapping* to obtain $\phi_{fg}[l, m]$ from $((\phi_{fg}[l, m]))_{2\pi}$
- Scaling of the unwrapped phase map $\phi_{fg}[l, m]$ to obtain the height map $\Delta h[l, m]$
- *Orthorectification* to correct layover based on the height information.

The images are formed using any SAR imaging algorithm appropriate to the collection scenario, such as the range-Doppler and polar format algorithms discussed previously. Because the height estimation depends on the difference in phase of the echo from each pixel at the two apertures, it is important to ensure that like pixels are compared. The slightly different geometries of the two offset apertures will result in slight image distortions relative to one another, so an image registration procedure is used to warp one image to align well with the other. Many registration procedures have been developed in the image processing literature. One that is popular in IFSAR uses a series of correlations between small subimages of each SAR map to develop a warping function. This concept is illustrated in Fig. 8.39. The actual resampling is typically done with simple bilinear interpolators. The procedure is described in detail by Jakowatz et al. (1996). Once the two images are registered, the wrapped phase map is easily computed as

$$((\phi_{fg}[l, m]))_{2\pi} = \arg\{f[l, m]g^*[l, m]\} \quad (8.80)$$

Local averaging, typically using a 3×3, 5×5, or 7×7 window, is often applied to the phase map at this point to smooth phase noise.

The two-dimensional phase unwrapping step to recover $\phi_{fg}[l, m]$ from $((\phi_{fg}[l, m]))_{2\pi}$ is the heart of IFSAR processing. Unlike many two-dimensional signal processing operations such as FFTs, two-dimensional phase unwrapping cannot be decomposed into one-dimensional unwrapping operations on the rows and columns. Two-dimensional phase unwrapping is an active research area; a thorough introduction is given by Ghiglia and Pritt (1998). Most unwrapping techniques can be classified as either *path following methods* or *minimum norm methods*; the latter are often based on fast transform techniques. An example of a minimum norm algorithm is the *discrete cosine transform* (DCT) method

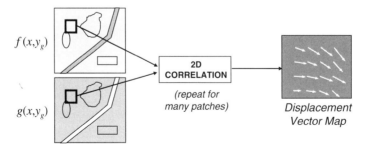

Figure 8.39 Generation of image pair warping function.

given in the paper by Ghiglia and Romero (1994). This technique finds an unwrapped phase function such that, when rewrapped, it minimizes the mean squared error between the gradients of the rewrapped phase function and the original measured wrapped phase.

The method begins by computing the wrapped gradients in the range (l) and cross-range (m) dimensions of the raw wrapped phase history data; these are then combined into a "driving function" $d[l, m]$

$$\Delta_{y_g}[l, m] = \begin{cases} \left(\left(\left((\phi_{fg}[l+1, m])\right)_{2\pi} - \left((\phi_{fg}[l, m])\right)_{2\pi}\right)\right)_{2\pi}, \\ \qquad 0 \leq l \leq L-2, \ 0 \leq m \leq M-1 \\ 0, \qquad \text{otherwise} \end{cases}$$

$$\Delta_x[l, m] = \begin{cases} \left(\left(\left((\phi_{fg}[l, m+1])\right)_{2\pi} - \left((\phi_{fg}[l, m])\right)_{2\pi}\right)\right)_{2\pi}, \\ \qquad 0 \leq l \leq L-1, \ 0 \leq m \leq M-2 \\ 0, \qquad \text{otherwise} \end{cases}$$

$$d[l, m] = (\Delta_{y_g}[l, m] - \Delta_{y_g}[l-1, m]) + (\Delta_x[l, m] - \Delta_x[l, m-1]) \quad (8.81)$$

Let $D[k, p]$ be the two-dimensional DCT of the driving function. The estimate of the unwrapped phase is then obtained as the inverse DCT of a filtered DCT spectrum

$$\hat{\phi}_{fg}[l, m] = \text{DCT}^{-1} \left\{ \frac{D[k, p]}{2 \left[\cos\left(\frac{\pi k}{M}\right) + \cos\left(\frac{\pi p}{N}\right) - 2\right]} \right\} \quad (8.82)$$

This function is then used in Eq. (8.77) to estimate the terrain height map $\Delta h[l, m]$.

As a simple demonstration of this algorithm, consider the simulated terrain profile of a hill shown in Fig. 8.40. This function was created as the outer product of two one-dimensional Hann window functions. A simulation of one-pass IFSAR data collection produces the wrapped interferometric phase function of Fig. 8.41a. In addition, noise has been added to a rectangular patch of the phase data to simulate a low-reflectivity or degraded area. Straightforward application of Eqs. (8.81) and (8.82) produces the unwrapped interferometric phase map estimate of Fig. 8.41b. The unwrapping is successful even in the noisy region. When converted back to terrain height, this phase map accurately reconstructs the hill height profile, although the height estimate is noisy in the degraded region.

IFSAR processing produces only relative height variations. Absolute height can be estimated by a variety of techniques. The most common is simply the use of surveyed reference points within an image; height relative to this point is then easily converted to absolute height. Another method, analogous to the use of multiple PRIs to resolve range and Doppler ambiguities, uses two interferometric phase maps having different scale factors and therefore different ambiguity intervals in height. Multiple maps with different scale factors

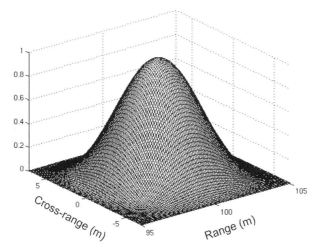

Figure 8.40 Height profile of simulated hill.

have been obtained using at least two different methods. In systems having a waveform bandwidth in excess of that required for the desired range resolution, the fast-time bandwidth can be split in half, and IFSAR processing completed separately for each half of the data. The effective center wavelength λ will be different for the two data sets, giving two different scale factors in Eq. (8.77). Another technique, applicable to radar systems having three or more receive

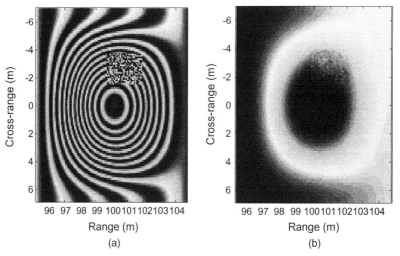

Figure 8.41 Example of DCT-based two-dimensional phase unwrapping on "hill" height map with noisy data patch. (*a*) Wrapped phase map $((\phi_{fg}))_{2\pi}$. (*b*) Estimated unwrapped phase $\hat{\phi}_{fg}$. (Images courtesy of Mr. Will Bonifant.)

Figure 8.42 Interferometric SAR height image of the football stadium area at the University of Michigan, Ann Arbor. (Image courtesy of General Dynamics Advanced Information Systems.)

apertures, forms one IFSAR height map using the first and second apertures, and another using the first and third apertures. The baselines for these two pairs are different, again giving rise to two different scaling factors in Eq. (8.77).

The last step in IFSAR processing is orthorectification, which corrects the displacement of image pixels due to layover. For each pixel in the SAR image $f[l, m]$, the corresponding height pixel $\Delta h[l, m]$ is used to estimate the layover $-\Delta h \tan \psi$ present in that pixel. The image is then resampled in the downrange dimension to shift each pixel by $+\Delta h \tan \psi$ to its correct downrange position.[†]

Figure 8.42 is an IFSAR image of the football stadium and surrounding area at the University of Michigan, Ann Arbor. The image clearly shows that the trees above the stadium in the image are taller than those to the left of the stadium; and that the stadium playing surface is actually below the level of the surrounding terrain. Also visible is the Crisler Arena at the lower middle of the image.

The relative phase of two SAR images can be used in other ways (Jakowatz et al., 1996). IFSAR presumes that there is no change in the imaged scene,

---

[†]If operated in squint mode, the image will also have layover in the cross-range dimension, so pixels must be shifted in both dimensions. See Jakowatz et al. (1996) for details.

so phase differences are due only to height variations viewed from slightly different aspect angles. Another application of growing interest is *coherent change detection* (CCD). CCD compares two images taken from the *same* trajectory at different times, from a few minutes apart to many hours or days apart. The pixel-by-pixel correlation coefficient will be close to one for pixels with the same complex reflectivity in both maps, and close to zero for pixels whose complex reflectivity has changed between the two collections. An image formed from these correlation coefficients can provide a very sensitive indicator of activity in an observed area. Finally, *terrain motion mapping* also computes correlation coefficients between two images. However, it is assumed that the reflectivity is unchanged between the two collections, so any complex image changes are due to *changes* in the height of the scene. The time lag between collections may be days to years. Terrain motion mapping has been used to study such phenomena as glacier movement, land subsidence, volcanic activity, and seismic activity.

8.7 Other Considerations

8.7.1 Motion compensation and autofocus

By assuming the radar platform's position in the x coordinate is $u = vt$, it has been implicitly assumed that the radar platform is flying a straight, level, and constant velocity path. In practice, this is never quite true. For airborne systems especially, atmospheric turbulence, minor maneuvers and course corrections, vibration, antenna gimbal transient motions, and similar effects all cause deviations of the SAR antenna phase center from this ideal. However, this assumption is built into the design of the image formation processor, which is based on Eq. (8.25) or one of its various approximations. To the extent that the actual $R(u)$ does not follow the model of $R(u)$ used in the processor design, the actual SAR data will not match the expected PSR. These differences, caused by deviations from the ideal flight path, manifest themselves as phase errors in the processing.

Specifically, suppose the transmitted waveform is of the form $A \exp[j 2\pi F t]$ and the range to some scatterer is R on a particular pulse. Then the received signal is

$$y(t) = A' \exp\left[j 2\pi F \left(t - \frac{2R}{c}\right)\right] \qquad (8.83)$$

where A' absorbs all of the range equation factors. Now suppose that the phase center of the radar antenna is, for whatever reason, displaced from the intended path by a distance ε; the received signal will instead be

$$\tilde{y}(t) = \tilde{A} \exp\left[j 2\pi F \left(t - \frac{2(R - \varepsilon)}{c}\right)\right] = \frac{\tilde{A}}{A} \exp\left[-j 4\pi \frac{\varepsilon}{\lambda}\right] y(t) \qquad (8.84)$$

The amplitude ratio in Eq. (8.84) will be very nearly unity, so the primary effect of motion errors is a phase rotation to the data. All of the fast time samples for a given pulse are rotated by the same phase factor.

TABLE 8.1 Effects of Various Phase Errors on the SAR Point Spread Response

Phase error class	Dominant effect on PSR
Low frequency	
Linear	Cross-range displacement
Quadratic	Mainlobe broadening, amplitude loss
Cubic	Mainlobe asymmetry and cross-range displacement
High frequency	
Deterministic periodic	Discrete "paired echo" high sidelobes
Random	Increased sidelobe level

The job of motion compensation is therefore to estimate ε and correct the data by a simple counter phase rotation

$$\hat{y}(t) = \exp\left[+j4\pi\frac{\varepsilon}{\lambda}\right]\tilde{y}(t) \approx y(t) \tag{8.85}$$

The difficult part is estimating ε to the required accuracy. A displacement of only $\lambda/4$ meters (0.3 inches at X band) corresponds to a two-way range change of $\lambda/2$ meters and therefore a 180° phase reversal. Thus, path deviations must be estimated to a small fraction of a wavelength to minimize phase error effects. Two factors ease this challenge somewhat. First, a displacement of the entire flight path has no effect; the same ε is added to the phase of all data samples, contributing a complex constant with no effect on focusing quality. Only variations in ε are significant. Second, any one pixel receives contributions from data only over the aperture time, on the order of a few tenths of a second to a few seconds in most SAR systems. Path drifts over longer time periods do not affect image focus.[†]

Phase errors are frequently categorized as either low- or high-frequency, based on the variation of the phase error in the slow time dimension (Lacomme et al., 2001). Low frequency errors are those that repeat on a time scale greater than the aperture time. Thus, any periodicity of low frequency errors is not apparent during formation of an image. High frequency errors are subdivided into deterministic and noise-like errors. In the former, periodic structure is evident during the aperture time. Low frequency errors produce net phase tapers across the aperture, affecting the resolution, gain, and cross-range accuracy of the PSR. For example, a linear phase taper is equivalent to an uncompensated Doppler shift and can produce a cross-range displacement of scatterers according to Eq. (8.36). Examples of the resolution and gain loss effects of quadratic phase errors were shown in Fig. 8.23 and Fig. 8.25a. High frequency errors affect primarily the side lobes of the PSR. Table 8.1 describes the major effect of the most important phase errors.

[†]Long term drift may be quite important, however, to image interpretation. Targets or land features detected in a SAR image may need to be precisely geolocated for targeting or for surveying purposes. It is then essential to have accurate information on the absolute position of the radar platform.

Phase errors are corrected by estimating the displacement $\varepsilon(u)$ at each aperture position u using a position measurement system, and in higher resolution systems by also using *autofocus* algorithms. Position estimates are developed using a motion compensation system with some or all of the elements in Fig. 8.43. Data from the aircraft or spacecraft's *inertial navigation system* (INS) provide the first level estimate of deviations from a straight line, constant-velocity flight path. The INS tracks the platform centroid, not the antenna phase center. Higher precision systems mount an additional strapdown *inertial measurement unit* (IMU) onto the antenna structure (Kennedy, 1988a). The IMU accounts more accurately for motions of the antenna relative to the airframe; these relative motions include not only intentional scanning but also vibration and airframe flexure. Inertial antenna mounts can also be used to stabilize the antenna as much as possible. Additional independent kinematic estimates such as radar-derived Doppler velocity estimates and *global positioning system* (GPS) data may also be used. In this system, the INS provides an initial attitude reference to the IMU. IMU data are sent to the *radar data processor* (RDP), which corrects for the lever arm displacement between the INS and IMU. The difference in INS and corrected IMU position estimates is input to the Kalman *transfer alignment* filter, whose output is fed back to the IMU to update its state. The Kalman filter typically has on the order of 20 states. The updated IMU state is fed to the motion compensation computer, which uses these data to compute the position error ε and the associated phase correction. The motion compensation computer typically also computes antenna steering commands to stabilize the line of sight.

A motion compensation error budget can be developed for the various INS/IMU errors. Table 8.2 is an example of such an analysis (Kennedy, 1988b).

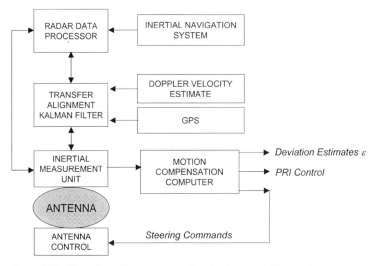

Figure 8.43 Generic motion compensation instrumentation package.

TABLE 8.2 Motion Compensation Error Budget

Error source	1σ Value		Root sum square (RSS) total	Quadratic phase error formula rads	Quadratic phase error rads
	INS	IMU			
Cross-range velocity error, v_e (m/s)	0.24	0.06	0.247	$\dfrac{\pi v v_e T_a^2}{\lambda R_0}$	0.43
LOS accelerometer bias, A_b (μg)	70	80	106.3	$\dfrac{\pi A_b T_a^2}{2\lambda}$	0.22
Platform tilt, α (mrad)	0.2	0.1	0.22	$\dfrac{\pi \alpha a g T_a^2}{2\lambda}$	0.45
Boresight accelerometer scale factor error, s (PPM)	150	150	212	$\dfrac{\pi s A T_a^2}{2\lambda}$	0.13
Total RSS quadratic phase error, radians					0.67

The parameter values used were an aperture time of 2 s, velocity of 152 m/s, standoff range of 37 km, gravitational acceleration $g = 9.8$ m/s^2, and a line-of-sight acceleration A of 3 m/s^2. These parameters imply a cross-range resolution of 1.8 m. In this example, the total error is 0.67 rad, well under the $\pi/2$ standard suggested earlier. Since the phase error contributions all scale as T_a^2/λ, the error margins get tighter as radar frequency increases and especially as resolution is made finer. Because of its dependence on range, the velocity error tends to dominate at shorter ranges; at longer ranges the other error effects become more significant. Another analysis of error contributions, in terms of longitudinal and latitudinal motions, is given by Lacomme et al. (2001).

In addition to estimates of the path deviation ε, motion compensation systems often provide two additional types of control data. The first is fine pulse-to-pulse PRI adjustments to provide uniform sample spacing along the velocity vector, i.e., in the x dimension. This technique eliminates the need to interpolate data in the slow-time dimension to obtain uniform sample spacing in higher resolution systems. The second, applicable primarily to squinted SAR and spotlight SARs, computes the range walk relative to a reference point on the ground and varies the time delay from the time of pulse transmission to the time when the A/D converter at the receiver output is triggered to begin fast-time sampling. The delay is chosen such that the echo from some reference point, generally in the middle of the imaged scene, always occurs in the same range bin. This technique, called *motion compensation to a point*, reduces or eliminates the computation required for range walk correction. Both techniques require more complicated receiver control.

8.7.2 Autofocus

In high resolution SAR systems, the motion compensation systems discussed previously often may not provide sufficient phase correction to achieve good image quality. Autofocus algorithms attempt to estimate residual phase errors present in the data after motion compensation and compensate them to improve image quality. A number of techniques have been suggested; see the book by Carrara et al. (1995) for an overview. Two common and representative methods are the map drift method and the phase gradient method. The most basic form of the map drift algorithm is designed specifically to correct quadratic phase errors. The algorithm divides the SAR data into two halves, corresponding to the first half of the synthetic aperture and the second half, and forms a lower-resolution image from each (Carrara et al., 1995; Elachi, 1988). If quadratic phase error is present, peaks in the images will be shifted away from their correct locations, but in opposite directions. Cross-correlating the two images produces a peak whose offset from the origin is directly proportional to the amount of residual quadratic phase error. The full data set is then phase-corrected with the conjugate of the estimated error, and the image reformed from the modified data. The algorithm is able to correct quadratic phase errors up to a few tens of radians. Extensions to the map drift algorithm which divide the data into more than two subapertures are capable of estimating higher-order polynomial phase error terms.

A very basic analysis of a simple map drift algorithm can be outlined as follows. Suppose a slow-time phase history sequence $y[m]$ is contaminated with a quadratic phase error sequence

$$y'[m] = y[m] \exp\{j4\alpha[m - (M-1)/2]^2/(M-1)^2\} \qquad 0 \le m \le M-1 \quad (8.86)$$

The parameter α is the maximum value of the quadratic phase error, which occurs at the two ends of $y'[m]$. Assume for convenience that M is even, and define the two half-sequences

$$\begin{aligned} y_1'[m] &= y'[m] & 0 \le m \le M/2 - 1 \\ y_2'[m] &= y'[m + M/2] & 0 \le m \le M/2 - 1 \end{aligned} \quad (8.87)$$

The phase error in $y_1'[m]$ can be decomposed into a quadratic component that is symmetric about the midpoint of $y_1'[m]$ and a linear component, which is just the straight line connecting the first and last sample of the phase curve in $y_1'[m]$; it is straightforward to show that this linear phase component is

$$\begin{aligned} \phi_{1_{\text{linear}}}[m] &= \alpha - \left(\frac{2}{M-2}\right)\left(1 + \frac{1}{(M-1)^2}\right)\alpha m \\ &\equiv \alpha - \alpha' m \end{aligned} \quad (8.88)$$

Similarly, the linear phase term of $y_2'[m]$ has an identical quadratic component, but a linear component of the opposite sign. Because each sequence is half of the original sequence $y'[m]$, their discrete time Fourier transforms will be similar in shape. The linear phase terms, however, will shift the two DTFTs $y_1'[m]$ and $y_2'[m]$ by $-\alpha'K/2\pi$ and $+\alpha'K/2\pi$ samples, respectively, where K is the DFT size. A cross-correlation of $Y_1'[k]$ and $Y_2'[k]$ will therefore produce a peak at lag $k_0 = \alpha' K/\pi$ radians. The quadratic phase error coefficient α can then be estimated from the spectral correlation peak as

$$\widehat{\alpha'} = \frac{\pi k_0}{K}$$
$$\hat{\alpha} = \left(\frac{M-2}{2}\right)\left(1 + \frac{1}{(M-1)^2}\right)^{-1} \widehat{\alpha'} \quad (8.89)$$

In practice, the estimate is not exact due to nonquadratic terms in the phase error and noise in the data. A corrected data sequence is then formed as

$$\hat{y}[m] = y'[m] \exp\{-j4\hat{\alpha}[m - (M-1)/2]^2/(M-1)^2\} \quad 0 \leq m \leq M-1 \quad (8.90)$$

The map drift algorithm is applied iteratively, typically requiring on the order of 3 to 6 iterations. The iteration is stopped when the peak estimated phase error is less than $\pi/2$ (or some other appropriate quality threshold).

Figure 8.44 illustrates the map drift algorithm on a real scene. The original image in part *a* of the figure was transformed to a simulated range-compressed phase history domain by assigning a random phase to each image pixel and then performing a Fourier transform in the cross-range dimension. A simulated phase error consisting of a quadratic term with a maximum of 5π radians across the aperture was applied in the cross-range dimension and the image reformed, giving the severely blurred image in Fig. 8.44*b*. Part *c* of the figure is the result after applying map drift autofocus to the blurred image; it is visually indistinguishable from the original. Finally, Fig. 8.44*d* is the difference image between the original and the autofocused image. While some correlation between the difference image and the original is visible on close inspection, the nearly uniform difference image confirms the quality of the phase error correction.

The *phase gradient algorithm* (PGA) (Carrara et al., 1995; Wahl et al., 1994) does not assume a polynomial form for the phase error and appears to offer excellent, robust performance. PGA uses the (blurred) image of the strongest scatterers in the SAR scene to estimate the actual cross-range phase function, and then multiplies the data by a compensating phase function. The algorithm estimates the first derivative of the cross-range phase error from a SAR image that is converted to (K_u, R) units by performing an FFT in cross-range. The phase error derivative is then integrated to get the actual phase error function, and a correction applied to the data. As with map drift, the phase gradient algorithm is usually applied iteratively.

A simplified analysis of the PGA algorithm gives some insight into its operation. PGA assumes that each cross-range row in the complex blurred image is dominated by the image of a single point scatterer. Consider an arbitrary range bin, and assume that the point scatterer in that range bin is located at $m = m_0$. The model of the complex image domain data in this range bin is then

$$f[k] = A\delta[k - k_0] + w[k] \quad (8.91)$$

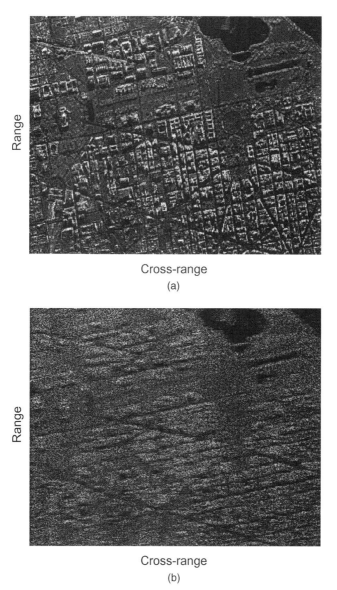

Figure 8.44 Illustration of map drift autofocus. (*a*) Original image. (*b*) Image formed from data with simulated noisy quadratic phase errors (*c*) Image after map drift autofocus. (*d*) Error image. (Original image courtesy of Sandia National Laboratories.)

where $w[k]$ is a random process representing other clutter scatterers. The dominant scatterer amplitude A and location m_0 will vary in different range bins, but for notational simplicity the dependence on l is not shown. If the cross-range compression algorithm is assumed to be approximately the Fourier transform of the range-compressed phase history data (implemented using a K-point DFT),

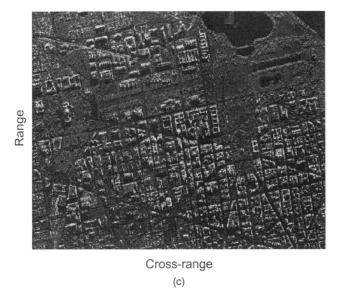

(c)

(d)

Figure 8.44 (*Continued*)

the corresponding phase history data $y[m]$ will be, ignoring the clutter, the inverse DFT of the cross-range image

$$y[m] = \text{IDFT}\{f[k]\} = \frac{A}{K}\exp\left[j\frac{2\pi}{K}k_0 m\right] \tag{8.92}$$

Equation (8.92) represents the phase history in the absence of phase errors. When phase errors $\phi[m]$ are present the data $y[m]$ are modulated by the phase error modulation function $\exp\{j\phi[m]\}$

$$y'[m] = \exp\{j\phi[m]\}y[m]$$
$$= \frac{A}{K}\exp\left\{j\frac{2\pi}{K}k_0 m\right\}\exp\{j\phi[m]\} \quad (8.93)$$

Denote the DFT of the phase error modulation function $\exp\{j\phi[m]\}$ as $e[k]$. The DFT of the phase history data including the phase errors is the actual observed complex cross-range image slice. Applying the modulation property of the DFT

$$f'[k] = \text{DFT}\{y'[m]\}$$
$$= A \cdot e[k - k_0] \quad (8.94)$$

Thus, the point target image is degraded into a replica of the DFT of the phase error function, centered at the point target location. $e[k]$ thus becomes the blurring function in the image domain due to the phase errors, and therefore isolated point targets carry the information in the shape of $e[k]$ and therefore on the phase error function $\phi[m]$. The blurred image data are the starting point for the PGA algorithm.

The algorithm begins by finding the peak amplitude of $|f'[k]|$; suppose this occurs at $k = k_p$. The cross-range slice is then circularly shifted left by k_p samples, giving the new sequence

$$f'_p[k] = A \cdot E[k - k_0 + k_p] \quad (8.95)$$

with corresponding IDFT

$$y'_p[m] = \frac{A}{K}\exp\{j\phi[m]\}\exp\{j2\pi(k_0 - k_p)m/K\} \quad (8.96)$$

The next step is to estimate the gradient of the phase of $y'_p[m]$. A simple maximum likelihood estimator is (Jakowatz and Wahl, 1993)

$$\widehat{\Delta\phi}[m] = \arg\{y'^*_p[m-1]y'_p[m]\}$$
$$\approx \phi[m] - \phi[m-1] \quad (8.97)$$

The phase error itself is estimated by integrating this gradient

$$\hat{\phi}[m] = \begin{cases} 0, & m = 0 \\ \sum_{q=1}^{m} \widehat{\Delta\phi}[q] & \text{otherwise} \end{cases} \quad (8.98)$$

The original slow-time phase history data are then corrected by compensating it with the estimated phase

$$\hat{y}[m] = y'[m]\exp\{-j\hat{\phi}[m]\} \quad (8.99)$$

Finally, the deblurred image slice is obtained by transforming back to the image domain

$$\hat{f}[k] = \text{DFT}\{\hat{y}[m]\} \tag{8.100}$$

If $\hat{\phi}[m]$ is a good estimate of $\phi[m]$, the result will be

$$\hat{f}[k] \approx A\delta[k - k_p] \tag{8.101}$$

Note that if the blur function $e[k]$ does not have its peak at the origin, $k_p \neq k_0$ and the corrected image slice will still be shifted by $(k_p - k_0)$ samples from the correct position. This does not hurt image quality, but does impact geolocation accuracy.

Since the image has many rows, each degraded by the same actual phase error function, an independent phase error estimate can be obtained by processing the data for each range bin separately. The signal-to-noise ratio of the phase estimate can then be improved by averaging over the range bins. Reintroducing the range bin index l

$$\widehat{\Delta\phi}[l, m] = \arg\{y'^*[l, m-1]y'[l, m]\}$$
$$\hat{\phi}[m] = \sum_{l=1}^{L-1} \widehat{\Delta\phi}[l, m] \tag{8.102}$$

Figure 8.45 illustrates this process on a simple synthetic image. The original image in part a of the figure is simply a collection of randomly located point scatterers of varying amplitude and phase in complex white Gaussian noise. A phase error consisting of quadratic, sinusoidal, and noise terms (see Fig. 8.46) was applied to the cross-range DFT of these data, and the image-reformed via a cross-range IDFT. The result, seen in part b, is severely blurred in cross-range. Application of the PGA algorithm produces the corrected image of part c. The smearing has been completely removed. Close inspection, however, shows that all of the scatterers have migrated three pixels to the left. This occurs because for the particular phase error function $\phi[m]$ used, the resulting blur function $e[k]$ has its peak at $k = 3$, rather than $k = 0$. Consequently, in the blurred image the peaks of each blurred scatterer are shifted three pixels right; that is, $k_p = k_0 + 3$. This adds a linear phase term in k corresponding to a $+3$ pixel shift to the phase function $\hat{\phi}[m]$ estimated by the PGA algorithm. When the phase of the original data is compensated, a -3 pixel shift results. This effect is clearly seen in the error image of Fig. 8.45d, which shows paired positive and negative differences three pixels apart for each scatterer.

The actual phase error function $\phi[m]$ applied in this example is shown in Fig. 8.46. The estimated phase error, and the difference between the two, is also shown. Only one iteration of PGA was needed because this simple simulation matches the algorithm assumptions very closely. The residual phase error is well within the $\pm\pi/2$ bounds denoted by the horizontal lines.

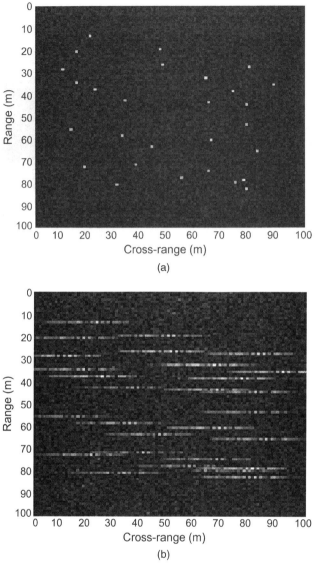

Figure 8.45 Illustration of phase gradient algorithm autofocus. (*a*) Synthetic image. (*b*) Image blurred by nonpolynomial phase error. (*c*) Image after 1 iteration of PGA autofocus. (*d*) Error image showing cross-range displacement.

A number of details in any practical implementation have not been addressed here. As mentioned earlier, the algorithm is typically applied iteratively. It is often applied to only a subset of the most energetic range bins, similar to the map drift algorithm. Furthermore, a window is used to select only a symmetric portion of the cross-range image slice around each peak magnitude pixel; this reduces the effects of noise and other bright pixels in the line. The window size

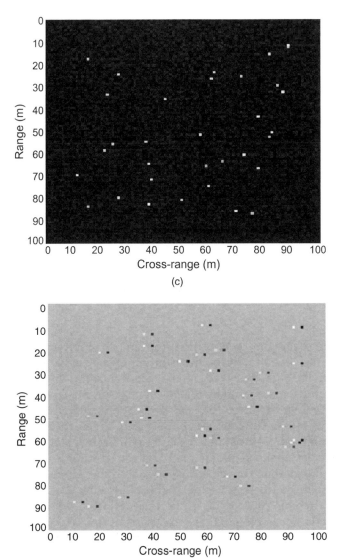

Figure 8.45 (*Continued*)

is usually selected based on an estimate of the width of the blur function $e[m]$ and is updated with each iteration. These and other details are addressed in the book by Jakowatz et al. (1996).

8.7.3 Speckle reduction

Like any coherent imaging system, SAR produces images contaminated by "speckle," an aptly-named multiplicative noise. Speckle is the natural result

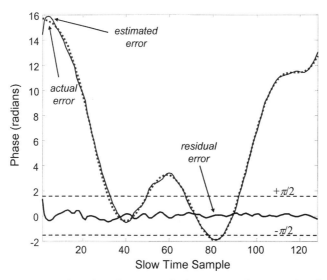

Figure 8.46 Actual and estimated phase error for example of Fig. 8.44.

of the coherent combination of echoes from many different scatterers to form an image pixel. If the amplitude distribution of the real and imaginary parts (I and Q channels) of the received signal is Gaussian and the phase distribution is uniform, conditions assured by the law of large numbers when many scattering centers are involved, then the pixel amplitude will be Rayleigh distributed as shown in Chap. 2. Thus, pixels representing areas with the same average RCS (mean echo amplitude) can give rise to different pixel amplitudes. These variations are not due to thermal, quantization, or other noise sources, but are nonetheless considered "noise" because of their effect on image quality.

Speckle is reduced through various forms of filtering and averaging schemes (Oliver and Quegan, 2004). One of the most effective is noncoherently integrating multiple uncorrelated images, or *looks,* of the same scene. This process reduces the pixel variance, reducing the amplitude variations among pixels representing the same RCS. Uncorrelated looks can be obtained using transmitter frequency or polarization agility. In many stripmap systems the maximum aperture time available exceeds that required to meet resolution goals. In this case the slow-time data can be divided into multiple subapertures, each long enough to form an image of the proper resolution. Images are calculated for each subaperture and combined. Typically, 4 to 10 looks might be used for speckle reduction. Figures 8.47*a* and *b* demonstrate the image enhancement obtained by integration of 10 simulated looks. Another source of independent looks can be multiple polarization channels. If multiple looks are not available, another method averages adjacent pixels of a high-resolution image, typically using a 3×3 or 5×5 window, to form one lower resolution but reduced-speckle pixel as illustrated in Fig. 8.47*c*. Figure 8.47*d* shows the result of a 3×3 median

Introduction to Synthetic Aperture Imaging 457

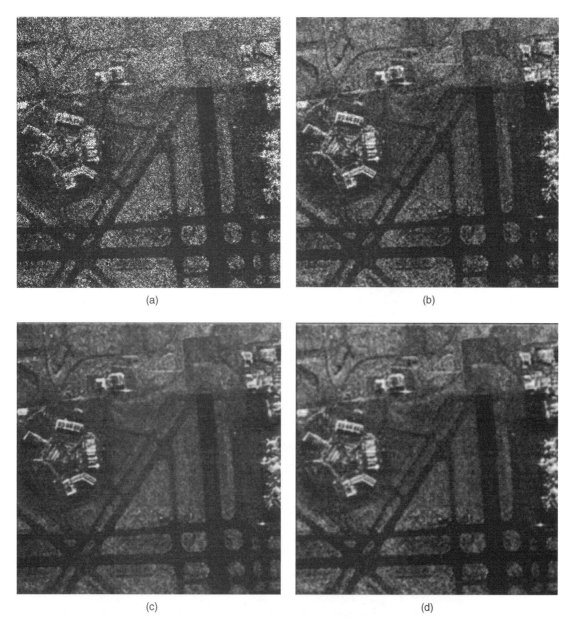

Figure 8.47 Speckle reduction. (*a*) Simulated single-look speckled image derived from Fig. 8.1*a*. (*b*) Full resolution image obtained by integration of 9 looks. (*c*) Reduced-resolution image obtained with 3×3 spatial filtering. (*d*) Image obtained with 3×3 median filter. (Original image courtesy of Sandia National Laboratories.)

filter applied to a single look image. This approach sacrifices less resolution than spatial averaging. Other techniques include a variety of adaptive filters and statistical methods.

References

Ausherman, D. A., "Digital vs. Optical Techniques in Synthetic Aperture Radar (SAR) Data Processing," *Optical Engineering.*, vol. 19(2), pp. 157–167, March/April 1980.

Ausherman, D. A., et al., "Developments in Radar Imaging," *IEEE Transactions on Aerospace & Electronic Systems*, vol. AES-20, no. 4, pp. 363–400, July 1984.

Bamler, R., "A Comparison of Range-Doppler and Wavenumber Domain SAR Focusing Algorithms," *IEEE Transactions on Geoscience and Remote Sensing*, vol. 30, no. 4, pp. 706–713, July 1992.

Brown, W. M., and L. J. Portello, "An introduction to synthetic-aperture radar," *IEEE Spectrum*, pp. 52–62, Sept. 1969.

Cafforio, C., C. Prati, and F. Rocca, "SAR Data Focusing Using Seismic Migration Techniques," *IEEE Transactions on Aerospace & Electronic Systems*, vol. AES-27, no. 2, pp. 194–207, March 1991.

Carrara, W. G., R. S. Goodman, and R. M. Majewski, *Spotlight Synthetic Aperture Radar*. Artech House, Norwood, MA, 1995.

Cumming, I. G., and F. H. Wong, *Digital Processing of Synthetic Aperture Radar Data*. Artech House, Norwood, MA, 2005.

Curlander, J. C., and R. N. McDonough, *Synthetic Aperture Radar: Systems and Signal Processing*. Wiley, New York, 1991.

Cutrona, L. J. et al., "On the Application of Coherent Optical Processing Techniques to Synthetic-Aperture Radar," *Proceedings of the IEEE*, vol. 54 (8), pp. 1026–1032, Aug. 1966.

Desai, M. D., and W. K. Jenkins, "Convolution Backprojection Image Reconstruction for Spotlight Mode Synthetic Aperture Radar," *IEEE Transactions on Image Processing*, vol. 1(4), pp. 505–517, Oct. 1992.

Dudgeon, D. E., and R. M. Mersereau, *Multidimensional Digital Signal Processing*. Prentice Hall, Englewood Cliffs, NJ, 1984.

Elachi, C., *Spaceborne Radar Remote Sensing: Applications and Techniques*. IEEE Press, New York, 1988.

Franceschetti, G., and R. Lanari, *Synthetic Aperture Radar Processing*. CRC Press, New York, 1999.

Freeman, A., et al., "The Myth of the Minimum SAR Antenna Area Constraint," *IEEE Transactions on Geoscience and Remote Sensing*, vol. 38, no. 1, pp. 320–324, Jan. 2000.

Ghiglia, D. C., and M. D. Pritt, *Two-Dimensional Phase Unwrapping: Theory, Algorithms, and Software*. Wiley, New York, 1998.

Ghiglia, D. C., and L. A. Romero, "Robust Two-dimensional Weighted and Unweighted Phase Unwrapping That Uses Fast Transforms and Iterative Methods," *Journal of Optical Society of America*, vol. 11, no. 1, pp. 107–117, Jan. 1994.

Gough, P. T., and D. W. Hawkins, "Unified Framework for Modern Synthetic Aperture Imaging Algorithms," *International Journal of Imaging Systems & Technology*, vol. 8, pp. 343–358, 1997.

Harger, R. O., *Synthetic Aperture Radar Systems: Theory and Design*. Academic Press, New York, 1970.

Jakowatz, C. V., et al., *Spotlight Mode Synthetic Aperture Radar*. Kluwer, Boston, 1996.

Jakowatz, C. V., Jr., and D. E. Wahl, "An Eigenvector Method for Maximum Likelihood Estimation of Phase Errors in SAR Imagery," *Journal of Optical Society of America*, vol. 10(12), pp. 2539–2546, Dec. 1993.

Kennedy, T. A., "The Design of SAR Motion Compensation Systems Incorporating Strapdown Inertial Measurement Units," *Proceedings of the IEEE 1988 National Radar Conference*, pp. 74–78, April 1988a.

Kennedy, T. A., "Strapdown Inertial Measurement Units for Motion Compensation for Synthetic Aperture Radars," *IEEE AESS Magazine*, vol. 3, no. 10, pp. 32–35, Oct. 1988b.

Kirk, J. C., Jr., "Discussion of Digital Processing in Synthetic Aperture Radar," *IEEE Transactions on Aerospace & Electronic Systems*, vol. AES-11, no. 3, pp. 326–337, May 1975.

Lacomme, P., J.-P. Hardange, J.-C. Marchais, and E. Normant, *Air and Spaceborne Radar Systems*. William Andrew Publishing, Norwich, NY, 2001.

Munson, D. C., Jr., J. D. O'Brien, and W. K. Jenkins, "A Tomographic Formulation of Spotlight-Mode Synthetic Aperture Radar," *Proceeding of the IEEE*, vol. 71(8), pp. 917–925, Aug. 1983.

Munson, D. C., Jr., et al., "A Comparison of Algorithms for Polar-to-Cartesian Interpolation in Spotlight Mode SAR," *Proceedings of the IEEE International Conference on Acoustics, Speech, and Signal Processing*, vol. 10, pp. 1364–1367, 1985.

Munson, D. C., Jr., and R. L. Visentin, "A Signal Processing View of Strip-Mapping Synthetic Aperture Radar," *IEEE Transactions on Acoustics, Speech, and Signal Processing*, vol. 27, no. 12, pp. 2131–2147, Dec. 1989.

Oliver, C., and S., Quegan, *Understanding Synthetic Aperture Radar Images*. SciTech Publishing, Raleigh, NC, 2004.

Raney, R. K., "A New and Fundamental Fourier Transform Pair," *Proceedings of IEEE 12th International Geoscience & Remote Sensing Symposium* (IGARSS '92), pp. 106–107, 26–29 May 1992.

Richards, M. A., "Nonlinear Effects in Fourier Transform Processing," Chap. 6 in *Coherent Radar Performance Estimation*, J. A. Scheer and J. L. Kurtz (eds.). Artech House, Norwood, MA, 1993.

Schleher, C. C., *MTI and Pulsed Doppler Radar*. Artech House, Norwood, MA, 1991.

Sherwin, C. W., J. P. Ruina, and R. D. Rawcliffe, "Some Early Developments in Synthetic Aperture Radar Systems," *IRE Transactions on Military Electronics*, vol. MIL-6, no. 2, pp. 111–115, April 1962.

Stimson, G. W., *Introduction to Airborne Radar*, 2d ed. SciTech Publishing, Mendham, NJ, 1998.

Soumekh, M., *Synthetic Aperture Radar Signal Processing With Matlab Algorithms*. Wiley, New York, 1999.

Sullivan, R. J., *Microwave Radar: Imaging and Advanced Concepts*. Artech House, Norwood, MA, 2000.

Wahl, D. E., P. H. Eichel, D. C. Ghiglia, and C. V. Jakowatz, Jr., "Phase Gradient Autofocus—A Robust Tool for High Resolution SAR Phase Correction," *IEEE Transactions on Aerospace & Electronics Systems*, vol. AES-30, no. 5, pp. 827–835, July 1994.

Wiley, C. A., "Pulsed Doppler Radar Methods and Apparatus," U. S. patent no. 3,196, 436, 1965 (originally filed, 1954).

Wiley, C. A., "Synthetic Aperture Radars—A Paradigm for Technology Evolution," *IEEE Transactions on Aerospace & Electronic Systems*, vol. AES-21, pp. 440–443, 1985.

Wu, C., K. Y. Liu, and M. Jin, "Modeling and a Correlation Algorithm for Spaceborne SAR Signals," *IEEE Transactions Aerospace & Electronics Systems*, vol. AES-18(5), pp. 563–574, Sept. 1982.

Chapter 9

Introduction to Beamforming and Space-Time Adaptive Processing

In Chap. 3, the concept of the radar datacube was introduced to describe the data collected in a coherent processing interval. Figure 3.1, repeated in part here as Fig. 9.1, illustrates the datacube $y[l, m, n]$, with its independent axes of fast time, slow time, and antenna phase center. The radar signal processing described up to this point has dealt almost entirely with a fast-time/slow-time matrix for a single phase center antenna; little use has been made of the third datacube dimension, other than to discuss the array factor of a phased array antenna pattern in Chap. 1.

Just as the temporal sampling of the slow-time axis enables analysis and processing of signals in a given range bin based on their temporal Doppler frequency content, the phase center axis enables analysis and processing of signals within a range bin based on their spatial frequency content, which is equivalent to the angle of arrival. In this chapter, a basic introduction to *beamforming* and *space-time adaptive processing* (STAP) is presented. Beamforming refers to the coherent combination of data from multiple phase centers to provide selectivity in the angle of arrival, i.e., to form and steer an antenna beam. STAP combines both spatial and temporal filtering on a moving radar platform to optimally discriminate targets from both clutter and jamming.

9.1 Spatial Filtering

9.1.1 Conventional beamforming

Consider a monochromatic plane wave with temporal variation $Ae^{j\Omega t}$ impinging on a uniform linear one-dimensional array as shown in Fig. 9.2. If the wave's

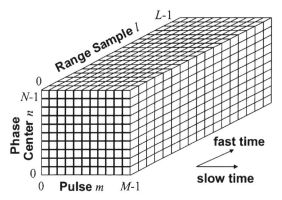

Figure 9.1 Radar datacube for a single CPI.

angle of arrival (AOA) relative to the array normal is θ radians, the signal observed at the nth array element is

$$\bar{y}_n(t) = A e^{j[\Omega(t - nd \sin\theta/c) + \phi_0]} \tag{9.1}$$

where the phase offset ϕ_0 accounts for the absolute phase at the $n = 0$ element. Now consider the single sample $y[n]$ formed from the individual array signals, sampled at a common time t_0

$$y[n] \equiv \bar{y}_n(t_0) = A e^{j[\Omega(t_0 - nd \sin\theta/c) + \phi_0]}$$
$$= \tilde{A} e^{-j\Omega nd \sin\theta/c} = \tilde{A} e^{-j 2\pi nd \sin\theta/\lambda} \quad n = 0, \ldots, N-1 \tag{9.2}$$

Assembling the N element samples into vector form gives a *snapshot* of the array at a fixed time

$$\begin{aligned}
\mathbf{y} &= [y[0] \quad y[1] \quad \cdots \quad y[N-1]]' \\
&= \hat{A} \begin{bmatrix} 1 & e^{-j 2\pi d \sin\theta/\lambda} & \cdots & e^{-j 2\pi (N-1)d \sin\theta/\lambda} \end{bmatrix}' \\
&= \hat{A} \begin{bmatrix} 1 & e^{-j K_\theta} & \cdots & e^{-j(N-1)K_\theta} \end{bmatrix}' \\
&\equiv \hat{A} \mathbf{a}_s(\theta)
\end{aligned} \tag{9.3}$$

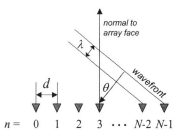

Figure 9.2 Wavefront impinging on uniform linear array.

where $K_\theta \equiv 2\pi d \sin\theta/\lambda$ is the spatial frequency in cycles and $\mathbf{a}_s(\theta)$ is the *spatial steering vector*. Thus, there is a one-to-one relationship between the AOA of a plane wave and the spatial frequency across the array face. Note that the range of K_θ is $\pm \pi d/\lambda$; the common element spacing of $d = \lambda/2$ gives $K_\theta \in (-\pi, +\pi)$. It is also useful to define $F_\theta \equiv K_\theta/2\pi$.

Conventional nonadaptive beamforming is implemented as a weighted sum of the element signals, $z = \mathbf{h}'\mathbf{y}$,[†] where $\mathbf{h}$ is a vector of complex weights

$$\mathbf{h} = [h_0 \quad h_1 \quad \cdots \quad h_{N-1}]' \tag{9.4}$$

A special case of interest occurs when $\mathbf{h}$ takes the form

$$\mathbf{h} = [w_0 \quad w_1 e^{+jK_\theta} \quad \cdots \quad w_{N-1} e^{+j(N-1)K_\theta}]'$$
$$= [w_0 \quad w_1 \quad \cdots \quad w_{N-1}]' \odot \mathbf{a}_s(\theta)$$
$$= \mathbf{w}' \odot \mathbf{a}_s(\theta) \tag{9.5}$$

The symbol $\odot$ represents the Hadamard (element-by-element) product of two vectors, i.e., if $\mathbf{a}$ and $\mathbf{b}$ are two N-element column vectors, then

$$\mathbf{a} \odot \mathbf{b} \equiv [a_0 b_0 \quad a_1 b_1 \quad \cdots \quad a_{N-1} b_{N-1}]' \tag{9.6}$$

Equation (9.5) represents $\mathbf{h}$ as the Hadamard product of two factors: a shape vector $\mathbf{w}$ that provides for any weighting for side lobe control, and a steering vector $\mathbf{a}_s(\theta)$ that provides maximum coherent integration for signals arriving from angle θ.

Suppose the weights are matched to an angle of θ_0; the array is said to be "steered" to θ_0. The response of a beamformer steered to θ_0 to an incoming wavefront at angle θ is

$$z(\theta) = \mathbf{h}'\mathbf{y} = \hat{A} \sum_{n=0}^{N-1} \alpha_n e^{-j(K_\theta - K_{\theta_0})n} \tag{9.7}$$

Note that $z(\theta)$ is just the discrete Fourier transform of the weight sequence $\{h_n\}$, shifted to a center spatial frequency of K_{θ_0} and multiplied by $\hat{A}$. When all of the weight amplitudes $\{w_n\} = 1$, this is a standard asinc pattern

$$z(\theta) = e^{j(N-1)(\pi d/\lambda)(\sin\theta - \sin\theta_0)} \left\{ \frac{\sin[N(\pi d/\lambda)(\sin\theta - \sin\theta_0)]}{\sin[(\pi d/\lambda)(\sin\theta - \sin\theta_0)]} \right\} \tag{9.8}$$

More commonly, the weights are chosen to reduce antenna side lobes at the cost of degraded resolution in the form of a wider mainbeam. Figure 9.3 illustrates the antenna pattern $|z(\theta)|$ for an array steered to $\theta_0 = 30°$ with $N = 11$ phase centers, both with and without Hamming weighting. In the latter case,

---

[†]Most STAP literature defines the filter according to $z = \mathbf{h}^H \mathbf{y}$. The convention $z = \mathbf{h}'\mathbf{y}$ is retained here for consistency with the discussion of vector matched filtering in Chap. 5. A consequence is that the forms for $\mathbf{h}$ obtained here are the conjugate of the results in common STAP literature.

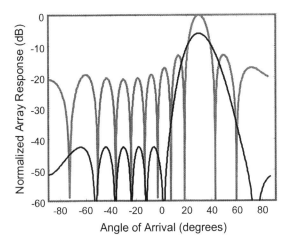

Figure 9.3 Antenna pattern with and without Hamming weighting. $N = 11$ phase centers, steering angle $\theta_0 = 30°$.

w is simply a vector of all 1s; in the former case, it is a Hamming window function.

The previous results can be obtained as the matched filter solution to a spatial filtering problem, using the same results developed for Doppler filtering in Chap. 5. Suppose it is desired to maximize the output $z(\theta)$ when the input is a snapshot of a plane wave having AOA θ, plus white noise. The math is exactly the same as used for vector matched filtering in Chap. 5, and the same results can be applied. Specifically

$$\mathbf{h} = \kappa \mathbf{S}_\mathbf{I}^{-1} \mathbf{t}^* \tag{9.9}$$

where $\mathbf{S}_\mathbf{I}$ = covariance matrix of the interference at the N phase center outputs
$\mathbf{t}$ = model of the desired target signal vector (i.e., steering vector)
κ = arbitrary constant

Recall that $\mathbf{S}_\mathbf{I} = \mathbf{S}_\mathbf{I}^H$ for covariance matrices. If the signal of interest is a monochromatic plane wave with AOA θ, the target model $\mathbf{t}$ is (letting $\hat{A} = 1$ without loss of generality) exactly the spatial steering vector $\mathbf{a}_s(\theta)$ defined in Eq. (9.3). When the interference is independent identically-distributed (i.i.d.) white noise with variance σ^2 at each element, $\mathbf{S}_\mathbf{I} = \sigma^2 \mathbf{I}$. Choosing $\kappa = 1/\sigma^2$ gives

$$\mathbf{h} = \mathbf{a}_s^*(\theta) = \begin{bmatrix} 1 & e^{+jK_\theta} & \cdots & e^{+j(N-1)K_\theta} \end{bmatrix}' \tag{9.10}$$

The filter output is then

$$z(\theta) = \mathbf{h}'\mathbf{y} = \sum_{n=0}^{N-1} y[n]e^{+jK_\theta n} = \sum_{n=0}^{N-1} y[n]e^{-j\widetilde{K}_\theta n} \qquad (9.11)$$

which shows that the antenna pattern is the *discrete-time Fourier transform* (DTFT) of the data snapshot, evaluated at $\widetilde{K}_\theta \equiv -K_\theta$. If the observed signal $y[n]$ is in fact a snapshot of a plane wave at AOA θ_0, $\mathbf{y}$ is of the form of Eq. (9.3) with $\theta = \theta_0$, and $z(\theta)$ is again given by Eq. (9.7). The peak output occurs when the filter-steering vector matches the actual steering vector of the input, i.e., at $z(\theta_0)$, and is just N times the amplitude of $y[n]$. A plot of $z(\theta)$ as a function of AOA θ gives the (unweighted) normalized antenna gain pattern of Fig. 9.3.

Note that, while the *fast fourier transform* (FFT) can be used to efficiently compute the antenna pattern for a given steering vector $\mathbf{t}$, the FFT output will be in terms of $\widetilde{K}_\theta$ and must be flipped to obtain z as a function of K_θ. The flipped pattern will be sampled at constant increments of spatial frequency $K_\theta = 2\pi d \sin\theta/\lambda$. Thus, the samples are not spaced uniformly in angle θ. If pattern samples at constant increments in θ are required, $z(\theta)$ must be computed explicitly using $z(\theta) = \mathbf{h}'\mathbf{y}$ and varying $\mathbf{y}$ for each desired value of AOA.

9.1.2 Adaptive beamforming

The vector matched filter approach leads directly to a means for designing array weight vectors $\mathbf{h}$ that can steer zeros of the antenna pattern, often called *nulls*, in specific directions to cancel interference sources. This capability is useful in combating *jammers*, which are hostile interfering signal sources that seek to degrade radar performance by any of a number of mechanisms, such as degrading the signal-to-noise ratio (SNR) by increasing the noise level, or creating false detections to overwhelm the radar with false targets. One of the most common forms of jamming is a simple noise jammer. This device radiates a relatively high-power waveform at the victim radar from a specific air-, space-, or ground-based platform. The jammer waveform is a random noise process that is white over the receiver bandwidth of the victim radar.[†] From the radar's point of view, the jammer signal is therefore a white noise process arriving from some specific AOA.

The antenna pattern that maximizes the output SNR in the presence of both white noise and jamming is still given by Eq. (9.9), however, the model for the interference covariance matrix $\mathbf{S_I}$ must be modified to incorporate a model for

---

[†]Effective jamming thus requires knowledge of the victim radar frequency and bandwidth. Unless the jammer has a priori knowledge of the precise portion of the band used by the radar, or can estimate it by detecting and analyzing the radar signal, it must spread its energy over a bandwidth wider than that actually used by the radar.

the noise jammer. The temporal variation of the jammer signal $J_n(t)$ received from AOA θ at each array phase center can be expressed as

$$J_n(t) = \sigma_J^2 w(t) e^{j[\Omega(t-nd\sin\theta/c)+\phi_0]} \tag{9.12}$$

where σ_J^2 is the jammer power and $w(t)$ is a unit variance zero mean white random process. A snapshot of the array response to the jammer gives

$$J[n] \equiv J_n(t_0) = \sigma_J w(t_0) e^{j[\Omega(t_0-nd\sin\theta/c)+\phi_0]}$$
$$= \sigma_J \hat{w}(t_0) e^{-j\Omega nd\sin\theta/c} = \sigma_J \hat{w}(t_0) e^{-j2\pi nd\sin\theta/\lambda}$$
$$= \sigma_J \hat{w}(t_0) e^{-jK_\theta n} \tag{9.13}$$

In vector form, this is

$$\mathbf{J} = \sigma_J \hat{w}(t_0) \begin{bmatrix} 1 & e^{-jK_\theta} & \cdots & e^{-j(N-1)K_\theta} \end{bmatrix}'$$
$$= \sigma_J \hat{w}(t_0) \mathbf{a}_s(\theta) \tag{9.14}$$

The covariance matrix of the jammer signals is then

$$\mathbf{S_J} = \mathbf{E}\{\mathbf{J}^*\mathbf{J}'\} = \sigma_J^2 \mathbf{E}\{|w(t_0)|^2\} \mathbf{a}_s^*(\theta)\mathbf{a}_s'(\theta)$$
$$= \sigma_J^2 \mathbf{a}_s^*(\theta)\mathbf{a}_s'(\theta)$$
$$= \sigma_J^2 \begin{bmatrix} 1 & e^{-jK_\theta} & \cdots & e^{-j(N-1)K_\theta} \\ e^{+jK_\theta} & 1 & e^{-jK_\theta} & e^{+j(N-2)K_\theta} \\ \vdots & \vdots & \ddots & \vdots \\ e^{+j(N-1)K_\theta} & e^{-j(N-2)K_\theta} & \cdots & 1 \end{bmatrix} \tag{9.15}$$

Thus, $\mathbf{S_J}$ is a Hermitian matrix, so that $\mathbf{S_J} = \mathbf{S_J}^H$. The total interference covariance matrix for the sum of receiver noise and some number P of mutually uncorrelated jammers is

$$\mathbf{S_I} = \sigma^2 \mathbf{I} + \sum_{p=0}^{P-1} \mathbf{S}_{\mathbf{J}_p} \tag{9.16}$$

The matched filter output can then be maximized by using the interference model of Eq. (9.16) in Eq. (9.9). As an example, consider a case with $N = 16$ antenna phase centers. Assume that two jammers are present, one at an AOA of $+18°$ with an SNR of $+50$ dB, and another at $-33°$ and an SNR of 30 dB. Figure 9.4 shows the resulting antenna patterns when the antenna is steered to $0°$ (sidelooking). In the left half of the figure, no adaptation was used; i.e., the beamformer weights were designed using $\mathbf{S_I} = \sigma^2 \mathbf{I}$. A standard pattern results. The two dotted vertical lines indicate the angles of arrival of the two jammers. While in the side lobes, both are on or near side lobe peaks and, given

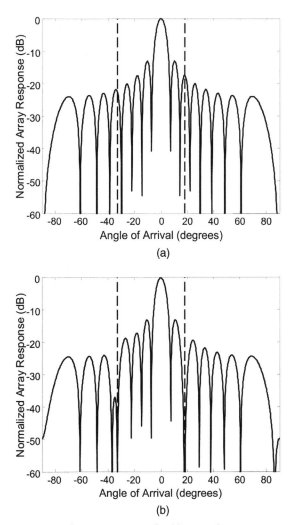

Figure 9.4 Antenna pattern for $N = 16$ phase centers. Two jammers present at indicated angles. (a) Without adaptation. (b) With adaptation.

their high power, will result in significant jamming energy in the beamformer output. In the right half of the figure, the beamformer weights were designed using a covariance matrix $\mathbf{S_I}$ computed via Eq. (9.16). The resulting pattern places deep nulls at the location of both jammers, effectively canceling them.

The *signal-to-interference ratio* (SIR) achieved by the optimum beamformer can be computed using the result of Eq. (5.7), namely

$$SIR = \frac{\mathbf{h}^H \mathbf{t}^* \mathbf{t}' \mathbf{h}}{\mathbf{h}^H \mathbf{S_I} \mathbf{h}} \qquad (9.17)$$

For the case where the interference is noise only, $\mathbf{S_I} = \sigma^2 \mathbf{I}$, $\mathbf{h} = \mathbf{a}_s^*(\theta)$, $\mathbf{t} = \tilde{A}\mathbf{a}_s(\theta)$, and since $\mathbf{a}_s'(\theta)\mathbf{a}_s^*(\theta) = \sum_n |\mathbf{a}_{sn}(\theta)|^2$, Eq. (9.17) reduces to

$$SIR = \frac{\left(|\hat{A}_s|^2 \sum_{n=0}^{N-1} |\mathbf{a}_{sn}(\theta)|^2\right)^2}{\sigma^2 |\hat{A}_s|^2 \sum_{n=0}^{N-1} |\mathbf{a}_{sn}(\theta)|^2} = \frac{|\hat{A}_s|^2 N^2}{\sigma^2 N} = N\frac{|\hat{A}_s|^2}{\sigma^2} \quad (9.18)$$

For the example of Fig. 9.4, the SIR gain relative to the single-element SIR of $|\tilde{A}_s|^2/\sigma^2$ for the case without jammers is therefore 16 (12.04 dB). The peak gain for the case with jammers is reduced by 0.2 dB while the SIR gain, computed using Eq. (9.16) in Eq. (9.17), is reduced slightly to 15.64 (11.94 dB), essentially because some of the *degrees of freedom* (DOF) are consumed canceling the jammers instead of providing coherent integration gain for the signal.

The variation in peak gain in the direction of the desired target signal $\mathbf{t}$ can be compensated by requiring that $\mathbf{h}'\mathbf{t} = 1$, resulting in what is often called a *distortionless* beamformer. Using this condition in Eq. (9.9) gives

$$\begin{aligned} 1 = \mathbf{h}'\mathbf{t} &= (\kappa \mathbf{S_I}^{-1}\mathbf{t}^*)'\mathbf{t} = \kappa \mathbf{t}^H (\mathbf{S_I}^{-1})'\mathbf{t} \\ &= \kappa \mathbf{t}^H (\mathbf{S_I}^{-1})^*\mathbf{t} = \kappa \mathbf{t}^H (\mathbf{S_I}^{-1}\mathbf{t}^*)^* \Rightarrow \\ \kappa &= \frac{1}{\mathbf{t}^H (\mathbf{S_I}^{-1}\mathbf{t}^*)^*} \quad \mathbf{h} = \frac{\mathbf{S_I}^{-1}\mathbf{t}^*}{\mathbf{t}^H (\mathbf{S_I}^{-1}\mathbf{t}^*)^*} \end{aligned} \quad (9.19)$$

Since κ is simply a scalar, this choice does not affect the shape of the antenna pattern $z(\theta)$, but merely scales it up or down in amplitude so that $z(\theta_0) = 1$, where θ_0 is the AOA of the target signal $\mathbf{t}$. This design approach can be extended to constrain the pattern gain at multiple AOAs; details are given by Van Trees (2002).

For each distinct jammer signal $\mathbf{J}_p$, the vector matched filter chooses $\mathbf{h}$ such that $\mathbf{h}'\mathbf{J}_p = 0$ (Guerci, 2003). If P jammers are present, this creates P such conditions. The distortionless constraint $\mathbf{h}'\mathbf{t} = 1$ adds a $(P+1)$th condition. There are therefore $P+1$ equations in the N unknowns of the filter vector $\mathbf{h}$. A solution will exist so long as $P+1 \leq N$. Thus, an N-phase center array can cancel up to $N-1$ jammers.

If a jammer is located in the mainbeam of the antenna pattern, the adaptive pattern is seriously degraded. In the previous example, the 3-dB beamwidth is approximately 6°. If the jammer at −33° is moved instead to −2° and the distortionless beamformer of Eq. (9.19) is applied, the pattern of Fig. 9.5 results. The two jammers are nulled, but the antenna pattern peak has been shifted from the target AOA of 0°. The distortionless constraint still guarantees that the antenna pattern gain is unity at 0°, but the peak is now +5.15 dB, occurring at $\theta = 3.3°$. In this example, the SIR gain is reduced to only 3.64 (5.61 dB).

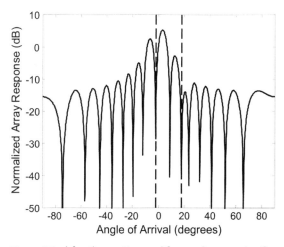

Figure 9.5 Adaptive pattern with one jammer in the main lobe of the adapted pattern.

9.1.3 Adaptive beamforming with preprocessing

The discussion so far has applied adaptive interference calculation directly to the individual array element signals, an approach called *element space* processing. Some systems perform fixed, conventional beamforming on the element signals first, and then apply adaptive processing to the individual beam outputs. This technique is called *beamspace processing*. Figure 9.3 was an example of forming a single fixed beam; a bank of such beams would provide coverage of the entire angle space. Beamspace processing tends to concentrate each interferer into the output of a small number, possibly only one, of beams, and can also reduce the dimensionality of the adaptive processing problem. As will be seen in Sec. 9.5, this reduction can lead to major savings in computational load.

Beams are formed as linear combinations of the element signals, as shown in the example of Eq. (9.7). A set of P beams formed from an N-element array can thus be represented as a $P \times N$ transformation matrix $\mathbf{T}$ acting on the $N \times 1$ spatial snapshot $\mathbf{y}$ to form a new $P \times 1$ vector of beam outputs $\tilde{\mathbf{y}}$ (Ward, 1994; Melvin, 2004)

$$\tilde{\mathbf{y}} = \mathbf{T}\mathbf{y} \qquad (9.20)$$

For instance, a set of conventional "DFT beams" at spatial frequencies K_{θ_0}, $K_{\theta_1}, \ldots, K_{\theta_{P-1}}$ can be formed with the transformation matrix

$$\mathbf{T} = \begin{bmatrix} 1 & e^{-jK_{\theta_0}} & e^{-j2K_{\theta_0}} & \cdots & e^{-j(N-1)K_{\theta_0}} \\ 1 & e^{-jK_{\theta_1}} & e^{-j2K_{\theta_1}} & \cdots & e^{-j(N-1)K_{\theta_1}} \\ \vdots & \vdots & \vdots & \ddots & \vdots \\ 1 & e^{-jK_{\theta_{P-1}}} & e^{-j2K_{\theta_{P-1}}} & \cdots & e^{-j(N-1)K_{\theta_{P-1}}} \end{bmatrix} \qquad (9.21)$$

If the $\{K_{\theta_p}\}$ are evenly spaced over the interval $(-\pi, +\pi)$, $\mathbf{T}$ computes the P-point DFT of the element snapshot.

To see how to apply adaptive processing to the preprocessed beamspace data, note that the snapshot $\mathbf{y}$ is now replaced with $\tilde{\mathbf{y}}$. This applies to whatever signals are present at the array face, so the transformation is applied to the interference as well as the target signals. The new interference covariance matrix becomes

$$\tilde{\mathbf{S}}_\mathbf{I} = \tilde{\mathbf{y}}^*\tilde{\mathbf{y}}'$$
$$= (\mathbf{Ty})^*(\mathbf{Ty})' = \mathbf{T}^*\mathbf{y}^*\mathbf{y}'\mathbf{T}'$$
$$= \mathbf{T}^*\mathbf{S}_\mathbf{I}\mathbf{T}' \qquad (9.22)$$

The adaptive weight vector and the filtered output are

$$\tilde{\mathbf{h}} = \kappa \tilde{\mathbf{S}}_\mathbf{I}^{-1}\tilde{\mathbf{t}}^*$$
$$\tilde{z}(\theta) = \tilde{\mathbf{h}}'\tilde{\mathbf{y}} \qquad (9.23)$$

where $\tilde{\mathbf{t}} = \mathbf{Tt}$ is the transformed target steering vector.

To illustrate this process, a DFT matrix $\mathbf{T}$ with $P = 10$ beams was applied to the same $N = 16$ element example used previously. In the absence of jammers, or when the jammers are located at $+18°$ and $-33°$, the resulting adapted antenna patterns are indistinguishable from Figs. 9.4a and b, respectively. When the jammer at $-33°$ is moved into the main lobe at $-2°$, the pattern of Fig. 9.6 results. This pattern still places nulls at both jammer locations, but details of the main lobe distortion differ from the element space example of Fig. 9.5.

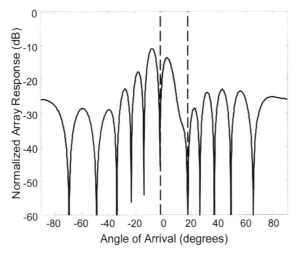

Figure 9.6 Adapted pattern using a 10-beam DFT fixed beamformer preprocessor. Compare to Fig. 9.5.

Note that the beamspace adaptive weight computation requires solving only a 10th-order system of equations, while the element space version requires the solution of a 16th-order system. The beamspace approach has the extra step of applying the beam formation preprocessor **T**, but in many cases the total computational load is less, especially if **T** has a structure that can be implemented efficiently, such as an FFT matrix.

The approach described here can be used with any linear transformation **T**. Additional examples include beams uniformly spaced in θ instead of K_θ, or approaches that form conventional beams and then combine sums of adjacent beams to achieve better cancellation. The technique can also be extended to include windowing for side lobe control and gain constraints similar to the distortionless constraint discussed previously.

9.2 Space-Time Signal Environment

In a multiphase center radar, filtering is possible in both Doppler shift and angle of arrival. It is therefore important to characterize the data in a given range bin in terms of Doppler and AOA. Figure 9.7 is a notional sketch of the general behavior of noise, jamming, clutter, and moving targets in the space-time environment.

Receiver noise has no structure in time or frequency, and therefore appears as a uniform noise floor throughout the angle-Doppler space. As described in the previous section, broadband noise jammers are localized in AOA but spread across the entire Doppler spectrum. This is reflected in Fig. 9.7 as a ridge of energy localized in AOA but spread across all Doppler shifts. Because the

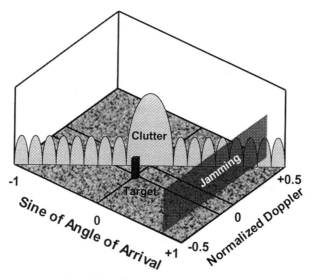

Figure 9.7 Space-time signal environment.

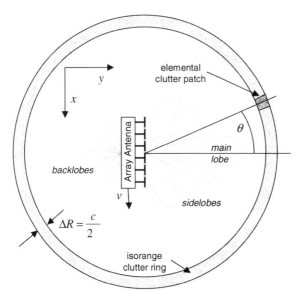

Figure 9.8 Clutter contributing to the angle-Doppler spectrum.

jamming energy is the same at all values of Doppler, discrimination against jamming must be based primarily on spatial filtering as described in the previous section.

Clutter is more complicated. Assume a platform with velocity v and a sidelooking radar,[†] as shown in Fig. 9.8; the height dimension is neglected for simplicity. In a given range bin, clutter scatterers anywhere on the isorange circle corresponding to the range of interest contribute to the total clutter return. (If the system is range ambiguous, there will be multiple isorange rings contributing to a given range bin.) A clutter scatterer directly on the radar boresight is at a squint angle of 90° with respect to the velocity vector; therefore the Doppler shift for that scatterer is zero. More generally, scatterers at an angle of θ radians with respect to the antenna boresight will have a Doppler shift of $(2v/\lambda)\sin\theta$ hertz. Note that there are two such patches that will produce the same Doppler shift, one in the radar look direction, and one behind it in the *back lobes*. The back lobes are often ignored due to low antenna gain in that direction, but in some systems they must be considered. If the back lobe return is ignored, there is a one-to-one relationship between AOA and Doppler shift for clutter

$$F_D = \frac{2v}{\lambda}\sin\theta \qquad (9.24)$$

[†]These results generalize readily to nonsidelooking cases, but for simplicity only the sidelooking case is considered in this chapter.

or in normalized units

$$f_D = \left(\left(\frac{2vT}{\lambda}\sin\theta\right)\right)_{1.0} \quad (9.25)$$

where T is the PRF and the notation $((\cdot))_{1.0}$ indicates arithmetic modulo 1.0 due to the use of discrete-time Fourier analysis. Echo from stationary ground clutter that has a Doppler shift of F_D hertz can therefore be presumed to be arriving at an angle of θ radians with respect to the sidelooking antenna boresight. Consequently, clutter echoes tend to fall on a diagonal ridge in a Doppler-$\sin\theta$ space, as diagrammed in Fig. 9.7. The amplitude of ground clutter from different AOAs is determined by the antenna gain in that direction, and is therefore largest near boresight, and lower in the side lobes and back lobes of the antenna.

Equation (9.25) can be rewritten as

$$\omega_D = 2\pi f_D = \left(\left(\frac{2vT}{d}K_\theta\right)\right)_{2\pi}$$
$$= ((\beta K_\theta))_{2\pi} \quad (9.26)$$

where $\beta \equiv 2vT/d$. Thus, β is the slope of the clutter ridge when plotted in (K_θ, ω_D) coordinates; it also represents the number of times the clutter ridge spans the range of $-\pi$ to $+\pi$ radians (-0.5 to $+0.5$ cycles) in Doppler as the AOA varies from $-\pi$ to $+\pi$.

Because F_D is proportional to $\sin\theta$ rather than θ itself, the clutter ridge is a straight line in Doppler-$\sin\theta$ space; however, if plotted as a function of θ instead of $\sin\theta$, the ridge curves visibly as the AOA approaches $\pm 180°$. This effect is visible in Fig. 9.9a which is a simulation of the angle-Doppler spectrum for a medium PRF radar. Two jammers are present, at approximately $-40°$ and $+60°$. The clutter ridge is diagonal through the center of the spectrum, but curves noticeably at $\pm 60°$. Also, the discrete time Doppler spectrum is periodic with period equal to the PRF. Thus, if the magnitude of F_D of Eq. (9.24) exceeds $PRF/2$, the clutter ridge will alias. This effect is seen in Fig. 9.9b that simulates a low PRF angle-Doppler spectrum.

The angle-Doppler characteristics of the echo from a (possibly) moving point target depend on both the radar platform motion and the target motion. Assuming the target *radar cross section* (RCS) is small enough that detection is likely only if it is in the radar mainbeam, the target can be presumed to be within a few degrees of the radar boresight. The Doppler shift will depend on the total radial velocity. Thus, if the target is stationary and directly on the boresight, the Doppler shift will be zero and it will fold in with the clutter. However, if the target is moving, it will separate from the clutter on the Doppler axis, as is shown in Fig. 9.7. This fact illustrates a key reason for the interest in space-time processing techniques, especially for the detection of relatively slow-moving ground targets ("slow movers"). As seen in Chap. 5, if only Doppler processing is used, the target Doppler shift must typically exceed the Doppler width of the clutter

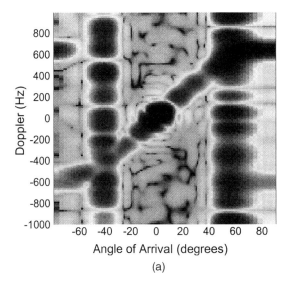

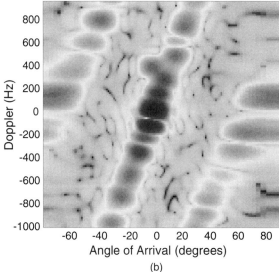

Figure 9.9 Simulation of angle-Doppler spectrum. Amplitude is in dB. (*a*) Illustration of clutter ridge curvature and two jammers. (*b*) Illustration of clutter ridge aliasing. (Figure courtesy of Dr. W. L. Melvin, GTRI.)

spectrum to achieve a signal-to-interference ratio adequate for detection. If the platform velocity is high or the mainbeam relatively wide, ground clutter can fill most of the Doppler spectrum, making detection of slow movers very difficult. Figure 9.7 shows that introducing spatial processing gives a second dimension in which to separate the target from the clutter. Any Doppler shift of the target echo causes it to compete with clutter arriving from a different AOA; the added capability of filtering based on AOA then allows separation of target and clutter having the same Doppler shift.

9.3 Space-Time Signal Modeling

Space-time adaptive processing applies vector matched filtering to the combined slow-time/phase center data set in each range bin. It is usually assumed that pulse compression has been applied prior to STAP processing. The two-dimensional slice of the datacube in range bin l_0, $y[l_0, m, n]$, is called a *space-time snapshot* (or just *snapshot*) of the data. To proceed, the $N \times M$ two-dimensional snapshot is converted to an $NM \times 1$ column vector by stacking the columns

$$\mathbf{y} = \begin{bmatrix} y[l_0, 0, 0] \\ y[l_0, 0, 1] \\ \vdots \\ y[l_0, 0, N-1] \\ y[l_0, 1, 0] \\ y[l_0, 1, 1] \\ \vdots \\ y[l_0, 0, N-1] \\ \vdots \\ y[l_0, M-1, 0] \\ y[l_0, M-1, 1] \\ \vdots \\ y[l_0, M-1, N-1] \end{bmatrix} \qquad (9.27)$$

This process of converting the data from a given range bin to a one-dimensional vector is illustrated in Fig. 9.10.

Next, the filter weight vector $\mathbf{h}$ must be designed using Eq. (9.9) or (9.19). The target model vector $\mathbf{t} = \mathbf{t}(f_D, \theta)$ must represent the expected signal from a target at some specific Doppler shift f_{Dt} and AOA θ_t of interest. Define a *temporal steering vector*

$$\mathbf{a}_t(f_{Dt}) = \begin{bmatrix} 1 & e^{-j2\pi f_{Dt}} & \cdots & e^{-j2\pi (M-1) f_{Dt}} \end{bmatrix}' \qquad (9.28)$$

This is simply the model for the slow-time data sequence corresponding to a target at normalized Doppler frequency f_{Dt}. The two-dimensional snapshot of the data from a target at Doppler f_{Dt} and AOA θ_t would have the temporal variation of $\mathbf{a}_t(f_{Dt})$ in each row, and the spatial variation of $\mathbf{a}_s(\theta_t)$ in each column. The snapshot therefore has the form

$$y[l_0, m, n] = \begin{bmatrix} a_{t0}(f_{Dt})\mathbf{a}_s(\theta_t) & a_{t1}(f_{Dt})\mathbf{a}_s(\theta_t) & \cdots & a_{t(M-1)}(f_{Dt})\mathbf{a}_s(\theta_t) \end{bmatrix} \qquad (9.29)$$

where $a_{tm}(f_{Dt})$ is the mth element of $\mathbf{a}_t(f_{Dt})$.

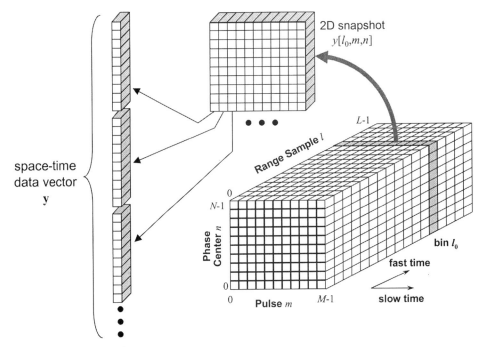

Figure 9.10 Mapping of a datacube range bin to a two-dimensional space-time snapshot, and then to a one-dimensional vector.

When this matrix is vectorized by stacking the columns, the result is the Kronecker product of the desired spatial and temporal steering vectors (Guerci, 2002; Melvin, 2004):

$$\mathbf{t} = \mathbf{a}_t(f_{Dt}) \otimes \mathbf{a}_s(\theta_t) = \begin{bmatrix} a_{t0}(f_{Dt})\mathbf{a}_s(\theta_t) \\ a_{t1}(f_{Dt})\mathbf{a}_s(\theta_t) \\ \vdots \\ a_{t(M-1)}(f_{Dt})\mathbf{a}_s(\theta_t) \end{bmatrix} \qquad (9.30)$$

Define $R = MN$. Next, a model for the $R \times R$ covariance matrix $\mathbf{S_I}$ of the interference is needed. The interference is the sum of receiver noise (**n**), jammer (**J**), and clutter (**c**) components. It is assumed that these are all uncorrelated with one another, with the result that the total interference covariance is the sum of the covariances of the three components, $\mathbf{S_I} = \mathbf{S_n} + \mathbf{S_J} + \mathbf{S_c}$.

Receiver noise, as usual, is assumed to be i.i.d. zero-mean complex Gaussian at each phase center and time sample, with the result that $\mathbf{S_n} = \sigma^2 \mathbf{I}_R$, an Rth-order identity matrix. Now consider a single noise jammer signal. The spatial variation was given in Eq. (9.14). The temporal variation can be modeled as

$$\mathbf{a}_{tJ} = \begin{bmatrix} a_{tJ0} & a_{tJ1} & \cdots & a_{tJ(M-1)} \end{bmatrix}' \qquad (9.31)$$

where the $\{a_{t_Jm}\}$ are uncorrelated i.i.d. random variables with equal power σ_J^2. Consequently, the covariance of the jammer temporal variation is

$$\mathbf{E}\{\mathbf{a}_{t_J}^* \mathbf{a}_{t_J}'\} = \sigma_J^2 \mathbf{I}_M \tag{9.32}$$

If the AOA of the jammer is θ_{J0}, the space-time data vector for the jammer component is

$$\mathbf{J} = \sigma_J^2 \, \mathbf{a}_{t_J} \otimes \mathbf{a}_{s_J}(\theta_{J0}) \tag{9.33}$$

and its covariance matrix can be shown to be (Ward, 1994)

$$\mathbf{S_J} = \mathbf{E}\{\mathbf{J}^* \mathbf{J}'\} = \sigma_J^2 \, \mathbf{I}_M \otimes \mathbf{a}_{s_J}^*(\theta_{J0})\mathbf{a}_{s_J}'(\theta_{J0}) \tag{9.34}$$

which is a block-diagonal matrix. If P multiple uncorrelated jammers are present, $\mathbf{S_J}$ is the sum of P terms of the form in Eq. (9.34), each with its own AOA θ_{Jp}.

The clutter signal is, as discussed previously, the sum of contributions from all of the clutter scatterers within the isorange ring of interest (or multiple rings, if the system is ambiguous in range). Strictly speaking, this is an integral of the pertinent scatterers, as with the angle-averaged reflectivity of Chap. 2. However, in STAP the integrated clutter is generally approximated as the sum of Q elemental clutter patches (see Fig. 9.8), each typically of an angular extent approximately equal to the radar beamwidth. For clutter patch q, the space-time data vector becomes

$$\mathbf{c}_q = \sigma_{c_q}^2 \mathbf{a}_{t_c}(f_{D_{cq}}) \otimes \mathbf{a}_{s_c}(\theta_{cq}) \tag{9.35}$$

where $\sigma_{c_q}^2$ represents the power of the qth clutter patch, which is determined by the radar range equation, and in particular is proportional to the antenna gain in the direction of that patch $G(\theta_q)$. The normalized Doppler shift and AOA of the clutter patch are related through Eq. (9.25). The total clutter vector is

$$\mathbf{c} = \sum_{q=0}^{Q-1} \mathbf{c}_q = \sum_{q=0}^{Q-1} \sigma_{c_q}^2 \mathbf{a}_{t_c}(f_{D_{cq}}) \otimes \mathbf{a}_{s_c}(\theta_{cq}) \tag{9.36}$$

The covariance matrix of the clutter is

$$\mathbf{S_c} = \mathbf{E}\{\mathbf{c}^* \mathbf{c}'\} = \sum_{q=0}^{Q-1} \sigma_{c_q}^2 \mathbf{c}_q^* \mathbf{c}_q'$$

$$= \sum_{q=0}^{Q-1} \sigma_{c_q}^2 \left[\mathbf{a}_{t_c}^*(f_{D_{cq}})\mathbf{a}_{t_c}'(f_{D_{cq}})\right] \otimes \left[\mathbf{a}_{s_c}^*(\theta_{cq})\mathbf{a}_{s_c}'(\theta_{cq})\right] \tag{9.37}$$

This is an $M \times M$ block matrix, where each "element" of the block matrix is the $N \times N$ cross-covariance of the spatial snapshots from two different PRIs. $\mathbf{S}_c$ can be factored as (Ward, 1994)

$$\mathbf{S_c} = \mathbf{C}\Sigma_c\mathbf{C}$$
$$\mathbf{C} = [\mathbf{c}_0 \quad \mathbf{c}_1 \quad \cdots \quad \mathbf{c}_{Q-1}] \quad (9.38)$$
$$\Sigma_c = \mathrm{diag}\left(\begin{bmatrix}\sigma_{c_q}^2 & \sigma_{c_q}^2 & \cdots & \sigma_{c_q}^2\end{bmatrix}\right)$$

The discussion of clutter so far assumes that it is uncorrelated in the spatial dimensions (range and cross-range) but perfectly correlated in slow time. It is indeed common to model the clutter as uncorrelated in space, assuming that clutter patches are separated by a distance on the order of a resolution cell. However, as discussed in Chap. 2 and again in the analysis of Doppler processing in Chap. 5, the clutter echo from any given ground patch cannot be reasonably modeled as constant in slow time over the time scale of a coherent processing interval. Natural clutter exhibits reflectivity fluctuations in time due to *internal clutter motion* (ICM, also called *intrinsic clutter motion*). This is simply the physical movement of scatterers due to wind or waves (if the surface of interest is a body of water). The radar system itself contributes sources of temporal modulation such as antenna scanning modulation or pulse-to-pulse instabilities. All of these temporal reflectivity fluctuations cause a broadening of the clutter power spectrum and cause decorrelation of the temporal snapshots.

ICM is easily incorporated into the model of Eq. (9.37) by replacing the temporal snapshot for patch q, $\sigma_{c_q}^2 \mathbf{a}_{t_c}(f_{D_{cq}})$, with a modified temporal snapshot that replaces the constant amplitude $\sigma_{c_q}^2$ with time-varying amplitudes $\sigma_{c_q}^2 \alpha_q$, where

$$\alpha_q = [\alpha_{0q} \quad \alpha_{1q} \quad \cdots \quad \alpha_{(M-1)q}]' \quad (9.39)$$

Let $\mathbf{A}_q = \mathbf{E}\{\alpha_q^* \alpha_q'\}$ be the covariance matrix of the temporal fluctuations α_q; $\mathbf{A}_q$ is an $M \times M$ Toeplitz matrix. The modified space-time data vector for the qth clutter patch is then

$$\mathbf{c}_q = \sigma_{c_q}^2 \left[\alpha \cdot \mathbf{a}_{t_c}\left(f_{D_{cq}}\right)\right] \otimes \mathbf{a}_{s_c}(\theta_{cq}) \quad (9.40)$$

and the clutter covariance matrix becomes

$$\mathbf{S_c} = \sum_{q=0}^{Q-1} \sigma_{c_q}^2 \left[\alpha \cdot \mathbf{a}_{t_c}^*(f_{D_{cq}})\mathbf{a}_{t_c}'(f_{D_{cq}})\right] \otimes \left[\mathbf{a}_{s_c}^*(\theta_{cq})\mathbf{a}_{s_c}'(\theta_{cq})\right] \quad (9.41)$$

An alternative approach for modeling ICM using covariance matrix tapers is described in Sec. 9.7.

Various models for the temporal correlation of the data are available in the literature. One that is popular in STAP simulations is the Billingsley model (Billingsley, 2001). This model assumes that the clutter temporal power spectrum is the sum of a triangular function and an impulse at the origin in Doppler frequency space

$$S_c(F) = \sigma_c^2 \left[\frac{\alpha}{\alpha+1} \delta_D(F) + \frac{1}{\alpha+1} \left(\frac{\beta\lambda}{4} \right) \exp\left(-\frac{\beta\lambda}{2} |F| \right) \right] \quad (9.42)$$

where α = ratio of the dc to ac components
β = parameter dependent primarily on wind conditions
α = function of both wind and radar frequency

The corresponding autocorrelation function is

$$s_c(\upsilon) = \sigma_c^2 \left(\frac{\alpha}{\alpha+1} + \frac{1}{\alpha+1} \frac{(\beta\lambda)^2}{(\beta\lambda)^2 + (4\pi\upsilon)^2} \right) \quad (9.43)$$

Experimental data are available to choose the parameters α and β to fit various scenarios distinguished by the type of clutter, radar wavelength, weather conditions, and so forth. Simple autoregressive filters can be used to implement the model in simulations (Mountcastle, 2004).

9.4 Processing the Space-Time Signal

9.4.1 Optimum matched filtering

Optimal space-time processing of the two-dimensional data snapshot consists of the following steps:

1. Form the interference covariance matrix $\mathbf{S_I}$. In practice, $\mathbf{S_I}$ must be estimated from the radar data, but discussion of this issue is deferred to Sec. 9.4.4.
2. Select a Doppler shift and angle of arrival at which to test for the presence of a target signal, and form the appropriate target space-time steering vector $\mathbf{t}$ using Eq. (9.30).
3. Compute the optimal filter weight $\mathbf{h}$ using either Eq. (9.9) or (9.19), depending on the normalization desired. For numerical and computational reasons, $\mathbf{h}$ is not computed by explicitly inverting $\mathbf{S_I}$ in practice; see Sec. 9.5.
4. Form the space-time data vector $\mathbf{y}$ for the range bin of interest as illustrated in Fig. 9.10 and described in Eq. (9.27).
5. Apply the weight vector to the data to obtain the *test statistic* $z = z(f_D, \theta) = \mathbf{h'y}$.

The test statistic can then be used for detection or for other purposes, such as angle estimation. If used for detection, typically $|z|$ or $|z|^2$ is computed and compared to an appropriate threshold.

The previous procedure computes the optimum (maximum SIR) test statistic for a single Doppler frequency and AOA, and in a single range bin. It therefore must be repeated for each Doppler and AOA of interest, and then for each range bin. Within a given range bin, the covariance matrix will remain constant and can be reused, but steps 2 through 5 must be repeated for each (f_D, θ) combination. However, $\mathbf{S_I}$ should ideally be recomputed for each range bin. Because the product $R = MN$ of the number of pulses in the *coherent processing interval* (CPI) and the number of phase centers in the antenna can easily be in the hundreds, the procedure above implies solving a system of hundreds of linear equations for each Doppler-AOA point of interest and each range bin. The computational load of STAP processing is immense. Algorithms for efficiently solving the equations and estimating the computational load are described in Sec. 9.5.

While the optimum weight vector is given by $\mathbf{S_I}^{-1}\mathbf{t}^*$, in practice it is often desirable to include windowing of the data to reduce side lobes, as was done in the spatial beamforming case in Fig. 9.3 and for Doppler processing in Chap. 5. Combined angle-Doppler weighting can be included by computing the weight vector as

$$\mathbf{h} = \kappa \mathbf{S_I}^{-1} \mathbf{t}_w^* \tag{9.44}$$

where the windowed target steering vector is

$$\mathbf{t}_w = (\mathbf{w}_f \otimes \mathbf{w}_\theta) \cdot \mathbf{t} \tag{9.45}$$

and $\mathbf{w}_f$ and $\mathbf{w}_\theta$ are the temporal and spatial weight vectors. That is, the conventional steering vector $\mathbf{t}$ is multiplied by a space-time window vector that is the Hadamard product of the desired Doppler and angle weighting functions.

9.4.2 STAP metrics

A common metric used to visualize STAP filtering performance is the *adapted pattern* (Ward, 1994). This is simply a two-dimensional plot of $|z(f_D, \theta)|^2 = \mathbf{h}^H \mathbf{t}^* \mathbf{t}' \mathbf{h}$ as f_D and θ are stepped over a regular grid. If the array has uniformly spaced phase centers and a constant PRI was used, the adapted pattern can be computed as the two-dimensional DFT of the weight vector $\mathbf{h}$ after it is remapped to a two-dimensional snapshot form, allowing the use of the FFT to efficiently compute the pattern.[†] However, the angle samples will be evenly spaced in K_θ rather than in θ itself. If it is preferred to display the pattern in terms of f_D and θ, one-dimensional FFTs can be applied to the temporal dimension of $\mathbf{h}$, but the angle dimension samples must be computed individually using one-dimensional DFTs at the non-uniformly spaced values of K_θ that correspond to uniformly spaced samples of θ.

Figure 9.11 illustrates the adapted pattern for an idealized example.[‡] The case considered corresponds to a sidelooking radar at an RF frequency of

---

[†]Note that the DFT will compute the pattern in terms of $\widetilde{K}_\theta$, not K_θ.

[‡]Figures 9.11 through 9.13 were generated using LL_STAP, a MATLAB™ program for demonstrating basic STAP patterns developed by Massachusetts Institute of Technology Lincoln laboratory (MIT/LL). LL_STAP is copyrighted 1995 by MIT/LL.

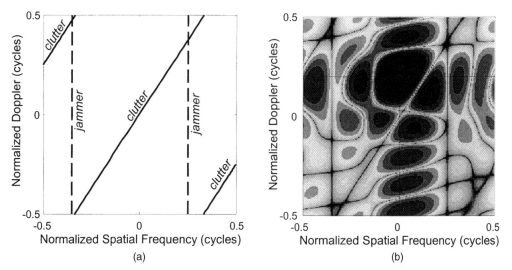

Figure 9.11 (*a*) Clutter and jamming loci for idealized example. (*b*) Adapted pattern. (See text for details.)

675 MHz. The platform velocity is 50 m/s, and the PRF is 200 pulses per second. There are eight phase centers spaced by $\lambda/2$ meters, and eight pulses in the CPI (that is, $M = N = 8$). Thus $\beta = 1.5$ in this example. Jammers are present at $+30°$ and $-44.43°$, corresponding to normalized spatial frequencies of $+0.25$ and -0.35 cycles. The CNR is $+40$ dB, as is the JNR for both jammers. Part *a* of the figure shows the loci of the jammer and clutter energy in angle-Doppler space. Part *b* of the figure shows the resulting adapted pattern when a target is present at an AOA of $0°$ and a normalized Doppler shift of 0.2 cycles. The adapted pattern clearly shows vertical nulls at the two jammer AOAs, and a three-part diagonal null corresponding to the clutter ridge. The adapted pattern has a strong peak at the target location of $0°$ and 0.2 cycles. Because no windowing was used in this example, high Doppler and spatial side lobes are evident as horizontal and vertical ridges extending from the peak. Figure 9.12 plots Doppler and spatial cuts through the adapted pattern taken at the target coordinates, showing the nulls in the spatial pattern at the jammer and clutter locations, and in the Doppler pattern at the clutter ridge location.

Another important class of metrics involves signal-to-interference ratio when the interference is noise only, and when it is noise plus clutter and/or jamming. Metrics in this class include SIR itself, and *SIR loss*, which is the reduction in SIR for the clutter + jamming case compared to the noise-only case. Each of these is a function of target angle and Doppler; typically they are plotted as a function of Doppler for a fixed AOA.

Consider a target that produces an SNR of χ_t for a single pulse and phase center, i.e., for a single sample of the two-dimensional snapshot. The optimum SNR is then

$$\chi_0 = MN\,\chi_c \qquad (9.46)$$

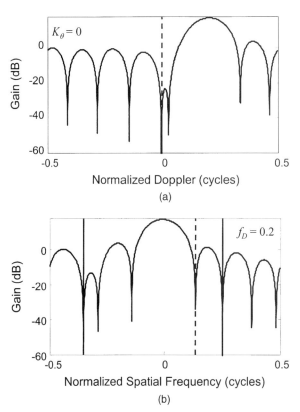

Figure 9.12 Pattern cuts from the adapted pattern of Fig. 9.11. (a) Doppler pattern. (b) Spatial pattern. Both cuts pass through the actual target location; cut lines are shown in Fig. 9.11b.

The factor MN represents the coherent integration gain achievable by combining MN samples. The SIR is, from Eq. (9.17)

$$SIR = \frac{\mathbf{h}^H \mathbf{t}^* \mathbf{t}' \mathbf{h}}{\mathbf{h}^H \mathbf{S_I} \mathbf{h}} = \frac{|z|^2}{\mathbf{h}^H \mathbf{S_I} \mathbf{h}} \tag{9.47}$$

Again from Chap. 5, when using the optimum weight vector, this becomes just

$$SIR_{\max} = \mathbf{t}' \mathbf{S_I}^{-1} \mathbf{t}^* \tag{9.48}$$

The SIR loss is then defined as

$$L_{SIR} = \frac{SIR}{\chi_0} \tag{9.49}$$

If the steering vector used in the STAP filter design exactly matches the target Doppler and angle, $SIR_{\max}$ is used in Eq. (9.49); however, this definition of L_{SIR} can be used with suboptimal filter designs as well, in which case Eq. (9.47) is used as the numerator of L_{SIR}. Note that L_{SIR} is defined here as a number less than one (negative dB), in keeping with common practice in the STAP literature.

L_{SIR} is a function of Doppler and angle of arrival. However, it is typically plotted as a function of Doppler, with the array steered in the correct target direction. Figure 9.13 is a plot of L_{SIR} versus Doppler for the interference scenario of Fig. 9.11. Each point on this curve is obtained by positing a target at the corresponding Doppler shift, and then evaluating Eq. (9.49) using the matched steering vector for that target Doppler. When the target velocity is far from the clutter ridge, for example $|f_D| > 0.3$, the loss is minimal, about 0.7 dB at the plot edges. As the target velocity nears the clutter at zero Doppler, the losses increase, reaching over 50 dB at $f_D = 0$.

This plot serves as the basis for two additional metrics (Ward, 1994). *Minimum detectable velocity* (MDV) is the velocity closest to the clutter notch at which an acceptable SIR loss is achieved. *Minimum detectable Doppler* (MDD) is the corresponding Doppler shift in either absolute or normalized units. Clearly, MDV is $\lambda/2$ times the MDD in hertz, or $\lambda/2T$ times the MDD in normalized frequency units of cycles. The frequency at which the SIR loss becomes acceptable is not necessarily symmetric about the clutter notch. Using normalized frequency units, define the upper and lower MDD as

$$MDD_+ = \min_{f_D}\{f_D \text{ such that } L_{SIR} \geq L_0, f_D > 0\}$$
$$MDD_- = \max_{f_D}\{f_D \text{ such that } L_{SIR} \geq L_0, f_D < 0\}$$
(9.50)

where L_0 is the maximum acceptable loss threshold. This definition assumes that the clutter notch is at $f_D = 0$; it is straightforward to generalize the

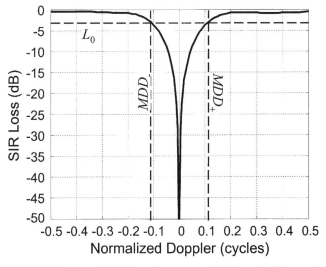

Figure 9.13 SIR loss versus Doppler for the scenario of Fig. 9.11.

definition to clutter notches at other frequencies. The MDD is then defined as the average of these two offsets, taking into account the fact that MDD_- is negative

$$MDD = \frac{1}{2}(MDD_+ - MDD_-) \qquad (9.51)$$

The choice of L_0 is a system design decision. Based on radar range equation considerations, $L_0 = -12$ dB would correspond to a 50 percent reduction in detection range, while $L_0 = -5$ dB would correspond to a 25 percent reduction. In Fig. 9.13, a value of $L_0 = -3$ dB has been selected. In this example, this choice results in $MDD_+ = 0.115$ and $MDD_- = -0.112$, giving $MDD = 0.1135$.

Another metric related to MDD is the *usable Doppler space fraction* (UDSF). This is the fraction of the Doppler space over which the SIR loss is acceptable, that is, $L_{SIR} > L_0$. UDSF is simply expressed in terms of the MDD in normalized frequency units:

$$UDSF = 1 - (MDD_+ - MDD_-)$$
$$= 1 - 2MDD \qquad (9.52)$$

In the example of Fig. 9.13, $UDSF = 0.773$, that is, the SIR loss is considered acceptable over 77.3 percent of the Doppler spectrum.

As a final example, Fig. 9.14 shows the adapted patterns obtained with the optimum filter for the two interference environments of Fig. 9.9. In the medium PRF case, there are now two vertical nulls in the Doppler dimension where the jammer energy was located. In addition, the STAP processing has implemented an S-shaped null to follow the clutter ridge and attenuate the clutter energy. The large response at approximately $F_D = 400$ Hz and $\theta = 0°$ is a target that was not visible in the original data. The two vertical ridges of energy in Doppler and angle are the side lobes of the target response. Similarly, the low PRF case shows the ability of STAP to implement a three-part null to remove aliased clutter, again revealing a hidden target.

9.4.3 Relation to displaced phase center antenna processing

Section 5.7 introduced displaced phased center antenna processing in the context of slow-moving target indication from a moving radar platform. DPCA processing combines data from multiple pulses and multiple antenna phase centers to form a clutter-cancelled output; as such, it appears to be related to STAP processing. To identify this connection, consider nonadaptive DPCA processing using two phase centers separated by a distance d_{pc}; thus $N = 2$ for DPCA processing. Assume that the DPCA condition is met with a time slip of M_s pulses; that is

$$M_s = \frac{d_{pc}}{2vT} \Rightarrow \beta = \frac{2vT}{d_{pc}} = \frac{1}{M_s} \qquad (9.53)$$

Introduction to Beamforming and Space-Time Adaptive Processing 485

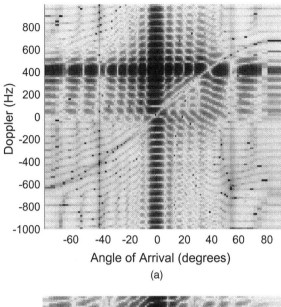

(a)

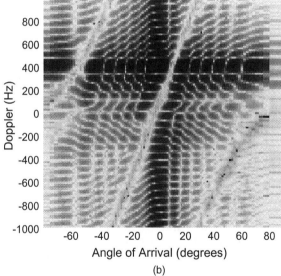

(b)

Figure 9.14 Result of optimal STAP processing of data in Fig. 9.9. (a) Medium PRF case with clutter and two jammers. (b) Low PRF case with aliased clutter. (Figure courtesy of Dr. W. L. Melvin, GTRI.)

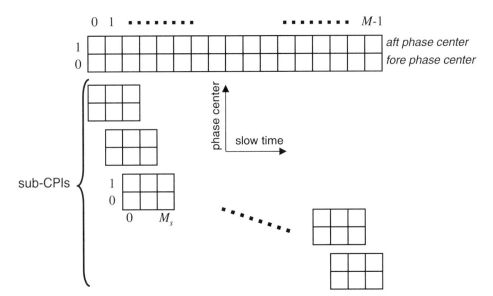

Figure 9.15 Schematic diagram of two-dimensional space-time snapshot for a DPCA processor, and its decomposition into sub-CPIs.

The space-time snapshot in two-dimensional form is therefore a $2 \times M$ array, with spatial index $n = 0$ corresponding to the fore phase center and $n = 1$ corresponding to the aft phase center, as shown in the upper portion of Fig. 9.15.

A nonadaptive DPCA processor is constrained to be of the form

$$z[m] = y_f[m] - y_a[m - M_s] \tag{9.54}$$

where $y_f[m]$ and $y_a[m]$ are the fore and aft phase center outputs, respectively. It is sufficient to work with a single "sub-CPI", i.e., an interval of M_s pulses, that spans the two pulses combined in the DPCA processor. The complete CPI snapshot is then constructed of a series of $M - M_s$ overlapping sub-CPIs, as shown in the bottom portion of Fig. 9.15 for the case $M_s = 2$. The weight vector $\mathbf{h}_1$ that implements Eq. (9.54) for a single sub-CPI, as well as the corresponding weight vector $\mathbf{h}$ for the entire CPI, are shown in Fig. 9.16 in two-dimensional form.

Viewed as a two-dimensional discrete function, the sub-CPI weight vector $\mathbf{h}_1$ can be written as

$$\mathbf{h}_1 \Rightarrow h_1[m,n] = \delta[m,n] - \delta[m - M_s, n-1] \tag{9.55}$$

The adapted pattern is the two-dimensional DTFT of this function

$$\begin{aligned} H_1(\omega_D, \tilde{K}_\theta) &= \sum_{m=0}^{M_s-1} \sum_{n=0}^{1} h_1[m,n] e^{-j(M_s \omega_D + \tilde{K}_\theta)} \\ &= 1 - \exp[-j(M_s \omega_D + \tilde{K}_\theta)] \\ &= 1 - \exp[-j(M_s \omega_D - K_\theta)] \end{aligned} \tag{9.56}$$

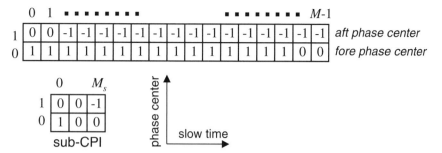

Figure 9.16 Full-CPI weight vector **h** and sub-CPI weight vector $\mathbf{h}_1$.

Note that this function has a zero at $\omega_D = K_\theta/M_s = \beta K_\theta$, which is exactly the Eq. (9.26) of the clutter ridge. Figure 9.17 shows $|H_1(f_D, K_\theta)|$ for the case $M_s = 2$. Clearly, the DPCA processor implements a null along the clutter ridge. Repeating the single sub-CPI DPCA process for the additional sub-CPIs within the total CPI simply provides coherent integration by a factor of $M - M_s$. Note also that the constrained form of the weight vector (Eq. (9.55)) prevents any combining of samples from different phase centers on the same pulse, so that the DPCA processor cannot provide any spatial beamforming capability. As a result, DPCA processing can cancel the clutter but not jammers.

The model of Eq. (9.56) implicitly assumes an omnidirectional antenna pattern for each of the antennas or subarrays that form the DPCA phase centers. A more realistic model would account for the subarray patterns. To do so, repeat the analysis, but with the output from each phase center filtered in angle by

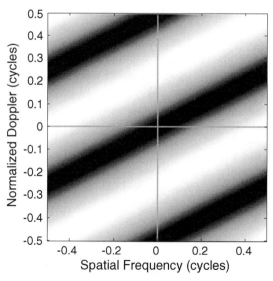

Figure 9.17 Adapted pattern for nonadaptive DPCA processor with $M_s = 2$.

the antenna pattern of the antenna or subarray that forms that phase center. Denoting the fore and aft subarray antenna patterns as $E_f(K_\theta)$ and $E_a(K_\theta)$, respectively, Eq. (9.56) can be generalized to

$$H_1(\omega_D, K_\theta) = E_f(K_\theta) - E_a(K_\theta)\exp[-j(M_s\omega_D - K_\theta)]$$
$$= E_f(K_\theta)\{1 - \exp[-j(M_s\omega_D - K_\theta)]\} \quad (9.57)$$

where the last step holds only if $E_f(K_\theta) = E_a(K_\theta)$. If the two patterns differ, the first line of Eq. (9.57) shows that there will not be a null at the clutter ridge location as desired. This emphasizes the need for carefully matched subarray patterns in DPCA processing.

Adaptive DPCA improves on these results by applying the vector matched filtering framework as described in Sec. 5.7.2. The adaptive DPCA processor is still constrained to a weight vector structure that prohibits combining of phase center outputs from the same pulse, and thus, like nonadaptive DPCA, provides no spatial beamforming. In addition, the use of the target model $\mathbf{t} = [1 \ \ 0]'$ provides a result that is optimized "on average" over all target Doppler frequencies, rather than for any specific target Doppler.

9.4.4 Adaptive matched filtering

In Chap. 6, it was seen that the knowledge of the interference power was required to set the detection threshold. Chapter 7 introduced *constant false alarm rate* (CFAR) techniques to estimate the noise level from the radar data. Exactly the same issue exists with STAP processing, which requires knowledge of the interference covariance matrix $\mathbf{S_I}$ to compute the optimal weight vector $\mathbf{h}$ and perform the matched filtering. Again, it is not realistic to assume that $\mathbf{S_I}$ can be known a priori, so it must be estimated from the radar data.

The most common approach to STAP when $\mathbf{S_I}$ is unknown is called the *sample matrix inverse* (SMI) method (Ward, 1994). This technique is exactly analogous to cell averaging CFAR. Figure 9.18 illustrates a datacube with a range bin of interest indicated, the *cell under test* (CUT), as well as a number of adjacent range bins. The data in these *reference cells* adjacent to the cell under test are assumed to consist of independent and identically distributed interference having the same statistics as the interference in the CUT. It is also assumed that the reference cells do *not* contain any target signals.[†] Consequently, they can be used to compute a sample mean estimate of $\mathbf{S_I}$, denoted by $\hat{\mathbf{S}}_\mathbf{I}$. Specifically, the covariance of the data from a single reference cell is

$$\hat{\mathbf{S}}_l = \mathbf{y}_l^* \mathbf{y}' \quad (9.58)$$

[†]If desired, guard cells can be included immediately adjacent to the CUT to prevent target contamination of the interference estimate.

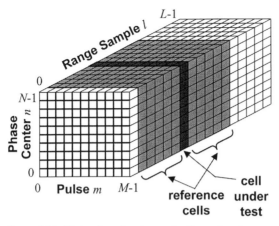

Figure 9.18 Datacube showing selection of reference cells for estimating interference covariance matrix.

If L_S reference cells are used, the estimate of $\mathbf{S}_I$ is simply

$$\hat{\mathbf{S}}_I = \frac{1}{L_s} \sum_{l=1}^{L_s} \hat{\mathbf{S}}_l \qquad (9.59)$$

Note that filter weight vectors computed using $\hat{\mathbf{S}}_I$ instead of $\mathbf{S}_I$ will be suboptimum because of the imperfect interference estimate.

The reduction in SIR due to the use of $\hat{\mathbf{S}}_I$ instead of $\mathbf{S}_I$ is denoted by L_{CFAR}. It has been shown by Reed, Mallet, and Brennan to be a beta-distributed random variable with expected value (Reed et al., 1974; Nitzberg, 1984)

$$\mathbf{E}(L_{\text{CFAR}}) = \frac{L_s + 2 - R}{L_s + 1} \qquad (9.60)$$

The quantity $\mathbf{E}(L_{\text{CFAR}})$ is plotted as a function of L_s/R, the size of the reference window relative to the number of DOF (snapshot size), in Fig. 9.19. This figure shows that the number of reference cells must be twice the number of DOF in the processor to limit the loss due to covariance estimation to 3 dB. To limit the loss to 1 dB requires $L_s > 5R$. The conclusion that the reference window size should usually be two to five times the snapshot size is known as the "Reed-Mallet-Brennan" or *RMB rule*. The shape of the curve is a very weak function of R; the example shown is for $R = 256$, corresponding for example to a system with $N = 8$ phase centers and $M = 32$ pulses in the CPI. These results apply only when there is no mismatch between the actual target steering vector and the model vector $\mathbf{t}$ used in the filter design; this also implies no use of windowing for reduced side lobes. Generalizations of Eq. (9.60) for mismatched and windowed target model vectors are discussed in the papers by Boroson (1980) and Kelly (1989).

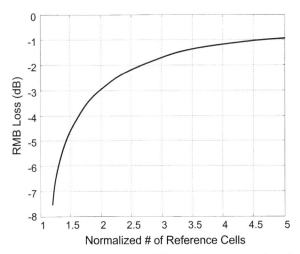

Figure 9.19 RMB estimate of SIR loss due to estimation of covariance matrix $\mathbf{S}_\mathbf{I}$.

Once the covariance matrix has been estimated, the weight vector is computed in the usual fashion. One particular choice of the scale factor κ of interest when using a square-law detector is the *adaptive matched filter* (AMF) (Kelly, 1986; Chen and Reed, 1991; Robey et al., 1992):

$$\kappa = \frac{1}{\sqrt{\mathbf{t}^H \left(\hat{\mathbf{S}}_\mathbf{I}^{-1}\right)^* \mathbf{t}}} \tag{9.61}$$

With this choice, the filter output becomes

$$|z|^2 = \frac{\left|\mathbf{t}^H \left(\hat{\mathbf{S}}_\mathbf{I}^{-1}\right)^* \mathbf{y}\right|^2}{\mathbf{t}^H \left(\hat{\mathbf{S}}_\mathbf{I}^{-1}\right)^* \mathbf{t}} \tag{9.62}$$

It can be shown that a threshold test applied to this test statistic exhibits CFAR behavior. It is also claimed that the AMF is more robust to targets in the training data than alternative tests such as the generalized likelihood ratio test (Steinhardt and Guerci, 2004).

The adapted pattern when using the SMI technique is subject to pattern degradations such as elevated side lobes and distorted mainbeams, especially when L_s is relatively small. If $L_s < R$, then $\hat{\mathbf{S}}_\mathbf{I}$ may also become nonsingular. A common extension of the basic SMI technique that addresses these issues is *diagonal loading*, in which a bias term is added to the diagonal elements of the estimated covariance matrix (Carlson, 1988)

$$\hat{\mathbf{S}}_\mathbf{I} \Rightarrow \hat{\mathbf{S}}_\mathbf{I} + \varepsilon \mathbf{I} \tag{9.63}$$

Since diagonal loading adds a factor that has the same structure as the covariance matrix of white noise, its effect is to increase the apparent noise floor of the data. The loading factor ε is typically set 10 to 30 dB above the actual noise level σ^2 (Kim et al., 1998). Diagonal loading tends to ensure nonsingularity of $\hat{\mathbf{S}}_\mathbf{I}$ and reduce distortion of the adapted pattern, but also reduces the depth of the nulls (Guerci, 2003).

9.5 Computational Issues in STAP

The filter $\mathbf{h}$ that maximizes SIR has been shown to be of the form

$$\mathbf{h} = \kappa \mathbf{S}_\mathbf{I}^{-1} \mathbf{t}^* \tag{9.64}$$

for *moving target indication* (MTI) filtering, adaptive beamforming, and now STAP. In practical problems, the interference covariance matrix $\mathbf{S}_\mathbf{I}$ is not known beforehand, so it must be estimated from the data; this in turn means that the filter design must be done in real time. As seen earlier, the distortionless case occurs when (using the estimated covariance matrix $\hat{\mathbf{S}}_\mathbf{I}$)

$$\mathbf{h} = \frac{\hat{\mathbf{S}}_\mathbf{I}^{-1} \mathbf{t}^*}{\mathbf{t}^H \left(\hat{\mathbf{S}}_\mathbf{I}^{-1} \mathbf{t}^* \right)^*} \tag{9.65}$$

Direct computation of $\mathbf{h}$ by inverting $\hat{\mathbf{S}}_\mathbf{I}$ may be practical for low-order problems such as MTI or adaptive beamforming with modest-sized antennas, but for STAP one rarely, if ever, actually computes $\hat{\mathbf{S}}_\mathbf{I}^{-1}$. This is because the large order of the matrix $\hat{\mathbf{S}}_\mathbf{I}$ makes the inversion numerically sensitive, computationally intensive, and ultimately superfluous, since the desired result is $\mathbf{h}$; $\hat{\mathbf{S}}_\mathbf{I}^{-1}$ is merely an intermediate factor. Note that solving Eq. (9.64) is equivalent to finding the solution to the set of linear equations

$$\hat{\mathbf{S}}_\mathbf{I} \mathbf{h} = \kappa \mathbf{t}^* \tag{9.66}$$

It is important for real-time implementation of adaptive algorithms in sonar, radar, and other high-rate applications to find a numerically robust and computationally efficient means of solving Eq. (9.66) without matrix inversion. One method popular in the radar and sonar arenas is the *Q-R method* or *Q-R decomposition*.

This technique can be applied in both *power domain* and *voltage domain* versions; both are described in this section, along with their computational requirements. The latter is used most commonly in practice due to its superior dynamic range properties, but introducing the power domain implementation first is more natural and eases the discussion of the voltage domain version.

The specific form of the equations for the distortionless STAP process considered here are as follows:

$$\hat{\mathbf{S}}_l = \mathbf{y}_l^* \mathbf{y}'$$

$$\hat{\mathbf{S}}_\mathbf{I} = \frac{1}{L_s} \sum_{l=1}^{L_s} \hat{\mathbf{S}}_l$$

$$\mathbf{h} = \frac{\hat{\mathbf{S}}_\mathbf{I}^{-1} \mathbf{t}^*}{\mathbf{t}^H \left(\hat{\mathbf{S}}_\mathbf{I}^{-1} \mathbf{t}^*\right)^*} = \frac{\mathbf{u}}{\mathbf{t}^H \mathbf{u}^*}$$

$$z = \mathbf{h}' \mathbf{t} \tag{9.67}$$

where $\mathbf{u} \equiv \hat{\mathbf{S}}_\mathbf{I}^{-1} \mathbf{t}^*$, implying

$$\hat{\mathbf{S}}_\mathbf{I} \mathbf{u} = \mathbf{t}^* \tag{9.68}$$

In most cases, the computations of Eq. (9.67) will be repeated for each of KP distinct target steering vectors $\mathbf{t}$ corresponding to K Doppler frequencies and P angles of arrival, and for some number L of range bins.

9.5.1 Power domain solution

Samples of the space-time data vector $\mathbf{y}$ are considered to be in voltage units, which they often literally are. The power domain approach begins by forming the estimated covariance matrix $\hat{\mathbf{S}}_\mathbf{I}$ using Eq. (9.67). Because this involves products of voltage-domain values, the covariance matrix values are considered to be in the power domain. The weight vector $\mathbf{h}$ is then found by applying the Q-R decomposition to $\hat{\mathbf{S}}_\mathbf{I}$ (Watkins, 1991; Cain, 1997). The technique begins by decomposing $\hat{\mathbf{S}}_\mathbf{I}$ into the product $\hat{\mathbf{S}}_\mathbf{I} = \mathbf{QR}$, where $\mathbf{Q}$ is a unitary matrix (meaning $\mathbf{QQ}^H = \mathbf{I}$ so that $\mathbf{Q}^{-1} = \mathbf{Q}^H$) and $\mathbf{R}$ is upper triangular. This factorization can be done using a series of Givens rotations (Watkins, 1991). Equation 9.68 then becomes

$$\mathbf{QRu} = \mathbf{t}^* \tag{9.69}$$

Now define $\mathbf{r} = \mathbf{Ru}$ so that Eq. (9.69) becomes

$$\mathbf{Qr} = \mathbf{t}^* \tag{9.70}$$

Because $\mathbf{Q}$ is unitary, Eq. (9.70) can be simply solved for $\mathbf{r}$ as

$$\mathbf{r} = \mathbf{Q}^H \mathbf{t}^* \tag{9.71}$$

Since $\mathbf{R}$ is upper triangular, the system $\mathbf{r} = \mathbf{Ru}$ can now be solved for $\mathbf{u}$ by simple back substitution. Finally, the filter vector $\mathbf{h}$ is computed and applied to the data using Eq. (9.67).

Introduction to Beamforming and Space-Time Adaptive Processing

To summarize, the algorithm steps are

1. Form the estimated covariance matrix $\hat{\mathbf{S}}_\mathbf{I}$.
2. Perform the Q-R decomposition of $\hat{\mathbf{S}}_\mathbf{I}$.
3. Compute $\mathbf{r} = \mathbf{Q}^H \mathbf{t}^*$.
4. Solve $\mathbf{r} = \mathbf{R}\mathbf{u}$ for $\mathbf{u}$ by back substitution.
5. Compute $\mathbf{h} = \mathbf{u}/\mathbf{t}^H \mathbf{u}^*$.
6. Compute the filter output $z = \mathbf{h}'\mathbf{t}$.

The covariance matrix is computed only once per range bin; if the interference environment permits, the same covariance matrix can be used for multiple range bins. Therefore, the Q-R decomposition is also required at most once per range bin. The computation in turn of $\mathbf{r}$, $\mathbf{u}$, $\mathbf{h}$, and z is repeated for each target steering vector $\mathbf{t}$ (AOA and Doppler frequency) of interest in each range bin.

9.5.2 Computational load of the power domain solution

It is fairly straightforward to count the arithmetic operations required to implement the power domain solution of the STAP equations as a function of the problem parameters. The parameters of interest are the data snapshot size $R = MN$, the number of snapshots L_s averaged to form the covariance matrix, the number of target steering vectors KP, and the number of range bins processed L. The results are derived in terms of estimated *real floating point operations* (RFLOP). In keeping with modern hardware and also for simplicity, it is assumed that real multiplications, divisions, additions, and subtractions all require the same amount of time, and so each is considered a single RFLOP. However, the basic computations are complex. Table 9.1 lists the factors used to convert from complex floating point operations to RFLOP. To convert to computational rate in RFLOPS (RFLOP per second), one must also take into account the time interval available for completing the computations before a new set of data snapshots is ready for processing. This is done in Sec. 9.5.4.

TABLE 9.1 Conversion from Complex to Real FLOP

Operation	Number of Real Floating Point Operations Assumed
Complex-complex multiply	4 real multiplies + 2 real adds = 6 RFLOP
Complex-complex add, subtract	2 real add, subtract = 2 RFLOP
Magnitude squared	2 real multiplies + 1 real add = 3 RFLOP
Real-complex multiply	2 real multiplies = 2 RFLOP
Complex divided by real	1 real inverse + 1 real-complex multiply = 3 RFLOP
Complex inverse	1 conjugate (not counted) + 1 magnitude-squared + 1 complex divided by real = 6 RFLOP
Complex-complex divide	1 complex inverse + 1 complex multiply = 12 RFLOP

The number of RFLOP in each algorithmic step of the power domain solution for all steering directions of interest in a single range bin can now be determined step-by-step:

Step 1a: *Formation of Covariance Matrix for lth Array Data Snapshot* ($\hat{\mathbf{S}}_l = \mathbf{y}_l^* \mathbf{y}'$). The outer product of a complex $R \times 1$ vector with itself is guaranteed to be Hermitian. Consequently, only the upper (or lower) half, including the diagonal, need be computed; the lower (upper) half can then be filled with copy and transpose operations, which are ignored in this computational estimate. Each element in the upper half of the matrix is a single complex multiply; there are $(R^2 - R)/2$ such elements. The R diagonal elements require only a magnitude squared instead of a general complex multiply. The result is a total of $3R^2$ RFLOP per range bin, or $3L_s R^2$ RFLOP for the L_s range bins required for step 1b.

Step 1b: *Form Averaged Covariance Matrix Over L_s Snapshots*, $\hat{\mathbf{S}}_\mathbf{I} = \frac{1}{L_s} \sum_{l=1}^{L_s} \hat{\mathbf{S}}_l$. The L_s covariance matrices from step 1a are averaged element-by-element; this requires $(L_s - 1)(R^2 + R)/2$ complex additions and $(R^2 + R)/2$ real-complex divides, for a total of $(L_s + 1/2)(R^2 + R) \approx L_s(R^2 + R)$ RFLOP to form the final estimated covariance matrix $\hat{\mathbf{S}}_\mathbf{I}$.

Step 2: *Perform a Q-R Decomposition of $\hat{\mathbf{S}}_\mathbf{I}$*. The next step is a Q-R decomposition of $\hat{\mathbf{S}}_\mathbf{I}$. This requires approximately $2R^3/3$ each of complex multiplications and additions (Watkins, 1991), for a total of $16R^3/3 = 5.33R^3$ RFLOP.

Step 3: *Compute* $\mathbf{r} = \mathbf{Q}^H \mathbf{t}^*$. This matrix-vector multiply requires R complex multiplications and $R - 1$ complex additions for each of the R elements of $\mathbf{r}$, a total of $8R^2 - 2R$ RFLOP. Repeating for each Doppler shift and AOA of interest gives $KP(8R^2 - 2R)$ RFLOP.

Step 4: *Solve for* $\mathbf{u}$ *in* $\mathbf{r} = \mathbf{Ru}$. Because $\mathbf{R}$ is triangular, this system can be solved by back substitution using approximately $R^2/2$ each of complex multiplications and additions (Watkins, 1991), for a total of $4R^2$ RFLOP per steering vector, and thus $4KPR^2$ RFLOP for all steering vectors.

Step 5: *Compute the Weight Vector* $\mathbf{h} = \mathbf{u}/\mathbf{t}^H \mathbf{u}^*$. The complex dot product of the two $R \times 1$ vectors $\mathbf{t}^H$ and $\mathbf{u}^*$ requires R complex multiplications and $R - 1$ complex adds. Dividing the resultant complex scalar into the vector $\mathbf{u}$ contributes R complex-complex divides. The total is $20R - 2$ RFLOP for computation of a single weight vector. The total for all steering directions is $KP(20R - 2)$ RFLOP.

Step 6: *Compute the Beamformer Output* $z = \mathbf{h}' \mathbf{t}$. This is another complex dot product of two $R \times 1$ vectors for each snapshot and therefore contributes $8R - 2$ RFLOP per steering vector, or $KP(8R - 2)$ RFLOP total.

It is useful to distinguish the number of RFLOP required to obtain the weight vectors, as opposed to those required to apply the weight vectors, since the weights are often updated at a less frequent rate than they are applied. The total number of RFLOP required per range bin to compute and factor $\hat{\mathbf{S}}_\mathbf{I}$, and then obtain the weight vectors $\mathbf{h}$ for each steering direction $\mathbf{t}$ in the previous sequence of calculations is the sum of the RFLOP counts in steps 1 through 5:

$$RFLOP = (5.33R^3 + 4L_s R^2) + (12KPR^2 + L_s R) + \cdots$$
$$\cdots + 18KPR - 2KP \text{ (power domain case)} \quad (9.72)$$

Terms are grouped by polynomial order in this equation.[†] Equation (9.72) shows that the computational load for computation of the weights is of order R^3, denoted as $O(R^3)$. The number of RFLOP required to apply the KP weight vectors per range bin to all L range bins (step 6) is $O(R)$

$$RFLOP = (8R - 2)KPL \text{ (power domain case)} \quad (9.73)$$

9.5.3 Voltage domain solution and computational load

One of the disadvantages of the power domain approach is the squaring effect implicit in the formation of the outer products $\mathbf{y}_l^* \mathbf{y}'$. For many real systems, the actual required computational rate for adaptive processing is very high; so much so that custom or semicustom computing hardware is often required. Such hardware is usually fixed-point. Squaring a signal doubles its dynamic range in dB, and therefore also doubles the number of bits required for its representation without overflow. This in turn increases design complexity and size as well as data rate in the processor. It is highly desirable *not* to square the signal unnecessarily.

It is thus of interest to apply the Q-R decomposition directly to the data vectors $\mathbf{y}_l$ and develop a similar solution technique for $\mathbf{h}$, called the *voltage domain* solution, that involves solving linear systems by forward and backward substitution, but avoids explicit computation of the estimated covariance matrix $\hat{\mathbf{S}}_\mathbf{I}$ (Watkins, 1991; Cain et al., 1997). The method begins by forming an $R \times L_s$ matrix $\mathbf{Y}$ by concatenating the data snapshots, $\mathbf{Y} = [\mathbf{y}_0 \; \mathbf{y}_1 \; \cdots \; \mathbf{y}_{L_s-1}]$. Note that $\mathbf{Y}^* \mathbf{Y}' = \hat{\mathbf{S}}_\mathbf{I}$; however, $\mathbf{Y}^* \mathbf{Y}'$ will not be explicitly computed, since that again implies squaring the data elements. Also note that the RMB rule implies that it will generally be the case that $L_s > R$.

Next, factor $\mathbf{Y}'$ into the product $\mathbf{Y}' = \mathbf{QR}$. $\mathbf{Y}'$ is used instead of $\mathbf{Y}$ because the nonsymmetric case of Q-R decomposition is conventionally defined for the situation where the number of rows exceeds the number of columns (Watkins, 1991);

[†] Note that the RMB rule implies that it will generally be the case that $L_s > R$, so a term of the form $L_s R^2$ is considered to be of order R^3, and so forth.

since typically $L_s > R$, **Y** will have more columns than row. **Q** will be an $L_s \times L_s$ unitary matrix and **R** is an $L_s \times R$ matrix of the form

$$\mathbf{R} = \begin{bmatrix} \hat{\mathbf{R}} \\ \mathbf{0} \end{bmatrix} \tag{9.74}$$

where $\hat{\mathbf{R}}$ is an $R \times R$ upper triangular matrix and **0** is an $(L_s - R) \times R$ zero matrix. It follows that

$$\hat{\mathbf{S}}_\mathbf{I} = \mathbf{Y}^*\mathbf{Y}' = \mathbf{R}^H \mathbf{Q}^H \mathbf{Q}\mathbf{R} = \mathbf{R}^H \mathbf{R} = \hat{\mathbf{R}}^H \hat{\mathbf{R}} \tag{9.75}$$

Using Eq. (9.75) in Eq. (9.68) gives

$$\hat{\mathbf{R}}^H \hat{\mathbf{R}} \mathbf{u} = \mathbf{t}^* \tag{9.76}$$

Define $\mathbf{r} \equiv \hat{\mathbf{R}}\mathbf{u}$. Then

$$\hat{\mathbf{R}}^H \mathbf{r} = \mathbf{t}^* \tag{9.77}$$

Because $\hat{\mathbf{R}}^H$ is lower triangular, Eq. (9.77) can be solved for **r**, given **t**, by forward substitution. Since $\hat{\mathbf{R}}$ is upper triangular, the relation $\mathbf{r} = \hat{\mathbf{R}}\mathbf{u}$ can then be solved for **u** by back substitution. This **u** vector is the same defined in the power domain solution; thus, the remaining steps are the same as in that case.

The computational load is counted in the same manner as in the power domain case. There are two differences in the voltage domain case. First, the computation of the covariance matrix $\mathbf{S}_\mathbf{I}$ in steps $1a$ and $1b$ of the power domain solution is not required. Second, the Q-R decomposition is applied to an $L_s \times R$ matrix instead of an $R \times R$ matrix. This requires approximately $L_s R^2 - R^3/3$ each of complex multiplies and additions, giving a total of $8L_s R^2 - 2.67R^3$ RFLOP. The total number of RFLOP required to obtain the weight vectors **h** for each steering direction **t** in the earlier sequence of calculations is

$$RFLOP = (8L_s R^2 - 2.67R^3) + 12KPR^2 + 18KPR - 2KP \text{ (voltage domain case)} \tag{9.78}$$

The number of RFLOP required to apply the weight vectors to all L snapshots is exactly the same as the power domain case

$$RFLOP = (8R - 2)KPL \text{ (voltage domain case)} \tag{9.79}$$

9.5.4 Conversion to computational rates

The results so far present the count of RFLOP, i.e., the number of operations to be carried out. Of greater interest in assessing the implementation requirements is the *rate* of floating point operations, expressed in RFLOP per second, or RFLOPS. This is a simple matter of dividing the appropriate RFLOP count (Eq. (9.72) or Eq. (9.78) for weight computation, Eq. (9.73) or Eq. (9.79) for weight application) by the time available for the computation. For weight computation, the user must specify the time interval T_u or corresponding rate F_u at which the weights are updated. For example, in the voltage domain case, Eq. (9.78) would become

$$RFLOPS \text{ (voltage domain, weight computation)} =$$
$$F_u[(8L_s R^2 - 2.67 R^3) + 12KPR^2 + 18KPR - 2KP] \quad (9.80)$$

The choice of the update rate F_u is based on a system engineering judgment of the frequency with which new weights are required to adapt to nonstationarity or nonhomogeneity of the interference environment; in addition, mode or parameter (e.g., PRF) changes require the computation of new weights. A typical weight update rate is on the order of $F_u = 20$ to 100 updates per second.

Weights are normally applied to each snapshot. If the time interval from the beginning of one CPI to the beginning of the next CPI is T_{CPI}, with corresponding rate F_{CPI}, T_{CPI} seconds are available to apply weights to each of the L snapshots in that CPI. Equation (9.79) thus becomes

$$RFLOPS \text{ (voltage domain, weight application)} = (8R - 2)KPLF_{CPI} \quad (9.81)$$

Similar equations result for the corresponding power domain computational rates.

Consider a modest-sized system with $N = 8$ phase centers and $M = 20$ pulses in the CPI, giving $R = MN = 160$ DOF. If the covariance estimation matrix loss is limited to 3 dB, the RMB rule requires $L_s = 2R = 320$ reference cells. Assume that the weights are updated at the rate of $F_u = 20$ updates per second. Assume also that $K = 20$ Doppler frequencies and $P = 12$ AOAs are tested in each range bin. Equation (9.80) then gives a computational rate of 2.6 GFLOPS to compute the adaptive weights.

To estimate the computational load for applying the weights, assume a range resolution of $\Delta R = 10$ m, implying a fast time sampling rate of 15 MHz, corresponding to a fast time sampling interval of $T_s = 66.67$ ns. If a range swath of $R_{swath} = 10$ km is required, there will be $L = 1000$ range bins. Assuming a duty cycle of 50 percent of the pulse repetition interval gives $PRF = 7.5$ kHz. The CPI rate is then $F_{CPI} = PRF/M = 375$ CPIs per second. Applying these values to Eq. (9.81) gives a computational load of 115 GFLOPS to apply the weights, and a total computational load of 117.6 GFLOPS for this system.

9.6 Reduced-Dimension STAP

It was shown in Sec. 9.1.3 that preprocessing of the element data could be used to reduce the dimensionality of the adaptive processing equations. The same technique can be applied to STAP processing. However, reduced-dimension processing is especially important in STAP because the snapshot dimensionality can be quite large, possibly in the hundreds. This aggravates two problems. First, the RMB rule states that the number of reference range bins required to maintain acceptable losses due to covariance estimation is on the order of $2R$ to $5R$, with the result that the training data are very unlikely to be statistically homogeneous as assumed in the SMI algorithm. In the example given earlier, the reference window would cover 3.2 km. More complex systems (more phase centers, longer CPIs) would have much longer reference windows. Second, the computational load is of order R^3. Reducing the dimensionality by only a factor of two produces nearly an order of magnitude reduction in computational load in solving the SMI equations. Increasing it by a factor of two for better performance (for instance, by reducing straddle losses or covariance matrix estimation losses) would similarly increase the load by an order of magnitude.

Because STAP operates on two data dimensions, preprocessing can be applied in the slow-time dimension, phase center dimension, or both. Figure 9.20 illustrates a taxonomy of reduced dimension STAP techniques based on the choice of preprocessing options (Ward, 1994). There are many variants of each general class. As an example, Fig. 9.21 illustrates a particular variant of the beamspace post-Doppler class of STAP processors (Melvin, 2004). DFTs in both dimensions are used to form a grid of fixed angle-Doppler bins. For a given target model vector **t**, adaptation is performed using only a small number of angle and Doppler bins around the target AOA and Doppler frequency. Beamspace post-Doppler architectures can isolate the interference in both angle and Doppler, and substantially reduce the order of the SMI equations. Typical implementations use three to five bins in each of the Doppler and angle dimensions, giving nine to 25 total degrees of freedom.

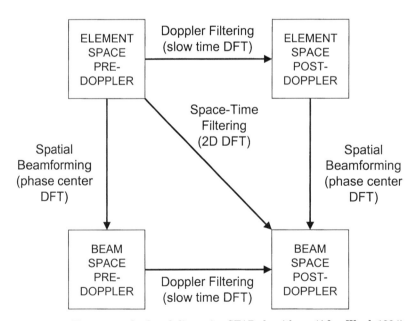

Figure 9.20 Taxonomy of reduced-dimension STAP algorithms. (After Ward, 1994)

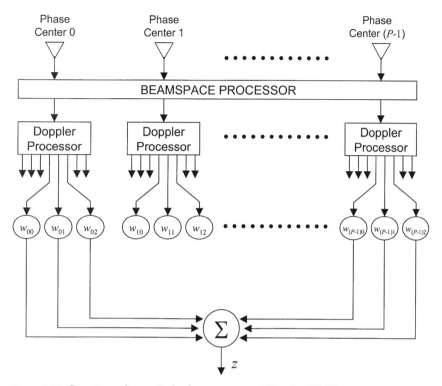

Figure 9.21 Structure of a particular beamspace post-Doppler STAP processor.

9.7 Advanced STAP Algorithms and Analysis

Only the most basic STAP algorithm, the sample matrix inverse approach based on the vector matched filter, has been introduced in this chapter, and only metrics related to the adapted pattern and signal-to-interference ratio have been used to evaluate its performance. A deeper understanding of the character of the interference and the behavior of STAP algorithms requires analysis of the eigenstructure of the signal environment, a topic beyond the scope of this book but thoroughly covered in texts devoted to STAP such as the books by Klemm (2002), Van Trees (2002), and Guerci (2003). The latter includes an extensive taxonomy of modern STAP algorithms proposed for radar data (Guerci, 2003, p. 112).

In particular, a number of ideas in advanced STAP are based on the realization that the covariance matrix of the clutter is not full rank; neither is that of the jammers. For instance, the rank of an ideal covariance matrix for J independent jammers is MJ (Guerci, 2003). The rank of the clutter-only covariance matrix under ideal conditions (no crab, constant velocity) can be estimated as (Brennan and Staudaher, 1992)

$$\mathrm{rank}\{\mathbf{S}_c\} \approx N + (M-1)\beta \qquad (9.82)$$

where β is the clutter ridge slope of Eq. (9.26). Because β is usually in the range of zero to three, this is well below the full rank $R = NM$. Thus, both the clutter and jamming signals can be represented with a relatively few basis vectors in the R-dimensional signal space, and eigenanalysis provides a convenient means for decomposing the signal components, defining a reduced-dimension representation, and analyzing the performance of the resulting algorithms.

This approach leads to algorithms that are fundamentally different from the SMI approach. Suppose that the combined clutter and jamming covariance matrix has Q dominant eigenvalues, with the remainder at or near the noise floor eigenvalue level. The *principal components* (PC) method forms the adaptive filter weight vector as a linear combination of the Q eigenvectors corresponding to the dominant eigenvalues, with weights related to the eigenvalue associated with each eigenvector used. The PC method and similar *subspace projection* techniques can construct high quality adapted patterns without the degradation of side lobes that often occurs with SMI techniques when estimated covariance matrices are used. Furthermore, because the filter vector is derived from only Q eigenvectors, these algorithms provide another means of dimensionality, and therefore computation, reduction. Algorithms based on projecting data into lower-dimensional spaces are generally called *reduced rank* STAP techniques (as opposed to the reduced dimension techniques of Sec. 9.6).

Even with effective dimensionality reduction, the nonstationarity and heterogeneity of the data due to error effects discussed in Sec. 9.8, and more fundamentally to variations in the physical clutter scene over the reference windows, remains a major limiting factor in STAP performance. Recently, significant research effort has been focused on *knowledge-aided* (KA) STAP (Weiner et al., 1998; Guerci, 2002). KA STAP attempts to use auxiliary sources of information to improve the interference covariance estimate for the cell under test. For example, map data can be used to identify changes in terrain type, roadways, and other variations in the characteristics of the clutter in the reference cells. Preprocessing algorithms can then excise some cells from the covariance matrix estimation process in an attempt to provide a more homogeneous set of training data and, therefore, an estimate of $\hat{\mathbf{S}}_I$ more consistent with the actual covariance matrix $\mathbf{S}_I$ in the CUT. The high level architecture of a KA STAP system is shown in Fig. 9.22. The knowledge-aided preprocessing edits or modifies the datacube to provide a modified datacube with more homogeneous statistics. Any of the conventional STAP algorithms can then be applied to cancel the now better-behaved clutter. In another example, *digital terrain elevation data* (DTED) maps can be used along with terrain type maps to predict clutter characteristics. In-band *electromagnetic interference* (EMI) can be identified using data on known emitters such as television stations and wireless services that may corrupt portions of the radar frequency band. The EMI can then be removed by slow time prefiltering of the data.

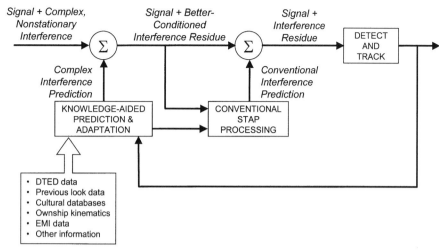

Figure 9.22 Basic architecture of a knowledge-aided STAP processor. (After Guerci, 2002.)

9.8 Limitations to STAP

In Chap. 5 it was seen that the performance of MTI is degraded by several factors, some internal to the radar system and some external to it. Internal factors include pulse-to-pulse variability in pulse amplitude, oscillator phase drifts, I/Q channel errors, and antenna scanning modulation. External factors are principally internal clutter motion and clutter heterogeneity.

All of these factors also degrade STAP performance, but additional complications exist due to the multi-phase center radar and the moving platform. For instance, mismatches will exist among the N channels of the receive array. These can be classified further as angle-independent mismatches, which are differences in the frequency responses $H_n(\Omega)$ of the channels, and angle-dependent mismatches. The latter arise from a variety of sources, including element placement errors, wideband dispersion, and mutual coupling of elements. Platform motion effects such as misalignment of the array face with the platform velocity vector due to platform crab angle creates additional degradations. Because adaptive weights are not generally updated every CPI, changing geometry between the moving platform and stationary or moving jammers can cause a jammer to move out of the adapted filter null until the weights are updated again, a phenomenon referred to as the "stale weights" problem.

All of these effects tend to increase the rank of the interference covariance matrix, a phenomenon called *interference subspace leakage* (ISL) because the expanded rank implies that the size of the subspace required to represent the interference is increased. The rank increase implies an increase in reference window size requirements. Many ISL effects can be modeled as *covariance matrix tapers* (CMTs) applied to the true covariance matrix. A CMT is an

$R \times R$ matrix $\mathbf{G}$ that combines with the ideal interference covariance matrix in a Hadamard product to give a modified and higher-rank covariance matrix

$$\tilde{\mathbf{S}}_\mathbf{I} = \mathbf{S}_\mathbf{I} \odot \mathbf{G} \tag{9.83}$$

$\mathbf{G}$, in turn, can often be modeled as the Hadamard product of several component CMTs representing different ISL effects (Guerci, 2003),

$$\mathbf{G} = \mathbf{G}_0 \odot \mathbf{G}_1 \odot \cdots \odot \mathbf{G}_{G-1} \tag{9.84}$$

for some number G of component effects. A one-dimensional example of this in adaptive beamforming defines an N-element CMT vector as samples of a sinc function (Mailloux, 1995; Zatman, 1995). Applying this CMT to the estimated covariance matrix widens the interference notch in the adapted pattern, providing increased immunity to the stale weight problems caused by changing jammer AOAs. As another example, in the space-time case, it can be shown that an appropriate CMT for angle-independent channel mismatch is the weighted sum of a matrix of all 1s and an identity matrix (Guerci, 2003). CMTs can also be used to represent the effects of internal clutter motion. A CMT based on the Billingsley model, discussed earlier, is obtained by sampling Eq. (9.43)

$$\mathbf{G}(m,n) = s_c(|m-n|T) \tag{9.85}$$

where T is the PRI.

However defined, CMTs can be used in at least two ways. The first uses the CMT in Eq. (9.83) to improve the model of the interference covariance in standard STAP algorithms. The second develops new algorithms that take account of the covariance structure, including the CMT component, to provide improved performance. Examples are given in the book by Guerci (2003).

As noted earlier, clutter heterogeneity is an especially significant concern in STAP because of the large reference window size required. Losses due to clutter heterogeneity can range from an insignificant 0.1 dB to as much as 16 dB for realistic scenarios (Melvin, 2000). Approaches to combating these losses include reduced dimension and reduced rank techniques described previously to limit reference window size, and knowledge-assisted algorithms to improve the homogeneity of the data used to estimate the clutter covariance in the cell under test.

References

Billingsley, J. B., *Radar Clutter*. Artech House, Norwood, MA, 2001.
Boroson, D. M., "Sample Size Consideration in Adaptive Arrays," *IEEE Transactions on Aerospace & Electronic Systems*, vol. AES-16(4), pp. 446–451, July 1980.
Brennan, L. E., and F. M. Staudaher, "Subclutter Visibility Demonstration," Technical Report RL-TR-92-21, Adaptive Sensors, 1992.
Cain, K. C., J. A. Torres, and R. T. Williams, "RT_STAP: Real-Time Space-Time Adaptive Processing Benchmark," MITRE Technical Report MTR 96B021, February 1997. Available at www.mitre.org/tech/hpc/pdf/mtr96b21_rtstap.pdf.

Carlson, B. D., "Covariance Matrix Estimation Errors and Diagonal Loading in Adaptive Arrays," *IEEE Transactions on Aerospace & Electronic Systems*, vol. AES-24(3), pp. 397–401, July 1988.

Chen, W. S., and I. S. Reed, "A New CFAR Detection Test for Radar," *Digital Signal Processing*, vol. 1, pp. 198–214. Academic Press, New York, 1991.

Guerci, J. R., "Knowledge Aided Sensor Signal Processing and Expert Reasoning," *Proceedings of Knowledge Aided Sensor Signal Processing and Expert Reasoning* (KASSPER) Workshop, April 2002.

Guerci, J. R., *Space-Time Adaptive Processing for Radar*. Artech House, Norwood, MA, 2003.

Kelly, E. J., "An Adaptive Detection Algorithm," *IEEE Aerospace & Electronic Systems Magazine*, vol. 28(1), pp. 115–127, March 1986.

Kelly, E. J., "Performance of an Adaptive Detection Algorithm: Rejection of Unwanted Signals," *IEEE Transactions on Aerospace & Electronic Systems*, vol. AES-25(2), pp. 122–133, March 1989.

Kim, Y. L., Su. U. Pillai, and J. R. Guerci, "Optimal Loading Factor for Minimal Sample Support Space-Time Radar," *Proceedings* of 1998 *IEEE International Conference on Acoustics, Speech, and Signal Processing* (ICASSP), vol. 4 , pp. 2505–2508, May 1998.

Klemm, R., *Principles of Space-Time Adaptive Processing*. Institution of Electrical Engineers (IEE), London, 2002.

Mailloux, R. J., "Covariance Matrix Augmentation to Produce Adaptive Array Pattern Troughs," *Electronics Letters*, vol. 31(10), pp. 771–772, 1995.

Melvin, W. L., "Space-Time Adaptive Radar Performance in Heterogeneous Clutter," *IEEE Transactions on Aerospace & Electronic Systems*, vol. AES-36(2)), pp. 621–633, April 2000.

Melvin, W. L., "A STAP Overview," *IEEE Aerospace & Electronic Systems Magazine,* vol. 19(1), pp. 19–35, Jan. 2004.

Mountcastle, P. D., "New Implementation of the Billingsley Clutter Model for GMTI Data Cube Generation," *Proceedings of IEEE 2004 Radar Conference*, pp. 398–401, April 2004.

Nitzberg, R., "Detection loss of the sample matrix inversion technique," *IEEE Transactions on Aerospace & Electronic Systems*, vol. AES-26(6), pp. 824–827, Nov. 1984.

Reed, I. S., J. D. Mallet, and L. E. Brennan, "Rapid Convergence Rate in Adaptive Arrays," *IEEE Transactions on Aerospace & Electronic Systems*, vol. AES-10(16), pp. 853–863, Nov. 1974.

Steinhardt, A., and J. Guerci, "STAP for RADAR: What Works, What Doesn't, and What's in Store," *Proceedings of IEEE Radar Conference*, pp. 469–473, April 2004.

Van Trees, H. L., *Optimum Array Processing: Part IV of Detection, Estimation, and Modulation Theory*. Wiley, New York, 2002.

Ward, J., "Space-Time Adaptive Processing for Airborne Radar," Technical Report 1015, Massachusetts Institute of Technology, Lincoln Laboratory, Dec. 13, 1994.

Watkins, D. S., *Fundamentals of Matrix Computations*. Wiley, New York, 1991.

Weiner, D. D., G. T. Capraro, C. T. Capraro, G. B. Berdan, and M. C. Wicks, "An Approach for Using Known Terrain and Land Feature Data in Estimation of the Clutter Covariance Matrix," *Proceedings of* IEEE 1998 National Radar Conference, pp. 381–386, Dallas, TX, May 1998.

Zatman, M., "Production of Adaptive Array Troughs Through Dispersion Synthesis," *Electronics Letters*, vol. 31(25), p. 2141, Dec. 1995.

Index

A/D conversion (*see* Analog-to-digital converters)
Adaptive displaced phase center antenna (*see* Displaced phase center antenna)
Albersheim's equation, 329–331, 336, 337
Aliased sinc function (*see* Asinc function)
Ambiguity
   Doppler, 121, 226, 228, 240
      (*See also* Doppler spectrum)
   range, 120, 126–128, 180–181, 280, 402, 472, 477
   resolution (*see* Ambiguity resolution)
   velocity, 240, 283–284
      (*See also* Ambiguity, Doppler)
Ambiguity function, 169–172
   complex ambiguity function, 170
   of Costas waveform, 222
   definition, 170
   ideal, 171–172
   of linear FM pulse, 194–196
   properties, 170–172
   of pulse burst, 183–187
   of simple pulse, 173–176
   of stepped frequency waveform, 211
   thumbtack, 172
Ambiguity resolution
   Chinese remainder theorem, 281–282
   coincidence algorithm, 282–283
   ghosts, 282
Analog-to-digital conversion, 5, 7, 145, 152–156, 198–200, 447
AN/FPS-108 radar, 9
Angle of arrival (AOA), 68, 137, 291–292, 462, 483
Antenna
   area constraint in SAR, 403
   array, 14–16, 461–463, 468, 469–470, 486, 501
   array element spacing, 137–138
   array factor, 16, 396
   beamwidth, 10, 12, 16, 22, 46, 60, 102–103, 125, 138–140, 402–403, 477
   boresight, 3, 10, 60, 77, 93, 104, 106, 125, 138, 390, 397, 400, 472–473
   effective aperture, 13
   fan beam, 23
   gain, 13
   grating lobes, 138
   nulls, 465, 470, 481, 484, 487–488, 491
   pencil beam, 12
   phase center, 13–14
   phase front, 13–14
   power pattern, 10, 139, 463
   side lobes, 12
   spatial spectra, 105
   voltage pattern, 10
Asinc function, 16, 132, 134, 182–183, 185–186, 208, 210, 255–256, 264, 267–268, 395, 436, 463
ASR-9 radar, 287
ASR-12 radar, 287
Atmospheric attenuation, 8–9, 56
Autofocus, 447–455
   map drift algorithm, 448–449
   phase gradient algorithm, 449–455
Automatic detection (*see* Constant false alarm rate detection)
Azimuth angle, 3

Bandwidth
   Doppler (*see* Doppler bandwidth)
   null-to-null, 121, 122
   Rayleigh, 122
   of a simple pulse, 122
Bandwidth-time product (*see* Time-bandwidth product)
Barker coded waveform (*see* Biphase coded waveform)
Beamforming, 41, 461–471
   adaptive, 41, 465–471
   beam space, 469
   degrees of freedom, 468
   distortionless, 468, 492
   element space, 469
   jammers, 465–467
   matched filter, 464–465
   number of jammers cancelled, 468
   nulls, 465

Beamforming (*Cont.*):
  preprocessing, 469–471
  signal-to-interference ratio (SIR), 467–468
Biphase coded waveform, 212–217
  ambiguity function of Barker code, 215–216
  Barker codes, 212–215
  combined Barker code, 214
  matched filter, 213–214
  minimum peak sidelobe codes (MPS), 216–217
  nested Barker code, 215
  pseudorandom sequences, 216

Cell averaging CFAR
  CFAR loss, 357–358
  concept, 350–351
  effect of false alarms, 349–350
  greatest of cell-averaging (*see* Greatest of cell-averaging CFAR)
  masking loss, 360–361
  order statistic (*see* Order statistic CFAR)
  performance, 354–357
  rank-based (*see* Order statistic CFAR)
  self-masking, 362–363
  smallest of cell-averaging (*see* Smallest of cell-averaging CFAR)
  target masking, 359–362, 364
  threshold multiplier, linear detector, 356–357
  threshold multiplier, square law detector, 354
Central reference point, 198, 412, 429
Chaff, 83
Channel mismatch (*see* I/Q errors)
Characteristic function (CF), 326
Chinese remainder theorem, 281–282
Chirp (*see* linear FM waveform)
Circular random process, 309
Circulator, 6
Clutter, 53, 82–86
  altitude line, 53–57, 226
  area, 83–84
  Billingsley model, 478–479, 502
  compound RCS models, 86–87
  constant gamma model, 83
  correlation, 85, 478–479
  edges, 363
  internal motion, 76, 85, 252–253, 478, 501–502
  log-normal, 72–76, 83, 87–88, 376
  main lobe, 226
  power spectrum, 479
    (*See also* Doppler spectrum)
  side lobe, 226
  volume, 66, 84
  Weibull, 70–76, 83, 86–87, 376
Clutter fill pulses, 128, 279
Clutter map, 284–285, 285
  CFAR, 377–379
  in moving target detector, 286
  threshold, 377
Coherent change detection, 444
Coherent processing interval, 40, 80, 176, 181, 187, 290, 410, 480, 486–487, 497, 501
Coherent signals, 19
Constant false alarm rate detection (CFAR), 49, 285, 293, 488
  adaptive CFAR, 374–375
  adaptive matched filter, 490
  cell-averaging (*see* Cell averaging CFAR)
  cell under test (CUT), 353, 488
  censored CFAR, 357
  CFAR loss, 357–358, 367–368, 370, 373–374
  CFAR window, 353, 376, 378, 381
  for clutter map, 377–379
  distribution-free, 379–381
  guard cells, 353
  lag window, 352
  lead window, 352
  log CFAR, 369–370
  order statistic (*see* Order statistic CFAR)
  reference cells, 353, 488
  reference window, 352–353
  sensitivity to threshold level, 348–350
  switching, 368
  trimmed mean, 357
  two-parameter, 375–376
Continuous wave, 1, 160
Correlation, 35–37
  autocorrelation, 35, 166, 191, 215–216, 218, 246–249, 274–276
  cross-correlation, 35–37, 162
  normalized, 36, 236, 247
Covariance matrix, 234–236, 464
  of clutter, 236, 290
  of jammers, 466
  of preprocessed data, 470
  rank, 499
  of space-time snapshot, 477–478
  taper, 478, 501
Cumulative distribution function, 371
Cumulative probability, 338–340
  of detection, 338
  of false alarm, 339

Data turning, 130–131, 259
Datacube, 116, 228, 461, 476, 488, 500
Decimation, 154

Index

Dechirp processing (*see* Stretch processing)
Deramp processing (*see* Stretch processing)
Detection, 3, 47–49, 120
   Bayesian approach, 297, 313
   in coherent receivers, 308–312
   coincidence detection (*see* Binary integration)
   complex constant-in-Gaussian example, 309–311
   constant false alarm rate (CFAR) (*see* Constant false alarm rate detection)
   detector loss, 317
   fluctuating targets, 331–336
   likelihood ratio test (*see* Likelihood ratio test)
   linear detector, 319
   Marcum's Q function, 316
   M of N detection (*see* Integration, binary)
   Neyman-Pearson criterion, 297–298, 325
   nonfluctuating targets, 324–329
   null hypothesis, 296
   real constant-in-Gaussian example, 302–305
   SNR loss, 317
   square law detector, 319
   sufficient statistic (*see* Sufficient statistic)
   Swerling 1 example, 333–334
   Swerling cases, 331–336
   threshold, 48, 299, 302
   uniformly most powerful test (UMP), 321
   unknown parameters, 312–321
   unknown phase example, 313–316
Digital IF (*see* Digital I/Q)
Digital I/Q, 152–156
   Lincoln Laboratory method, 155–156
   Rader's method, 152–154
Digital sinc function (*see* Asinc function)
Digital terrain elevation data (DTED), 500
Dirichlet function (*see* Asinc function)
Discrete cosine transform, 440–441
Displaced phase center antenna processing (DPCA), 287–293
   adaptive, 289–293, 488
   relation to STAP, 484–488
   time slip, 289, 484
Doppler
   ambiguity, 121, 226, 228, 240
   mismatch, 167–176, 181–182, 199, 205
   tolerance, 176, 197, 205, 215
   unambiguous, 118, 239–240
Doppler bandwidth, 123–126
   due to intrinsic clutter motion, 124
   due to platform motion, 124–125, 403
Doppler beam sharpening (DBS), 389, 401, 411–421
   azimuth dechirp, 421
   beam sharpening ratio, 414
   cross-range/Doppler mapping, 412
   geometric distortion, 413–414
   quadratic phase errors, 416–421
   range curvature, 414
   range migration correction, 421
   range walk limit, 414
   secondary range compression, 421
Doppler interpolation
   via DFT, 264–265
   quadratic, 265–269
Doppler shift, 92–100
   classical approximation, 92, 95–97
   relativistic, 92
   spatial Doppler, 97–100, 128, 147
   stop-and-hop assumption, 97–99
Doppler spectrum
   Billingsley model, 478–479
   characteristics, 225–228
   clear region, 225
   clutter region, 225
   effect of platform motion, 125–126, 472
   meteorological model, 273
   relation to angle of arrival, 473
   skirt region, 225
   unambiguous width, 118, 239–240, 273
Downchirp, 189
Duplexer, 6
Dwell, 176
Dynamic range, 5, 71, 141

Electromagnetic interference (EMI), 348, 500
Elevation angle, 3
Error function, 303, 304, 311
   approximations, 343–344
   complementary, 303, 304, 311

False alarm, 48, 381
Fast Fourier transform, 5, 183, 258–259, 286, 412, 424, 436, 449, 465, 471, 480
Fast time, 39, 48, 116, 121–122, 182–183, 253, 313, 321, 392, 407–410, 425, 444, 461
FFT (*see* Fast Fourier transform)
FIR filter, 202, 233
Fourier transform
   continuous, 25
   discrete, 26, 129–135, 183, 210–211, 253, 258–268, 283, 290, 448–453, 469–470, 480
   discrete time, 26, 129, 183, 256, 465
   fast (*see* Fast Fourier transform)
Frequency agility, 79
Frequency-modulated continuous wave (FMCW), 198

Ghosts, 283
Global positioning system, 446
Greatest-of cell averaging CFAR, 366–367
  CFAR loss, 367
  clutter edge performance, 367
  target masking performance, 367
  threshold multiplier, 367
Ground moving target indication (GMTI), 287
Ground plane, 404

Hadamard product, 463, 480, 502
Hermitian, 37, 466
History of radar (*see* Radar history)
Hypothesis testing, 296–297

I channel (*see* In-phase channel)
IF sampling (*see* Digital I/Q)
IIR filter, 234, 377
Imaging, 45–47
  (*See also* Synthetic aperture radar)
Improvement factor, 244, 246–249
  autocorrelation method, 246–248
  frequency domain method, 245–246
  suboptimum filter loss, 249–250
  vector method, 248–249
Inertial measurement unit (IMU), 446
Inertial navigation system (INS), 446
In-phase channel, 17, 145–146, 149
Integration, 33–35
  binary, 323, 324, 338–341
  coherent, 33, 179, 323
  noncoherent, 34, 323, 324
Integration gain, 34
  coherent, 323
  noncoherent, 329, 331
Interferometric SAR, 436–444
  absolute height estimation, 441–443
  baseline, 437
  coherent change detection, 444
  concept, 436–439
  interferometric phase difference, 438
  layover, 438
  one-pass, 439
  orthorectification, 443
  phase estimation, 440
  phase unwrapping, 440–441
  pixel phase, 437
  processing steps, 439–443
  registration, 440
  terrain motion mapping, 444
  two-pass, 439
Internal clutter motion (*see* Clutter)

Interpolation, 5, 27
  Doppler (*see* Doppler interpolation)
  fast time (*see* Interpolation, range)
  range, 289, 414, 425
  spatial frequency, 433, 434–436
Intrinsic clutter motion (*see* Clutter, internal motion)
I/Q errors, 145–151
  amplitude-only, 147
  correcting, 149–150
  image component, 147
  phase only, 148
I/Q imbalance (*see* I/Q errors)

Jamming, 53, 92, 466

Kronecker product, 476

Lagrange multipliers, 299
Least of cell-averaging CFAR (*see* Smallest of cell-averaging CFAR)
Likelihood ratio test (LRT), 298–301, 308, 325
  generalized likelihood ratio test, 313, 321, 490
  log likelihood ratio test, 300, 309, 325
Linear FM waveform, 43, 111, 159, 428
  ambiguity function, 194–196
  definition, 188
  range-Doppler coupling, 197–198
  Rayleigh resolution, 191, 196
  stationary phase spectrum approximation
Linear time-invariant system, 242
Loss
  amplitude, 416, 445
  antenna, 13
  array steering, 208
  atmospheric (*see* Atmospheric attenuation)
  brightness, 416
  CFAR (*see* Constant false alarm rate detection)
  conversion, 20
  covariance matrix estimation, 489
  detector, 317, 344–345
  mismatch, 215
    (*See also* Doppler mismatch)
  ohmic, 88
  processing (*see* Processing loss)
  in processing gain (*see* Loss in processing gain)
  propagation (*see* Atmospheric attenuation)
  receiver, 90
  resolution, 416
  SIR loss in STAP, 481–484
  SNR, 317, 344–345
  straddle, 131–136, 259–261, 268–269

Loss in processing gain (LPG), 204, 257–258
Low Earth orbit, 397

Magnitude function approximations, 344–345
Marcum's Q function (*see* Detection)
Masking, 201
Matched filter, 44, 161–169, 310, 464, 479
   all-range, 165–166
   in beamforming, 464, 468
   continuous, 161–162
   for moving targets, 167–169
   for moving target indication (MTI), 234–238, 261–262, 289–293
   peak SNR, 162
   for pulse burst, 177–178
   pulse-by-pulse processing (*see* Pulse-by-pulse processing)
   range resolution, 168
   for simple pulse, 163–165
   in space-time adaptive processing, 479–480
   vector formulation, 234–235, 464–465
   whitening, 163
Maximum likelihood estimate
   of clutter edge location, 374–375
   of interference power, 351–352
   of phase gradient, 452
Minimum detectable Doppler (*see* Space-time adaptive processing)
Minimum detectable velocity (*see* Space-time adaptive processing)
Moore's law, 6, 389
Motion compensation, 444–447
   autofocus (*see* Autofocus)
   global positioning system (GPS), 446
   inertial measurement unit (IMU), 446
   inertial navigation system (INS), 446
   Kalman filter, 446
   phase errors, 445
   PRI control, 447
   to a point, 447
Moving target detector (MTD), 286–287, 378
Moving target indication, 225, 228–253
   blind speeds, 239–240
   clutter attenuation, 244, 245
   combined with pulse Doppler, 279
   concept, 228–230
   figures of merit, 244–251
   FIR filter type, 233
   ground MTI, 287
   improvement factor (*see* Improvement factor)
   interference covariance matrix, 236
   limitations (*see* Moving target indication limitations)
   in moving target detector, 286
   N-pulse canceller, 233
   optimum filter, 235–239
   signal model, 238–239
   as slow time highpass filter, 229–230
   subclutter visibility, 245
   surface MTI, 287
   three-pulse canceller, 230–232
   transient effects, 279
   two-pulse canceller, 230–232
   visibility factor, 251
Moving target indication limitations, 251–253
   amplitude instability, 251–252
   intrinsic clutter motion, 252
   phase instability, 252
   platform motion, 253
MTI (*see* moving target indication)
Multipath, 108–109
Munson, David, 389

NEXRAD, 278–279
Neyman-Pearson criterion (*see* Detection)
Noise, 88–92
   effective temperature, 91
   figure, 91
   power spectrum of, 88
   in quadrature receivers, 89
   types, 20, 88
Noise equivalent bandwidth, 89–90
Null hypothesis (*see* Detection)
Nyquist rate, 31
   in 3 dB beamwidths, 140
   in angle of arrival, 139
   in Doppler, 129
   in fast time, 121–123
   in range, 121–123
   in sine of angle of arrival, 139
Nyquist theorem, 27–31

Order statistic, 370
Order statistic CFAR, 370–375
   CFAR loss, 373–374
   performance, 371–373
   threshold multiplier, 370–372
   in two-parameter CFAR, 376
Oscillator
   local, 17
   stable local, 19

Phase history, 407
Phenomenology, 40
Pilot signal, 149
Point scatterer response (PSR) (*see* Stripmap SAR)

Polar format algorithm, 433–436
    angular-range interpolation, 436
    keystone grid, 435
    polar-to-rectangular interpolation, 434–436
    radial-keystone interpolation, 435–436
    rectangular format algorithm, 436
Polarization scattering matrix, 65
Polyphase coded waveform, 218–221
    ambiguity function, 219
    Frank codes, 218
    P3 codes, 219
    P4 codes, 219
    precompression bandlimiting, 221
Principle of stationary phase, 192–193, 205, 221, 424
Probability density functions
    beta, 489
    chi-square, 72–76, 81, 86
    circular Gaussian, 309
    Erlang, 327, 354
    exponential, 71–76, 81, 83, 86, 331
    Gamma, 327
    Gaussian, 300
    log-normal, 70–76, 83, 87–88
    Rayleigh, 324, 326
    Rician, 316, 324, 326, 361
    Weibull, 70–76, 83, 86–87
Probability of detection, 4, 297
Probability of false alarm, 4, 297
Probability of miss, 297
Processing loss (PL), 204, 257–258
Projection-slice theorem, 430–431
Pulse burst waveform, 176, 183–188
    ambiguity function, 183–187
    Doppler response, 181–183
    relation of ambiguity function to slow time spectrum, 187–188
Pulse cancellers (see Moving target indication)
Pulse-by-pulse processing, 178, 209–210
Pulse compression, 42–43, 188
Pulse Doppler processing, 225, 253–273
    as bandpass filter bank, 261–263
    combined with MTI, 279
    DTFT of moving target, 255–256
    fine Doppler estimation (see Doppler interpolation)
    high PRF operation, 280
    low PRF operation, 280
    as matched filter, 262
    medium PRF operation, 280
    modern spectral estimation in, 279–271
    in moving target detector, 286
    PRF regimes, 280
    transient effects, 279

Pulse pair processing, 273–279
    autocorrelation method, 274–276
    frequency domain method, 276–277
    spectral subtraction, 277
    spectral width, 274
Pulse repetition frequency, 9, 115
Pulse repetition interval, 9, 115

Q channel (see Quadrature channel)
Q function (see Detection)
Quadrature channel, 17, 145–146, 149
Quantization, 140–145
    linear model, 142
    McClellan-Purdy model, 143–145
    rounding, 141
    saturation, 143
    sign-magnitude encoding, 140
    truncation, 141
    two's complement encoding, 140
    underflow, 143

Radar
    AN/FPS-108, 9
    antenna (see Antenna)
    applications, 2
    ASR-9, 287
    ASR-12, 287
    bistatic, 2
    Chain Home, 1
    continuous wave, 1, 160
    frequency bands, 7–8
    imaging (see Synthetic aperture radar)
    literature, 49–51
    monostatic, 2
    NEXRAD, 278–279
    receiver (see Receiver)
    range equation (see Radar range equation)
    SCR-270, 1
    SIR-C radar, 388–389
    synthetic aperture (see Synthetic aperture radar)
    transmitter, 7
Radar cross section (RCS), 64–88, 100–109, 473
    convolutional models, 100–108
    of clutter (see Clutter)
    common pdfs
        for RCS, 72–74
        for voltage, 75–76
    correlation, 75–79
        in angle, 78
        in frequency, 78
        pulse-to-pulse, 80–81
    definition, 64
    meteorological, 66

spatial models, 100–109
spectral model, 109–111
statistical models (*see* Target models)
Swerling models (*see* Swerling models)
typical values, 65
Radar history, 1–2
   Chain Home, 1
   Hülsmeyer, Christian, 1
   Page, Robert, 1
   Radiation Laboratory, 2
   SCR-270, 1
   Taylor, Albert, 1
   Watson-Watt, Sir Robert, 1
   Young, Leo, 1
Radar range equation, 54–63
   for area clutter, 60–63
   for distributed targets, 57–58
   for point scatterers, 59
   for volume clutter, 59–60
Range
   ambiguity (*see* Ambiguity)
   bins, 118
   cells, 118
   gates, 118
   unambiguous, 120, 239
Range-Doppler algorithm, 421–428
   frequency domain implementation, 423
   narrowband approximation, 425
Range-Doppler coupling (*see* linear FM waveform)
Range migration, 99, 401, 406
   range curvature, 406, 407
   range walk, 406
Range profile, 210, 429
Range side lobe control, 201–206
   by matched filter design, 202–204
   by waveform spectrum shaping, 204–206
Range window, 198
Real-beam imaging, 390
Receiver, 6
   effective temperature, 91
   superheterodyne, 20
Receiver operating characteristic (ROC), 303, 305
Reed-Mallet-Brennan rule, 489
Reflectivity, 101
   angle-averaged effective, 106
   baseband complex, 101
   effective, 102
   projections, 107–108
   range-averaged effective, 103
Region of interest (ROI), 397
Replication, 30, 130

Residual video phase (*see* Spotlight SAR)
Resolution
   cell, 22
   cross-range, 22, 390–391, 395, 396, 397
   range, 9, 21
   Rayleigh, 168, 191, 205, 210
   velocity, 168
RMB rule (*see* Reed-Mallet-Brennan rule)

Sampling
   in angle, 138–140
   array element spacing, 137–138
   criteria, 116
   in Doppler frequency, 118, 129–131
   in fast time, 121
   Nyquist theorem, 27–31
   in slow time, 123–128
   spatial, 118
   temporal, 118
Schwarz inequality, 161, 171, 234
Sensitivity time control (STC), 107
Shnidman's equation, 336–338
Shuttle imaging radar – C (*see* SIR-C radar)
Sigma-nought (σ^0) (*see* Clutter)
Signal-to-clutter ratio, 84, 225, 230, 234, 245–246
Signal-to-interference ratio, 4, 53, 163, 322, 381, 467–468, 481–484, 489, 491
Signal-to-noise ratio, 20, 33–34, 84, 91, 131, 161–164, 203–204, 277, 305–208, 323, 329–330, 332–336, 337, 339, 348, 357–364, 397, 465
Signal-to-quantization noise ratio, 142
SIR-C radar, 388–389, 410
Slant plane, 404
Slant range, 404
Slow time, 40, 116, 123, 187, 225–227, 253–255, 273–276, 285, 290, 399–401, 407, 410, 412, 416–421, 424–425, 448, 456, 461, 475, 498, 500
Smallest-of cell averaging CFAR, 364–366
   CFAR loss, 367
   clutter edge performance, 365
   target masking performance, 365
   threshold multiplier, 365
Snapshot
   space-time, 475, 486
   spatial, 462
Space-time adaptive processing, 287, 471–502
   adapted pattern, 480, 488
   Billingsley model, 478, 502
   clutter model, 472–473
   clutter ridge slope, 473, 487
   column stacking, 475

Space-time adaptive processing (*Cont.*):
  computational load, 491–497
  covariance matrix (*see* Covariance matrix)
  covariance matrix taper (CMT), 478, 501
  diagonal loading, 490
  distortionless, 492
  interference subspace leakage, 501
  knowledge-aided, 500
  limitations, 501–502
  metrics, 480–484
  minimum detectable Doppler (MDD), 483–484
  minimum detectable velocity (MDV), 483
  power domain solution, 491, 492–495
  principle components method, 500
  Q-R decomposition, 491, 492, 495–496
  rank of covariance matrix, 499
  reduced dimension, 497–498
  reduced rank, 500
  Reed-Mallet-Brennan rule, 489
  relation to DPCA, 484–488
  sample matrix inverse, 488
  signal model, 471, 475–479
  SIR loss, 481–483
  subspace projection, 500
  temporal correlation, 478–479
  test statistic, 479
  usable Doppler space fraction (UDSF), 484
  voltage domain solution, 491, 495–496
Spatial frequency, 23–24, 433, 434–436
Spotlight SAR, 389, 397–399
  central reference point, 429
  convolution-backprojection, 436
  cross-range resolution, 397
  data model, 428–433
  limitations to polar format model, 433
  mapping rate, 403–404
  polar format algorithm (*see* Polar format algorithm)
  polar format data, 432
  projection, 429
  rectangular format algorithm, 436
  residual video phase, 429
  sampling requirements, 432–433
Staggered PRF
  block-to-block, 240
  dwell-to-dwell, 240, 272–273
  pulse canceller frequency response, 242–243
  pulse-to-pulse, 240
  stagger ratio, 241
  staggers, 241
STC (*see* Sensitivity time control)
Steering vector
  space-time, 476
  spatial, 453
  temporal, 475
Stepped frequency waveform, 206–211
  ambiguity function, 211
  array steering, 208
  definition, 208
  pulse-by-pulse processing, 209–210
Straddle loss, 131–136
  definition, 134
  in Doppler, 259–260
  sampling rate for specified limit, 134–135
Stretch processing, 198–201, 428–429
Stop-and-hop assumption (*see* Doppler shift)
Stripmap SAR, 389, 392–397
  ω-k algorithm, 410
  cross-range chirp, 406
  data set size, 409–410
  depth of focus, 426–428
  geometry, 404–407
  lower bound on cross-range resolution, 396
  mapping rate, 403
  point scatterer response (PSR), 404, 407–410
  range migration algorithm, 410, 426
  space-bandwidth product, 406
  subswaths, 426
  waveform, 407
Sufficient statistic, 301–304, 310–311, 314, 321, 326
Surface moving target indication (SMTI), 287
Swerling models, 79–82, 324, 354
Synthetic aperture radar, 45–47, 385–389
  antenna area constraint, 403
  aperture time, 393, 395
  array synthesis, 392
  autofocus (*see* Autofocus)
  beamwidth of synthetic array, 394–395
  cross-range resolution (*see* Resolution)
  Doppler beam sharpening (*see* Doppler beam sharpening)
  Doppler viewpoint, 399–401
  interferometric SAR (*see* Interferometric SAR)
  layover, 438
  looks, 456
  mapping rate, 403–404
  motion compensation (*see* Motion compensation)
  multilook integration, 456
  point scatterer response (PSR) (*see* Stripmap SAR)
  quadratic phase errors, 401
  range curvature (*see* Range migration)

range-Doppler algorithm, 401
range migration (*see* Range migration)
range walk (*see* Range migration)
sidelooking, 393, 402
speckle, 386, 455–456
speckle reduction, 455–458
spotlight (*see* Spotlight SAR)
squinted SAR, 394
stripmap (*see* Stripmap SAR)
swath constraint, 403
swath length, 401–403
synthetic aperture size, 392
Synthetic array radar (*see* Synthetic aperture radar)

Target masking (*see* Masking)
Target models, 67–82, 334–338
    fluctuating, 331–338
    nonfluctuating, 324, 341
    Rician, 324, 326
    Swerling (*see* Swerling models)
Target visibility (*see* MTI visibility factor)
Terrain motion mapping, 444
Test statistic, 479, 490
Threshold detection (*see* Detection)
Time-bandwidth product, 188–190, 199, 201–202

Tracking, 49
Transmit/receive switch, 6

Unambiguous range (*see* Range)
Upchirp, 189

Vector representation, 32

Walker, Jack, 389
Waveform
    biphase coded (*see* Biphase coded waveform)
    Costas frequency codes, 222–223
    linear FM (LFM) (*see* Linear FM waveform)
    nonlinear FM (NLFM), 160, 204–207, 221
    phase coded, 211–221
    polyphase coded (*see* Polyphase coded waveform)
    pulse burst (*see* Pulse burst waveform)
    simple pulse, 173–176
    stepped frequency (*see* Stepped frequency waveform)
Wavefront (*see* Antenna phase front)
Whitening, 163
Wiley, Carl, 389

Zero padding, 130
Zero velocity filter, 285

ABOUT THE AUTHOR

Mark A. Richards, Ph.D., is a principal research engineer and adjunct professor at the Georgia Institute of Technology. He has over 20 years experience in academia, industry, and government in radar signal processing and embedded computing. He has served as a program manager in the Defense Advanced Research Projects Agency; the General Chair of the IEEE 2001 Radar Conference, and as an associate editor of the *IEEE Transactions on Image Processing* and the *IEEE Transactions on Signal Processing*. Dr. Richards teaches frequently in graduate and professional education courses in radar signal processing, radar imaging, and related topics. He lives in Marietta, Georgia.

McGraw-Hill's
Digital Engineering Library

A powerful online repository of engineering content from the field's premier publisher

UNPARALLELED CONTENT AND FUNCTIONALITY

✓ 4,000 articles derived from 150+ McGraw-Hill titles

✓ 12 major engineering disciplines

✓ 500+ Topics

✓ Easy and effective search or topical browse

✓ By the most trusted and respected authors

✓ Subscription or pay-per-view option

No other online engineering resource is this **AUTHORITATIVE!**

Nothing else is this **COMPREHENSIVE!**

SIGN-UP FOR A FREE 30-DAY TRIAL:

www.digitalengineeringlibrary.com

(CLICK ON "SUBSCRIPTION INFORMATION")